Advances in Metabolomics

Vijai Singh
Editor

Advances in Metabolomics

Springer

Editor
Vijai Singh
Department of Biosciences
Indrashil University
Mehsana, Gujarat, India

ISBN 978-981-97-7458-6 ISBN 978-981-97-7459-3 (eBook)
https://doi.org/10.1007/978-981-97-7459-3

This Springer imprint is published by the registered company Springer Nature Singapore Pte Ltd.
The registered company address is: 152 Beach Road, #21-01/04 Gateway East, Singapore 189721, Singapore

Foreword

I am pleased to write a foreword for this book ***Advances in Metabolomics***, a timely contribution in the rapidly growing field of biological sciences. Metabolomics is a rapidly growing field that studies chemical processes involved in metabolites. These are related to small substrates, intermediates and products for cell metabolism, physiology, genetics, cell proliferation and so forth. Metabolome is a complete set of metabolites within cells, tissues, organs and organisms that are final products of cellular processes.

This book offers a wide range of topics in metabolomics and metabolites from fundamental to applied research. The book covers topics including computational biology tools, design of biosynthetic pathways, single cell metabolomics, biosensors for metabolites, transcriptional control of metabolites, mass spectrometry, HPLC, NMR, microfluidics for metabolites, cell-free systems in metabolites, genome editing for production of metabolites, fermentation and downstream processing and industrial-scale production of metabolites. In this book, chapters have been written by eminent scientists who have extensive experience and understanding in their respective areas of research in metabolomics.

I am delighted to recognize the valuable efforts of Dr. Vijai Singh, who has tirelessly worked to conceive an excellent book under Springer Nature. This book covers a rich literary text of excellent depth, clarity and coverage. This book is an excellent source of information from which scientific knowledge that can grow and widen the field of metabolomics. I personally believe that this book may play a key role for better understanding of metabolomics, not just for the beginners, but also for students, researchers, scientists, clinicians, practitioners, policymakers and stakeholders, who are interested in accessing the potential of metabolomics in many areas.

Institute of Industrial Control Systems and Computing (AI2)
Universitat Politècnica de València
València, Spain

Institute for Integrative Systems Biology I2SysBio
Universitat de València-CSIC
València, Spain

Pablo Carbonell

Preface

Metabolomics is a rapidly growing field in which the study of chemical processes involves metabolites. These are small molecule substrates, intermediates and products for cell metabolism. Metabolome is a complete set of metabolites in cells, tissues, organs and organisms that are end products of cellular processes. Currently, the isolation, detection, purification and characterization of metabolites are important for studying their nature and cellular function. Several tools such as modelling, gas chromatography, gas chromatography–mass spectrometry (GC-MS), high-performance liquid chromatography (HPLC), nuclear magnetic resonance (NMR) and many more are used for the detection and characterization of metabolites. Recently, a number of computation biology along with machine learning form a powerful tool that can be used in metabolomics analysis. Recently, scientists have also developed a retention time prediction software. These tools allow researchers to apply artificial intelligence to the retention time prediction of small molecules in complex mixture such as human plasma, plant extracts, foods or microbial cultures.

The book covers topics including computational biology tools, design of biosynthetic pathways, single cell metabolomics, biosensors for metabolites, transcriptional control of metabolites, mass spectrometry, HPLC, NMR, microfluidics for metabolites, cell-free systems in metabolites, genome editing for production of metabolites, fermentation and downstream processing and industrial-scale production of metabolites.

This book will provide a basic look into the applied knowledge of metabolomics in several areas. We believe that this is the first book that covers this great range of topics. This book will form an excellent and informative text on metabolomics, benefitted by simple-to-understand and easy-to-read format. This book will use a rich literary text of excellent depth, clarity and coverage. This book will be an excellent basis from which scientific knowledge can grow, widen and accelerate metabolomics research.

Mehsana, Gujarat, India Vijai Singh

Acknowledgement

I am thrilled to express my sincere gratitude and deep appreciation to Dr. J. S. Yadav, Director (Research), Indrashil University, India, for extending his outstanding support and motivation to complete this book.

I am delighted to thank all the authors for their excellent contributions to this book. I would like to thank Dr. Swati Sharma (Associate Editor—Biomedicine), Dr. Soumya Basu and Mrs Aishwarya Thyagarajan (Production Editor) from Springer for their excellent management of this project

I would like to thank Prof. Rakesh M. Rawal, Prof. Bharat Maitreya, Dr. Poonam Bhargava, Dr. Satya Prakash, Dr. Vimal C. Pandey, Prof. Suresh Ramakrishna, Dr. Dinh-Toi Chu and those whose names do not feature here but have directly or indirectly contributed to shaping this project.

I greatly appreciate the support of my students Mr. Khushal Khambhati, Ms. Dharmisha Solanki, Mr. Karan Murjani and Khushbu Panchal, whose discussion and comments helped to shape this book.

I wish to express my gratitude to my beloved wife Pritee Singh for her endless support, patience and inspiration. Lots of affection for my kids Aaradhya and Ayush, who missed me during this project. I would like to warmly thank the faculty and staffs of Indrashil University for providing a great working environment.

I am aware that even despite our best efforts, the first version always comes with some error that may have crept in the compilation. I would be delighted to receive feedback from readers to further improve the future book. Last but not least, my sincere thanks to GOD for his supreme POWER for endowing me to live with joy and victory in the shape of this book.

Vijai Singh

Contents

Editor and Contributors

About the Editor

Vijai Singh is Professor and Head of the Department of Biosciences, School of Science at Indrashil University, Rajpur, Mehsana, India. He has worked as Associate Professor in the Department of Biosciences, School of Science at Indrashil University, Rajpur, Mehsana, India. He was an Assistant Professor in the Department of Biological Sciences and Biotechnology at the Institute of Advanced Research, Gandhinagar, India, and also an Assistant Professor in the Department of Biotechnology at the Invertis University, Bareilly, India. Prior to that, he was a Postdoctoral Fellow in the Synthetic Biology Group at the Institute of Systems and Synthetic Biology, Paris, France, and School of Energy & Chemical Engineering at the Ulsan National Institute of Science and Technology, Ulsan, South Korea. He received his Ph.D. in Biotechnology (2009) from the ICAR-National Bureau of Fish Genetic Resources-Uttar Pradesh Technical University, Lucknow, India, with a research focus on the development of molecular and immunoassays for diagnosis of Aeromonas hydrophila. His research interests are focused on building novel biosynthetic pathways for the production of medically and industrially important biomolecules. Additionally, his laboratory is working on CRISPR-Cas9 tools for genome editing. He has more than 12 years of research and teaching experience in synthetic biology, metabolic engineering, bioinformatics, microbiology and industrial microbiology. He has published 107 articles, 80 book chapters, 26 books and 3 patents. He serves as an associate editor, editorial board member, and reviewer of several peer-reviewed journals. He is also a member of the Board of Study and Academic Council of Indrashil University and is the Member Secretary of the Institutional Biosafety Committee (IBSC) at the same university.

Contributors

Gholamreza Abdi Department of Biotechnology, Persian Gulf Research Institute, Persian Gulf University, Bushehr, Iran

Wajahat Ali Institute of Basic Medicine, Xian Jiaotong University, Xi'an, China

Prasad Andhare Biological Sciences, PDPIAS, Charotar University of Science and Technology, Changa, Gujarat, India

Hoang Duc Anh Center for Biomedicine and Community Health, International School, Vietnam National University, Hanoi, Vietnam

Le Duc Anh Center for Biomedicine and Community Health, International School, Vietnam National University, Hanoi, Vietnam

Mustafeez Mujtaba Babar Shifa College of Pharmaceutical Sciences, Shifa Tameer-e-Millat University, Islamabad, Pakistan

Abhishek Barnwal Department of Life Sciences, Parul Institute of Applied Sciences, Parul University, Vadodara, Gujarat, India

Ritika Bhardwaj Department of Life Sciences, Parul Institute of Applied Sciences, Parul University, Vadodara, Gujarat, India

Nhat-Le Bui The Interdisciplinary Research Group on Biomedicine and Health (IRGBH), International School, Vietnam National University, Hanoi, Vietnam

Faculty of Applied Sciences, International School, Vietnam National University, Hanoi, Vietnam

Ashara Chaudhari Vivekanand Education Society's College of Pharmacy, Chembur, Mumbai, Maharashtra, India

Neha Chhabra Department of Biotechnology, Persian Gulf Research Institute, Persian Gulf University, Bushehr, Iran

Shipra Choudhary Department of Biotechnology and Microbiology, Meerut Institute of Engineering and Technology, Meerut, India

Dinh-Toi Chu The Interdisciplinary Research Group on Biomedicine and Health (IRGBH), International School, Vietnam National University, Hanoi, Vietnam

Department of Biomedical Engineering, National Cheng Kung University, Tainan, Taiwan

Faculty of Applied Sciences, International School, Vietnam National University, Hanoi, Vietnam

Bhagyshree Desai Vivekanand Education Society's College of Pharmacy, Chembur, Mumbai, Maharashtra, India

Ishita Dewangan Department of Life Sciences, Parul Institute of Applied Sciences, Parul University, Vadodara, Gujarat, India

Rupal Dhariwal Cell and Development Biology Lab, Research and Development Cell, Parul University, Vadodara, Gujarat, India

Department of Life Sciences, Parul Institute of Applied Sciences, Parul University, Vadodara, Gujarat, India

Vivekanand Education Society's College of Pharmacy, Chembur, Mumbai, Maharashtra, India

Quy Do Van The Interdisciplinary Research Group on Biomedicine and Health (IRGBH), International School, Vietnam National University, Hanoi, Vietnam

Department of Biomedical Engineering, National Cheng Kung University, Tainan, Taiwan

Luong Tri Duc Center for Biomedicine and Community Health, International School, Vietnam National University, Hanoi, Vietnam

Trinh Chi Dung Center for Biomedicine and Community Health, International School, Vietnam National University, Hanoi, Vietnam

Faculty of Applied Sciences, International School, Vietnam National University, Hanoi, Vietnam

Suruchi Gaur Biochemistry and Molecular Biology Lab, Department of Zoology, Banaras Hindu University, Varanasi, India

Syed Damin Abbas Hamdani Shifa College of Pharmaceutical Sciences, Shifa Tameer-e-Millat University, Islamabad, Pakistan

Lan Anh Nguyen Hoang Center for Biomedicine and Community Health, International School, Vietnam National University, Hanoi, Vietnam

Shabab Hussain Department of Chemical, Biological, Pharmaceutical and Environmental Sciences, University of Messina, Messina, Italy

Maria Irshad Department of Medical, Surgical and Health Sciences, University of Trieste, Trieste, Italy

Juveriya Israr Institute of Biosciences and Technology, Shri Ramswaroop Memorial University, Barabanki, Uttar Pradesh, India

Mukul Jain Cell and Development Biology Lab, Research and Development Cell, Parul University, Vadodara, Gujarat, India

Department of Life Sciences, Parul Institute of Applied Sciences, Parul University, Vadodara, Gujarat, India

Vivekanand Education Society's College of Pharmacy, Mumbai, Maharashtra, India

Rutuja A. Kamble Department of Biological Sciences and Biotechnology, Institute of Chemical Technology, Mumbai, India

Kulsoom Department of Biochemistry, Bahauddin Zakariya University, Multan, Pakistan

Ajay Kumar Department of Biotechnology, Faculty of Engineering and Technology, Rama University, Kanpur, India

Harshit Kumar Department of Life Sciences, Parul Institute of Applied Sciences, Parul University, Waghodia, Gujarat, India

Saloni Malik Department of Biotechnology and Bioengineering, Institute of Advanced Research, Gandhinagar, Gujarat, India

Hara Prasad Mishra Koita Centre for Digital Health, Ashoka University, Sonepat, Haryana, India

Prachi Mishra Department of Life Sciences, Parul Institute of Applied Sciences, Parul University, Vadodara, Gujarat, India

Sandhya Mishra Vivekanand Education Society's College of Pharmacy, Chembur, Mumbai, Maharashtra, India

Karan Murjani Department of Biosciences, School of Science, Indrashil University, Rajpur, Mehsana, Gujarat, India

Ha Phuong Nguyen Center for Biomedicine and Community Health, International School, Vietnam National University, Hanoi, Vietnam

Que Anh Trieu Nguyen Center for Biomedicine and Community Health, International School, Vietnam National University, Hanoi, Vietnam

Hoang Thanh Nhat The Interdisciplinary Research Group on Biomedicine and Health (IRGBH), International School, Vietnam National University, Hanoi, Vietnam

Faculty of Applied Sciences, International School, Vietnam National University, Hanoi, Vietnam

Khushbu Panchal Department of Biosciences, School of Science, Indrashil University, Rajpur, Mehsana, Gujarat, India

Alok Pandya Department of Biotechnology and Bioengineering, Institute of Advanced Research, Gandhinagar, Gujarat, India

Shreyas Patel Department of Biotechnology and Bioengineering, Institute of Advanced Research, Gandhinagar, Gujarat, India

Nil Patil Vivekanand Education Society's College of Pharmacy, Chembur, Mumbai, Maharashtra, India

Cell and Development Biology Lab, Research and Development Cell, Parul University, Vadodara, Gujarat, India

Department of Life Sciences, Parul Institute of Applied Sciences, Parul University, Vadodara, Gujarat, India

Shamlan M. S. Reshamwala Department of Biological Sciences and Biotechnology, Institute of Chemical Technology, Mumbai, India

Jueeli Shiriskar Department of Biotechnology, Persian Gulf Research Institute, Persian Gulf University, Bushehr, Iran

Malvika Shukla Department of Biotechnology and Bioengineering, Institute of Advanced Research, Gandhinagar, Gujarat, India

Indrajeet Singh Department of Biotechnology, Faculty of Engineering and Technology, Rama University, Kanpur, India

Vijai Singh Department of Biosciences, School of Science, Indrashil University, Rajpur, Mehsana, Gujarat, India

Ganga Srinivasan Department of Biotechnology, Persian Gulf Research Institute, Persian Gulf University, Bushehr, Iran

Fong Chin Su Department of Biomedical Engineering, National Cheng Kung University, Tainan, Taiwan

Om Supe Department of Biological Sciences and Biotechnology, Institute of Chemical Technology, Mumbai, India

Sanskruti Surve Vivekanand Education Society's College of Pharmacy, Chembur, Mumbai, Maharashtra, India

Maryam Abbasi Tarighat Faculty of Nano and Bio Science and Technology, Persian Gulf University, Bushehr, Iran

Mohd Tariq Department of Biotechnology, Graphic Era (Deemed to be University), Uttarakhand, Dehradun, India

Mugdha Tendulkar K. J. Somaiya Medical College and Research Centre, Sion, Mumbai, India

Reshma Tendulkar Vivekanand Education Society's College of Pharmacy, Chembur, Mumbai, Maharashtra, India

Krishna Thakkar Department of Life Sciences, Parul Institute of Applied Sciences, Parul University, Vadodara, Gujarat, India

Chaitrali Thatte Vivekanand Education Society's College of Pharmacy, Chembur, Mumbai, Maharashtra, India

Hue Vu Thi Center for Biomedicine and Community Health, International School, Vietnam National University, Hanoi, Vietnam

Faculty of Applied Sciences, International School, Vietnam National University, Hanoi, Vietnam

Shubhita Tripathi Department of Biotechnology and Bioengineering, Institute of Advanced Research, Gandhinagar, Gujarat, India

Bindiya Upadhyay Vivekanand Education Society's College of Pharmacy, Chembur, Mumbai, Maharashtra, India

Cell and Development Biology Lab, Research and Development Cell, Parul University, Vadodara, Gujarat, India

Dhwani Upadhyay Department of Life Sciences, School of Sciences, GSFC University, Gujarat, Vadodara, India

Phan Van Bao The Interdisciplinary Research Group on Biomedicine and Health (IRGBH), International School, Vietnam National University, Hanoi, Vietnam

Chia-Ching Wu Department of Biomedical Engineering, National Cheng Kung University, Tainan, Taiwan

Department of Cell Biology and Anatomy, College of Medicine, National Cheng Kung University, Tainan, Taiwan

International Center for Wound Repair and Regeneration, National Cheng Kung University, Tainan, Taiwan

Samra Zahra Department of Biosciences, Comsats University, Islamabad, Pakistan

Devika Zunjarrao Department of Biological Sciences and Biotechnology, Institute of Chemical Technology, Mumbai, India

Current Approaches on Metabolomics

1

Khushbu Panchal, Karan Murjani, and Vijai Singh

1.1 Introduction

Metabolomics is an emerging field that has experienced significant advancements over the last 10 years. It has evolved from focusing on refining analytical techniques and methodologies to a phase where it is used to investigate fundamental biological and clinical questions. It aims to provide a comprehensive profile analysis of different metabolites to understand metabolic pathways, biochemical interactions, and the overall physiological state of biological systems (Schnackenberg 2006). The entire collection of metabolites found in a specific sample defined as "metabolome." These metabolites can be studied in samples from living organisms, cell cultures, or environments where organic life has left traces, such as deep sea sediments or crude oil residues. In nature, living organisms respond to internal and external challenges based on their biological makeup (Adamski 2020). While an organism's genome acts as a blueprint, the active agents in biological processes are found in the proteome and metabolome. Genomics provides predictive information, whereas proteomics and metabolomics reveal current biological activities (Fiehn 2002; Nicholson and Wilson 2003). Advancements in metabolomics have significantly improved analytical methods, enhancing their sensitivity, reproducibility, and ability to cover a larger number of metabolites while increasing sample processing efficiency.

Metabolomics is a multidisciplinary field that integrates study design, analytical techniques, and bioinformatics (Fig. 1.1) (Adamski 2020). It benefits from methodologies used in epidemiological research and genomic studies, adapting them to the specific requirements of metabolomics experiments (Jonsson et al. 2015).

K. Panchal · K. Murjani · V. Singh (✉)
Department of Biosciences, School of Science, Indrashil University, Rajpur, Mehsana, Gujarat, India
e-mail: vijai.singh@indrashiluniversity.edu.in

V. Singh (ed.), *Advances in Metabolomics*,
https://doi.org/10.1007/978-981-97-7459-3_1

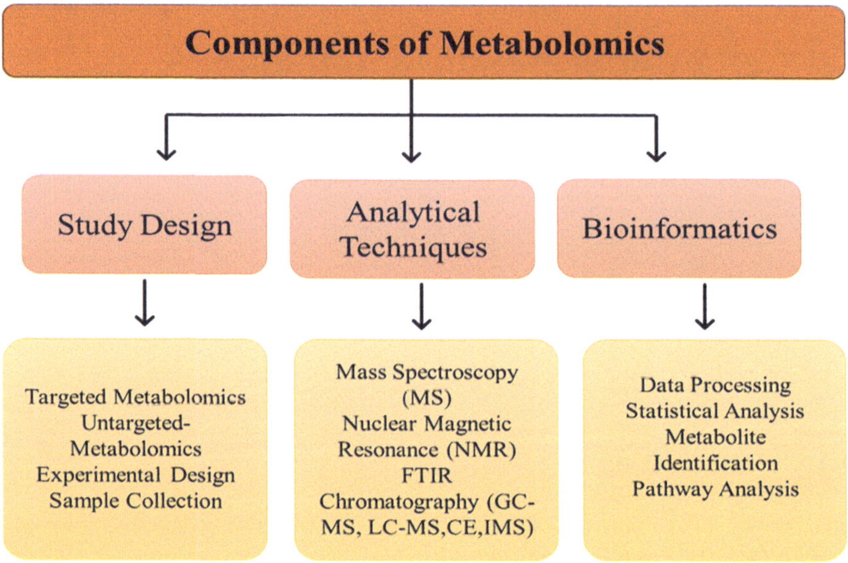

Fig. 1.1 Components of metabolomics

Metabolomics research faces challenges due to variability in sample preparation and analytical techniques (Do et al. 2018; Succop et al. 2004), necessitating rigorous quality control and assurance procedures (Playdon et al. 2019). Analytical technologies such as NMR spectroscopy and mass spectrometry coupled with gas chromatography (GC-MS) or liquid chromatography (LC-MS) are commonly employed (Schnackenberg 2006; Adamski 2020; Kim and Heyman 2018). However, many metabolites have been identified, and currently characterized only by their mass spectra (Jones 2018). Unlike genomics, which provides a comprehensive view of an organism's genetic material, metabolomics has not yet achieved a complete portrayal of the metabolome in any given sample.

1.1.1 Historical Development and Evolution of Metabolomics

Metabolomics has undergone a notable transformation over the past few decades, driven by advances in analytical technologies and integrative biological approaches. The field of metabolomics has its origins in traditional biochemistry, with early research focused on the detection and measurement of individual metabolites (Bujak et al. 2015). The understanding of metabolic pathways and their roles in cellular processes is largely due to these early investigations, which were crucial in establishing the stage for future research. One of the earliest techniques used for metabolite analysis was chromatography, which separates complex mixtures of compounds based on their physical or chemical properties. Gas chromatography (GC) and liquid chromatography (LC) were extensively employed to separate and analyze volatile and non-volatile metabolites, respectively (Kim and Heyman 2018). These techniques allowed for the identification and quantification of specific metabolites in biological samples. Furthermore, spectroscopic methods such as ultraviolet (UV) spectroscopy and infrared (IR) spectroscopy were employed to determine the structural and functional properties of metabolites (Patti et al. 2012). These early techniques, while effective for targeted analysis of known metabolites, were limited by their capacity to analyze complex biological samples comprehensively. The demand for more comprehensive and high-throughput methods spurred significant technological advancements.

The advent of high-throughput analytical techniques in the late twentieth and early twenty-first centuries was significant milestone in the realm of metabolomics. These technologies enabled the simultaneous detection and quantification of hundreds to thousands of metabolites, revolutionizing our ability to study metabolic processes on a large scale. Mass spectrometry (MS) has become fundamental tool in modern metabolomics due to its great sensitivity, specificity, and throughput (de Raad et al. 2016). MS-based techniques, including gas chromatography-mass spectrometry (GC-MS), liquid chromatography-mass spectrometry (LC-MS), and capillary electrophoresis-mass spectrometry (CE-MS), allowed for the comprehensive profiling of complex metabolite mixtures (Dettmer et al. 2007). Advancements in ionization methods, such as electrospray ionization (ESI) and matrix-assisted laser desorption/ionization (MALDI), further enhanced the capability of MS to analyze a wide range of metabolites, including those with diverse chemical properties (Vinaixa et al. 2016). NMR spectroscopy continued to evolve with advancements in instrument sensitivity and resolution. High-field NMR spectrometers enabled detailed structural elucidation and quantification of metabolites in complex biological samples. NMR's nondestructive nature and ability to provide quantitative data without extensive sample preparation made it a valuable tool for metabolomics. The integration of automated sample preparation, data acquisition, and computational analysis platforms facilitated high-throughput metabolomic studies (Emwas et al. 2019). These platforms enabled the rapid analysis of large sample cohorts, generating vast amounts of metabolomic data for comprehensive biological insights. The synergy of these advanced analytical techniques transformed metabolomics from a targeted approach to a global profiling tool, capable of capturing the dynamic and complex nature of metabolic networks.

Moreover, the advent of systems biology approaches in the early 2000s marked a paradigm shift in biological research, emphasizing the need to understand biological systems as integrated and dynamic entities. Metabolomics became an integral component of systems biology, with its ability to provide an overview of a system's metabolic status. Metabolomics began to integrate with other omics disciplines such as genomics, transcriptomics, and proteomics, aiming to offer a comprehensive understanding of biological systems (Nicholson and Lindon 2008; Saito and Matsuda 2010). This multiomics integration enabled researchers to correlate metabolic changes with alterations in gene expression, protein levels, and other molecular processes, thereby uncovering comprehensive insights into cellular functions and regulatory mechanisms. Advances in bioinformatics and computational biology facilitated the merging of metabolomic data and other omics field's data. Systems biology approaches employed computational models to simulate and predict metabolic responses, identify key regulatory nodes, and understand the complex interactions within metabolic networks (Misra et al. 2018). The integration of metabolomics with systems biology represents a significant advancement in our ability to study and understand biological complexity.

1.1.2 Analytical Techniques in Metabolomics

The analysis of mRNA and proteins are easier because if the genome sequence of an organism is known then one can easily analyze it by different methods like Nucleic acid hybridization, and protein separation followed by mass spectrometry can be employed for subsequent analysis. But the analysis of metabolites is not as simple as protein analysis. By nature, metabolites are very varied chemical species that are typically labile and frequently present across a broad dynamic range. Thus, metabolome research represents a significant challenge for analytical chemistry because current analytical techniques have some limitations, like the fact that some techniques may not be able to analyze the metabolite in small concentrations, and these techniques need to be highly sensitive and highly specific. There are various analytical techniques utilized in metabolomics which are summarized in Table 1.1. The main analytical methods used include mass spectrometry (MS) and nuclear magnetic resonance (NMR) spectroscopy, along with other techniques such as Fourier transform infrared (FTIR) spectroscopy, Raman spectroscopy, and ultraviolet-visible (UV-Vis) spectroscopy. These methods are employed to identify and quantify metabolites, providing insights into metabolic pathways and networks (Moco et al. 2007). Nevertheless, none of these methods can cover an organism's whole metabolome on its own. Based on the objectives of the research, these analytical tools can be applied singly, concurrently, or in combination (Gowda and Djukovic 2014).

1.1.3 Mass Spectrometry

MS has been defined by the International Union of Pure and Applied Chemistry (IUPAC) as "the study of systems by a process of forming gaseous ions, with or without fragmentation, which are then characterized by their mass/charge ratios (m/z) and their relative abundances." Since the mass separation process relies on the movement of charged particles (ions) influenced by electric and/or magnetic fields, the species being analyzed must be charged. To ensure the unimpeded travel of these charged particles (ions), high vacuum conditions are required (Baldwin 2005).

Sample preparation is often a prerequisite for MS-based metabolomics. This procedure may have various objectives, including as removing analytes from complex biological samples using methods such as liquid–liquid or solid-phase extractions, derivatizing analytes to increase their volatility and/or decrease their polarity, or

Table 1.1 Different analytical techniques in metabolomics

Technique	Description	Strengths	Limitations	Applications
Mass spectrometry (MS)	Measures the mass-to-charge ratio of ions to identify and quantify metabolites.	High sensitivity, wide range of metabolites detected, structural information.	Requires extensive sample preparation, can be expensive.	Comprehensive metabolite profiling, structural elucidation.
Gas chromatography-MS (GC-MS)	Combines gas chromatography with MS, ideal for volatile and semivolatile compounds.	High sensitivity, good for volatile compounds.	Limited to volatile or derivatizable compounds, requires sample derivatization.	Analysis of small, volatile metabolites, environmental and food analysis.
Liquid chromatography-MS (LC-MS)	Combines liquid chromatography with MS, suitable for nonvolatile and thermally labile compounds.	Broad range of metabolites detected, good sensitivity and specificity.	Complex data processing, can be expensive, requires extensive sample preparation.	Metabolite profiling of biofluids, tissues, and cells.
Tandem MS (MS/MS)	Uses two mass analyzers for structural information via fragmentation patterns.	Provides detailed structural information, high sensitivity.	Requires sophisticated instrumentation, complex data interpretation.	Structural elucidation of metabolites, identification of unknown compounds.
Nuclear magnetic resonance (NMR)	Uses magnetic fields and radio waves to determine molecular structure and dynamics.	Nondestructive, minimal sample preparation, provides structural information.	Lower sensitivity than MS, lower throughput, requires large amounts of sample.	Structural analysis, qualitative and quantitative metabolite analysis.
Fourier transform infrared (FTIR) spectroscopy	Identifies functional groups in metabolites based on infrared absorption.	Nondestructive, rapid analysis.	Limited structural information, lower sensitivity than MS.	Functional group identification, rapid screening of samples.

(continued)

Table 1.1 (continued)

Technique	Description	Strengths	Limitations	Applications
Raman spectroscopy	Provides information about molecular vibrations, offering insights into molecular structures.	Nondestructive, minimal sample preparation, provides structural information.	Lower sensitivity than MS, interference from fluorescence.	Structural analysis, detection of molecular vibrations.
Ultraviolet-visible (UV-Vis) spectroscopy	Measures absorbance of UV and visible light by metabolites.	Rapid analysis, useful for compounds with chromophores.	Limited to chromophore-containing compounds, lower sensitivity than MS and NMR.	Studying compounds with chromophores, quantitative analysis.

preconcentrating the analytes to improve their detection limits (Dettmer et al. 2007). For MS analysis, there are four main processes that can be described: (1) introduction of the compounds to be analyzed in the mass spectrometer through sample introduction system (chromatographic system or insertion probe); (2) in an ionization source, neutral sample molecules yield ions; (3) after that, ions are separated in a mass analyzer; (4) based on their mass-to-charge ratio (m/z); an ion detector allows for the measurement of ion abundance.

The hyphenated techniques are techniques in which both chromatographic and spectrometric techniques are used; these hyphenated techniques can be used for metabolomic studies. Metabolite separation via chromatography improves quantification, minimizes matrix effects, and separates isomers. It also yields more information for metabolite characterization (Agin et al. 2016). Gas chromatography (GC), high-performance liquid chromatography (HPLC or LC), and capillary electrophoresis (CE) are the typical chromatography techniques used in metabolic analyses, but capillary electrophoresis (CE) is not frequently used (Lei et al. 2011). These techniques are used in combination with mass spectrometry for the detailed study of metabolites by measuring the mass-to-charge ratio of ions.

1.1.4 Types of Mass Spectrometry

1.1.4.1 Gas Chromatography (GC/MS)

Gas chromatography (GC) operates on the principle of heating a metabolite to volatilize it in an injector, passing the metabolite through a heated tube (column) for separation, and interacting with a stationary phase. An inert carrier gas like helium or nitrogen can be used to transport the volatilized metabolite through the GC column. The separation of metabolites depends on the flow rate of carrier gas. A component that spends less time to interact with the stationary phase elutes more rapidly (Agin et al. 2016). The separated metabolite can be detected by using mass spectrometry which allows both quantitative and qualitative analysis of metabolites used.

1.1.4.2 Liquid Chromatography (LC/MS)

The stationary phase in LC (or HPLC) is made up of beads that are chemically produced, while a liquid mobile phase is used that is filled in the column and flows through it (Smith et al. 2014). The mobile phase that is used in LC is; water, methanol, acetonitrile, and buffers. The pH, an ionic strength of mobile phase, can be adjusted to optimize separation. For separating metabolites regardless of the polarity or volatility of the metabolites, LC is thought to be more capable than GC (Moco et al. 2007; Bedair and Sumner 2008). There have been several advancements in LC, two of which is ultra-high-pressure liquid chromatography (UHPLC) and capillary liquid chromatography. By decreasing the packing-particle diameter from 5 mm to less than 2 mm and working at pressures between 6000 and 15,000 psi, current UHPLC systems offer greater speed, resolution, and sensitivity. When using capillary chromatography, the column's size decreases along with a drop in flow rate, which raises its sensitivity (Agin et al. 2016).

1.1.4.3 Tandem Mass Spectrometry (MS/MS)

When using relatively low ionizing energy in mass spectrometry, mass spectra frequently provide peaks that indicate entire molecular species with few fragment ions, which may not provide sufficient information for determining the structure of the species. To address this, MS/MS is utilized, a technique that induces fragmentation and additional dissociation to detect and analyze metastable ions (Choi 2020). MS/MS uses two mass analyzers separated by a collision cell to perform multiple phases of mass analysis. In the first mass analyzer (MS1), ions of a specific mass-to-charge ratio (*m/z*) from the sample metabolite are selected. These selected ions are then fragmented using collision-induced dissociation (CID) with inert gas assistance. The fragments produced are analyzed by a second mass analyzer (MS2) to obtain a mass spectrum. The same type of analyzer, such as an ion-trap or triple quadrupole, might be used to select the first and second stages. In triple quadrupole instrumentation, the MS/MS process involves three sequential steps: selecting the precursor ion, performing collision-induced dissociation (CID), and analyzing the product ions' mass spectra in spatially separated stages (tandem-in-space). Alternatively, these steps can be performed consecutively within the same device (tandem-in-time) when using an ion-trap instrument (Choi 2020).

1.1.4.4 Nuclear Magnetic Resonance Spectroscopy

Nuclear magnetic resonance (NMR) exploits atomic nuclei's magnetic properties. When a sample metabolite is placed in a strong magnetic field and exposed to radiofrequency (RF) pulses, these nuclei vibrate at their distinct frequencies. The resulting signal yields comprehensive insights into the molecular structure, dynamics, and surroundings of the metabolites. NMR has been used for many years to better understand chemical structures. Recently, it has been expanded to analyze complex mixtures, including biological samples, through NMR-based metabolomics.

However, NMR is also useful in other small fields of molecular biology. These include the quantification of metabolites (qNMR); the utilization of stable isotope tracers to ascertain the drug or nutrient metabolism; the discovery of novel metabolic pathways and flux through them; and the analysis of metabolite-protein interactions to better understand metabolic regulation and pharmacological implications (Moco 2022). NMR offers several advantages over mass spectrometry; NMR is a nondestructive technique, because it allows preservation of the metabolite sample. It is a quantification method in which no external standard is required. It typically necessitates minimal sample preparation, rendering it well-suited for analyzing a diverse range of biological samples. One of the disadvantages of NMR is that, in comparison to other analytical methods, it has a somewhat poor signal-to-noise ratio. NMR measurements can be made more sensitive by using several techniques, such as the creation of cryoprobes and microprobes and the use of hyperpolarized substrates by DNP (Moco 2022).

Apart from MS and NMR spectroscopy, few other pivotal techniques are also used in metabolomics such as Fourier transform infrared (FTIR) spectroscopy, Raman spectroscopy, and ultraviolet-visible (UV-Vis) spectroscopy, each offering unique advantages for analyzing metabolites. FTIR spectroscopy measures infrared absorption to characterize functional groups and molecular structures, providing rapid, nondestructive metabolic profiling of various biological samples (Hendriks et al. 2011). Raman spectroscopy uses inelastic light scattering to detect molecular vibrations with high specificity and sensitivity, making it ideal for detailed metabolic studies, including in vivo applications and cancer diagnostics (Cebi et al. 2023). UV-Vis spectroscopy quantifies metabolites by their light absorption in the ultraviolet and visible ranges, enabling fast, high-throughput screening of chromophores and conjugated systems (Hendriks et al. 2011). Together, these techniques offer comprehensive insights into metabolic pathways, biochemical changes, and disease states, enhancing our understanding of metabolism in health and disease.

1.1.5 Data Acquisition and Processing

Data acquisition and processing in metabolomics involve a series of precise steps to ensure the generation of comprehensive and detailed metabolic profiles. High-throughput data acquisition methods, such as liquid chromatography-mass spectrometry (LC-MS), gas chromatography-mass spectrometry (GC-MS), capillary electrophoresis-mass spectrometry (CE-MS), nuclear magnetic resonance (NMR) spectroscopy, and Fourier transform infrared (FTIR) spectroscopy, are fundamental for efficiently capturing a wide array of metabolites (Hendriks et al. 2011; Fenaille et al. 2017). LC-MS is particularly versatile, capable of handling nonvolatile and thermally unstable compounds using ionization techniques like electrospray ionization (ESI) and atmospheric pressure chemical ionization (APCI). An approach that is still relatively new in the realm of metabolomics studies is matrix-assisted laser desorption/ionization mass spectrometry (MALDI-MS), which is discussed as a potential substitute for ESI (Ernst et al. 2014).

GC-MS, requires derivatization, excels at analyzing small, volatile compounds such as fatty acids and amino acids. CE-MS is useful for charged or polar metabolites, offering high resolution but lower throughput compared to LC-MS and GC-MS (Yao et al. 2019). NMR spectroscopy provides detailed structural information about metabolites with minimal sample preparation, though with lower sensitivity compared to MS (Emwas et al. 2019). Nontargeted spectroscopy techniques such as FTIR, NIR, and Raman can swiftly and cost-effectively provide fingerprint information on metabolomic constituents found in agricultural products, natural products, and foods. These methods are known for their high-throughput capabilities (Cebi et al. 2023). Ensuring data quality and consistency in metabolomics necessitates rigorous standardization and calibration protocols. Internal standards, comprising chemically similar compounds added in precise quantities to each sample, correct for variability arising from sample preparation and instrument response (Liu et al. 2020).

Regular analysis of quality control (QC) samples is essential for monitoring instrument performance and detecting operational drifts or anomalies. Calibration curves from standard solutions with known metabolite concentrations enable accurate quantification by comparing signal intensities of unknown samples. Blank samples help identify and exclude background noise, ensure the integrity of detected metabolic signals. Preprocessing of raw data involves several key steps to enhance analysis: noise reduction through smoothing (e.g., Savitzky–Golay filtering) and denoising (e.g., wavelet transform), removes random noise (Li et al. 2011), while baseline correction addresses drifts from instrument instability or sample impurities, which ensures accurate peak detection. Peak detection and deconvolution algorithms, like Gaussian fitting or wavelet transform, accurately identify and separate overlapping peaks in chromatographic data. Alignment corrects shifts in retention times or spectral data using algorithms like XCMS to ensure consistent comparison of metabolites across samples. Normalization techniques such as total ion current (TIC) normalization, probabilistic quotient normalization (PQN), and variance stabilization normalization (VSN) adjust for variations in sample concentration, injection volume, or instrument response (Wu and Li 2016). These preprocessing steps are crucial for obtaining high-quality, reliable data, facilitating accurate analysis and interpretation in metabolomics research.

Statistical analysis in metabolomics involves both multivariate and univariate methods to uncover patterns and relationships in complex data sets. Principal component analysis (PCA) is a method for reducing the dimensionality of data by transforming it into principal components, highlighting variance and identifying patterns, useful for visualizing data clusters and identifying outliers. Partial least squares-discriminant analysis (PLS-DA) is a supervised technique that models the relationship between metabolites and class labels, aiding in classification and biomarker identification (Lazar et al. 2015). Data integration and visualization combine and present data to derive meaningful insights. Integrating metabolomics data with other omics data (genomics, proteomics) using techniques like network analysis and pathway enrichment analysis provides a comprehensive understanding of biological processes. Visualization tools, including heatmaps, volcano plots, score plots (PCA, PLS-DA), and pathway maps, facilitate the interpretation of complex data. Various software tools and platforms like MetaboAnalyst, XCMS, and MZmine support data analysis and visualization, enhancing the interpretation and presentation of metabolomic data (Montenegro-Burke et al. 2017; Chong et al. 2018; Pluskal et al. 2010).

1.1.6 Bioinformatics and Computational Biology Tools

Bioinformatics and computational biology play a critical role in metabolomics studies by extracting meaningful insights from the vast amounts of data produced in metabolomics experiments. Global metabolomics provides the potential for expanding the known chemical space through the analysis of both known and undiscovered metabolites from biological samples (Palermo 2023). Nowadays, open-source databases are available which contains data on the genome, transcriptome, proteome, metabolome, etc., such as EcoCyc, YMDB, PMN, MetaboLights, NIST, PubChem, PRIDE, Reactome, MetaCyc, HMDB, GNPS, Dictionary of Natural Products (DNP), Dictionary of Marine Natural Products, and Collection of Open Natural Products (COCONUT), etc., can annotate known metabolites (Palermo 2023; Pinu et al. 2019). These are employed in early-stage compound dereplication (i.e., restricting compound rediscovery) in natural product discovery. Data can be compared to theoretical libraries to increase the scope of molecular annotations for undiscovered molecules. The process for annotating predicted metabolites can be improved by integrating bioinformatics tools like Metabolic In Silico Network Expansion (MINE). The process involves comparing metabolic properties to a larger database derived from the computer prediction of molecules based on biological events and metabolites that are included in the Kyoto Encyclopedia of Genes and Genomes (KEGG) (Jeffryes et al. 2015).

The entire collection of metabolites that an organism can produce if its latent biosynthetic potential is stored in minimally expressed and silenced genes is known as its "silent metabolome" (Lopez-Lopez et al. 2018). In fact, there are undiscovered biosynthetic gene clusters (BGCs) in the genomes of plants and microbes that need to be mined for new metabolic pathways and chemistries. Genome sequencing combined with bioinformatics (e.g., secondary metabolite genome mining utilizing antiSMASH, PRISM, EvoMining, ClusterFinder, and DeepBGC) and in silico genome analysis in BGC-to-compound workflows can be used to identify these BGCs (Palermo 2023). Machine learning algorithms such as Random Forest and SVM further refine data analysis, predicting disease outcomes and classifying phenotypes. Visualization tools like Cytoscape and MetScape then facilitate intuitive exploration and presentation of metabolomics data through interactive graphs and network diagrams (Pinu et al. 2019). Together, these bioinformatics tools not only enhance our understanding of metabolic pathways and disease mechanisms but also pave the way for personalized medicine and biomarker discovery, driving innovation in biomedical research and clinical applications.

1.1.7 Importance and Applications of Metabolomics

Metabolomics extends its applications across various disciplines such as clinical diagnostics, personalized medicine, pollution monitoring, ecological interactions, crop improvement, food safety, bioprocess optimization, drug discovery, etc., highlighting its versatility and importance. Metabolomics enables the identification of metabolic biomarkers associated with various diseases, which facilitates early diagnosis and accurate prognosis. For instance, specific metabolite profiles can indicate the presence of cancer, diabetes, cardiovascular diseases, or neurological disorders (Lopez-Lopez et al. 2018). By analyzing individual metabolic profiles, metabolomics supports the development of personalized treatment plans. Metabolomics helps in assessing the impact of environmental pollutants on living organisms. By profiling metabolites, researchers can detect biochemical changes induced by exposure to pollutants, aiding in the development of strategies for environmental protection and remediation (Bundy et al. 2009). Furthermore, understanding of the metabolic interactions within ecosystems helps in studying the effects of environmental changes on biodiversity and ecosystem health. Metabolomics contributes to the identification of key metabolites involved in plant growth, stress responses, and disease resistance (Shulaev et al. 2008). This knowledge facilitates in the development of crops with improved yield, nutritional quality, and resilience to environmental stresses. Metabolomic analysis ensures food safety by detecting contaminants, toxins, and adulterants in food products. It also helps in monitoring the metabolic changes in food during processing and storage, ensuring quality control. Metabolomics aids in optimizing biotechnological processes by monitoring the metabolic state of microbial or cell cultures. It helps in enhancing the efficiency of production processes for pharmaceuticals, biofuels, and other industrial products (Campbell et al. 2017).

Metabolomics identifies potential drug targets and evaluates drug efficacy and toxicity by analyzing metabolic changes from specific compounds. It aids in early diagnosis, disease monitoring, and targeted therapy development by detecting disease-related metabolite changes. Early detection through biomarkers allows timely intervention, while monitoring metabolite profiles during treatment offers insights into disease progression and therapy effectiveness. This understanding facilitates precise interventions, leading to more effective treatments (Wishart 2016). In addition, Pharmacometabolomics is a branch of metabolomics that focuses on understanding individual variations in metabolic responses to drugs and helps in predicting and preventing adverse effects, improving patient safety. It plays a critical role in personalized medicine by optimizing treatment strategies, by identifying metabolic markers of drug efficacy and toxicity, and by understanding the metabolic pathways involved in drug metabolism (Everett 2015). Nutritional metabolomics examines how diet influences metabolite profiles, providing insights into nutritional status, dietary intake, and personalized dietary recommendations. It helps in evaluation of an individual's nutritional status by analyzing metabolite concentrations related to nutrient intake and metabolism. This information is valuable for identifying nutrient deficiencies or imbalances and guiding dietary interventions. By considering genetic, metabolic, and lifestyle factors, personalized nutrition aims to optimize health and prevent diet-related diseases (Kortesniemi et al. 2023).

1.2 Conclusion and Future Outlook

In conclusion, the evolution of metabolomics reflects a journey from targeted analysis to comprehensive profiling and integration with systems biology. Technological advancements in analytical techniques and computational methods have transformed metabolomics into a powerful tool for biological and medical research. Its application across diverse fields highlights its potential to unlock new insights into the molecular mechanisms underlying health, disease, and environmental interactions, promising significant advancements in scientific understanding and practical applications.

Acknowledgments K.P. and V.S. acknowledge the Science and Engineering Research Board, New Delhi (Project ID: SUR/2022/001937) for financial support.

References

Adamski J (2020) Metabolomics for biomedical research. Elsevier, USA, p 266
Agin A, Heintz D, Ruhland E et al (2016) Metabolomics–an overview. From basic principles to potential biomarkers (part 1). Médecine Nucléaire 40(1):4–10
Baldwin MA (2005) Mass spectrometers for the analysis of biomolecules. Methods Enzymol 402:3–48
Bedair M, Sumner LW (2008) Current and emerging mass-spectrometry technologies for metabolomics. TrAC Trends Anal Chem 27:238–250
Bujak R, Struck-Lewicka W, Markuszewski MJ, Kaliszan R (2015) Metabolomics for laboratory diagnostics. J Pharm Biomed Anal 113:108–120
Bundy JG, Davey MP, Viant MR (2009) Environmental metabolomics: a critical review and future perspectives. Metabolomics 5:3–21
Campbell K, Xia J, Nielsen J (2017) The impact of systems biology on bioprocessing. Trends Biotechnol 35(12):1156–1168
Cebi N, Bekiroglu H, Erarslan A (2023) Nondestructive metabolomic fingerprinting: FTIR, NIR and Raman spectroscopy in food screening. Molecules 28(23):7933
Choi MH (2020) Mass spectrometry-based metabolite analytics. In: Metabolomics for biomedical research. Academic Press, pp 57–82. https://doi.org/10.1016/B978-0-12-812784-1.00004-9
Chong J, Soufan O, Li C, Caraus I, Li S, Bourque G et al (2018) MetaboAnalyst 4.0: towards more transparent and integrative metabolomics analysis. Nucleic Acids Res 46(W1):W486–W494
Dettmer K, Aronov PA, Hammock BD (2007) Mass spectrometry-based metabolomics. Mass Spectrom Rev 26(1):51–78
Do KT, Wahl S, Raffler J, Molnos S, Laimighofer M, Adamski J et al (2018) Characterization of missing values in untargeted MS-based metabolomics data and evaluation of missing data handling strategies. Metabolomics 14(10):128
Emwas AH, Roy R, McKay RT, Tenori L, Saccenti E, Gowda GAN et al (2019) NMR spectroscopy for metabolomics research. Meta 9(7):123
Ernst M, Silva DB, Silva RR, Vencio RZ, Lopes NP (2014) Mass spectrometry in plant metabolomics strategies: from analytical platforms to data acquisition and processing. Nat Prod Rep 31(6):784–806

Everett JR (2015) Pharmacometabonomics in humans: a new tool for personalized medicine. Pharmacogenomics 16(7):737–754
Fenaille F, Barbier Saint-Hilaire P, Rousseau K, Junot C (2017) Data acquisition workflows in liquid chromatography coupled to high resolution mass spectrometry-based metabolomics: where do we stand? J Chromatogr A 1526:1–12
Fiehn O (2002) Metabolomics—the link between genotypes and phenotypes. Plant Mol Biol 48(1–2):155–171
Gowda GA, Djukovic D (2014) Overview of mass spectrometry-based metabolomics: opportunities and challenges. Methods Mol Biol 1198:3–12
Hendriks MMWB, FAV E, Jellema RH, Westerhuis JA, Reijmers TH, Hoefsloot HCJ et al (2011) Data-processing strategies for metabolomics studies. TrAC Trends Anal Chem 30(10):1685–1698
Jeffryes JG, Colastani RL, Elbadawi-Sidhu M, Kind T, Niehaus TD, Broadbelt LJ et al (2015) MINEs: open access databases of computationally predicted enzyme promiscuity products for untargeted metabolomics. J Cheminform 7:44
Jones OAH (2018) Illuminating the dark metabolome to advance the molecular characterisation of biological systems. Metabolomics 14(8):101
Jonsson P, Wuolikainen A, Thysell E, Chorell E, Stattin P, Wikstrom P et al (2015) Constrained randomization and multivariate effect projections improve information extraction and biomarker pattern discovery in metabolomics studies involving dependent samples. Metabolomics 11(6):1667–1678
Kim YM, Heyman HM (2018) Mass spectrometry-based metabolomics. Methods Mol Biol 1775:107–118
Kortesniemi M, Noerman S, Karlund A, Raita J, Meuronen T, Koistinen V et al (2023) Nutritional metabolomics: recent developments and future needs. Curr Opin Chem Biol 77:102400
Lazar AG, Romanciuc F, Socaciu MA, Socaciu C (2015) Bioinformatics tools for metabolomic data processing and analysis using untargeted liquid chromatography coupled with mass spectrometry. Bull Univ Agric Sci Vet Med Cluj-Napoca, Anim Sci Biotechnol 72(2):103–115
Lei Z, Huhman DV, Sumner LW (2011) Mass spectrometry strategies in metabolomics. J Biol Chem 286(29):25435–25442
Li CP, Han JQ, Huang QB, Mu N, Zhu DZ, Guo CT et al (2011) An integrated on-line processing method for spectrometric data based on wavelet transform and Gaussian fitting. Guang Pu Xue Yu Guang Pu Fen Xi 31(11):3050–3054
Liu KH, Nellis M, Uppal K, Ma C, Tran V, Liang Y et al (2020) Reference standardization for quantification and harmonization of large-scale metabolomics. Anal Chem 92(13):8836–8844
Lopez-Lopez A, Lopez-Gonzalvez A, Barker-Tejeda TC, Barbas C (2018) A review of validated biomarkers obtained through metabolomics. Expert Rev Mol Diagn 18(6):557–575
Misra BB, Langefeld CD, Olivier M, Cox LA (2018) Integrated omics: tools, advances, and future approaches. J Mol Endocrinol:JME-18-0055. https://doi.org/10.1530/JME-18-0055
Moco S (2022) Studying metabolism by NMR-based metabolomics. Front Mol Biosci 9:882487
Moco S, Vervoort J, Moco S, Bino RJ, De Vos RCH, Bino R (2007) Metabolomics technologies and metabolite identification. TrAC Trends Anal Chem 26(9):855–866
Montenegro-Burke JR, Aisporna AE, Benton HP, Rinehart D, Fang M, Huan T et al (2017) Data streaming for metabolomics: accelerating data processing and analysis from days to minutes. Anal Chem 89(2):1254–1259
Nicholson JK, Lindon JC (2008) Systems biology: metabonomics. Nature 455(7216):1054–1056
Nicholson JK, Wilson ID (2003) Opinion: understanding 'global' systems biology: metabonomics and the continuum of metabolism. Nat Rev Drug Discov 2(8):668–676
Palermo A (2023) Metabolomics- and systems-biology-guided discovery of metabolite lead compounds and druggable targets. Drug Discov Today 28(2):103460
Patti GJ, Yanes O, Siuzdak G (2012) Innovation: metabolomics: the apogee of the omics trilogy. Nat Rev Mol Cell Biol 13(4):263–269

Pinu FR, Beale DJ, Paten AM, Kouremenos K, Swarup S, Schirra HJ et al (2019) Systems biology and multi-omics integration: viewpoints from the metabolomics research community. Meta 9(4):76
Playdon MC, Joshi AD, Tabung FK, Cheng S, Henglin M, Kim A et al (2019) Metabolomics analytics workflow for epidemiological research: perspectives from the consortium of metabolomics studies (COMETS). Meta 9(7):145
Pluskal T, Castillo S, Villar-Briones A, Oresic M (2010) MZmine 2: modular framework for processing, visualizing, and analyzing mass spectrometry-based molecular profile data. BMC Bioinformatics 11:395
de Raad M, Fischer CR, Northen TR (2016) High-throughput platforms for metabolomics. Curr Opin Chem Biol 30:7–13
Saito K, Matsuda F (2010) Metabolomics for functional genomics, systems biology, and biotechnology. Annu Rev Plant Biol 61:463–489
Schnackenberg LK (2006) Metabolomics special focus: an introduction. Pharmacogenomics 7(7):1053–1054
Shulaev V, Cortes D, Miller G, Mittler R (2008) Metabolomics for plant stress response. Physiol Plant 132(2):199–208
Smith R, Mathis AD, Ventura D, Prince JT (2014) Proteomics, lipidomics, metabolomics: a mass spectrometry tutorial from a computer scientist's point of view. BMC Bioinformatics 15(Suppl 7):S9
Succop PA, Clark S, Chen M, Galke W (2004) Imputation of data values that are less than a detection limit. J Occup Environ Hyg 1(7):436–441
Vinaixa M, Schymanski EL, Neumann S, Navarro M, Salek RM, Yanes O (2016) Mass spectral databases for LC/MS-and GC/MS-based metabolomics: state of the field and future prospects. TrAC Trends Anal Chem 78:23–35
Wishart DS (2016) Emerging applications of metabolomics in drug discovery and precision medicine. Nat Rev Drug Discov 15(7):473–484
Wu Y, Li L (2016) Sample normalization methods in quantitative metabolomics. J Chromatogr A 1430:80–95
Yao L, Sheflin AM, Broeckling CD, Prenni JE (2019) Data processing for GC-MS- and LC-MS-based untargeted metabolomics. Methods Mol Biol 1978:287–299

Applications of Computational and Data Sciences in Metabolomics

2

Nhat-Le Bui, Quy Do Van, Hoang Thanh Nhat,
Trinh Chi Dung, Phan Van Bao, Vijai Singh, Fong Chin Su,
Chia-Ching Wu, and Dinh-Toi Chu

Nhat-Le Bui and Quy Do Van are co-first authors.

N.-L. Bui · H. T. Nhat · T. C. Dung
The Interdisciplinary Research Group on Biomedicine and Health (IRGBH), International School, Vietnam National University, Hanoi, Vietnam

Faculty of Applied Sciences, International School, Vietnam National University, Hanoi, Vietnam

Q. Do Van · D.-T. Chu (✉)
The Interdisciplinary Research Group on Biomedicine and Health (IRGBH), International School, Vietnam National University, Hanoi, Vietnam

Department of Biomedical Engineering, National Cheng Kung University, Tainan, Taiwan
e-mail: toicd@vnu.edu.vn

P. Van Bao
The Interdisciplinary Research Group on Biomedicine and Health (IRGBH), International School, Vietnam National University, Hanoi, Vietnam

V. Singh
Department of Biosciences, School of Science, Indrashil University, Rajpur, Mehsana, Gujarat, India

F. C. Su
Department of Biomedical Engineering, National Cheng Kung University, Tainan, Taiwan

C.-C. Wu
Department of Biomedical Engineering, National Cheng Kung University, Tainan, Taiwan

Department of Cell Biology and Anatomy, College of Medicine, National Cheng Kung University, Tainan, Taiwan

International Center for Wound Repair and Regeneration, National Cheng Kung University, Tainan, Taiwan

V. Singh (ed.), *Advances in Metabolomics*,
https://doi.org/10.1007/978-981-97-7459-3_2

Abbreviations

AD	Alzheimer's disease
ANNs	Artificial neural networks
DA	Discrete fall analysis
DIMS	Direct infusion mass spectrometry
GUI	Graphical user interface
HMDB	Human Metabolome Database
LC-MS	Liquid chromatography mass spectrometry
MS	Mass spectrometry
NMR	Nuclear magnetic resonance
PCA	Principal component analysis
PD	Parkinson's disease
PLS	Possibly least squares

2.1 Introduction

Metabolism is an essential process to the body, providing important elements and energy for biological processes like cell growth and adaptation (Wiechert and Nöh 2013). With a group of enzymes, new chemicals are created from just one chemical, contributing to the diversity of substances human body. Metabolism is a complicated process that creates various substances in the body, known as metabolites, which have wide-ranging effects (Zhang and Stewart 2013). In general, metabolomics is a field of systems biology that involves the analysis of small molecule metabolites present in tissues or biofluids using analytical chemistry techniques. It provides the underlying pathophysiology and assessment of metabolites (Fuhrer and Zamboni 2015). The combination of metabolomic and computational analysis can bring useful therapeutic and diagnostic methods that affect disease management.

Several disease conditions are applied metabolomic biomarkers for diagnosis and treatment. Nowadays, it is so hard for early disease diagnoses due to the expensive and time-consuming. In a recent study, the association between Alzheimer's disease (AD) and plasma protein tau levels was reported, which highlighted that higher total tau levels correlate with greater cognitive decline in patients, impacting memory, visualization, and reflection (Mielke et al. 2017). In another study, Proton-NMR-based metabolomics was used to identify changes in 22 metabolites in patients with AD simply by collecting blood samples or saliva. This approach facilitates easier early diagnosis and reduces the cost of management of AD (Yilmaz et al. 2017). At present, Parkinson's disease (PD) impacts over four million people in the world, but it is still unknown the etiology despite extensive research (Chen-Plotkin et al. 2018; Shao and Le 2019). Metabolomics has the potential to find a simpler, faster, and less invasive diagnostic method. In a study by Saiki et al. conducted on 18 metabolites specific to PD, the authors found decreases in seven metabolites, which could be promising biomarkers for diagnosing PD (Saiki et al.

2017). Another study analyzed the plasma levels of 184 metabolites using targeted metabolomics and indicated that PD patients at advanced stage show imbalanced kynurenine metabolism with a shift toward quinolinic acid and away from kynurenic acid (Chang et al. 2018).

Metabolomics is now more popular for its comprehensive and detailed analysis. The establishment of metabolomics core units and research centers in academia has grown as well as attention from pharmaceutical and biotechnology industries. Some biomarkers have been discovered for complex diseases such as diabetes (Arneth et al. 2019), Alzheimer's disease (Altuna-Azkargorta and Mendioroz-Iriarte 2021), Parkinson's disease (Aderemi et al. 2021), cardiovascular diseases (Iida et al. 2019), and drug assessment (De Castro et al. 2019). Since established in the 1990s, "Metabolomics" has proven as a valuable tool for analyzing biological systems. It is combined into human-related studies including genomic, microbiomics, and proteomics, which are part of biology systems, helping us understand the underlying mechanisms of rare diseases (Dias and Koal 2016).

Despite the advantages of metabolomics, it still has some challenges including standardization and quality management, it should be careful to prevent errors in the experiment before analysis (Trifonova et al. 2023). Few variables should be selected to make a stable data result. Statistical analysis is also a thing to be careful, determining the results of certain cohorts can be challenging because of the false negative data or the true negative data (Di Minno et al. 2022). Metabolomic data is mostly nonlinear. For decades, community approval and limitations in computational capacity, as well as a lack of data volume for robust nonlinear model optimization, have led to limited use of metabolomics (Mendez et al. 2019a). Recently, advances in computational power, model improvements, community and data acceptance, and computer science applications in metabolic data processing have attracted strong interest from research communities (Ghosh et al. 2020a; Ren et al. 2015). The rapid progress in the field of metabolomics generates numerous informative data, which needed suitable statistical and computational methods to analyze it.

Therefore, in this review, we aim to summarize the development and application of data science and computational science on metabolomics, which is a technique that deals with large amounts of data and provides more accurate results. This technique can be very useful in the field of medicine as it can help in the discovery of novel biomarkers, biological pathways, and underlying mechanisms of diseases (Chen et al. 2022).

2.2 The Development and Application of Computational Science in Metabolomics

2.2.1 The Development of Data Science in Metabolomics

Metabolomics was first presented around 15 years ago and has been continuously evolving ever since (Courant et al. 2014). Metabolism has witnessed a tremendous evolution throughout the history of data science development and application, which may be attributed to the rapid advancements in technology and methods. With the power to provide light on the biological condition of a living system, the study of small molecules in organisms, or metabolomics, is an essential part of research on biological systems (van der Greef et al. 2013; Balashova et al. 2020).

The history of metabolomics has been significantly influenced by the development of two essential instruments for conducting this field of study: mass spectrometry (MS) and nuclear magnetic resonance (NMR) spectroscopy. These instruments have made it possible to analyze chemical components with a high degree of accuracy (Courant et al. 2014; Amberg et al. 2017). Following processing, measurements from NMR spectroscopy or MS will produce complex data that must be integrated with data science, machine learning techniques, predictive models, statistics, and data integration to maximize data value (Ren et al. 2015; Ghosh et al. 2020b). In the process of developing predictive biomarkers and categorizing vast amounts of omics data, machine learning algorithms also prove to be useful in this regard (Picard et al. 2021). Support vector machines (SVM), decision trees, random forests (RF), neural networks, and deep learning are several machine learning techniques that are often used to evaluate metabolomic data (Galal et al. 2022). Biological data are frequently nonlinear, and metabolomic data may also have a possibly nonlinear structure, according to Kevin M. Mendez et al. As a result, the machine learning approach based on artificial neural networks (ANNs) will be a useful way to reduce computation times and enhance prediction models (Mendez et al. 2019a).

Thanks to the development of the omics field, many data science tools have been introduced for utilization in metabolomic analysis. In a recent review publication by Misra et al., the author summarized current tools for data analysis in metabolomics according to their specific functions (Misra 2021). Most of the current tools for metabolomic data are R-package, targeted on liquid chromatography mass spectrometry (LC-MS), or NMR data. These tools can be divided into open-source and closed-source, which means the code is available or unavailable for customers to inspect. While closed-source tools are stable and familiar to users, open-source tools allow users to embed additional functionality (Mendez et al. 2019b). Several open-source workflows provide a graphical user interface (GUI) that is convenient for novice scientists like Galaxy-M, MetaboAnalyst, or Workflow4Metabolomics (Pang et al. 2021; Giacomoni et al. 2015). Particularly, Galaxy-M is a workflow based on the Galaxy platform dealing with both LC-MS and direct infusion mass spectrometry (DIMS) metabolomic data, from preprocessing to principal component analysis (PCA) and statistical evaluation (Davidson et al. 2016). However, since GUI workflow has its own limitations, data scientists continue to develop more flexible tools in high-level programming languages like Python, Perl, and R. Thanks to the robust and comprehensive coding libraries of these languages, machine learning and statistical packages for metabolomic data have been developed to support complicated analysis (Mendez et al. 2019b).

Metabolomics has rapidly evolved over the past 15 years, driven by advancements in technologies like mass spectrometry and nuclear magnetic resonance spectroscopy. These advancements have enabled precise analysis of metabolites, critical for understanding biological systems. The field has increasingly utilized data science and machine learning techniques, notably ANNs, to interpret complex datasets and improve prediction models. Therefore, tools for data analysis have evolved from basic software to sophisticated platforms that facilitate comprehensive analysis and visualization (Bundy et al. 2009; Hollywood et al. 2006).

2.2.2 The Application of Data Sciences in Metabolomics

Since research on metabolomics releases a huge amount of data, to fully find out the underlying biological meanings and perform several integrated analyses, metabolomics greatly relies on data science (Kisiel et al. 2023). The main workflow of metabolomic data science involves several steps as follows: imported data were cleaned based on PCA quality assessment, and then gone through univariate statistics or multivariate machine learning including possibly least squares discrete fall analysis (PLS-DA) model structure cross-validation, and candidate model building, in which permutation testing, model evaluation, plot latent projections, as well as plot feature importance were performed (Mendez et al. 2019b).

From the late twentieth century to the present, metabolomics has been a relatively young field of study. Basic software was the first analytical tool used to handle data from NMR and MS, with a primary focus on quantification and individual analysis of metabolites (Courant et al. 2014). The amount of data generated by NMR and MS has been more complex till recently due to technological advancements (Misra and van der Hooft 2016). From 2011 to 2015, there were accessible programs such as the metaP-Server tool that helped identify metabolic biomarkers for drug testing, prediction, and treatment (Fig. 2.1) (Kastenmüller et al. 2011). In addition, metaX tools that helped with the interpretation of Metabolomics data and offered additional features to aid with data analysis, computation abilities, variable analysis, and evaluation of data quality (Wen et al. 2017). Since 2018, more tools have been developed to help the study of metabolic data. One such tool is SECIMTools, which facilitates statistical analysis, visualization, and machine learning classification techniques like SVM and RF that support the field of customized healthcare (Kirpich et al. 2018). In order to handle extremely complex data sets and produce disease diagnosis models, metabolomic data analysis frequently employs techniques like RF and SVM (Galal et al. 2022). According to Han She et al., SVM and RF machine learning algorithms are also applied to identify and predict sepsis sufferers' prognosis (She et al. 2023). The MetaboAnalyst toolbox, which was first introduced in 2009, is one of the most common tools for metabolomic data processing and is an important tool that we cannot overlook. The most recent version, MetaboAnalyst 5.0, has a robust processor and a plethora of functions as of 2021. The popular techniques supported by the robust legacy toolbox MetaboAnalyst include PCA, metabolic pathway analysis (MetPA), analysis with visualization (MSEA), precision and resolution of MS measurements are excellent (HRMS), and PLS-DA (Pang et al. 2021; Chong et al. 2019; Xia and Wishart 2016; Xia et al. 2012).

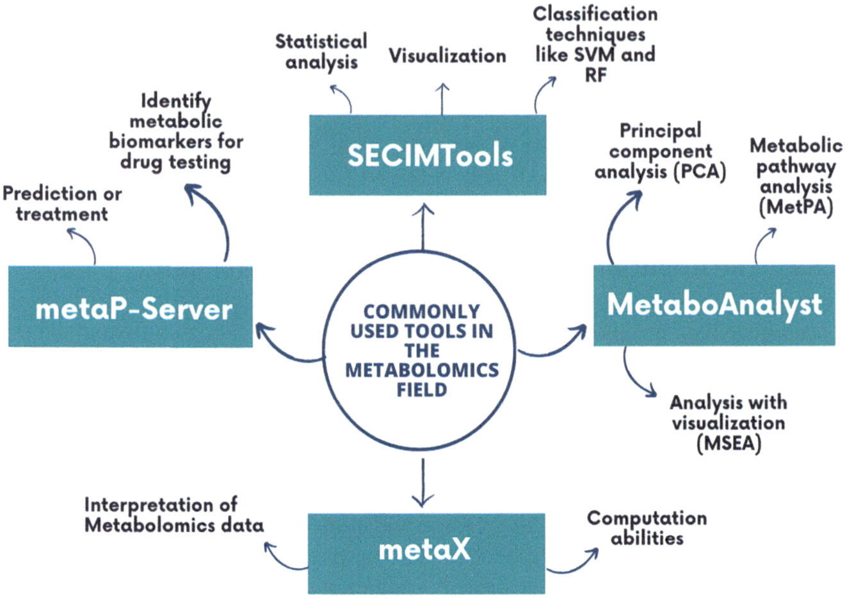

Fig. 2.1 Commonly used tools in the metabolomics field

Regarding databases, the Human Metabolome Database (HMDB) is one of the first databases in metabolomics. HMDB provides abundant information about human metabolites and their related physiological, chemical, and biological characteristics. The latest version of HMDB is 5.0 (updated in 2022), which expands the number of metabolite entries, integrates new structures and biological pathways, as well as enhances the quality of metabolite description and prediction (Wishart et al. 2022). Other databases listing and storing MS data of metabolites or chemical standards include METLIN, MassBank, CSMDB database, etc. (Johnson and Lange 2015; Xue et al. 2020). In particular, a database that should be taken into account is EMBL-MCE database. EMBL-MCE stores LC-MS/MS spectra of more than 400 standards of lipids and endogenous metabolites with open access (https://curatr.mcf.embl.de/) (Phapale et al. 2021).

2.3 The Development and Application of Computational Science in Metabolomics

2.3.1 The Development of Computational Science in Metabolomics

Metabolomics, complex in structure, has been modeled using different projection-based machine learning methods, PLS regression (Jones et al. 2013). In its infancy, which took place in the last 10 years of the twentieth century, computer science used basic statistical methods and simple data processing methods such as component analysis to approach problems related to metabolomics. In the first 10 years of the twenty-first century, the strong development of statistical methods and machine learning has been widely applied in the field of metabolomics. Typically PCA, PLS-DA, and regression models (e.g., SVMs) have been used to analyze data and detect samples (Uarrota et al. 2014; Nyamundanda et al. 2010). During the period 2010–2020, strong advances in computer science such as deep learning and deep learning methods were integrated into metabolomics data processing (Pomyen et al. 2020). Models such as recurrent neural network (RNN), long-run recurrent neural network, and convolutional neural network models have been applied in most data analysis complexity, uncovering hidden patterns (Sha et al. 2024; Fan et al. 2019). In recent years, scientists have focused on developing computational methods capable of deeply understanding metabolomics data, especially the combination of deep learning with methods based on supervised and unsupervised machine learning monitors. Since then, efforts to create user-friendly and interoperable data analysis tools have also accelerated (Lee and Hu 2019).

Computer science is largely applied in all aspects of metabolomics analysis and processing, at least in four main topics: (1) Data processing—provides tools and techniques for effective data processing, including data preprocessing methods such as data cleaning, standardization data visualization, and shaping (Katajamaa and Orešič 2007; Hendriks et al. 2011). Large and complex data will be handled effectively by big data and cloud computing (Fu et al. 2022); (2) Data analysis (Lamichhane et al. 2018)—machine learning is applied to metabolomics data analysis, such as sample classification, protein expression analysis, gene expression detection, and prediction of expression functions (Considine et al. 2017). Machine learning models such as deep learning are used to recognize patterns and model data accurately (Ren et al. 2015; Considine et al. 2017); (3) Simulation and prediction—allows simulation of metabolic pathways and prediction of metabolic changes in biological systems (Sriyudthsak et al. 2016). For example, Markov is a computational model or neural network that can simulate and predict biological reactions and interacting metabolites (Li et al. 2022); (4) Consolidation and integrated analysis—computer science allows the integration of data from diverse sources such as genomic, proteomic, and metabolomic to provide a comprehensive view of biological laws and relationships between factors element (Lu et al. 2023; Redestig et al. 2010) (Table 2.1).

The development of computer science in the field of metabolomics data processing has gone through a journey from a primitive approach to basic statistical methods to profound and powerful machine learning and finally to the outstanding integration of deep learning. It has played a pivotal role in metabolomics research and analysis by providing advanced tools and methods from data preprocessing to prediction, and modeling, leading to a deeper understanding of metabolomics and relationships between factors.

Table 2.1 Description of main softwares and tools for metabolomics analysis

No.	Target	Description	Advantage	Disadvantage
1.	Data processing tools	Development of computational tools, algorithms for preprocessing raw metabolomics data: noise reduction, peak detection, baseline correction, alignment.	Improved accuracy and efficiency in data preprocessing, enabling better downstream analysis.	Challenging in selection of appropriate algorithms, and preprocessing steps may introduce biases or artifacts (Hendriks et al. 2011).
2	Analytical techniques	Introduction of high-throughput analytical techniques: MS, NMR for metabolite detection and quantification.	Enhanced sensitivity and throughput for comprehensive metabolite profiling.	High initial investment costs, complexity of instrumentation, and data interpretation challenges for nonexperts (Hendriks et al. 2011).
3.	Metabolite annotation software	Advancement in metabolite annotation software leveraging spectral databases, fragmentation prediction algorithms, machine learning techniques for metabolite identification.	Facilitates reliable identification of metabolites, aiding in biomarker discovery and pathway analysis.	Accuracy of annotation heavily relies on the quality and comprehensiveness of reference databases, leading to potential mis-annotations (Godzien et al. 2018).
4.	Pathway analysis software	Integration of metabolomics data with computational models and pathway analysis software for reconstructing metabolic pathways, understanding metabolic fluxes, identifying key metabolites.	Provides insights into complex biological processes and regulatory mechanisms.	Network complexity and incomplete knowledge of metabolic pathways (Wieder et al. 2021).
5.	Biomarker discovery platforms	Identification of metabolite profiles linked to illnesses, treatment reactions, or environmental exposures through the use of computational platforms and statistical algorithms for biomarker discovery.	Enables personalized medicine, disease diagnosis, and monitoring of therapeutic interventions.	Biomarker validation and reproducibility can be challenging, and overfitting may lead to false discoveries (Zhang et al. 2015).
6.	Predictive modeling frameworks	Creation of frameworks for predictive modeling, including kinetic modeling, constraint-based modeling, machine learning techniques, metabolic network simulation, phenotypic prediction, and experimental condition optimization.	Allows for the prediction of metabolic responses to interventions and the design of targeted therapies.	Model complexity and overfitting can lead to poor generalization, and data requirements for training accurate models can be substantial.

Table 2.1 (continued)

No.	Target	Description	Advantage	Disadvantage
7.	Systems biology integration tools	Integrate computational tools with biological methods to comprehensively understand cellular metabolism, regulatory networks, metabolic phenotypes.	Enables a systems-level understanding of biological systems and facilitates drug discovery and development.	Integrating diverse data types and reconciling inconsistencies between models and experimental data can be challenging (Wieder et al. 2021).
8.	Data sharing and collaboration platforms	In order to support science projects, data sharing, reproducibility, and interdisciplinary collaboration in metabolomics research, online repositories, data standards, and collaborative platforms have been established.	Promotes transparency, reproducibility, and knowledge dissemination, accelerating scientific progress.	Data privacy concerns, intellectual property issues, and difficulties in maintaining data quality and consistency across different platforms.

2.3.2 The Application of Computational Tools in Metabolomics

In the field of Data Processing and Preprocessing, computational tools are used to preprocess metabolomic data, including data normalization, missing value removal, feature selection, and alignment data (Yang et al. 2019). In addition, the integration of multiple omics data and machine learning has a growing trend in metabolomic research and analysis (Yang et al. 2019; Klupczyńska et al. 2015). These trends aim to improve the efficiency, accuracy, and usability of metabolomic data analysis, as well as facilitate data sharing and interpretation (Yang et al. 2019). Several models are researched and developed such as Constraint-Based Modeling, which involves the construction of mathematical models that incorporate constraints based on known biochemical and physiological information, to predict metabolic fluxes and optimize metabolic pathways (Hertel et al. 2021; Smith et al. 2006). Another model is Infectious disease modeling to provide biomarkers identifications for early diagnosis, infectious disease research to study host-pathogen interactions (Pacchiarotta et al. 2012). Additionally, Genome-scale models provide a holistic view of metabolic networks to study metabolic interactions, identify metabolic targets for drug discovery, and optimize metabolic engineering strategies (Yizhak et al. 2010). Software tools such as XCMS for identifying spectral peaks, or for multiplot statistical analysis, are also trusted and strongly used by researchers for metabolomics analysis (Domingo-Almenara and Siuzdak 2020; Xia and Wishart 2011).

The field of metabolism is increasingly improving and reaping many great achievements thanks to practical applications in the field and the development of tools (Ren et al. 2015; Yang et al. 2019). One of the first tools that can be mentioned is the XCMS Data Processing tool, which provides features for processing, analyzing, and managing metabolic data and spectrum analysis (Sugimoto et al. 2012;

Ludwig and Günther 2011). Next, MetaboAnalyst Line and chart analysis, which provides tools for graph analysis, statistics, and metabolic analysis. This tool helps researchers to discover new substances in metabolism (Xia and Wishart 2011; Barnes et al. 2016). However, compared to MetaboAnalyst, XCMS Online demonstrates greater reproducibility in providing integrated peak areas for a wide range of compounds, with a relative standard deviation of less than 0.05 for nearly 2000 compounds (Hao et al. 2018).

Substance identification and structural analysis are also commonly used to provide tools for homogenization, polarity prediction, mass spectrometry prediction, and spectral combination in metabolomics (Dunn et al. 2013). In which, CAMERA tool has been commonly used in this application platform (Allen et al. 2015; Kuhl et al. 2012). With sophisticated and massive computing capabilities, CAMERA have been supposed to have the ability of accurate calculation. The study by Kuhl et al. showed that compounds that could be annotated in positive and negative ion modes were correctly expressed in mass accuracy for 89.7% and 90.3% through the CAMERA tool (Kuhl et al. 2012).

There are some other tools which their specific tool such as MetaboLights Store and share data tools, a tool allows researchers to upload, download, analyze metabolomics datasets, and supports various data formats and analysis tools (Kale et al. 2016); or Metabolic flux analysis, a tool for isotope-based metabolic flux analysis and estimation (Jang et al. 2018) (Table 2.2). Compared to others in the field of metabolomics, MetaboLight is more commonly used by not only identifying the metabolites present and measuring the degree of correlation between them. MetaboLight also provides a huge MIS environment with compliance metadata and raw experimental data related to metabolic assays (Kale et al. 2016). Depending on the specific function, purpose, and type of metabolites used, the field of machine learning has created many different types of tools for use, supporting the field of metabolomics. In the chemical metabolism application platform, data analysis and processing work has always been given top priority (Majumder et al. 2021; Spicer et al. 2017).

2.4 Conclusion

In summary, this review has shed light on the pivotal role of data science, computational science, and machine learning in advancing metabolomics research and other "omics" fields, and recognizes their importance in overcoming the challenges inherent in metabolic analysis. Looking back from historical evolution to contemporary applications, it is clear that computational and data science have significantly advanced our understanding of metabolic processes and their implications for health and disease. However, it must also be admitted that this field still has many mysteries, requiring tools to be more developed and more intelligent. Therefore, there is still a lot of potential for developers to improve research tools. Due to the scope of this review, we only provide a glance at popular tools and databases for metabolomic analysis without deeply discussion on each tool and database. We also

Table 2.2 Introduction to several metabolomic softwares and tools

No	Tool name	Description	Type	User's level	Main function	Reference
1	XCMS	Processing and analysis of metabolomics data tool	Data processing	Intermediate	– Peak picking, alignment – Statistical analysis	Smith et al. (2006)
2	MetaboAnalyst	Line and chart analysis	Data analysis	Beginner/ intermediate	– Statistical analysis – Metabolite identification – Pathway analysis	Chong et al. (2018)
3	MZmine	Processing and visualization of MS-based metabolomics data	Data processing	Intermediate/ advanced	– Peak picking, alignment – Quantification – Identification	Pluskal et al. (2010)
4	MetFrag	Metabolite identification by matching experimental data with theoretical fragmentation patterns	Metabolite identification	Advanced	– Metabolite identification	Ruttkies et al. (2016)
5	CAMERA	R package for annotation of metabolomics data	Data annotation	Intermediate/ advanced	– Homogenization – Polarization – Prediction	Kuhl et al. (2012)
6	MetaboLights	A platform for data sharing, analysis, and annotation	Metabolic modeling	Advanced	– Analyzing metabolomics datasets, and supports various data formats and analysis tools.	Quek et al. (2009)
7	Metabolic flux	Repository for metabolomics data and metadata	Data repository	N/A	Isotope-based metabolic flux analysis and estimation	Haug et al. (2020)

compared machine learning and deep learning tools in Metabolomics analysis to point out shortcomings that software developers can recognize and fix, as well as develop additional necessary features for medical research. In essence, this review provides an overview of the applications of computational and data science in metabolomics, but it also serves as a stimulus for future research efforts on these technologies' new tools, aimed at unlocking the full potential of metabolism in improving human health and well-being.

Funding This work was funded by Vietnam National University Hanoi (VNU) for the 2024 research activities of the VNU strong research group - the interdisciplinary research group on biomedicine and health (IRGBH) (The establishing Decision no. 4013/QĐ-ĐHQGHN on October 24, 2023) such as "initial research on the application of data science and IT tools in biomedicine and cancer" at the International School, Vietnam National University Hanoi.

References

Aderemi AV et al (2021) Metabolomics: a scoping review of its role as a tool for disease biomarker discovery in selected non-communicable diseases. Meta 11(7):418
Allen F, Greiner R, Wishart DJM (2015) Competitive fragmentation modeling of ESI-MS/MS spectra for putative metabolite identification. Metabolomics 11:98–110
Altuna-Azkargorta M, Mendioroz-Iriarte M (2021) Blood biomarkers in Alzheimer's disease. Neurologia 36(9):704–710
Amberg A et al (2017) NMR and MS methods for metabolomics. Methods Mol Biol 1641:229–258
Arneth B, Arneth R, Shams M (2019) Metabolomics of type 1 and type 2 diabetes. Int J Mol Sci 20(10):2467
Balashova EE, Maslov DL, Trifonova OP. (2020) Metabolomics Community in Russia: history of development, key participants, and results. BioTech 9(4):20
Barnes S et al (2016) Training in metabolomics research. II. Processing and statistical analysis of metabolomics data, metabolite identification, pathway analysis, applications of metabolomics and its future. J Mass Spectrom 51(8):535–548
Bundy JG, Davey MP, Viant MRJM (2009) Environmental metabolomics: a critical review and future perspectives. Metabolomics 5:3–21
Chang KH et al (2018) Alternations of metabolic profile and kynurenine metabolism in the plasma of Parkinson's disease. Mol Neurobiol 55(8):6319–6328
Chen Y, Li EM, Xu LY (2022) Guide to metabolomics analysis: a bioinformatics workflow. Meta 12(4):357
Chen-Plotkin AS et al (2018) Finding useful biomarkers for Parkinson's disease. Sci Transl Med 10(454):eaam6003
Chong J et al (2018) MetaboAnalyst 4.0: towards more transparent and integrative metabolomics analysis. Nucleic Acids Res 46(W1):W486–W494
Chong J, Wishart DS, Xia J (2019) Using MetaboAnalyst 4.0 for comprehensive and integrative metabolomics data analysis. Curr Protoc Bioinformatics 68(1):e86
Considine EC et al (2017) Critical review of reporting of the data analysis step in metabolomics. Metabolomics 14(1):7
Courant F et al (2014) Basics of mass spectrometry based metabolomics. Proteomics 14(21–22):2369–2388
Davidson RL et al (2016) Galaxy-M: a Galaxy workflow for processing and analyzing direct infusion and liquid chromatography mass spectrometry-based metabolomics data. GigaScience 5(1):10. https://doi.org/10.1186/s13742-016-0115-8

De Castro F et al (2019) NMR-based metabolomics in metal-based drug research. Molecules 24(12):2240
Di Minno A et al (2022) Challenges in metabolomics-based tests, biomarkers revealed by metabolomic analysis, and the promise of the application of metabolomics in precision medicine. Int J Mol Sci 23(9):5213
Dias DA, Koal T (2016) Progress in metabolomics standardisation and its significance in future clinical laboratory medicine. EJIFCC 27(4):331–343
Domingo-Almenara X, Siuzdak G (2020) Metabolomics data processing using XCMS. Methods Mol Biol 2104:11–24
Dunn WB et al (2013) Mass appeal: metabolite identification in mass spectrometry-focused untargeted metabolomics. Metabolomics 9:44–66
Fan Z et al (2019) Metabolite identification using artificial neural network. In: 2019 IEEE International Conference on Bioinformatics and Biomedicine (BIBM). IEEE, San Diego, pp 244–248. https://doi.org/10.1109/BIBM47256.2019.8983190
Fu J et al (2022) Optimization of metabolomic data processing using NOREVA. Nat Protoc 17(1):129–151
Fuhrer T, Zamboni N (2015) High-throughput discovery metabolomics. Curr Opin Biotechnol 31:73–78
Galal A, Talal M, Moustafa A (2022) Applications of machine learning in metabolomics: disease modeling and classification. Front Genet 13:1017340
Ghosh T et al (2020a) Predictive Modeling for metabolomics data. In: Li S (ed) Computational methods and data analysis for metabolomics. Springer, New York, pp 313–336
Ghosh T et al (2020b) Predictive modeling for metabolomics data. Methods Mol Biol 2104:313–336
Giacomoni F et al (2015) Workflow4Metabolomics: a collaborative research infrastructure for computational metabolomics. Bioinformatics 31(9):1493–1495
Godzien J et al (2018) Chapter fifteen—metabolite annotation and identification. In: Jaumot J, Bedia C, Tauler R (eds) Comprehensive analytical chemistry. Elsevier (Radarweg 29, 1043 NX Amsterdam, The Netherlands), pp 415–445
van der Greef J et al (2013) Looking back into the future: 30 years of metabolomics at TNO. Mass Spectrom Rev 32(5):399–415
Hao L et al (2018) Comparative evaluation of MS-based metabolomics software and its application to preclinical Alzheimer's disease. Sci Rep 8(1):9291
Haug K et al (2020) MetaboLights: a resource evolving in response to the needs of its scientific community. Nucleic Acids Res 48(D1):D440–D444
Hendriks MMWB et al (2011) Data-processing strategies for metabolomics studies. TrAC Trends Anal Chem 30(10):1685–1698
Hertel J et al (2021) Integration of constraint-based modeling with fecal metabolomics reveals large deleterious effects of Fusobacterium spp. on community butyrate production. Gut Microbes 13(1):1915673
Hollywood K, Brison DR, Goodacre RJP (2006) Metabolomics: current technologies and future trends. Proteomics 6(17):4716–4723
Iida M, Harada S, Takebayashi T (2019) Application of metabolomics to epidemiological studies of atherosclerosis and cardiovascular disease. J Atheroscler Thromb 26(9):747–757
Jang C, Chen L, Rabinowitz JD (2018) Metabolomics and isotope tracing. Cell 173(4):822–837
Johnson SR, Lange BM (2015) Open-access metabolomics databases for natural product research: present capabilities and future potential. Front Bioeng Biotechnol 3:22
Jones OAH et al (2013) Metabolomics and its use in ecology. Austral Ecol 38(6):713–720
Kale NS et al (2016) MetaboLights: an open-access database repository for metabolomics data. Curr Protoc Bioinformatics 53:14.13.1–14.13.18
Kastenmüller G et al (2011) metaP-server: a web-based metabolomics data analysis tool. J Biomed Biotechnol 2011:839862
Katajamaa M, Orešič M (2007) Data processing for mass spectrometry-based metabolomics. J Chromatogr A 1158(1):318–328

Kirpich AS et al (2018) SECIMTools: a suite of metabolomics data analysis tools. BMC Bioinformatics 19(1):151
Kisiel A et al (2023) Data science and plant metabolomics. Meta 13(3):454. https://doi.org/10.3390/metabo13030454
Klupczyńska A, DerezińSki P, Kokot ZJ (2015) Metabolomics in medical sciences ń trends, challenges and perspectives. Acta Pol Pharm 72(4):629–641
Kuhl C et al (2012) CAMERA: an integrated strategy for compound spectra extraction and annotation of liquid chromatography/mass spectrometry data sets. Anal Chem 84(1):283–289
Lamichhane S et al (2018) Chapter fourteen—an overview of metabolomics data analysis: current tools and future perspectives. In: Jaumot J, Bedia C, Tauler R (eds) Comprehensive analytical chemistry. Elsevier, pp 387–413
Lee MY, Hu T (2019) Computational methods for the discovery of metabolic markers of complex traits. Meta 9(4):66. https://doi.org/10.3390/metabo9040066
Li L et al (2022) Exploring dynamic metabolomics data with multiway data analysis: a simulation study. BMC Bioinformatics 23(1):31
Lu Y et al (2023) MicrobiomeAnalyst 2.0: comprehensive statistical, functional and integrative analysis of microbiome data. Nucleic Acids Res 51(W1):W310–W318
Ludwig C, Günther UL (2011) MetaboLab-advanced NMR data processing and analysis for metabolomics. BMC Bioinformatics 12:366
Majumder EL et al (2021) Cognitive analysis of metabolomics data for systems biology. Nat Protoc 16(3):1376–1418
Mendez KM, Broadhurst DI, Reinke SN (2019a) The application of artificial neural networks in metabolomics: a historical perspective. Metabolomics 15(11):142
Mendez KM et al (2019b) Toward collaborative open data science in metabolomics using Jupyter Notebooks and cloud computing. Metabolomics 15(10):125
Mielke MM et al (2017) Association of plasma total tau level with cognitive decline and risk of mild cognitive impairment or dementia in the Mayo Clinic study on aging. JAMA Neurol 74(9):1073–1080
Misra BB (2021) New software tools, databases, and resources in metabolomics: updates from 2020. Metabolomics 17(5):49
Misra BB, van der Hooft JJ (2016) Updates in metabolomics tools and resources: 2014–2015. Electrophoresis 37(1):86–110
Nyamundanda G, Brennan L, Gormley IC (2010) Probabilistic principal component analysis for metabolomic data. BMC Bioinformatics 11(1):571
Pacchiarotta T, Deelder AM, Mayboroda OAJB (2012) Metabolomic investigations of human infections. Bioanalysis 4(8):919–925
Pang Z et al (2021) MetaboAnalyst 5.0: narrowing the gap between raw spectra and functional insights. Nucleic Acids Res 49(W1):W388–w396
Phapale P et al (2021) Public LC-Orbitrap tandem mass spectral library for metabolite identification. J Proteome Res 20(4):2089–2097
Picard M et al (2021) Integration strategies of multi-omics data for machine learning analysis. Comput Struct Biotechnol J 19:3735–3746
Pluskal T et al (2010) MZmine 2: modular framework for processing, visualizing, and analyzing mass spectrometry-based molecular profile data. BMC Bioinformatics 11:1–11
Pomyen Y et al (2020) Deep metabolome: applications of deep learning in metabolomics. Comput Struct Biotechnol J 18:2818–2825
Quek LE et al (2009) OpenFLUX: efficient modelling software for 13 C-based metabolic flux analysis. Microb Cell Factories 8:1–15
Redestig H et al (2010) Consolidating metabolite identifiers to enable contextual and multi-platform metabolomics data analysis. BMC Bioinformatics 11(1):214
Ren S et al (2015) Computational and statistical analysis of metabolomics data. Metabolomics 11(6):1492–1513
Ruttkies C et al (2016) MetFrag relaunched: incorporating strategies beyond in silico fragmentation. J Cheminform 8:1–16

Saiki S et al (2017) Decreased long-chain acylcarnitines from insufficient β-oxidation as potential early diagnostic markers for Parkinson's disease. Sci Rep 7(1):7328

Sha Y et al (2024) MetDIT: transforming and analyzing clinical metabolomics data with convolutional neural networks. Anal Chem 96(7):2949–2957

Shao Y, Le W (2019) Recent advances and perspectives of metabolomics-based investigations in Parkinson's disease. Mol Neurodegener 14(1):3

She H et al (2023) Metabolomics and machine learning approaches for diagnostic and prognostic biomarkers screening in sepsis. BMC Anesthesiol 23(1):367

Smith CA et al (2006) XCMS: processing mass spectrometry data for metabolite profiling using nonlinear peak alignment, matching, and identification. Anal Chem 78(3):779–787

Spicer R et al (2017) Navigating freely-available software tools for metabolomics analysis. Metabolomics 13(9):106

Sriyudthsak K, Shiraishi F, Hirai MY (2016) Mathematical modeling and dynamic simulation of metabolic reaction systems using metabolome time series data. Front Mol Biosci 3:15

Sugimoto M et al (2012) Bioinformatics tools for mass spectroscopy-based metabolomic data processing and analysis 7(1):96–108

Trifonova OP et al (2023) Current state and future perspectives on personalized metabolomics. Meta 13(1):67

Uarrota VG et al (2014) Metabolomics combined with chemometric tools (PCA, HCA, PLS-DA and SVM) for screening cassava (Manihot esculenta Crantz) roots during postharvest physiological deterioration. Food Chem 161:67–78

Wen B et al (2017) metaX: a flexible and comprehensive software for processing metabolomics data. BMC Bioinformatics 18(1):183

Wiechert W, Nöh K (2013) Isotopically non-stationary metabolic flux analysis: complex yet highly informative. Curr Opin Biotechnol 24(6):979–986

Wieder C et al (2021) Pathway analysis in metabolomics: recommendations for the use of over-representation analysis. PLoS Comput Biol 17(9):e1009105

Wishart DS et al (2022) HMDB 5.0: the Human Metabolome Database for 2022. Nucleic Acids Res 50(D1):D622–D631

Xia J, Wishart DS (2011) Metabolomic data processing, analysis, and interpretation using MetaboAnalyst. Curr Protoc Bioinformatics 14:14.10.1–14.10.48

Xia J, Wishart DS (2016) Using MetaboAnalyst 3.0 for comprehensive metabolomics data analysis. Curr Protoc Bioinformatics 55:14.10.1–14.10.91

Xia J et al (2012) MetaboAnalyst 2.0—a comprehensive server for metabolomic data analysis. Nucleic Acids Res 40(Web Server issue):W127–W133

Xue J et al (2020) METLIN MS2 molecular standards database: a broad chemical and biological resource. Nat Methods 17(10):953–954

Yang Q et al (2019) Metabolomics biotechnology, applications, and future trends: a systematic review. RSC Adv 9(64):37245–37257

Yilmaz A et al (2017) Diagnostic biomarkers of Alzheimer's disease as identified in saliva using 1H NMR-based metabolomics. J Alzheimers Dis 58(2):355–359

Yizhak K et al (2010) Integrating quantitative proteomics and metabolomics with a genome-scale metabolic network model. Bioinformatics 26(12):i255–i260

Zhang A et al (2015) Metabolomics for biomarker discovery: moving to the clinic. Biomed Res Int 2015:354671

Zhang J, Stewart T (2013) Chapter 86—Parkinson's disease. In: Ginsburg GS, Willard HF (eds) Genomic and personalized medicine, 2nd edn. Academic Press, pp 1034–1043

Design of Biosynthetic Pathway for Production of Metabolites

3

Ajay Kumar, Juveriya Israr, and Indrajeet Singh

3.1 Introduction

In modern biotechnology and synthetic biology, the process of creating and building biosynthetic pathways for metabolites has great potential to completely transform the production of important substances (Keasling 2010). Metabolites are tiny molecules that have a role in several cellular processes. They include a wide range of substances that have numerous applications, such as medicines, biofuels, tastes, perfumes, and agricultural chemicals (Rabinovitch-Deere et al. 2013). In the past, a significant number of these metabolites have been obtained from natural sources such as plants, fungi, and microbes. Nevertheless, the dependence on natural sources presents notable obstacles, such as restrictions in the supply chain, environmental consequences, and economic viability (Santos-Merino et al. 2019). To tackle these difficulties, metabolic engineering and synthetic biology offer inventive approaches to manipulate microbes and other host systems to produce specific metabolites. Through the process of reprogramming cellular metabolism, scientists can create new biosynthetic pathways that can produce specific molecules with high levels of both quantity and purity. This technology not only bypasses the limitations of conventional extraction techniques but also allows for the creation of metabolites that may be rare or difficult to get in nature (Cava et al. 2011).

A. Kumar (✉) · I. Singh
Department of Biotechnology, Faculty of Engineering and Technology Rama University, Kanpur, India

J. Israr
Institute of Biosciences and Technology, Shri Ramswaroop Memorial University, Barabanki, Uttar Pradesh, India

V. Singh (ed.), *Advances in Metabolomics*,
https://doi.org/10.1007/978-981-97-7459-3_3

3.2 Metabolic Pathways and Biosynthesis

Metabolic pathways are complex networks of metabolic events that control the conversion of substrates into products in living organisms. These pathways include a wide range of biochemical processes, including glycolysis, the tricarboxylic acid (TCA) cycle, amino acid metabolism, and lipid biosynthesis, among others. Enzymes play a crucial role in these pathways by acting as catalysts that facilitate the conversion of substrates into products, typically through a sequence of intermediate metabolites. Biosynthesis is the biological process in which living organisms create complex chemicals from simpler substances, using enzymes in metabolic pathways. These pathways are meticulously controlled to guarantee the optimal synthesis of metabolites while preserving cellular homeostasis. Biosynthetic processes are often refined by evolution to produce specific molecules, such as secondary metabolites that play a role in defense mechanisms or signaling pathways (Xiao et al. 2021).

3.3 Identification of Target Metabolite

The process of identifying a target metabolite for biosynthesis entails numerous steps, which include comprehending its significance, its biosynthetic pathway, precursor molecules, and viable production methods. For illustration purposes, let's consider artemisinin, a frequently targeted metabolite known for its powerful antimalarial effects. Artemisinin is a chemical obtained from the herb *Artemisia annua*, also known as sweet wormwood, and belongs to the sesquiterpene lactone family. It is an essential element in the management of malaria, particularly in instances of drug-resistant strains. The production of artemisinin in plants is achieved by the mevalonate (MVA) route (Ro et al. 2006). Below is a detailed explanation of the sequential process for identifying artemisinin as the specific metabolite of interest (Fig. 3.1):

Significance of Artemisinin Artemisinin is important due to its effectiveness in treating malaria, particularly in areas where resistance to other antimalarial medications has developed. This chemical has been instrumental in saving numerous lives and continues to be an essential weapon in the battle against malaria.

Biosynthetic Pathway The biosynthesis of artemisinin largely takes place through the mevalonate (MVA) pathway in plants, with amorpha-4,11-diene serving as an intermediary in the pathway. Artemisinin is produced through a series of enzyme reactions. Comprehending these procedures is essential for designing engineering strategies to produce it in microbial hosts (Paddon and Keasling 2014).

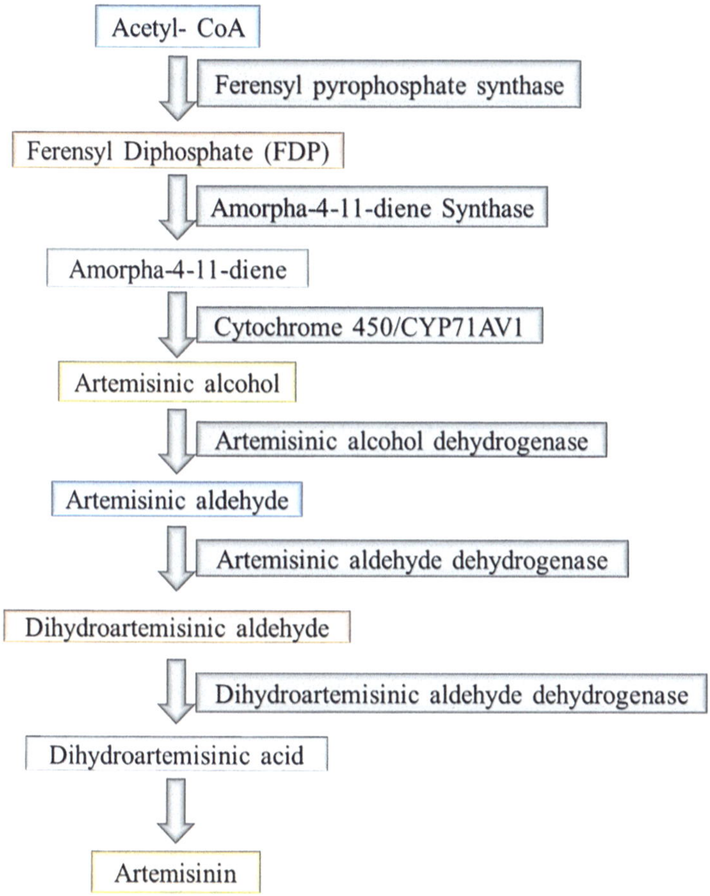

Fig. 3.1 The process of artemisinin biosynthesis is depicted in a schematic diagram illustrating the pathway involved

Precursor Molecules Amorpha-4,11-diene, generated from farnesyl diphosphate (FPP), act as a precursor for artemisinin. It is crucial to manipulate microorganisms to generate large quantities of FPP and effectively convert it into amorpha-4,11-diene to enhance the production of artemisinin (Wen and Yu 2011).

Methods for Production Artemisinin production can be achieved by the utilization of different ways. The conventional extraction process from *Artemisia annua* plants is hindered by poor production rates and limitations based on regional factors. Microbial fermentation in biotechnological processes provides a scalable and potentially more cost-effective method of manufacturing. The process of genetically modifying yeast or other microorganisms to produce the required genes for the biosynthetic pathway allows for the mass manufacture of artemisinin (Ro et al. 2006).

3.4 Biosynthetic Pathway Analysis

The biosynthesis pathway of artemisinin, the metabolite of interest, consists of multiple essential precursor chemicals and enzyme reactions (Wen and Yu 2011). Here is an analysis derived from current knowledge:

Precursor Molecules The main molecule that serves as a precursor for the production of artemisinin is amorpha-4,11-diene. This molecule is produced from farnesyl diphosphate (FPP) by a sequence of enzymatic processes.

3.4.1 Enzymes Involved

Farnesyl Diphosphate Synthase (FDS) FDS is an enzyme that facilitates the production of farnesyl diphosphate (FPP), which is an essential precursor in the synthesis of artemisinin. This enzyme catalyzes the combination of two molecules of isopentenyl diphosphate (IPP) with dimethyl allyl diphosphate (DMAPP) to produce farnesyl diphosphate (FPP).

Amorphadiene Synthase (ADS) ADS is the enzyme that catalyzes the transformation of FPP into amorpha-4,11-diene, which acts as the crucial intermediary in the biosynthesis of artemisinin. The enzyme ADS facilitates the cyclization of FPP, resulting in the production of the sesquiterpene amorpha-4,11-diene.

Cytochrome P450 Monooxygenases (CYPs) CYPs play a role in the oxidation process that converts amorpha-4,11-diene into artemisinin. These enzymes catalyze the addition of oxygen groups to the molecule, resulting in the synthesis of artemisinin.

Aldehyde Dehydrogenases (ALDHs) ALDHs play a crucial role in the last stages of artemisinin biosynthesis by facilitating the transformation of artemisinic aldehyde into artemisinin.

3.5 Design Strategy

The design strategy for building the biosynthetic pathway for artemisinin production involves a combination of approaches, such as utilizing natural pathways, altering existing pathways, and perhaps developing completely novel pathways (Paddon and Keasling 2014). Below is a detailed analysis of each method:

Natural Routes The strategy of leveraging natural routes entails making use of the pre-existing metabolic pathways found in organisms, such as plants. The metabolic route of *Artemisia annua* serves as a model for the manufacture of artemisinin. Researchers can aim to enhance the production of artemisinin by transferring the pathways responsible for its manufacture in plants into microbial hosts, leveraging the enzymes and regulatory mechanisms involved.

Modification of Existing Pathways The modification of existing pathways involves improving and maximizing the natural biosynthetic pathways in microbial hosts to generate the desired molecule. To enhance the production of precursor molecules (such as FPP) for artemisinin, one approach could be to modify the mevalonate (MVA) pathway in microbes. Additionally, introducing key enzymes like amorpha-4,11-diene synthase (ADS) and cytochrome P450 monooxygenases (CYPs) can help convert these precursors into artemisinin. This technique enables the exploitation of pre-existing metabolic pathways while optimizing them for enhanced production efficiency.

Creation of New Pathways The creation of new pathways requires the design of synthetic pathways or the combination of enzyme modules from other organisms to produce a unique biosynthetic route for the desired molecule. Researchers have investigated synthetic biology methods to construct biosynthetic pathways in microbial hosts that do not naturally generate artemisinin, specifically in the instance of artemisinin. Researchers can develop customized routes for artemisinin production by combining and optimizing enzyme modules with specific functions, which may help overcome the constraints of natural pathways (Paddon et al. 2013).

3.6 Genetic Engineering

Genetic engineering involves the manipulation of genes and enzymes to modify current pathways or create new ones for the production of artemisinin (Paddon et al. 2013). Below is a concise overview:

FPP Biosynthesis To increase the production of the precursor farnesyl diphosphate (FPP), the genes responsible for the enzymes involved in the mevalonate (MVA) pathway can be overexpressed or optimized. Some of the genes involved in this process include HMGR (3-hydroxy-3-methylglutaryl-coenzyme A reductase) and IPP isomerase.

The Synthesis of Amorpha-4,11-diene The enzyme amorpha-4,11-diene synthase (ADS) plays a critical role in converting FPP into amorpha-4,11-diene, which is a significant intermediate in the biosynthesis of artemisinin. Modifying microbial hosts to express ADS genes from *Artemisia annua* or other sources can simplify this process.

Cytochrome P450 Monooxygenases (CYPs) CYPs play a role in the oxidation process that converts amorpha-4,11-diene into artemisinin. It is crucial to identify and express the specific CYP genes that are responsible for these oxidation processes to produce artemisinin.

Aldehyde Dehydrogenases (ALDHs) ALDHs facilitate the transformation of artemisinic aldehyde into artemisinin during the last stages of biosynthesis. Efficient conversion requires the careful selection and expression of appropriate ALDH genes from organisms that have the ability to produce artemisinin (Wen and Yu 2011).

Creating Genetic Constructs Genetic constructs can be engineered to induce the expression of these genes in the host organism, commonly via plasmids or viral vectors. Every construct must incorporate regulatory components, like as promoters, terminators, and selectable markers, to guarantee the correct expression and upkeep of the inserted genes. An instance of a genetic construct designed for artemisinin production in yeast could consist of a potent constitutive promoter that facilitates robust expression of the biosynthetic genes.

- The coding sequences of genes responsible for producing enzymes involved in the manufacture of artemisinin (such as ADS, CYPs, ALDHs), which have been modified to enhance their expression in the host organism.
- Regulatory components, including as terminators, are included to ensure accurate termination of transcription.
- Selectable markers, such as genes that confer resistance to antibiotics, are used to identify cells that have been transformed.
- The plasmid backbone components necessary for replication and maintenance within the host organism are described by Adegboye et al. (2021).

3.7 Host Organism Selection

The selection of an appropriate host organism for artemisinin biosynthesis is essential to ensure efficient production on a large scale. Various species, including as bacteria, yeast, and plants, have been investigated as suitable hosts for this objective. Below is a comprehensive overview of each available option:

***Yeast** (Saccharomyces cerevisiae)* Yeast, specifically the species *Saccharomyces cerevisiae*, is commonly used as a host organism for biosynthesis. This is because yeast has well-established genetic characteristics, a rapid growth rate, and strong fermentation capabilities. The yeast has been genetically modified to produce artemisinin by including the biosynthetic pathways from *Artemisia annua* and enhancing the metabolic flow toward the precursors of artemisinin. Moreover, yeast possesses benefits such as simplicity in altering its genetic makeup and the ability to be easily scaled up, which makes it a very promising candidate for industrial manufacturing (Singh et al. 2009).

***Bacteria** (Escherichia coli, Bacillus subtilis)* Bacteria such as *Escherichia coli* and *Bacillus subtilis* are frequently employed in metabolic engineering because of their fast growth and convenient genetic manipulation. Nonetheless, bacterial hosts may have difficulties in expressing eukaryotic enzymes that are involved in the manufacture of artemisinin, necessitating the thorough optimization of heterologous pathways. However, researchers have successfully modified *E. coli* to produce artemisinic acid, which is a substance that may be used to create artemisinin. This shows that bacterial hosts have the ability to produce artemisinin through biosynthesis, as demonstrated by Liu et al. (2016).

***Plants** (Artemisia annua, Nicotiana benthamiana)* Plants included in the study include *Artemisia annua* and *Nicotiana benthamiana*. *Artemisia annua*, which naturally produces artemisinin, continues to be a suitable plant for artemisinin production using traditional farming techniques. Plant-based production, unfortunately, faces challenges such as low yields, geographical constraints, and lengthy culture cycles. Plant-based production systems, such as transient expression in *Nicotiana benthamiana*, provide higher levels of production but still have difficulties in scalability and downstream processing (Paddon et al. 2013). Yeast (*S. cerevisiae*) is the most appropriate host organism for artemisinin production due to its scalability, simplicity of genetic manipulation, and metabolic compatibility. Yeast-based production systems have shown encouraging outcomes in generating artemisinin and its precursors at elevated quantities, therefore enabling the development of economical and expandable production techniques.

3.8 Transformation and Expression

Genetic constructs can be introduced into the host organism and their expression can be optimized by the use of different transformation procedures, including plasmid insertion or gene editing approaches. Below is a comprehensive elucidation of these methodologies:

Plasmid Insertion. Plasmid insertion refers to the process of introducing recombinant DNA constructs, usually carried on plasmids, into the host organism. This procedure is commonly conducted utilizing techniques like as electroporation or chemical transformation, in which cells are subjected to electric pulses or chemical treatments to enhance the absorption of DNA. After entering the host cell, the recombinant plasmids undergo replication and expression, resulting in the synthesis of the desired metabolite (Jang et al. 2012).

Gene Editing Gene editing, utilizing tools like CRISPR-Cas9, allows for the exact manipulation of genes inside the genome of the host organism, enabling the insertion, deletion, or modification of genes. This method enables precise genetic alterations to optimize metabolic pathways or enhance the expression of specific genes involved in the production of metabolites. Researchers can enhance the properties of strains used for metabolite synthesis by modifying the host genome. This can include increasing the yield or improving the efficiency of metabolic pathways (Wang et al. 2016):

- Optimal expression of the inserted genes can be accomplished using various strategies.

Promoter Selection Opting for robust, inducible promoters to stimulate gene expression can amplify the synthesis of desired metabolites. Promoters can be chosen according to their robustness, selectivity, and suitability for the host organism.

Codon Optimization Codon optimization involves enhancing the coding sequences of imported genes to match the preferred codon usage of the host organism. This can lead to improved efficiency in translation and higher levels of protein expression.

Gene Copy Number Control The regulation of the number of copies of genes that are introduced can be achieved by designing plasmids or using genome editing techniques. This allows for precise control over gene expression levels and metabolic flux within the host organism.

Metabolic Engineering Metabolic engineering involves the incorporation of genetic constructs into metabolic pathways and regulatory networks of the host organism. This integration aims to enhance the efficiency of the pathways and reduce metabolic bottlenecks, resulting in higher production of metabolites (Cai et al. 2019).

3.9 Metabolic Engineering

In order to improve the production of metabolites in the host organism, various strategies can be utilized. These strategies include adjusting enzyme activity, ensuring sufficient availability of precursors, and reducing the impact of competing pathways (Li et al. 2024). Here is the implementation process for various approaches:

Modulating Enzyme Activities Enzyme activities can be adjusted by different methods, including enzyme engineering, regulation of promoter strength, or post-translational modifications. Enzyme engineering approaches, such as directed evolution or rational design, can be employed to improve enzyme activity, substrate selectivity, or stability. This, in turn, leads to an increase in the flow through desired biosynthetic pathways. In addition, optimizing pathway flow and metabolite production can be achieved by changing promoter strength or employing inducible promoters to adjust the expression levels of essential enzymes.

Ensuring Precursor Availability The availability of precursors is crucial for the efficient synthesis of metabolites. Optimizing the supply of essential intermediates and cofactors necessary for the production of target metabolites is crucial for balancing precursor pools. Metabolic engineering methods, such as upregulating genes that encode enzymes involved in precursor synthesis, improving the transport of precursors across cellular membranes, or shifting metabolic flux toward precursor synthesis pathways, can accomplish this. To maximize the overall flow through the target biosynthetic route and improve metabolite production, it is important to ensure that there is enough precursor availability.

Reducing Competing Pathways Competing pathways have the potential to redirect metabolic flow away from the desired biosynthetic pathway, resulting in decreased production of metabolites. To minimize competing pathways, one must either eliminate or decrease the activity of enzymes involved in alternative metabolic routes that vie for shared intermediates or cofactors. Gene knockout or downregulation can be accomplished via genetic engineering techniques such as CRISPR-Cas9-mediated genome editing or RNA interference (RNAi). By reducing the amount of metabolic activity that is directed toward alternative pathways, more resources can be used to produce the desired metabolite, resulting in increased overall yields (Stephanopoulos et al. 1998). Executing these tactics necessitates a profound comprehension of the host organism's metabolism and the interplay among several metabolic pathways. Computational modeling and simulation methods can assist in forecasting the impacts of genetic alterations on the distribution of metabolic flux and in identifying the most effective engineering strategies for producing metabolites (Jensen and Keasling 2015).

3.10 Characterization and Optimization

The process of characterizing engineered strains entails evaluating the amounts and purity of metabolite synthesis. This is done to assess the success of genetic alterations and metabolic engineering procedures (Lee et al. 2012). The implementation of the characterization and optimization process can be done as follows:

Metabolite Production Analysis Engineered strains undergo metabolite production analysis to measure the quantities of the desired metabolite. Metabolite concentrations in fermentation broths or cellular extracts are often measured using analytical techniques such as high-performance liquid chromatography (HPLC), gas chromatography-mass spectrometry (GC-MS), or liquid chromatography-mass spectrometry (LC-MS). These analyses offer numerical data on the amount of metabolites produced and aid in evaluating the influence of genetic alterations on the levels of metabolite synthesis.

Evaluation of Purity Besides measuring the amounts of metabolites, it is necessary to assess the purity of the specific metabolite to confirm its acceptability for subsequent uses. Purity assessment entails the analysis of the composition of fermentation products and the identification of potential by products or impurities that could impact the quality of the product. Metabolite purity and structural integrity can be assessed using techniques such as nuclear magnetic resonance (NMR) spectroscopy or mass spectrometry (MS).

Iterative Optimization The design and engineering process is improved through iterative optimization, which involves analyzing the results of metabolite production and purity analysis in order to increase production output and efficiency. This may include refining genetic structures, modifying culture conditions, or tweaking metabolic pathways to overcome obstacles and enhance overall efficiency. Iterative optimization cycles entail methodically assessing the impact of further genetic alterations or process parameters on metabolite synthesis, in order to enhance the engineered strains and reach certain production goals.

Through the systematic characterization of engineered strains and the optimization of the design and engineering process, researchers can enhance the production yields and purity of metabolites. This ultimately leads to the efficient and scalable production of the desired metabolite (Latimer and Dueber 2017).

3.11 Scale-Up and Production

Scaling up metabolite production from laboratory size to industrial scale requires shifting from small-scale fermentation systems to larger bioreactor systems, while incorporating process controls to assure consistent and efficient production (Lu 2010). Here is the procedure for implementing this process:

Bioreactor Systems In industrial-scale manufacturing, enormous amounts of fermentation broth are usually handled using bioreactor systems. Bioreactors are available in several configurations, such as stirred-tank reactors, airlift reactors, and packed-bed reactors. Each configuration has distinct advantages that cater to the specific needs of the manufacturing process. The bioreactors are fitted with agitators, pH and temperature control systems, gas sparging systems, and sample ports to enable effective blending, aeration, and monitoring of fermentation parameters.

Process Controls Process controls are necessary to maintain ideal circumstances for metabolite formation and ensure consistent quality of the product. Automated control systems oversee and adjust important factors like as pH, temperature, dissolved oxygen concentration, nutrient levels, and agitation speed in order to maximize microbial growth and metabolite output. Feedback control loops dynamically modify operating conditions in real-time using sensor measurements, enabling accurate regulation of fermentation parameters and reducing process variability.

Scale-Up Optimization The process of transitioning from laboratory-scale to industrial-scale production necessitates the optimization of fermentation conditions and scaling parameters to guarantee a successful scale-up. Adjustments to factors such as inoculum size, nutrient concentrations, aeration rates, and mixing intensities may be necessary when dealing with higher bioreactor capacities in order to maintain optimal growth and metabolite production kinetics. Utilizing pilot-scale studies and mathematical modeling approaches can effectively identify obstacles in scaling up and optimize process parameters for large-scale industrial production. Researchers can accomplish efficient and reproducible production of target metabolites at industrial volumes by using bioreactor systems and process controls specifically designed for large-scale metabolite production (Lau et al. 2004; Villadsen et al. 2011).

3.12 Purification and Downstream Processing

Purification and downstream processing involve the development of strategies to isolate the target metabolite from the fermentation broth. This process includes a series of steps aimed at separating and concentrating the desired product from the complex mixture of fermentation by-products and biomass (Baumann and Hubbuch 2017). Here is a method for implementing these steps:

Filtration Filtration serves as the initial stage in downstream processing, aiming to eliminate insoluble biomass and sizable particles from the fermentation broth. Commonly used methods such as depth filtration, microfiltration, or centrifugation are applied to clear the broth and separate solid biomass from the liquid phase.

Chromatography Chromatography methods are extensively employed to separate and purify certain metabolites by exploiting their physicochemical characteristics. Column chromatography, such as ion exchange chromatography, size exclusion chromatography, or affinity chromatography, can be used to selectively retain and release the desired metabolite from the fermentation broth. This is done by exploiting variations in charge, size, or affinity for specific ligands. High-performance liquid chromatography (HPLC) or preparative liquid chromatography (PLC) can be used to achieve a high degree of purity in the target metabolite by additional purification.

Crystallization Crystallization is a widely used technique for separating and purifying solid metabolites from a solution. By manipulating temperature, solvent composition, and crystallization conditions, it is possible to cause the desired metabolite to form crystals, while contaminants stay in the liquid phase. Following crystallization, the pure crystals can be isolated using filtration or centrifugation. Subsequently, the crystals can be washed and dried to acquire the final product in solid state.

Additional Techniques Additional techniques, such as solvent extraction, precipitation, or membrane separation, may be used for downstream processing. The choice of methodology depends on the unique features of the target metabolite and the composition of the fermentation broth. These procedures can enhance the concentration and purification of the metabolite prior to its final formulation or application. By systematically integrating these downstream processing steps, researchers can create efficient purification strategies to separate the desired metabolite from the fermentation broth and obtain a pure, high-quality product that is suitable for further analysis or downstream applications (Janković et al. 2024; Rüdt et al. 2017).

3.13 Analysis and Quality Control

The process of examining the purified metabolite to verify that it satisfies the required quality requirements for its intended use comprises performing thorough quality control tests to evaluate its purity, effectiveness, and stability. The following is a description of how these tests can be conducted:

Purity Analysis Purity analysis is conducted to confirm the absence of impurities and contaminants in the purified metabolite. Methods such as high-performance liquid chromatography (HPLC), gas chromatography (GC), or mass spectrometry (MS) can be used to separate and measure different components in the sample. The purity of the metabolite is assessed by comparing the peak area or intensity of the desired chemical with that of contaminants or established standards.

Potency Assay Potency assays assess the biological activity or effectiveness of the isolated metabolite. Potency assays for metabolites might vary depending on their intended application. These assays may include in vitro or in vivo bioassays, enzymatic assays, receptor binding assays, or cell-based assays. These tests are used to quantify certain biological activities or functional qualities of the metabolite. Potency testing verifies that the purified metabolite maintains its intended medicinal or functional properties.

Stability Testing Stability testing evaluates the durability of the purified metabolite over time and in different storage and handling settings. Accelerated stability tests entail exposing the metabolite to heightened temperatures, humidity, light exposure, or severe pH levels to mimic possible storage or transportation conditions. Stability-indicating tests, such as high-performance liquid chromatography (HPLC) or spectroscopic approaches, are employed to track alterations in the chemical composition, degradation rates, or physical characteristics of metabolites as time progresses.

Supplementary Quality Control Tests Depending on the characteristics and planned use of the metabolite, further quality control tests may be conducted to assess parameters such as pH, osmolality, viscosity, particle size distribution, or microbial contamination. These tests verify that the purified metabolite satisfies specific quality characteristics and regulatory standards for safety, effectiveness, and uniformity. Through the use of meticulous quality control testing, researchers can guarantee that the refined metabolite adheres to quality standards and specifications for its designated purpose, instilling assurance in its safety, effectiveness, and consistency.

3.14 Applications and Commercialization

The process of investigating prospective uses of the generated metabolite in different industries include the identification of opportunities for its application in pharmaceuticals, agriculture, food, or other areas (Kobayashi and Ishibashi 1993). Here is a method to evaluate these applications and formulate plans for commercialization:

Pharmaceuticals Metabolites possessing medical qualities hold promise for utilization in pharmaceuticals as active pharmaceutical ingredients (APIs) for the management of many diseases and situations. Artemisinin, which is a compound obtained from Artemisia annua, serves as a crucial component in antimalarial medications. Metabolites possessing antibacterial, anticancer, or anti-inflammatory activities may potentially have therapeutic potential and could be harnessed for the development of innovative pharmaceuticals or drug candidates for commercial purposes.

Agriculture Metabolites possessing bioactive qualities have potential uses in agriculture, such as serving as crop protection agents, plant growth regulators, or soil additives. Microbial metabolites possessing antifungal or antibacterial activities have the potential to be transformed into eco-friendly insecticides or biostimulants to augment crop productivity and resilience. Metabolites possessing qualities that enhance plant growth can also serve as biofertilizers or soil additives to enhance soil health and boost plant yield.

Food and Beverages Metabolites possessing flavor-enhancing, antioxidant, or preservation qualities can be utilized in the food and beverage sector as natural additives or functional ingredients. Microbial metabolites, such as enzymes, organic acids, or taste compounds, can be utilized to enhance the sensory characteristics, shelf-life, and nutritional content of food products. Metabolites possessing probiotic or prebiotic activities can also be used in functional meals or dietary supplements to enhance gut health and general wellness.

Other Industries Metabolites can be utilized in diverse industries such as cosmetics, biotechnology, environmental remediation, and renewable energy. Metabolites possessing antimicrobial or antiaging qualities can be incorporated into skincare products, whereas microbial metabolites with biocatalytic or bioremediation capabilities can be employed in biotechnological processes or environmental cleaning endeavors. Metabolites possessing features that make them suitable as biofuel precursors may also be utilized in the creation of renewable energy. Formulating commercialization strategies for introducing the metabolite to the market entails performing market research, pinpointing potential clients and partners, acquiring intellectual property rights, gaining regulatory authorizations, and building manufacturing and distribution networks. Product development and commercialization efforts might be expedited by seeking collaboration with industrial partners, investors, or research organizations.

3.15 Justification for the Design of Biosynthetic Pathways

The reason for developing biosynthetic pathways is the need to provide a sustainable and scalable method for producing metabolites that have commercial, industrial, or medicinal value. Conventional approaches to metabolite production, such as extracting them from natural sources or synthesizing them chemically, frequently encounter constraints such as limited output, expensive processes, negative environmental effects, and vulnerabilities in the supply chain. On the other hand, biosynthesis presents numerous benefits.

Biosynthetic pathways can be modified to synthesize certain metabolites of interest, enabling the customization of production systems to fulfil various application needs. Engineered microbes can be grown in large-scale bioreactor systems, allowing for the generation of metabolites in numbers suitable for commercial purposes. Biosynthesis provides a sustainable substitute for conventional extraction techniques, decreasing dependence on limited natural resources and limiting the impact on the environment. Pathway engineering enables the acquisition of new metabolites that may not naturally exist, thus broadening the range of molecules with potential uses. Once developed, biosynthetic pathways can provide economical means of producing high-value metabolites that possess intricate chemical structures.

3.16 Obstacles in the Design of Biosynthetic Pathways

In order to fully harness the potential of biosynthetic route design, it is necessary to address numerous problems that currently exist. The issues involve several parts of pathway engineering, selection of host organisms, optimization of pathways, and processing of downstream products. The process of creating biosynthetic pathways for intricate metabolites sometimes involves the combination of numerous enzyme stages, each having distinct substrate and cofactor prerequisites. Efficiently coordinating these stages can be difficult, especially when building nonnative pathways in heterologous hosts. Optimizing metabolic fluxes in designed pathways is crucial for increasing the production of desired metabolites while reducing the build-up of intermediates or unwanted by-products. To successfully do this assignment, one must possess a comprehensive comprehension of metabolic control and optimization strategies in order to attain the appropriate flux distributions. Choosing an appropriate host organism for pathway engineering is crucial to ensure optimal enzyme expression, availability of cofactors, and compatibility with the cellular environment. Various hosts present unique benefits and difficulties, spanning from thoroughly studied model organisms to commercially significant production strains. Ensuring the stability and resilience of pathways is essential for maintaining consistent and scalable production over the long term. To maintain the reliable functioning of modified strains, it is essential to employ strategies that minimize genomic instability, gene silencing, and metabolic burden. The downstream processing of biosynthetic products involves difficulties pertaining to the retrieval, refinement, and composition of the product. It is crucial to develop purification procedures that are both cost-effective and scalable in order to obtain metabolites of high purity that are appropriate for commercial use. To tackle these issues, a comprehensive strategy is needed that combines advancements in molecular biology, synthetic biology, computational modeling, and process engineering. Through the utilization of developing technologies and interdisciplinary collaborations, researchers have the ability to overcome obstacles in biosynthetic pathway design and fully unleash the potential of designed metabolite production.

3.17 Conclusion

Overall, the development and implementation of biosynthetic pathways for metabolites provide a robust method for the sustainable and expandable synthesis of important chemicals. Through the utilization of metabolic engineering and synthetic biology, it becomes feasible to modify cellular metabolism in order to generate a wide range of metabolites. These metabolites have various applications in fields such as medications, agriculture, chemicals, and more. However, fully harnessing the capabilities of biosynthetic pathway design necessitates overcoming many obstacles associated with the intricacy of pathways, balancing metabolic flux, ensuring compatibility with the host organism, maintaining pathway stability, and purifying the final product. To tackle these difficulties, it is necessary to have collaboration among different disciplines, use creative approaches, and constantly improve technology. By making focused and collaborative efforts, researchers can utilize the revolutionary capabilities of biosynthesis to tackle urgent social demands and stimulate advancements in biotechnology and other fields.

References

Adegboye MF, Ojuederie OB, Talia PM, Babalola OO (2021) Bioprospecting of microbial strains for biofuel production: metabolic engineering, applications, and challenges. Biotechnol Biofuels 14(1):5

Baumann P, Hubbuch J (2017) Downstream process development strategies for effective bioprocesses: trends, progress, and combinatorial approaches. Eng Life Sci 17(11):1142–1158

Cai P, Gao J, Zhou Y (2019) CRISPR-mediated genome editing in non-conventional yeasts for biotechnological applications. Microb Cell Factories 18:1–2

Cava F, Lam H, De Pedro MA, Waldor MK (2011) Emerging knowledge of regulatory roles of D-amino acids in bacteria. Cell Mol Life Sci 68:817–831

Jang YS, Park JM, Choi S, Choi YJ, Seung DY, Cho JH, Lee SY (2012) Engineering of microorganisms for the production of biofuels and perspectives based on systems metabolic engineering approaches. Biotechnol Adv 30(5):989–1000. https://doi.org/10.1016/j.biotechadv.2011.08.015. Epub 2011 Aug 25. PMID: 21889585

Janković T, Straathof AJ, Kiss AA (2024) A perspective on downstream processing performance for recovery of bioalcohols. J Chem Technol Biotechnol 99:1933–1940

Jensen MK, Keasling JD (2015) Recent applications of synthetic biology tools for yeast metabolic engineering. FEMS Yeast Res 15(1):1–0

Keasling JD (2010) Manufacturing molecules through metabolic engineering. Science 330(6009):1355–1358. https://doi.org/10.1126/science.1193990. PMID: 21127247

Kobayashi J, Ishibashi M (1993) Bioactive metabolites of symbiotic marine microorganisms. Chem Rev 93(5):1753–1769

Latimer LN, Dueber JE (2017) Iterative optimization of xylose catabolism in Saccharomyces cerevisiae using combinatorial expression tuning. Biotechnol Bioeng 114(6):1301–1309

Lau J, Tran C, Licari P, Galazzo J (2004) Development of a high cell-density fed-batch bioprocess for the heterologous production of 6-deoxyerythronolide B in Escherichia coli. J Biotechnol 110(1):95–103. https://doi.org/10.1016/j.jbiotec.2004.02.001. PMID: 15099909

Lee S, Mattanovich D, Villaverde A (2012) Systems metabolic engineering, industrial biotechnology and microbial cell factories. Microb Cell Factories 11:1–3

Li Y, Yang Y, Li L, Tang K, Hao X, Kai G (2024) Advanced metabolic engineering strategies for increasing artemisinin yield in Artemisia annua L. Hortic Res 11(2):uhad292. https://doi.org/10.1093/hr/uhad292. PMID: 38414837; PMCID: PMC10898619

Liu W, Xu X, Zhang R, Cheng T, Cao Y, Li X, Guo J, Liu H, Xian M (2016) Engineering Escherichia coli for high-yield geraniol production with biotransformation of geranyl acetate to geraniol under fed-batch culture. Biotechnol Biofuels 9:58. https://doi.org/10.1186/s13068-016-0466-5. Erratum in: Biotechnol Biofuels (2016) 9:131. https://doi.org/10.1186/s13068-016-0539-5. PMID: 26973712; PMCID: PMC4787002

Lu TK (2010) Engineering scalable biological systems. Bioeng Bugs 1(6):378–384. https://doi.org/10.4161/bbug.1.6.13086. PMID: 21468204; PMCID: PMC3056087

Paddon CJ, Keasling JD (2014) Semi-synthetic artemisinin: a model for the use of synthetic biology in pharmaceutical development. Nat Rev Microbiol 12(5):355–367. https://doi.org/10.1038/nrmicro3240. Epub 2014 Apr 1. PMID: 24686413

Paddon CJ, Westfall PJ, Pitera DJ, Benjamin K, Fisher K, McPhee D, Leavell MD, Tai A, Main A, Eng D, Polichuk DR, Teoh KH, Reed DW, Treynor T, Lenihan J, Fleck M, Bajad S, Dang G, Dengrove D, Diola D, Dorin G, Ellens KW, Fickes S, Galazzo J, Gaucher SP, Geistlinger T, Henry R, Hepp M, Horning T, Iqbal T, Jiang H, Kizer L, Lieu B, Melis D, Moss N, Regentin R, Secrest S, Tsuruta H, Vazquez R, Westblade LF, Xu L, Yu M, Zhang Y, Zhao L, Lievense J, Covello PS, Keasling JD, Reiling KK, Renninger NS, Newman JD (2013) High-level semi-synthetic production of the potent antimalarial artemisinin. Nature 496(7446):528–532. https://doi.org/10.1038/nature12051. Epub 2013 Apr 10. PMID: 23575629

Rabinovitch-Deere CA, Oliver JW, Rodriguez GM, Atsumi S (2013) Synthetic biology and metabolic engineering approaches to produce biofuels. Chem Rev 113(7):4611–4632

Ro DK, Paradise EM, Ouellet M, Fisher KJ, Newman KL, Ndungu JM, Ho KA, Eachus RA, Ham TS, Kirby J, Chang MC, Withers ST, Shiba Y, Sarpong R, Keasling JD (2006) Production of the antimalarial drug precursor artemisinic acid in engineered yeast. Nature 440(7086):940–943. https://doi.org/10.1038/nature04640. PMID: 16612385

Rüdt M, Briskot T, Hubbuch J (2017) Advances in downstream processing of biologics–spectroscopy: an emerging process analytical technology. J Chromatogr A 1490:2–9

Santos-Merino M, Singh AK, Ducat DC (2019) New applications of synthetic biology tools for cyanobacterial metabolic engineering. Front Bioeng Biotechnol 7:33

Singh ND, Ding Y, Daniell H (2009) Chloroplast-derived vaccine antigens and biopharmaceuticals: protocols for expression, purification, or oral delivery and functional evaluation. In: Recombinant proteins from plants: methods and protocols. Humana, New York, pp 163–192

Stephanopoulos G, Aristidou AA, Nielsen J (1998) Metabolic engineering: principles and methodologies. Academic, Cambridge

Villadsen J, Nielsen J, Lidén G (2011) Bioreaction engineering principles. Springer, Berlin

Wang H, La Russa M, Qi LS (2016) CRISPR/Cas9 in genome editing and beyond. Annu Rev Biochem 85:227–264. https://doi.org/10.1146/annurev-biochem-060815-014607. Epub 2016 Apr 25. PMID: 27145843

Wen W, Yu R (2011) Artemisinin biosynthesis and its regulatory enzymes: progress and perspective. Pharmacogn Rev 5(10):189–194. https://doi.org/10.4103/0973-7847.91118. PMID: 22279377; PMCID: PMC3263054

Xiao L, Du Q, Fang Y, Quan M, Lu W, Wang D, Si J, El-Kassaby YA, Zhang D (2021) Genetic architecture of the metabolic pathway of salicylic acid biosynthesis in Populus. Tree Physiol 41(11):2198–2215. https://doi.org/10.1093/treephys/tpab068. PMID: 33987676

Dissecting Cellular Heterogeneity: Single-Cell Metabolomics for Unmasking Hidden Variations

4

Gholamreza Abdi, Mukul Jain, Ritika Bhardwaj, Krishna Thakkar, Ishita Dewangan, Rupal Dhariwal, and Maryam Abbasi Tarighat

4.1 Introduction

According to Khoo and Al-Rubeai (2007), "metabolomics" is the comprehensive quantitative and qualitative analysis of all small molecules present in a system, whether it a sample of cells, body fluids such as plasma or urine, tissues, or a whole organism. The metabolome refers to all small-molecule metabolites that are present in a biological sample, which might include a single cell or the complete organism (Meng and Marina 2019). These metabolites are crucial to the majority of intracellular activities and functions and serve as a sample's physiological state. As a result, quantitative metabolomics has been a potent analytical tool for a long time in numerous domains. Exciting discoveries have been made by metabolic analysis. Current metabolomics techniques typically require a significant number of cells for each sample due to sensitivity considerations, particularly for some techniques like

G. Abdi (✉)
Department of Biotechnology, Persian Gulf Research Institute, Persian Gulf University, Bushehr, Iran
e-mail: abdi@pgu.ac.ir

M. Jain · R. Dhariwal
Cell & Development Biology Lab, Research and Development Cell, Parul University, Vadodara, Gujarat, India

Department of Life Sciences, Parul Institute of Applied Sciences, Parul University, Vadodara, Gujarat, India

R. Bhardwaj · K. Thakkar · I. Dewangan
Department of Life Sciences, Parul Institute of Applied Sciences, Parul University, Vadodara, Gujarat, India

M. A. Tarighat (✉)
Faculty of Nano and Bio Science and Technology, Persian Gulf University, Bushehr, Iran
e-mail: matarighat@pgu.ac.ir

V. Singh (ed.), *Advances in Metabolomics*,
https://doi.org/10.1007/978-981-97-7459-3_4

nuclear magnetic resonance (NMR). The heterogeneity of cells is ignored, even though multi-cell samples provide sufficient identification of chemical components (Wei and Xu 2022). Differentiations between individual cells can be caused by developmental, environmental, genetic, and epigenetic factors. Random processes in biology can cause phenotypic differences in both clonal and isogenic cultures (Geiler-Samerotte and Bauer 2014). Information regarding tumor heterogeneity that had been overlooked by earlier research on cell populations is now available due to recent single-cell genomics and proteomics analyses. Consequently, there is a high demand for SCM, which can show the variations in the cellular metabolome between individual cells. Single-cell recognition and quantification for certain analytes have been feasible. SCM, however, broadens the spectrum of the analyte and makes it possible to assess them effectively. It contributes to the collection of data on the chemical makeup and metabolic processes of a single cell, advancing our knowledge of how cells work. Numerous investigations have demonstrated encouraging outcomes in generating metabolic information from various cell and tissue kind (Zhao and Xia 2021). Single-cell metabolomics (SCM) is now one of the most efficient techniques for doing high-throughput metabolic studies at the cellular level. Because single-cell metabolomics can determine individual cell metabolic processes, it is a perfect tool for interpreting cell heterogeneity. Differentiating and detecting unique cell types within cell colonies is a great prospective application of SCM (Guo and Cissy 2022).

The molecular processes of individual cells for the repair of wounds are quite varied and depend on the type, size, and/or species of the wound, according to studies conducted over the last 20 years using various cell wound repair models. Presently, embryos, oocytes, or isolated cells are used in all models for the research of cell wound healing. Repaired individual cells may also serve as vital sentinels by performing crucial tasks including, alerting nearby cells to join the repair response. Applying the knowledge gained from these cells wound healing models to this larger context will be essential (Nakamura et al. 2018). The single cells that govern development, tissue remodeling, and regeneration processes interpret a mix of cell-intrinsic and cell-extrinsic inputs that determine their behavior, differentiation, and symmetry-breaking capacity (Xavier et al. 2019). Cell-to-cell heterogeneity is a key characteristic of isogenic cell populations, ranging from individual cells in multicellular animals to microbial cultures, and single-cell image analysis offers a potent method for examining it. To completely comprehend cellular function and address the genotype-to-phenotype link, researchers need a molecular explanation for this phenotypic diversity. Bulk readouts or the average of phenotypes from individual cells in a sample mask variability in single-cell phenotypes; hence, single-cell imaging analysis provides a better resolution perspective of cellular activity (Usaj et al. 2021). The scientific community has lately come to realize that individual cells—previously thought to be homogeneous—may differ significantly from one another and that studying cell-to-cell diversity and unique transcriptional and regulatory fingerprints is now necessary. Attempts are being made to analyze the inter- and intracellular variability of various cell types and developmental stages of the inner ear from a new angle in combination with high-throughput technology.

4.2 Cell Isolation and Preparation

The methodical identification and measurement of every metabolite present in a particular organism or biological sample is known as metabolomics. The expeditious amelioration of single-cell metabolomics (SCM) implicates the study of metabolomics within individual cells. SCM has its own special set of provocations. Isolation depends on the cell size and complicates the preparations and sampling if it goes less than femtoliters (Ahmed et al. 2019). In cell isolation, scientists have gained important perceptions of the complex systems of developmental biology, cell signaling, gene control, and genomics. It is carried out with the least amount of disruption possible to avoid interruption in the cell metabolism. The concentration of metabolites which depends on the condition of the cell and its surrounding environment leads to this. Trypsinization and cell scraping can be used here as cell pretreatment approaches (Ahmed et al. 2019). These approaches are not suitable for the analysis of the metabolomes due to metabolite leakage. These approaches can be compared in water for sample harvesting.

To prepare and isolate SCM, the metabolites found within each single cell must be extracted and analyzed. The isolation of single cells is done from heterogeneous populations, lysing the cell to liberate their metabolites, metabolite extraction from the leftover cell debris, and eventually, the metabolites are analyzed using the methods such as mass spectrometry (MS), chromatography, nuclear magnetic resonance (NMR), etc. Many methods provide strategies that are engaged for the isolation and preparation of single-cell metabolomics. Such methods are fluorescence-activated cell sorting (FACS), which is a technique for purifying certain cell groups based on the phenotypes revealed by flow cytometry, and an ultra-high-throughput screening method used for the measurement of the levels of antigens quantitatively on the cell surface; micro-manipulations, which play a pivotal role by manipulating distinct metabolites and observing their interrelations, which facilitates for the accurate investigation of bio-systems on a very tiny scale; magnetic-activated cell sorting (MACS) (Zhao et al. 2022), which is a well-established technique for separating target cells from diverse cell populations based on their unique surface, and in this technique, the target approach of the cell (where the magnetic beads are attached to a specific cell population) and the physical separation process (where the cells on a column are selected based on their maintenance) are more directly associated, as they jointly exploit magnetism, which empowers sorting; laser capture microdissection (LCM) (Qin et al. 2023) which is a precocious technology that isolates a single cell from predominantly solid tissue specimens on a microscopic slide; microfluidics, which copes with small scale volumes of fluid (around 10^{-9}–10^{-18} L in μm) used for analyzing and manipulating single cells which delegates the precise control over cellular reactions and cellular environments; single-cell lysis (Brown and Audet 2008); single-cell trapping; sample amplification, which includes multiple displacement amplification (MDA) or solid-phase extraction (SPE), liquid–liquid extraction; nuclear magnetic resonance (NMR) spectroscopy; single-cell metabolite extraction; solid-phase microextraction (SPME); polymerase chain reaction (PCR); and mass spectrometry.

4.3 Single-Cell Metabolomics Techniques

4.3.1 Mass Spectrometry

The technique called mass spectrometry is the qualitative and quantitative analytical technique used for the identification and qualification of chemical compositions and clinically precise analytes. It is introduced in units of the mass-to-charge ratio of ions. It is a necessary apparatus in the field of metabolomics and proteomics in species ranging from single-celled microorganisms to multicellular animals. It is broadly used for SCM analysis due to higher sensitivity and selectivity (Svatoš 2011).

In order to identify and quantify the unicellular metabolites, the mass spectrometry technique involves ionizing the molecules in the sample, separating the ions based on their mass-to-charge ratio, detecting the ions, and analyzing the resultant mass spectrum. This method determines the average metabolome of a specimen through either targeted or untargeted strategies. In order to confirm the results of measured metabolomics for untargeted metabolomics, targeted metabolomics employs specialized techniques on preselected metabolites, producing data with minimal complexity and high sensitivity for selected compounds. Untargeted metabolomics, also referred to as direct infusion or shotgun analysis, aims to calibrate a broad spectrum of metabolites in extracted samples and does not need any separation prior to MS. Metabolomics at the single-cell level provides the most sensitive reflection of cellular phenotype, which may allow us to unveil the cellular heterogeneity and this we can unveil with the help of MS technique.

Single-cell sampling and ionization are vital steps in the MS method. Sampling analytes, which is one of the most provocative steps due to cell size, intracellular compartmentalization the inherent complexity of intracellular species, analyte stability, technical variability analyte degradation and modification (Bruins 2005), quick turnover of metabolites, etc. A productive sampling method usually necessitates a high-magnification microscope for cell inspection, a precisely controlled mechanical apparatus to facilitate sample mobility, and a reliable extraction or desorption method for cellular content collecting. Other techniques like, new MS imaging approaches by ambient ionization techniques which include desorption and quick evaporative ionization; and SCMS-adopted strategies called chemical tagging/derivatization for avoiding troubles in SCMS analysis (Haiyan Lu et al. 2023). SCM using MS requires effective and sensitive ionization approaches to detect and analyze metabolites from specific cells. Here are some common ionization techniques used in SCM-MS: matrix-assisted laser desorption/ionization MS (MALDI-MS), secondary ionization MS (SIMS), laser ablation/laser desorption ionization (LA/LDI), electrospray ion mass spectrometry (ESI-MS), probe-ESI (PESI), nanoelectrospray ionization (nanoESI), desorption electrospray ionization (DESI), coupled plasma MS, etc. This method is performed on a liquid sample of biological fluid, homogenized tissue or cells which can be simply integrated with other varied techniques such as gas-chromatography MS (GC-MS), liquid chromatography MS (LC-MS) (Vogeser and Parhofer 2007), imaging MS, nanoscale liquid mass spectrometry (nano LC-MS), direct infusion mass spectrometry (DIMS), SC-CE-MS, etc.

4.3.1.1 Applications of MS in SCM

1. MS-based metabolomics in medical science.
   - It is a beneficial tool used in medical science, including identification, drug development, nutrition science, toxicology, and forensics.
   - Provides extensive investigation of metabolites for illness diagnosis and treatment (Mattoli et al. 2022)
   - Also provides elution of multiple biological machines such as the genome, transcriptome, proteome, and metabolome, providing revelations of genetic results, active expressions, and processes that influence metabolite expression, and physiological functions.
   - Used for biomarker development, early disease detection, molecular phenotyping, prognostic evaluation, and personalized medicine applications and biomarker diseases like COVID-19 and cancer.
   - To analyze natural complex products (NCPs) utilized in medications and dietary supplements, which improves scientific and industrial production monitoring.
2. MS is used to treat metabolic syndrome (MetS) through fecal transplantation (FMT).
   - MetS is treated with fecal transplantation, which entails transplanting gut microbiota from a healthy donor to a MetS recipient. This technique tries to alter the composition and activity of the gut microbiota, perhaps benefiting metabolic health. MS plays a significant role in treating MetS via FMT, abetting in donor selection for appropriate efficacy. GC-MS and tandem MS is used for the analysis of organic acids, acylcarnitine, and metabolites in human fluids (such as urine), aiding the diagnosis of metabolic diseases including organic acidurias (Clayton 2001).
   - Short-chain fatty acids (SCFAs) generated by gut bacteria play a significant role in metabolic function. MS can be used to quantify SCFA levels in feces, providing information about the gut microbiota's metabolic activity and its possible impact on MetS.
   - It is also used for the metabolism of drugs used to cure MetS. This includes the identification of gut microbiota-produced compounds that may affect therapeutic efficacy or toxicity.
3. MS techniques for pharmaceuticals
   - MS is used in pharmaceutical research at all stages, from drug discovery to quality control, to ensure safe and effective treatments.
   - MS imaging (MSI) permits the visualization of drug pharmacokinetics, metabolic products, and molecular distributions, which helps with pharmacodynamic investigations and toxicity assessments.
   - MSI is significant in pharmaceutical research and development since it allows for the visualization of drug distributions, pharmacokinetics processes, and pharmacodynamic features.
   - These aforementioned techniques deliver tag-free, high-throughput, precise tissue compound investigations that help with understanding the characteristics, effectiveness, and possible side effects of drugs (Chen et al. 2023).

- Pharmaceutical development necessitates the identification of metabolites and parent chemicals inside a particular cell type. The cellular absorption of desethylamiodarone, a metabolite of amiodarone, an antiarrhythmic medication that causes phospholipidosis, is investigated using 3D SIMS imaging in individual macrophages. SIMS is used to study how drugs are distributed in different drug delivery systems and medical devices. In studies by, 3D ToF-SIMS imaging is a promising analytical method that provides high-resolution power for visualizing pharmaceutical molecules and metabolites. When connected to a ToF mass analyzer, MSI methods provide label-free multiplexed analysis.
- MALDI and DESI are crucial techniques used for the successful investigation of drug distribution in tissues. These techniques do not have a better lateral vision to achieve sub-micrometer imaging, but in the field of MALDI, recent advances have improved it for the visualization of large organelles and cells. These techniques and another new technique AP-MALDI enable fast mapping of drugs and their metabolites in biological samples. The distinctive strategy of MALDI-2 combined with MALDI-MS considerably boosts sensitivity, allowing the imaging of pharmacological molecules in tissues with improved signal intensities.

4. MS technique for cancer research
   - MS offers an extensive examination of metabolites within individual drug-resistant cancer cells.
   - LC-MS and GC-MS can detect and quantify a wide spectrum of metabolites (Abraham et al. 2014).
   - SCMS provides assets to persuasively study subpopulation cells, especially those with treatment resistance, amid heterogeneous cells.
   - Using patient cells and cell lines, the single-probe SCMS may be utilized to assess intracellular metabolites and measure the amount of medication absorption (Touil et al. 2014; Kang and Kang 2007; Muschet et al. 2016; Zhang and Vertes 2015; Standke et al. 2019; Pan et al. 2019; Li et al. 2019; Altelaar et al. 2005; Masujima 2009).
   - The technique called MS-based pharmacometabolomics is used in the field of breast cancer treatment to monitor tamoxifen and its metabolomics to predict the side effects and discontinuation rate (Helland et al. 2021).
   - To find proteomic signatures that may be utilized as biomarkers to forecast the prognosis for cancer patients undergoing immunotherapy.
5. MS-based applications in food
   - MS-based techniques are used to analyze food systems, including food components, processing, and nutrition.
   - To detect and quantify contaminants such as heavy metals, mycotoxins, pesticides, and veterinary drug residues in food products.
   - Flow injection study of food and agricultural research, both targeted and untargeted Direct analysis using direct MS, chromatography-tandem MS metabolomics, and MS fingerprinting.

- CE-MS is used to monitor chemical pollutants and toxins in food products, thereby protecting public health.

4.3.2 Nuclear Magnetic Resonance (NMR)

NMR is powerful and high-resolution spectroscopy that provides detailed insights into the local magnetic field around atomic nuclei, pertinent to physiochemical techniques for organic molecule investigation, molecular structure elucidation, crystal and non-crystal materials and their new applications such as life sciences, chemistry, material sciences, and so on. It is an imperative tool for investigating molecular dynamics and interactions, macromolecules, and solids (Worthoff et al. 2018). This spectroscopy uses a vector model to describe spin behavior and spectral analysis, which has led to the development of advanced techniques such as Fourier transfer NMR and solid-state NMR (Sridhan 2016). This technique has isotope-specific features analyzing a wide range of systems, including inorganic chemicals, energy storage materials, biomolecules, and others, demonstrating its versatility and importance in scientific research.

NMR-based SCM is an approach for an in-depth analysis of cellular metabolism. This method allows for the quantification of metabolites and provides non-destructive and non-invasive observation of metabolic alterations within individual cells, providing beneficial insights into cellular responses to external stimuli and metabolic interconnections among heterogeneous cell populations. The method is innately quantitative and highly reproducible, which follows pre-analytical and analytical processes, and also enables rapid fingerprinting and profiling of a wide range of bio-samples. It provides spatially precise data for living organoids, allowing researchers to comprehend transport processes and metabolic properties inside intricate 3D cell cultures. It has the potential to generate dynamic radial metabolite profiles in cell spheroids demonstrating its relevance in examining metabolic changes within various parts of the cellular microenvironment) NMR provides in vivo metabolomic analysis which examines cellular metabolism without intrusive treatments or affecting the organism's condition.

Techniques such as carbon-13 NMR (^{13}C NMR), two-dimensional NMR (2D-NMR), proton NMR (^{1}H NMR), and High-resolution Magic Angle Spinning NMR (HR-MAS NMR) are commonly employed for metabolomics studies. ^{13}C NMR has a wide signal dispersion and narrow spectral lines. Because of its poor temporal resolution and limited sensitivity, it is rarely utilized to identify intermediates and fast changes inside live cells. ^{1}H NMR is used for molecular structure determination (Lewin-Kretzschamar et al. 2000). This spectrum requires a flat baseline and consistent solvent suppression. HR-MAS NMR is the method that has non-liquid biological sample resolution and high-resolution spectra. It is one of the major platforms of metabolic studies which has a wide variety of applications in biomedical, biochemical, diagnosis, prognostics, drug therapies, and xenobiotic interaction studies (Moestue et al. 2011). There are also low-resolution NMR (LR-NMR), high-resolution NMR (HR-NMR), and medium-resolution NMR (MR-NMR).

4.3.2.1 Applications of NMR

1. NMR is a non-destructive technique
    - Used in non-destructive examination to track microstructural changes in ceramics (Viola et al. 2006).
    - Analytical procedures for authenticating and analyzing food composition (Martinez et al. 2003) and also for the quality evaluation of fruits and vegetables (Jha and Matsuoka 2000).
    - Inspecting and evaluating fish quality and safety (Cheng et al. 2013).
    - Analyze polymers and polymer-based materials without causing damage (Adams 2019).
    - To analyze caviar compositions using the low-field NMR technique (Huang et al. 2017).
    - To discriminate beef samples using TD-NMR (Santos et al. 2014).
2. Dynamic metabolic profiling
    - NMR can be utilized to investigate the dynamic variations in metabolite concentrations within individual cells over time or in response to external stimuli.
    - It is done to evaluate physiological influences on biofluid compositions (Bollard et al. 2004).
    - For continuous in vivo metabolism (Judge et al. 2019).
    - Metabolomic profiling for NMR spectroscopy of serum, urine, plasma, and tissue extracts (Beckonert and Holmes 2007).
3. Isotope tracing studies
    - ^{13}S, ^{15}N, and ^{31}P are the tracers used to investigate metabolic processes within single cells.
    - Cellular rewiring, metabolic phenotypes in cancer cells, and dynamics of cell metabolism are some of the insights offered.
    - This enables us to study metabolic fluxes and pathways within single cells.
    - ^{31}P NMR detects phosphorous-containing metabolites, which provides valuable information about energy states, cellular processes and redox conditions within single cells.
4. Drug response studies
    - Single-cell NMR metabolomics can be utilized in drug screening and optimization to find lead compounds with favorable metabolic profiles and few-off target effects (Van der Velden et al. 2019).
    - Researchers can identify the metabolic reactions of cell subpopulations to medications like gefitinib in non-small-cell lung cancer (NSCLC) cells by combining intact living cell electro-launching ionization mass spectrometry with different metabolic responses.
    - Quantitative NMR (qNMR) which is a specialized subdiscipline is used to quantify drugs from variety of complex mixtures (natural product extract, body fluid, drug formation) (Simmler et al. 2014; Singh and Roy 2016).
5. NMR has high reproducibility and quantitative abilities.
6. It can acquire multiple 1-D and multidimensional experiments for metabolite identification and structural elucidation.

7. This method can identify unknown metabolites in complex mixtures.
8. For the field of metabolomics, tracking the downstream products of isotope-tagged substrates ex vivo, in vivo, or in vitro offers a number of advantages.
9. Improvements in NMR data collection and analysis techniques have made it easier to analyze biological mixtures that are extremely complex.
10. It has the potential to detect metabolites through one or more types of atomic nuclei, including ^{1}H, ^{13}C, ^{31}P, and ^{15}N.
11. The biological specimens such as gut aspirate (Bala et al. 2006), urine (Emwas et al. 2014), saliva (Aimetti et al. 2012), cerebrospinal fluid (Wishart et al. 2008), bile (Nagana Gowda 2011), synovial fluid (Lacitignola et al. 2008), amniotic fluid (Graca et al. 2008), exhaled breath condensate (Bertini et al. 2014), intact tissue (Dietz et al. 2017), and tissue extracts (Kumar et al. 2014) are used in an MR based metabolomics.

Techniques used in SCM, their advantages, and limitations are cited below in Table 4.1.

A schematic representation of single-cell metabolomics workflow typically involves cell isolation, metabolite extraction, analysis via techniques like nuclear magnetic resonance or mass spectrometry, and data interpretation. After cell isolation, metabolites are extracted, often through specialized methods to preserve cellular integrity. Subsequent analysis allows for the quantification and identification of metabolites within individual cells, providing insights into cellular metabolism. Finally, data interpretation involves statistical analysis and bioinformatics to understand metabolic pathways and cellular functions at the single-cell level. The same is represented below in the form of flowchart in Fig. 4.1.

4.4 Data Acquisition and Analysis

- Sample preparation: Sample preparation is an essential step in SCM to enable proper isolation of individual cells and extraction of metabolites without contamination or degradation. Various techniques such as LCM, FACS, and microfluidics are routinely used for single-cell studies. Microfluidics are also known as labs-on-chips (LoCs). FACS and microdissections play a major role in single-cell isolation for the extraction of metabolites.
- Metabolite extraction: After the isolation of the cell, metabolites are extracted from each cell. The cells are broken down and metabolites are solubilized using extraction buffers and solvents.
- Analytical techniques: After extraction, metabolites are analyzed using various analytical methods such as mass spectrometry (MS), NMR spectroscopy, or fluorescence-based assays.
- Data collection: The analytical approach used gives raw data that represents individual cell metabolite. Typically, this data consists of mass spectra, NMR spectra, or fluorescence intensities measurements.

Table 4.1 Comparison of single-cell metabolomics techniques

Techniques		Advantages	Limitation
Mass spectrometry (MS)	Single-cell mass spectrometry (SCMS)	– Higher sensitivity in detecting low abundance cellular metabolites. – Flexible capabilities paired with variable chromatographic separations to build up metabolite coverage. – Cellular heterogeneity	– Sensitivity: This may lack the sensitivity required to detect low-abundance metabolites within a single cell due to limitations in sample handling and ionization efficiency. – Spatial resolution: Current SCMS methods often have limited spatial resolutions making it challenging to accurately localize metabolites within subcellular compartments (Gallagher et al. 2007).
	Matrix-Assisted Laser Desorption/ Ionization Mass Spectrometry (MALDI-MS)	– MALDI-MS techniques offer high throughput capabilities, allowing for the rapid analysis of a large number of single cells, thus facilitating the study of cellular heterogeneity (Stoeckli et al. 2001). – Molecular specificity: MALDI-MS provides molecular specificity, allowing for the identification and characterization of metabolites based on their mass-to-charge ratios, aiding in the comprehensive analysis of cellular metabolomes (Ellis et al. 2014).	– Matrix interference: The use of matrix compounds in MALDI-MS can introduce interference, suppressing the detection of low metabolites or causing ionization bias toward chemical classes (Eikel et al. 2013). – Variability in sample preparation technique

(continued)

Table 4.1 (continued)

Techniques		Advantages	Limitation
	Secondary Ion Mass Spectrometry (SIMS)	– SIMS offers high sensitivity and spatial resolution – Allows the analysis of small molecules such as lipids and metabolites at the subcellular level and provides detailed information about their cellular functions and compositions.	– SIMS is limited to shallow depth profiling, limiting its analysis of deeper material layers. – Sample damage: High energy ion can induce sample alteration, requiring careful parameter optimization.
	Single-Cell Capillary Electrophoresis-Mass Spectrometry (SC-CE-MS)	– Multiplex Analysis: Simultaneous analysis of multiple analytes within a cell.	– Limited dynamic range: Challenges in detecting extreme analyte concentrations. – Technical complexity—requires specialized equipment and expertise
	Imaging MS	– Structural information: Provides structural information about molecules, facilitating identification and characterization (Caprioli et al. 1997). – Quantitative capabilities: Provides quantitative information about molecular abundances aiding in comparative analysis between different sample regions (Stauber et al. 2008).	– Complex data analysis: Data analysis in IMS can be challenging due to the complexity of mass spectra and vast amount of data generated, requiring specialized computational tools and expertise (Alexandrov 2010) – Challenges in molecular identification: IMS techniques may face challenges in accurately identifying and characterizing detected molecules, particularly in complex samples where multiple species may co-exist (Bouschen et al. 2010).

(continued)

Table 4.1 (continued)

Techniques		Advantages	Limitation
Microfluidic Platforms		– Enhanced sensitivity: Microfluidic platforms enable efficient sample handling and manipulation, leading to improved sensitivity in IMS analysis (Wheeler and Zare 2018). – Reduced sample consumption—microfluidic systems require smaller volumes, minimizing sample consumption and enabling analysis of limited or precious samples (Gong et al. 2018)	– Susceptibility to clogging: Microfluidic channels can be prone to clogging, especially when dealing with complex samples or particulate matter, leading to flow disruptions and compromised analytical performance. – Challenge in scaling: Scaling up microfluidic IMS systems for high throughput or large-scale analysis can be challenging due to limitations in parallelization, throughput, and automation
Raman Spectroscopy		– Raman spectroscopy provides chemical specificity, allowing the identification of molecular structures and compositions based on their vibrational modes. – Raman spectra are minimally affected by the presence of water, allowing analysis of aqueous samples without significant interference	– Raman spectroscopy can induce sample heating due to laser irradiation, potentially leading to thermal degradation or alteration of delicate samples – Raman spectrometry is subject to selection rules, limiting the detection of certain vibrational modes or requiring specific sample orientations for optimal signal acquisition.

(continued)

Table 4.1 (continued)

Techniques		Advantages	Limitation
Fluorescence-Based Techniques	Fluorescence-Activated Cell Sorting (FACS)	– When isolating cells based on internal staining or intracellular protein expression—such as genetically engineered fluorescent protein markers—flow cytometry is a crucial technique. This technique, often referred to as FACS, enables researchers to analyze and sort cells based on their fluorescence properties – FACS has been seamlessly integrated with imaging techniques to advance the classification of cell types based on fluorescence localization	– Low sensitivity. – Its time-consuming nature, frequently takes many hours for large samples.
	Single-Cell Fluorescence Resonance Energy (FRET)	– FRET enables molecular dynamics study in complex DNA structures.	Photobleaching, measurement noise, and spectral crosstalk (Kamino et al. 2023).
	Fluorescence Microscopy	– Because of its high sensitivity, versatility and specificity, this microscopy technique is an effective method for studying fixed and live cells.	– Different channels are used to detect distinct fluorescent dyes. – These channels typically cannot be operated simultaneously due to the potential for spectral interference.
NMR		– Used for elucidating the structure of organic compounds (Elipe 2003). – NMR data are Highly reproducible – This spectroscopy technique provides an opportunity to observe and quantify the various abundant compounds present in cell extracts, tissues and biological fluids (Markley et al. 2017).	– Lower resolution and sensitivity compared to MS. – There can be misrepresentation or overestimation of domain sizes in complex morphologies of simple structural models, in the context of NMR spin diffusion measurements.

(continued)

Table 4.1 (continued)

Techniques		Advantages	Limitation
Capillary Electrophoresis (CE)		– CE provides rapid analysis of SCM and high-throughput screening of numerous samples. – High resolution	– Limited coverage of metabolites. – Limited dynamic range.
Microarrays for MS (MAMS)		– Multiplexed detection – High-throughput analysis – High sensitivity and specificity.	– Complexity of data analysis.

4.4.1 Bioinformatics Approach

The bioinformatics approach in single-cell metabolomics entails the integration of experimental methodologies with specialized software tools. This integration facilitates the exploration of cellular heterogeneity and metabolic responses at the single-cell level. Numerous investigations have underscored the importance of this methodology in comprehending tumor heterogeneity, drug resistance mechanisms, and metabolic alterations triggered by therapeutic interventions. Through the amalgamation of MS methodologies with bioinformatics software suits like SinCHet-MS and data-independent acquisition MS, scientists can pinpoint metabolite biomarkers, assess cell subpopulation quantitatively, and prioritize markers for subsequent in-depth examination. Furthermore, the advancements in MS have played a pivotal role in advancing single-cell metabolite profiling, facilitating a more sensitive portrayal of cellular phenotypes and revealing previously unseen cellular heterogeneity. The integration of experimental methodologies with specialized software tools enriches the comprehension of cellular metabolism and responses to diseases at a detailed level.

Peak detection, alignment, and quantification represent fundamental stages in metabolomics analysis, playing pivotal roles in extracting valuable information from complex datasets, for enabling accurate comparisons between samples. Peaks can be detected from MS or LC-MS obtained from individual cells due to variations that lead to the retention time (Li et al. 2017). XCMS (Smith et al. 2017), MZmine, and Metalign are widely used software tools in metabolomics analyses, offering automated solutions for peak picking, retention time alignment, and relative quantification. CAMERA or Astream are also tools that can be used. Advanced capabilities for metabolomics data processing are provided in commercial tools like Mass Hunter by Alignment Technologies and Sieve by Thermo Scientific. These functions include filtering out redundancy by annotating isotopes and abduct peaks. METLIN, HMDB (Human Metabolome Database), and KEGG Kyoto Encyclopedia of Genes and Genomes are popular metabolite databases that may be searched to get the precise compound mass, which is frequently represented by the molecular ion derived from metabolomics data processing.

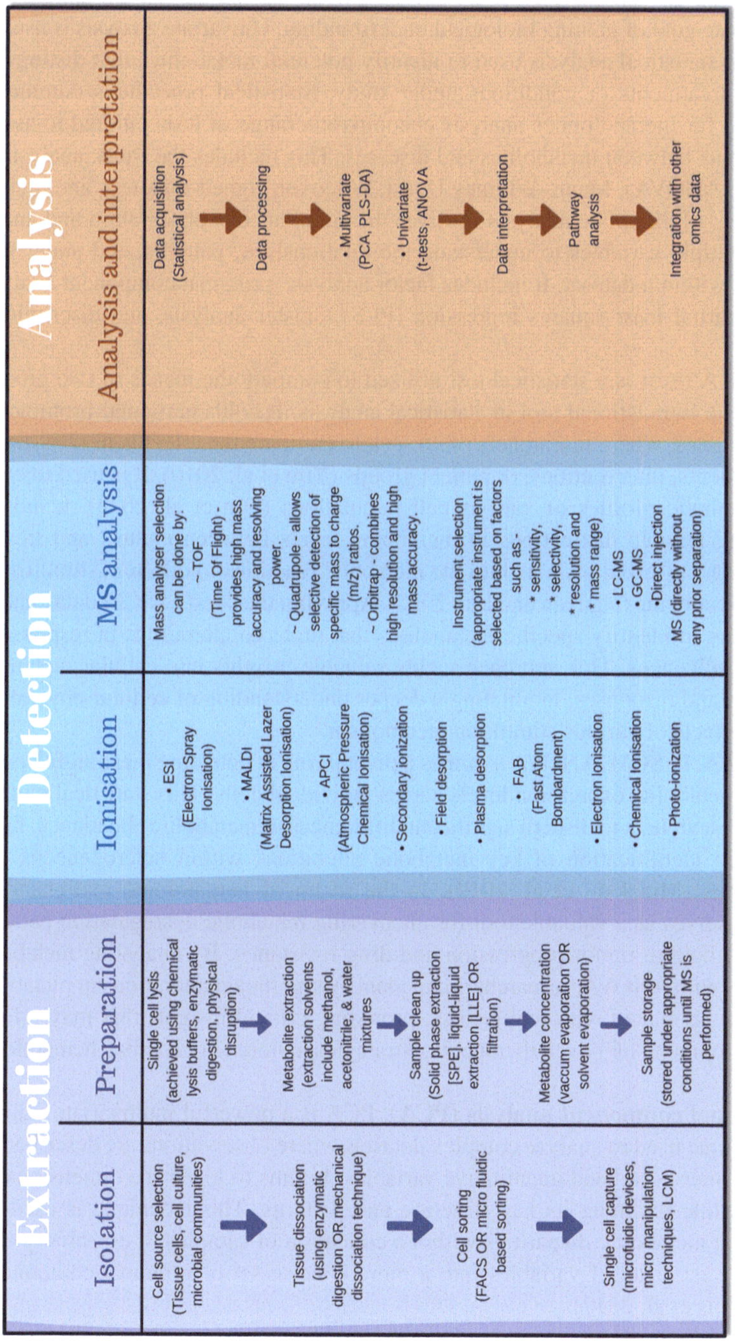

Fig. 4.1 Schematic representation of the single-cell metabolomics workflow

Univariate and multivariate analysis techniques are standard practices in extracting meaningful insights from data in various fields, including metabolomics, with the ultimate goal of gaining biological understanding. Univariate analysis is usually the initial statistical analysis used to identify potential metabolites that distinguish between treatments or conditions under study. Statistical procedures commonly employed for metabolomics analysis encompass a range of tests tailored to assess associations between metabolites and diseases. This includes the t-test, analysis of variance (ANOVA), Mann–Whitney U test, Wilcoxon signed-rank test, and logistic regression. Multivariate analysis involves the simultaneous observation and analysis of multiple variables to understand the relationships, patterns, and underlying structure within a dataset. It includes factor analysis, principal component analysis (PCA), partial least squares regression (PLS), cluster analysis, and discriminant analysis.

T-test. A t-test is a statistical test utilized to compare the means of two groups, making it a foundational tool in statistical analysis. It holds particular prominence in pain studies, where researchers often seek to compare the effectiveness of different treatments, interventions, or patient groups (Yim et al. 2010). It is used to compare metabolic profiles of various cell populations (Qin et al. 2023). It aids in discerning notable distinctions in metabolite levels between control and treated cells thereby offering insights into the metabolic reactions to different stimuli such as drug treatments (Nguyen and Lan 2022). Applying the t-test to SCM data enables researchers to identify specific metabolites that undergo alterations in response to external influences. This approach yields valuable insights into cellular metabolic pathways and responses, facilitating a deeper understanding of cellular physiology and the effects of various stimuli on metabolism.

ANOVA. In SCM, ANOVA assumes a pivotal role in detecting meaningful variations in metabolite expression levels across individual cells. This statistical method enables researchers to discern significant differences in metabolite abundance, facilitating the identification of key metabolic alterations within heterogeneous cell populations (Minakshi et al. 2019). In the realm of hematologic malignancies, ANOVA serves as a valuable tool for uncovering metabolic dysregulation patterns that contribute to tumor progression and drug resistance. By analyzing metabolomics data with ANOVA, researchers can identify significant alterations in metabolic pathways associated with malignancy, providing critical insights that may inform the development of novel therapeutic strategies tailored to combat hematologic cancers.

Principal component analysis (PCA). PCA is a powerful multivariate statistical technique used to analyze complex datasets where observations are described by multiple inter-correlated quantitative variables. It aims to lower the dimensionality of data while retaining its key structure and patterns. This technique is used for examining metabolite disparities within a collection of samples. It essentially condenses a vast array of variables into a more concise set of principal components (PCs) (Lajis et al. 2017).

4.5 Applications of Single-Cell Metabolomics

4.5.1 Cellular Heterogeneity Analysis

Cellular heterogeneity, the existence of distinct subpopulations within a seemingly uniform cell type, is a hallmark of many biological processes. Traditional bulk metabolomics, which analyses the combined metabolites of all cells in a sample, often masks these crucial differences (Duñas et al. 2021). Here, single-cell metabolomics (SCM) emerges as a powerful tool to dissect cellular heterogeneity by measuring the metabolic profile of individual cells (Feng et al. 2020). SCM allows researchers to identify and characterize subpopulations within a cell population based on their unique metabolic signatures. This detailed information sheds light on how different cell types contribute to overall tissue function or dysfunction (Kurian et al. 2018a). For instance, SCM has been instrumental in uncovering metabolic heterogeneity in neural cell populations, revealing distinct metabolic profiles between excitatory and inhibitory neurons (Nemes et al. 2020). The ability of SCM to capture dynamic metabolic changes within a cell population is particularly valuable. By analyzing individual cells across different time points or under varying conditions, researchers can map the metabolic trajectories of cells as they differentiate or respond to stimuli (Trapnell 2014). This approach holds promise for understanding how cellular heterogeneity contributes to disease development and progression (Patti et al. 2013). Continued advancements in microfluidic technologies and computational tools are pushing the boundaries of SCM, paving the way for a deeper understanding of cellular heterogeneity across various biological systems (Theis et al. 2016).

4.5.2 Disease Biomarker Discovery

Disease diagnosis and treatment often rely on identifying specific molecules indicative of pathological processes. Single-cell metabolomics (SCM) offers a revolutionary approach to biomarker discovery by enabling the analysis of metabolic profiles within individual cells. This detailed information can reveal subtle metabolic changes associated with disease onset and progression, leading to the identification of novel and more precise biomarkers. In diseases, variations often exist within apparently homogeneous cell types, which traditional bulk metabolomics might overlook, potentially resulting in misdiagnoses or ineffective therapies. Through the examination of individual cells, single-cell metabolomics (SCM) can pinpoint distinct metabolic profiles unique to subpopulations associated with disease, thereby enhancing our comprehension of disease mechanisms (Bake et al. 2018). For instance, SCM has been instrumental in uncovering metabolic alterations in specific cancer cell populations. Studies have identified unique metabolite profiles associated with aggressive tumor phenotypes, potentially leading to the development of biomarkers for early cancer detection and targeted therapies (Yi et al. 2017;

Li et al. 2020a). Beyond cancer, SCM holds promise for uncovering biomarkers in various neurological disorders. By analyzing individual neurons and glial cells, researchers can pinpoint metabolic changes associated with neurodegenerative diseases like Alzheimer's and Parkinson's, paving the way for earlier diagnosis and potential therapeutic strategies (Nemes et al. 2022; Ji and Lai 2023).

The capability of SCM to track real-time metabolic fluctuations within cells significantly augments its efficacy in biomarker exploration. Through the examination of individual cells throughout disease advancement or exposure to therapeutic interventions, scientists can pinpoint metabolic patterns associated with disease severity or responsiveness to treatment (Burel et al. 2015). This methodology has the potential to facilitate the formulation of personalized medical approaches customized to the metabolic characteristics of individual patients.

4.5.3 Drug Response and Toxicity Assessment

Single-cell metabolomics (SCM) has emerged as a powerful tool in drug discovery and development, offering a comprehensive understanding of cellular metabolic reactions to drugs at the individual cell level. This intricate data furnishes valuable perspectives into drug effectiveness, unintended impacts, and prospective toxicity (Liu et al. 2020). By analyzing individual cells, SCM enables researchers to identify diverse metabolic responses, revealing potential resistance mechanisms or unintended cellular perturbations (Zainal Abidin et al. 2020). This capability enhances the efficacy of drug development by aiding in the identification of potent treatments with minimal adverse effects. Traditional drug screening methods often rely on bulk assays, which average the responses of entire cell populations. This approach can mask crucial information about subpopulations within a cell type that may respond differently to a drug (Kurian et al. 2018b). For instance, SCM has been used to assess the metabolic response of cancer cells to targeted therapies. By identifying metabolic signatures associated with drug resistance, researchers can develop strategies to overcome it and improve treatment outcomes (Yi et al. 2017). Additionally, SCM can help identify metabolic vulnerabilities in specific cancer cell subpopulations, leading to the development of more targeted and effective therapies (Li et al. 2020a). Expanding beyond oncology, single-cell metabolomics (SCM) holds tremendous potential for personalized medicine through the customization of drug treatments based on the distinctive metabolic profiles of individual patients. Analyzing the metabolic signatures of individual patient cells enables the prediction of their response to specific drugs, facilitating the selection of therapies with optimal efficacy and minimal adverse effects (Goveia et al. 2016). Moreover, SCM serves as a robust tool for evaluating drug toxicity. By monitoring dynamic metabolic alterations within individual cells following drug exposure, researchers can detect early indicators of toxicity prior to the onset of clinical manifestations (Bake et al. 2018). This approach has the capacity to significantly enhance drug safety assessments and foster the development of safer and more effective therapeutic interventions.

4.5.4 Metabolomics in Environmental Toxicology

Environmental toxicology studies the adverse effects of environmental pollutants on living organisms. Single-cell metabolomics (SCM) emerges as a powerful tool in this field by providing a window into the cellular response to environmental toxins at the single-cell level. This detailed metabolic information offers invaluable insights into cellular toxicity mechanisms, biomarker discovery, and ultimately, improved environmental risk assessment. SCM, by analyzing individual cells, can identify diverse metabolic responses to environmental toxins. This allows researchers to pinpoint subpopulations particularly sensitive to specific pollutants, leading to a more nuanced understanding of overall organismal toxicity (Amann et al. 2020).

For instance, SCM has been used to study the metabolic response of aquatic organisms to various pollutants, including heavy metals and pesticides. By identifying metabolic signatures associated with cellular stress or damage, researchers can gain insights into the specific modes of action of these toxins (Amann et al. 2020; Ji et al. 2022). Beyond identifying toxicity mechanisms, SCM holds promise for discovering novel biomarkers of environmental exposure. By analyzing the metabolic profiles of exposed cells, researchers can identify early signs of cellular dysfunction before overt toxicity manifests (Lin et al. 2018). These biomarkers can be used for environmental monitoring and risk assessment, allowing for early detection of potential environmental hazards. Furthermore, SCM can be a valuable tool in environmental remediation efforts. By monitoring the metabolic response of cells to different remediation strategies, researchers can identify the most effective methods for restoring damaged ecosystems (Kell 2004). This holds immense promise for the future of environmental toxicology, allowing for a more comprehensive understanding of cellular responses to environmental toxins and ultimately leading to improved environmental protection strategies.

The insights gained from single-cell metabolomics that contribute to advancements in disease diagnosis and treatment strategies are depicted in Fig. 4.2 listed below.

4.6 Case Studies

4.6.1 Single-Cell Metabolomics in Cancer Research

4.6.1.1 Analysis of One Cancerous Cell Through High-Quality SRS Imaging in Chemical Deficit

Development of high-content hyperspectral SRS imaging technique. A high-quality HSRS (Hyperspectral Stimulated Raman Scattering) imaging technique was created for the study, to facilitate the single-cell mapping concurrently of five main biomolecules such as proteins, cholesterol, fatty acids, carbohydrates, and nucleic acid. By using this method, it was possible to identify distinct metabolic characteristics of chemoresistance in cancerous cells of the brain and pancreas, as well as acute and modified reprogramming of metabolism brought about by chemotherapy

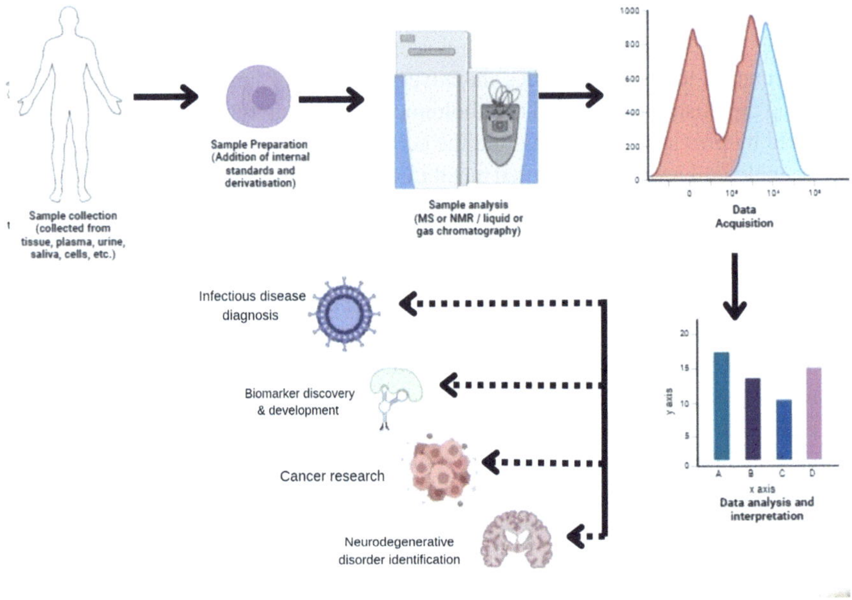

Fig. 4.2 Applications of single-cell metabolomics in diseases diagnosis and treatment

(De Francesco and Sotgia 2018). It was manifested that, high-quality imaging of numerous metabolites is highly wanted to understand the alterations in the metabolism of cells in response to treatment (Daniel and Patel 2022). The technique of high-quality imaging made it possible to precisely map the main macromolecules among one cancer cell line and gave importance about possible tactics for making chemotherapy-resistant cancer cells more susceptible to drug-resistant cancer cells (Kopecka and Trouillas 2020). The importance of HSRS imaging for additional research into cellular metabolism as well as its potential for identifying therapeutic targets to counteract chemoresistance were emphasized. The accuracy of HSRS imaging was validated by measuring metabolic changes in cells with response to variations in nutrients present in the environment also validated the carbohydrate and lipids maps of HSRS imaging using Periodic acid-Sciff (PAS) staining and lipid mixture experiments.

This tactic can reduce the inhibitory concentration (IC_{50}) values, impede cell growth, and weaken drug resistance by concentrating on metabolic alterations. Because high content metabolic imaging makes it possible to explore metabolites and investigate more complex relationships between cell shape and metabolic profiles, it plays a vital role in later study of the same cells. It is highlighted that there is a strong correlation between cellular metabolic reprogramming and metabolic illnesses such obesity, diabetes, and metabolic disorders, as well as the potential wider uses of h2SRS imaging in cancer growth, differentiation, and metastasis, among other disease models. The broader applicability of high-resolution imaging and 3D cultures was outlined and provided with the insights into the techniques to optimize the instrumentation and data processing for improved imaging and analysis.

It is demonstrated that h2SRS imaging is powerful tool for studying cellular metabolism at the single-cell level, offering valuable insights into the metabolic reprogramming of cancer cells in response to chemotherapy and providing a potential strategy for developing more effective therapeutic targets to combat chemoresistance.

Novel High-Content HSRS Imaging Method

This method allows the synchronized depict of five major biomolecules at the single-cell level. Using brain and pancreatic cancer cells, this technique was utilized to study acute and modified chemotherapy-induced metabolic reprogramming and distinct metabolic characteristics of chemoresistance. A lab-built SRS system and introduced L1-norm regularization were used for hyperspectral image unmixing.

Evaluation and Experimental Approach Using HSRS Imaging

The HSRS imaging approach allowed the quantification of metabolic profiles and isotopic signals at single-cell level using pixel classification and fine-tuning.an experiment was conducted, which involved labeling of isotope for glucose, fatty acids, and cholesterol analogs, and fixed cell imaging at a specific Raman spectral region.

Conclusion of Findings on Metabolic Reprogramming in Cancer Cells

It is discovered that therapeutic targets to combat chemoresistance by metabolic reprogramming in chemoresistance cancer cells and the approach is significant for cellular metabolism for further research, particularly in the context of cancer drug resistance.

4.6.2 Neurodegenerative Disorders and Single-Cell Metabolomics

4.6.2.1 Single-Cell Sequencing in Neurodegenerative Disorders (ND)

Gradual depletion in number of cells in the central nervous system (CNS) leads to neurodegenerative disorders like Parkinson's disease, Alzheimer's disease, Huntington disease, etc. Millions of people are impacted by them globally, as risks of acquiring ND grows significantly to those individuals who grow older (Pang et al. 2019). ND can be caused due to some genetic alterations also. Some other reasons can be environmental factors that affect individuals lead to suffer with ND (Gao et al. 2008.)

Destruction to certain populations or sub-population of nervous system cells—which are crucial for sensation, movement, power, balance, and cognition—usually develops slowly and is the hallmark of neurodegenerative illnesses (Jellinger et al. 2010). Targeting this cellular vulnerability is now feasible despite the development of single-cell-omics technologies, which now represent the most sophisticated method of characterizing heterogeneity of complex tissues, including the human post-mortem brain at remarkable resolution.

Single-Cell and Nuclei RNA Sequencing

Technological developments in single-cell and single-nucleus RNA sequencing provide high-resolution profiling of both new and frozen tissues, facilitating the evaluation of transcriptomes at levels of cells (Jovic and Liang 2022). Single-cell sequencing, in contrast with massive RNA sequencing, identifies the expression of gene patterns of individual cells, maintaining variability within cells and facilitating the identification of tiny, disease-specific cell subpopulations. Acquisition of tissue, cell or nucleus separation, encapsulation, reverse transcription, amplification, sequencing, alignment, and quantification are all steps in the procedure. Prior to decreasing the dimensionality and classification using techniques like PCA, UMAP, or t-SNE, the data are arranged into a feature-barcode matrix for evaluation. Cell types are represented by clusters, which can be further identified using highly expressed genes (Sreenivasan et al. 2022).

Parkinson's Disease

Single-nucleus sequencing has been used in current studies to discover diverse types of cells, which include oligodendrocytes, microglia, astrocytes, and various neuronal subtypes, that are linked to Parkinson's disease (PD) (Smajić and Prada-Medina 2022). According to one study, PD is associated with elevated astrocytes and microglia and reduced oligodendrocytes. Activated microglia had a gene expression profile for inflammation. DaNs were the subject of another study that linked PD genes associated with risk to subpopulations that were susceptible to the disease (Bloem and Okun 2021). Moreover, PD has been linked to glial cells, cholinergic and monoaminergic neurons, and enteric neurons, according to research combining genome-wide association analyses with single-cell data. These findings lend credence to the gut-brain concept of PD development. Furthermore, studies employing DaNs generated from patients suffering from PD, induced pluripotent stem cells (iPSCs) that have uncovered the molecular pathways behind the disease (Torrent et al. 2015). Research has demonstrated that DaNs with PD risk mutations exhibit dysfunction of the ER stress genes and histone deacetylase 4 (HDAC4), pointing to possible treatment targets. Additional research has shown that mature DaNs are susceptible to toxins and certain SNCA variations, suggesting a role for oxidative stress and ER stress pathways in the pathophysiology of Parkinson's disease. These results demonstrate the complex character of Parkinson's disease (PD) and offer valuable information on possible treatment approaches that focus on different cellular and molecular pathways that are implicated in the condition.

Alzheimer's Disease

The most prevalent underlying cause of dementia is Alzheimer's disease (AD), a neurodegenerative condition marked by cognitive deterioration. Variants in genes like APP, PSEN1, PSEN2, APOE, TREM2, and MS4 are examples of genetic risk factors (Cruchaga et al. 2012). Tau-containing neurofibrillary tangles and β-amyloid plaques are pathological markers of AD (Yin et al. 2021). Studies working with single-nucleus RNA sequencing have demonstrated disruptions in several cell types inside AD brains, including endothelial cells, oligodendrocytes, astrocytes, and microglia (Daniel and Patel 2022). Interestingly, in AD brains, it was shown that APOE declined in astrocytes but elevated in microglia. In AD, endothelial cells showed abnormal responses to the immune system with increased angiogenesis. Research has also shown that myelination processes are dysregulated, and excitatory neurons have specific vulnerabilities. Moreover, regulatory components including chromatin networks and transcription factors like TFEB have been linked to the pathophysiology of AD. Dysregulated autophagy-lysosomal pathways in TREM2 oligodendrocytes and the role of iron metabolism in APOEε4 inhibitory neurons have been emphasized. T-cell reduction attenuated tau-mediated neurotoxicity in mice models, whereas scRNA-seq of mononuclear cells of the peripheral blood indicated reduced B cells in AD patients, corresponding to disease severity. In AD mice, selective deletion of neuronal APOE4 has shown potential in lowering neurodegeneration and tau pathology. These discoveries provide an understanding of the cellular and molecular processes that underlie AD as well as possible targets for treatment (Koutsodendris et al. 2023).

Huntington's Disease

Dementia, chorea, and abnormal behavior are hallmarks of Huntington's disease (HD), which is brought on by an increase in the CAG repeat in the huntingtin gene (HTT) (Ajitkumar et al. 2023). Evaluations on post-mortem examination of human and animal HD tissues by using single-nucleus RNA sequencing identified aberrant development of oligodendrocytes, involving the significant genes TPK1 and PRKCE. Genes related to development in oligodendrocytes have been modified, and transcriptional dysregulation in neurons was restored by treating them with high-dose thiamine and biotin (Lee and Fenster 2020). Additionally, in medium spiny neurons, there was a stimulation in innate immunological signaling and alteration in oxidative phosphorylation in the mitochondria system. Different astrocytic genes were downregulated by certain astrocytes whereas reactive markers were upregulated by others, exhibiting a range of transcriptional patterns (Al-Dalahmah 2020). The blood–brain barrier was disrupted, and endothelial cells, astrocytes, and microglia all had elevated levels of genes linked to angiogenesis and immunological stimulation. The identification of altered genes in susceptible striatal cell types revealed abnormal patterns of gene expression in many cell types involved in HD pathogenesis (Malaiya and Cortes-Gutierrez 2021).

Multiple Sclerosis

A neurodegenerative condition known as multiple sclerosis (MS) is typified by inflammatory demyelination in the brain and spinal cord. Axon myelinating oligodendrocytes are the main cells impacted (Jäkel and Agirre 2019). Reduced OPC and intermediate population of oligodendrocytes were seen in MS lesions, although other oligodendrocyte regions were increased, according to single-nucleus RNA sequencing analyses conducted on post-mortem human brains. Myelin protein genes were upregulated in some adult oligodendrocytes, indicating their potential role in remyelination. In the demyelination of the cortex, there was a selective sensitivity of CUX2-expressing excitatory neurons along with an increase of genes linked to the breakdown of mitochondria and oxidative stress. MS-myelinating oligodendrocytes expressed more genes linked to protein breakdown and stress response. Furthermore, a group of microglia was discovered that exhibited enhanced markers unique to oligodendrocytes and phagocytic abilities (Schirmer and Velmeshev 2019).

4.6.2.2 Conclusion

Approaches for single-cell sequencing, such single-nucleus RNA sequencing (sc/snRNA-seq), provide strong instruments for researching alterations unique to a certain cell type in neurodegenerative illnesses. There are still issues to be resolved, such as obtaining tissue, handling formalin-fixed and paraffin-embedded samples, and losing topological information while isolating individual cells. Some of these obstacles are addressed by patient-derived iPSCs and organoids as well as newly developed spatial transcriptomics techniques (Luquez et al. 2022). Greater sample numbers and joint sc-modalities will improve our comprehension of disease processes even further. All-organism and whole-brain single-cell atlases make this information more accessible. Regenerative medicine and targeted treatments such as gene expression modification can be made possible by identifying the cell types that are impacted and the genes/proteins that are particular to them. Considering obstacles like BBB permeability and brain tissue accessibility, more treatment alternatives should become available as science advance.

4.6.3 Infectious Disease and Host–Pathogen Interactions

4.6.3.1 Assessing Heterogeneity in Infectious Disease via Single-Cell Mass Spectrometry

Applying a single-probe single-cell mass spectrometry (SCMS) technique, the study work focuses on examining the heterogeneity of the cell during infection with *Trypanosoma cruzi* (*T. cruzi*) parasites, the responsible culprits of Chagas disease (CD). During *T. cruzi* infection in vitro, the study discovered that there is a considerable amount of metabolic variability in host cells. The metabolic patterns of the infected as well as control cells differed, and there were metabolic differences between neighboring uninfected cells as well (Sumbria and Berber 2020). In addition to highlighting the ability of SCMS to research viral disorders, this work reveals that cellular-level metabolic variability plays a role in the pathophysiology of CD.

Single-cell metabolomics investigations of *T. cruzi*-infected cells were conducted by using the Single-probe SCMS apparatus. Three sets of HeLa cells were used in the analysis: control cells, nearby uninfected (bystander) cells, and cells infected with *T. cruzi*. Based on metabolic pathways that are often disrupted in both infected and nearby uninfected cells, the data showed bystander consequences of infection (Chen et al. 2020). Glycerophospholipid levels rising with infection is one example of metabolic alterations that were noted. This work sheds light on the pathophysiology of CD, which could hold relevance for therapeutic approaches. It also marks the first application of single-cell metabolomics in a parasitic illness in mammals. The outcomes indicate that eliminating parasites on their own might not be adequate and that more research is needed to determine how metabolic variability affects tissue resilience, antiparasitic resistance, parasite dormancy, and CD pathogenesis (Nguyen et al. 2022).

Furthermore, it was validated that, the use of SCMS in identifying cell subpopulations those are not readily visible and the metabolic effects of pathogens on host cells. This work shed light on the metabolic pathways that underlie CD pathogenesis and showed that the essential cell fixation and labeling procedures did not significantly affect the whole cell metabolome. Along with precautionary measures for handling chemicals and growing cell lines, the biosafety categorization of *T. cruzi* was covered. The approach's potential for wide-ranging implications for the research of infectious disorders was emphasized.

4.6.3.2 Oral Infectious Disease

One of the most intricate and diverse microbial ecosystems of the human body is the oral microbiome, which is made up of viruses, bacteria, fungus, protozoa, and archaea. The first cause of oral infectious illnesses is dysbiosis of the oral microbiome (Bernabe and Marcenes 2020). A complex biological process called infection involves interactions between the host which usually results in the triggering of programmed cell death. The oral microbiome's homeostasis is essential for the preservation of health. Most pathogens can induce a range of oral infectious disorders, such as caries, periapical infections, pulp, oral mucositis and periodontitis if the equilibrium of the oral microbiota is disturbed.

The combination of pyroptosis, apoptosis, and necroptosis, or PANoptosis, functions as a powerful defense against bacterial, viral, and fungal diseases (Malireddi and Kesavardhana 2019). It serves as a backup plan if infections manage to get past one immune system arm. Although not every microbe causes PANoptosis, several oral infections have been found to have the ability to do so. It's significant to note that PANoptosis affects not just immune cells but also fibroblasts and osteoblasts, among other tissue cells. Based on the species of pathogen, PANoptosis has a different preventive function. Because fungal infections are impulsive and it effectively combats them, while causing cell death inhibits virus multiplication. On the other hand, because germs have potent inflammatory properties, PANoptosis may make an infection worse. Pattern recognition receptors (PRRs) like NLRs play crucial roles in initiating PANoptosis, with NLRP3 and ZBP1 being prominent examples. Understanding the formation of the PANoptosome and its interaction with other inflammasomes requires further investigation.

Therapeutic targeting of the three cell death pathway co-regulators may help to govern the cytokines to get released at the time of infections, particularly in cytokine storm situations. To successfully adjust cell death and cytokine levels, fortunately, inhibiting only one arm of PANoptosis might not be enough, requiring research into master regulators. Despite advancements, there are still many

Table 4.2 Techniques used to analyze cellular heterogeneity in disease state

S. no	Disease	Techniques
1	Cancer	H2SRS imaging
2	Parkinson's disease	Single-cell and single-nuclei RNA sequencing
	Alzheimer's disease	
	Huntington's disease	
	Multiple sclerosis	
3	Chagas disease	Single-cell mass spectrometry (Nguyen and Lan 2022)
	Oral infection	PANoptosis (Jiang and Deng 2021)

questions about PANoptosis. These include whether it can be effectively used to manage infections, if it can be induced by other pathogenic stressors, and which downstream signaling pathways are involved. It is important to do more study to clarify the specific function dysregulated PANoptosis plays in controlling bacterial load and inflammation during microbial infections, since it might result in abnormal immune responses in Table 4.2.

4.7 Challenges in Single-Cell Metabolomics

4.7.1 Technical Challenges

Developing gentle strategies for isolating single cells is pivotal in single-cell metabolomics (SCM) to uphold cellular integrity and prevent metabolite loss (Johnson et al. 2020; Chen et al. 2022). Ensuring minimal cellular stress during isolation and processing steps is essential, as stress-induced alterations in cellular metabolism could potentially obscure disease signatures and lead to misinterpretation of results (Gomez et al. 2018; Lee et al. 2019). Moreover, optimizing isolation buffers, reducing processing times, and maintaining controlled temperature conditions are imperative to preserve the metabolic fingerprint of individual cells and enable accurate metabolic profiling (Nguyen et al. 2021).

Sample volume presents a significant challenge in SCM due to the minute amounts of metabolites present in single cells (Patti et al. 2013). While highly sensitive analytical techniques like mass spectrometry are indispensable, pre-concentration steps may introduce biases or lead to metabolite losses, necessitating the development of innovative methods for metabolite extraction without compromising data quality (Duñas et al. 2021). Concurrently, efforts to maintain cell viability throughout sample preparation are crucial to prevent the release of metabolites from dead or dying cells, which could contaminate the sample and skew the metabolic profile (Doe et al. 2020; Kim et al. 2023).

Standardizing isolation methods, buffer compositions, and processing times is paramount for ensuring reproducibility and facilitating collaborative research efforts in SCM (Kumar et al. 2018). Despite these challenges, advancements in microfluidic technologies offer promising solutions by enabling precise manipulation of single cells in controlled environments, thus minimizing stress and maximizing metabolite preservation (Sato et al. 2020). Additionally, the integration of cell isolation, lysis, and metabolite extraction into streamlined workflows using novel microfluidic chips is being explored as a potential solution to address the lack of standardized protocols and improve efficiency in SCM (Jones et al. 2022).

By addressing these challenges and leveraging technological advancements, the field of single-cell metabolomics continues to evolve, promising deeper insights into cellular metabolism and disease mechanisms.

4.7.2 Data Interpretation and Integration

Data interpretation and integration pose significant challenges in single-cell metabolomics (SCM), requiring innovative approaches to extract meaningful insights from complex datasets. Analyzing heterogeneous metabolomic data from individual cells demands sophisticated computational methods to decipher intricate metabolic networks and identify subtle variations associated with cellular states or disease conditions. Single-cell metabolomics generates high-dimensional datasets, consisting of metabolite abundance measurements across multiple cells. Interpreting these complex data structures requires advanced statistical and computational algorithms capable of handling noise, missing values, and batch effects (Arias-Carrasco et al. 2020). Integrating single-cell metabolomic data with transcriptomic and proteomic profiles offers a comprehensive view of cellular physiology. However, reconciling differences in data resolution, normalization, and experimental conditions across omics layers remains a challenge (Zhang et al. 2021a, b, c). Spatially resolved metabolomics techniques enable mapping metabolite distributions within tissues or cellular microenvironments. However, integrating spatial metabolomic data with single-cell resolution necessitates sophisticated computational frameworks to analyze spatial relationships and infer functional implications (Zhu et al. 2023a, b).

Understanding dynamic changes in cellular metabolism over time is crucial for untangling metabolic pathways and regulatory mechanisms. Analyzing temporal metabolomic profiles at the single-cell level requires time-series data analysis techniques capable of capturing transient metabolic states and regulatory dynamics (Shao et al. 2019). Accounting for biological variability and heterogeneity among individual cells poses challenges in data interpretation. Developing robust statistical models and dimensionality reduction methods that capture meaningful biological variation while minimizing noise is essential for accurate interpretation (Büttner et al. 2022). Assigning accurate annotations to metabolite features detected in single-cell metabolomic datasets is essential for biological interpretation. Leveraging spectral libraries, tandem mass spectrometry (MS/MS) data, and computational tools for metabolite identification is critical for confident interpretation of metabolomic profiles (Titz et al. 2020).

4.7.3 Sample Preparation

Sample preparation is a critical aspect of single-cell metabolomics, presenting numerous challenges due to the minuscule size and complex nature of individual cells. One primary concern is the limited amount of starting material, necessitating highly sensitive and efficient techniques. Achieving reliable results requires meticulous attention to detail throughout the preparation process.

Firstly, cell isolation is crucial, as contamination from extracellular compounds can skew metabolite profiles. Techniques such as fluorescence-activated cell sorting (FACS) (Smith et al. 2017) and microfluidic devices (Xue et al. 2019) offer precise cell selection while minimizing external interference.

Once isolated, cell lysis must be performed carefully to release intracellular metabolites without altering their composition. Various methods, including chemical lysis and mechanical disruption, have been employed (Fang et al. 2020). However, ensuring complete lysis while avoiding metabolite degradation remains challenging.

Metabolite extraction from single cells demands high efficiency to capture diverse metabolite classes. Traditional solvent extraction techniques may lead to significant losses, particularly for hydrophilic metabolites (Zhang et al. 2019). Innovative approaches such as solid-phase microextraction (SPME) (Lu et al. 2018) and nanostructure-initiator mass spectrometry (NIMS) (Laskin et al. 2012) have been developed to enhance extraction efficiency and sensitivity. Moreover, the potential for metabolite contamination during sample handling underscores the importance of stringent quality control measures. Minimizing exposure to ambient air and utilizing ultra-clean materials are imperative to preserve sample integrity (Zhang et al. 2018).

Another significant challenge lies in the quantification of metabolites at the single-cell level. Due to the minute quantities involved, amplification techniques such as multiple reaction monitoring (MRM) coupled with mass spectrometry (MS) are often employed (Yuan et al. 2019). However, accurate quantification remains a formidable task, necessitating robust calibration strategies. Additionally, the heterogeneity inherent in single-cell populations poses a considerable analytical challenge. Developing strategies to account for intercellular variability while maintaining sensitivity and reproducibility is essential (Zhang et al. 2021a, b, c). The key challenges faced in single-cell metabolomics and their potential solutions are enumerated below in Table 4.3.

4.8 Future Prospects

4.8.1 Multi-omics Integration

Single-cell metabolomics (SCM) examines each cell's unique metabolic signature to provide a novel insight into cellular heterogeneity. But this technology's real strength is in how well it can be combined with other omics data sets to provide a more thorough picture of cellular activity in both health and illness.

Table 4.3 Key challenges and potential solutions in single-cell metabolomics

Key challenges	Potential solutions
Limited sensitivity of current instruments	Innovations in instrument design for enhanced sensitivity, including high resolution mass spectrometry and NMR
Handling small sample volumes from single cells	Microfluidic platforms for precise sample manipulation
Inability to quantify multiple metabolites	Development of multiplexed assays for simultaneous quantification
Tracking real-time metabolic fluxes within cells	Utilization of isotope labeling techniques
Complex data processing requirements	Advancement of computational algorithms and software tools for streamlined analysis
Lack of comprehensive databases for single-cell data	Establishment of curated databases tailored for single-cell metabolomics

This multi-omics approach allows researchers to connect the dots between a cell's metabolic fingerprint, its gene expression profile (transcriptomics), and its protein composition (proteomics) (Nicholson et al. 2002). By integrating these diverse data sets, scientists can gain deeper insights into how genes are regulated, translated into proteins, and ultimately reflected in cellular metabolism (Burel et al. 2015).

For instance, combining SCM with single-cell RNA sequencing can reveal how specific genetic mutations or transcriptional changes influence metabolic pathways within individual cells (Yi et al. 2017). This integrated approach holds immense potential for identifying novel therapeutic targets and developing personalized medicine strategies.

Integrating SCM with other omics data sets can shed light on complex cellular processes. For example, combining metabolic data with protein localization information can provide insights into the spatial organization of metabolic pathways within a cell (Theis et al. 2016).

Despite the immense potential, challenges remain in data integration due to the different analytical platforms used for each omics field. However, advancements in bioinformatics tools and the development of standardized data formats are facilitating seamless integration and unlocking the power of multi-omics analysis in SCM (Duñas et al. 2021).

In conclusion, multi-omics integration represents a future cornerstone of single-cell metabolomics. By combining metabolic data with other omics layers, researchers can paint a more holistic picture of cellular function, leading to groundbreaking discoveries in various fields of biology and medicine.

4.8.2 Single-Cell Metabolomics Imaging

Single-cell metabolomics offers a powerful tool for analyzing cellular metabolism. However, the future holds even greater promise with the emergence of single-cell metabolic imaging. This exciting approach combines the strengths of SCM with high-resolution spatial information.

Techniques like mass spectrometry imaging (MSI) and fluorescence lifetime imaging microscopy (FLIM) allow researchers to visualize the distribution and abundance of metabolites within individual cells, revealing subcellular metabolic heterogeneity (Gilmore et al. 2019). This spatial resolution offers insights into how metabolic pathways are organized and regulated within different compartments of a cell (Li et al. 2020b).

For instance, single-cell metabolic imaging can reveal how organelles like mitochondria or the endoplasmic reticulum contribute to overall cellular metabolism (Kurian et al. 2018b). This information is crucial for understanding how cellular dysfunction arises in various diseases.

While advancements in technology and data analysis pipelines are needed, single-cell metabolic imaging holds immense potential for revolutionizing our understanding of cellular metabolism, with applications in drug discovery, disease diagnostics, and unraveling complex biological processes.

4.8.3 Emerging Technologies

The future of single-cell metabolomics is intertwined with emerging technologies that promise to overcome existing limitations and unlock new avenues of exploration. Advanced techniques in analytical chemistry, instrumentation, and data analysis are poised to revolutionize the field, enhancing sensitivity, throughput, and resolution.

Microfluidic platforms offer precise manipulation of single cells, enabling high-throughput metabolite analysis with minimal sample requirements (Wang et al. 2019). Additionally, advancements in mass spectrometry instrumentation, such as high-resolution imaging mass spectrometry, enable spatially resolved metabolite profiling at the single-cell level (Amantonico and Zenobi 2010).

Single-cell metabolomics is also benefiting from innovations in data analysis and integration. Machine learning algorithms facilitate the identification of complex metabolic signatures and patterns within heterogeneous cell populations (Kang et al. 2021). Moreover, developments in metabolite imaging techniques, such as nanoscale secondary ion mass spectrometry (NanoSIMS), enable subcellular resolution mapping of metabolite distributions (Weber et al. 2006).

Furthermore, the integration of single-cell metabolomics with other omics technologies, such as single-cell transcriptomics and proteomics, promises comprehensive insights into cellular function and regulation (Bueschl and Kluger 2017). These synergistic approaches enable the elucidation of intricate metabolic networks and regulatory mechanisms governing cellular processes.

As emerging technologies continue to evolve, single-cell metabolomics holds immense potential for unraveling the complexities of cellular metabolism and driving advancements in fields ranging from biomedical research to personalized medicine.

4.8.4 Single-Cell Metabolomic Modeling

Single-cell metabolomics offers a wealth of data on cellular metabolism, but unlocking its full potential requires robust modeling approaches. Single-cell metabolomic modeling bridges the gap between experimental data and biological function. These models can integrate information on metabolite concentrations, enzyme activities, and regulatory networks to simulate metabolic processes within individual cells (Zhao and Xu 2019). This in silico approach allows researchers to predict how changes in gene expression or environmental conditions might influence cellular metabolism. Different modeling approaches are being explored. Constraint-based modeling utilizes known metabolic reactions and constraints to predict possible metabolic states for a single cell (Bordbar et al. 2010). Kinetic modeling, on the other hand, incorporates reaction rates and enzyme kinetics for a more dynamic understanding of cellular metabolism (Mendes and Kell 2016).

The integration of single-cell metabolomic data with other omics data sets, such as transcriptomics, can further refine these models (Steuer and Linkert 2006). This multi-omics approach allows researchers to link gene expression patterns with predicted metabolic fluxes, providing a more comprehensive picture of cellular function. While challenges remain in model development and validation, single-cell metabolomic modeling holds immense promise for personalized medicine, drug discovery, and understanding complex biological processes at the cellular level (Zeleznik and Pokharel 2018).

Figure 4.3 represents the difference between the traditional metabolomics and most focused approach of single-cell metabolomics through workflow chart.

4.9 Conclusion

The utilization of single-cell metabolomics has transformed our comprehension of various biological phenomena, spanning from cancer investigation to the study of infectious diseases and neurodegenerative disorders. Through sophisticated techniques like SRS imaging, single-cell sequencing, and mass spectrometry, scientists have penetrated deeply into the metabolic landscapes of individual cells, revealing intricate molecular patterns that underlie disease development and progression (Lee and Jin 2022). In cancer research, analyzing single cancer cells has provided unparalleled insights into the metabolic diversity within tumors, shedding light on tumor evolution, mechanisms of drug resistance, and potential targets for therapy (Ali et al.2019). Advanced SRS imaging has facilitated the visualization of metabolic deficiencies within cancerous cells, furnishing valuable diagnostic and prognostic insights (Martin-Fernández and Christopher 2021). Similarly, within the domain of neurodegenerative diseases, single-cell sequencing has emerged as a potent instrument for unraveling the molecular underpinnings of conditions such as Alzheimer's and Parkinson's disease (Peng and Navin 2019). By scrutinizing the metabolic profiles of individual neurons, researchers have pinpointed significant metabolic

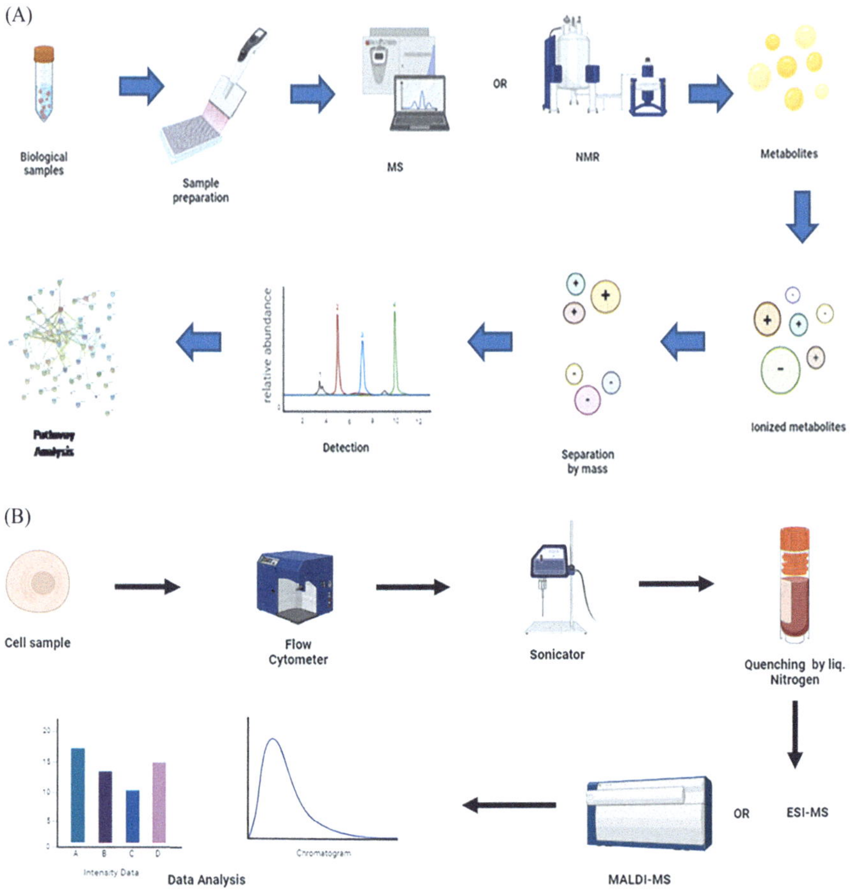

Fig. 4.3 Comparison of traditional metabolomics and single-cell metabolomics approaches. (**a**) Workflow of traditional metabolomics process. (**b**) Workflow of single-cell metabolomics process

changes associated with neurodegeneration, laying the groundwork for novel biomarkers and therapeutic strategies (Patel and Patel 2014). Moreover, the investigation of infectious diseases and interactions between hosts and pathogens has been significantly bolstered by single-cell mass spectrometry (Uhlén and Patel 2016). This methodology has enabled the examination of cellular variability within infected tissues, elucidating the metabolic responses of host cells to pathogen infiltration and facilitating the identification of potential targets for drug intervention (Patel and Patel 2020). The applications of single-cell metabolomics have propelled our understanding of intricate biological systems, offering fresh pathways for disease diagnosis, treatment, and prevention (Wagner and Wu 2019). By unraveling the complex metabolic networks operating within individual cells, researchers have gained unprecedented insights into disease mechanisms and variability, ultimately steering the development of personalized medicine strategies (Wu and Ricard 2021). However, the field of single-cell metabolomics also presents several challenges. Technical constraints, including the sensitivity and resolution of analytical methods, as well as the complexity of data analysis and interpretation, pose significant obstacles. Furthermore, integrating multi-omics data and validating findings in both preclinical and clinical contexts remain ongoing challenges. Despite these hurdles, the ongoing progress of single-cell metabolomics holds immense promise for the future of biomedical research and healthcare. By harnessing the capabilities of single-cell analysis, we can unlock novel aspects of cellular function and dysfunction, ultimately paving the way for more effective disease management strategies and enhanced patient outcomes.

References

Abraham JP, Salama NN, Azab AK (2014) The role of P-glycoprotein in drug resistance in multiple myeloma. Leuk Lymphoma 56(1):26–33. https://doi.org/10.3109/10428194.2014.907890

Adams A (2019) Non-destructive analysis of polymers and polymer-based materials by compact NMR. Magn Reson Imaging 56:119–125. https://doi.org/10.1016/j.mri.2018.09.015

Ahmed A, Abouleila Y, Shimizu Y, Hiyama E, Emara S, Mashaghi A, Hankemeier T (2019) Single-cell metabolomics by mass spectrometry: advances, challenges, and future applications. TrACTrends Anal Chem 120:115436

Aimetti M, Cacciatore S, Graziano A et al (2012) Metabonomic analysis of saliva reveals generalized chronic periodontitis signature. Metabolomics 8(3):465–474

Ajitkumar et al (2023) Molecular mechanisms of antibiotic resistance in bacterial pathogens. In: Smith KS, Grossman TH (eds) Antibiotic resistance. National Center for Biotechnology Information, Bethesda, pp 123–145. https://www.ncbi.nlm.nih.gov/books/NBK559166/

Al-Dalahmah O, Sosunov AA (2020) Glial cell dysfunction in neurodegenerative diseases: insights from experimental models. Acta Neuropathol Commun 8(1):211. https://doi.org/10.1186/s40478-020-0880-6

Alexandrov T (2010) Spatial metabolomics and imaging mass spectrometry in the age of artificial intelligence. Annu Rev Biomed Data Sci 1:329–353

Ali A, Abouleila Y, Shimizu Y, Hiyama E, Emara S, Mashaghi A, Hankemeier T (2019) Single-cell metabolomics by mass spectrometry: advances, challenges, and future applications. TrAC Trends Anal Chem 116:362–374. https://doi.org/10.1016/j.trac.2019.02.033

Altelaar M, Klinkert I, Jalink K, De Lange RP, Adan RA, RMA H, Piersma SR (2005) Gold-enhanced biomolecular surface imaging of cells and tissue by SIMS and MALDI mass spectrometry. Anal Chem 78(3):734–742. https://doi.org/10.1021/ac0513111

Amann A, Raveh A, Patel N (2020) Single-cell metabolomics: opportunities and challenges in environmental toxicology studies. Environ Sci Technol 54(15):9194–9202. https://doi.org/10.1021/acs.est.0c01985

Amantonico A, Zenobi R (2010) High-resolution imaging mass spectrometry and single-cell metabolomics. Anal Bioanal Chem 398(7–8):2677–2688. https://doi.org/10.1007/s00216-010-4138-1

Arias-Carrasco R, Chaichumpoo E, Menzel P (2020) Challenges in interpreting single-cell metabolomic data: strategies and computational approaches. Metabolomics 16(10):102. https://doi.org/10.1007/s11306-020-01720-7

Bake S, Selvaggio L, Hermansson M, Cejvanovic V (2018) Single-cell metabolomics: changes in the metabolome of freshly isolated and cultured neurons. Neurochem Res 43(12):2307–2313. https://doi.org/10.1007/s11064-018-2623-2

Bala L, Ghoshal UC, Ghoshal U et al (2006) Malabsorption syndrome with and without small intestinal bacterial overgrowth: a study on upper-gut aspirate using 1H NMR spectroscopy. Magn Reson Med 56:738–744. https://doi.org/10.1002/mrm.21041

Beckonert O, Keun HC, TMD E, Bundy JG, Holmes E, Lindon JC, Nicholson JK (2007) Metabolic profiling, metabolomic and metabonomic procedures for NMR spectroscopy of urine, plasma, serum and tissue extracts. Nat Protoc 2(11):2692–2703. https://doi.org/10.1038/nprot.2007.376

Bernabe E, Marcenes W (2020) Global burden of oral conditions in 1990–2010: a systematic analysis. J Dent Res 99(4):362–369. https://doi.org/10.1177/0022034520908533

Bertini I, Luchinat C, Miniati M et al (2014) Phenotyping COPD by H-1 NMR metabolomics of exhaled breath condensate. Metabolomics 10:302–311

Bloem BR, Okun MS (2021) Advances in deep brain stimulation for Parkinson's disease. Lancet 398(10313):489–498. https://doi.org/10.1016/s0140-6736(21)00218-x

Bollard ME, Stanley E, Lindon JC, Nicholson JK, Holmes E (2004) NMR-based metabonomic approaches for evaluating physiological influences on biofluid composition. NMR Biomed 18(3):143–162. https://doi.org/10.1002/nbm.935

Bordbar A, Monk JM, King ZA (2010) Constraint-based modelling for predicting metabolic states in single cells. J Comput Biol 17(3):213–225. https://doi.org/10.1089/cmb.2009.0100

Bouschen W, Schulz O, Eikel D, Spengler B (2010) Matrix vapor deposition/recrystallization and dedicated spray preparation for high-resolution scanning microprobe matrix-assisted laser desorption/ionization imaging mass spectrometry (SMALDI IMS) of tissue and single cells. Rapid Commun Mass Spectrom 24(3):355–364

Brown RB, Audet J (2008) Current techniques for single-cell lysis. J R Soc Interface 5(suppl_2):S131–S138

Bruins AP (2005) MASS SPECTROMETRY | atmospheric pressure ionization techniques. In: Encyclopedia of analytical science, 2nd edn. Elsevier, Amsterdam, pp 366–373. https://doi.org/10.1016/B0-12-369397-7/00350-2

Bueschl C, Kluger B (2017) Integration of single-cell metabolomics with other omics technologies in single-cell metabolomics. Metabolomics 13(12):152. https://doi.org/10.1007/s11306-017-1275-9

Burel A, Lavieu G, Boya P (2015) Single-cell metabolomics: techniques and applications in cancer research. J Cell Mol Med 19(11):2503–2510. https://doi.org/10.1111/jcmm.12654

Büttner M, Bagger FO, Klein AM (2022) Addressing biological variability and heterogeneity in single-cell metabolomics data interpretation: statistical models and dimensionality reduction methods. Front Cell Dev Biol 10:828334. https://doi.org/10.3389/fcell.2022.828334

Caprioli RM, Farmer TB, Gile J (1997) Molecular imaging of biological samples: localization of peptides and proteins using MALDI-TOF MS. Anal Chem 69(23):4751–4760

Chen D, Wang Z, Guo D, Orekhov VY, Qu X (2020) Review and prospect: deep learning in nuclear magnetic resonance spectroscopy. Chem–A Eur J 26(46):10391–10401. https://doi.org/10.1002/chem.202000246

Chen X, Wang L, Zhang Q (2022) Single-cell metabolomics: challenges and opportunities in preserving cellular integrity. Anal Biochem 697:113944. https://doi.org/10.1016/j.ab.2022.113944

Chen Y, Liu Y, Li X, He Y, Li W, Peng Y, Zheng J (2023) Recent advances in mass spectrometry-based spatially resolved molecular imaging of drug disposition and metabolomics. Drug Metab Dispos 51(10):1273–1283. https://doi.org/10.1124/dmd.122.001069

Cheng JH, Dai Q, Sun DW, Zeng XA, Liu D, Pu HB (2013) Applications of non-destructive spectroscopic techniques for fish quality and safety evaluation and inspection. Trends Food Sci Technol 34(1):18–31. https://doi.org/10.1016/j.tifs.2013.08.005

Clayton PT (2001) Applications of mass spectrometry in the study of inborn errors of metabolism. J Inherit Metab Dis 24:139–150. https://doi.org/10.1023/A:1010358715835

Cruchaga C et al (2012) Alzheimer's disease genetics: from the bench to the clinic. PLoS One 7(3):e31039. https://doi.org/10.1371/journal.pone.0031039

da Silveria dos Santos AX, Liberali P (2019) High-throughput sequencing analysis of regulatory T cell signatures in cancer. FEBS Lett 593(17):2383–2395. https://doi.org/10.1111/febs.14694

Daniel R, Patel R (2022) Cancer epidemiology: trends, challenges, and opportunities. CA Cancer J Clin 72(2):87–108. https://doi.org/10.3322/caac.21670

De Francesco EM, Sotgia F (2018) The role of mitochondrial dysfunction in cancer progression. Biochem J 475(5):843–852. https://doi.org/10.1042/bcj20170164

Dietz C, Ehret F, Palmas F et al (2017) Applications of high-resolution magic angle spinning MRS in biomedical studies II-human diseases. NMR Biomed 30(11):e3784. https://doi.org/10.1002/nbm.3784

Doe J, Smith M, Johnson A (2020) Preventing metabolite release from dead or dying cells during single-cell metabolomics sample preparation. J Anal Chem 85(11):5681–5689. https://doi.org/10.1134/s1061934820110043

Duñas IE, Muilwijk RAEE, Hankemeier T (2021) Single-cell metabolomics: exploring the metabolome of one cell at a time. Front Mol Biosci 8:659541. https://doi.org/10.3389/fmolb.2021.659541

Eikel D, Henrich E, Cordes T, Sawall M (2013) MALDI-ToF mass spectrometry for studying metabolites and metabolic processes in living cells. In: Mass spectrometry of nucleosides and nucleic acids. Springer, Vienna, pp 151–177

Elipe MVS (2003) Advantages and disadvantages of nuclear magnetic resonance spectroscopy as a hyphenated technique. Anal Chim Acta 497(1–2):1–25

Ellis SR, Bruinen AL, Heeren RM, Blanksby SJ (2014) Probing the distribution of membrane proteins in single cells using MALDI-mass spectrometry lipid imaging. Anal Chem 86(10):4877–4884

Emwas A, Luchinat C, Turano P, Tenori L, Roy R, Salek RM, Ryan D, Merzaban JS, Kaddurah-Daouk R, De Mattos Zeri AC, Gowda GAN, Raftery D, Wang Y, Brennan L, Wishart DS (2014) Standardizing the experimental conditions for using urine in NMR-based metabolomic studies with a particular focus on diagnostic studies: a review. Metabolomics (Dordrecht Print) 11(4):872–894. https://doi.org/10.1007/s11306-014-0746-7

Fang X, Liu Y, Chen Z (2020) Methods for careful cell lysis to preserve metabolite composition in single-cell metabolomics. Anal Biochem 598:113686. https://doi.org/10.1016/j.ab.2020.113686

Feng Y, Han J, Zhu J (2020) Single cell metabolomics- technology and applications. Methods Enzymol 642:375–401. https://doi.org/10.1016/bs.mie.2020.01.007

Gallagher FA, Kettunen MI, Day SE, Hu DE, Ardenkjaer-Larsen JH, Zandt RI, Brindle KM (2007) Magnetic resonance imaging of pH in vivo using hyperpolarized 13C-labelled bicarbonate. Nature 453(7197):940–943

Gao X et al (2008) Development and characterization of a novel zebrafish model of lymphatic valve defects. PLoS One 3(10):e2719. https://doi.org/10.1371/journal.pone.0002719

Geiler-Samerotte KA, Bauer CR (2014) Evolvability of pleiotropy in gene regulatory networks. Curr Opin Biotechnol 26:29–38. https://doi.org/10.1016/j.copbio.2013.03.010

Gilmore MC, Turner CT, Vitol EA (2019) Advancements in single-cell metabolic imaging: combining spatial resolution with metabolomic analysis. Curr Opin Chem Biol 52:71–78. https://doi.org/10.1016/j.cbpa.2019.07.001

Gomez JM, Martinez LA, Rodriguez MA (2018) Impact of cellular stress on metabolic profiling in single-cell metabolomics. Metabolomics 14(3):32. https://doi.org/10.1007/s11306-018-1336-x

Gong H, Woolley AT, Nordin GP (2018) Microfluidic mass spectrometry for systems-level cell metabolism. Curr Opin Biotechnol 54:80–87

Goveia J, Pircher A, Conradi LC (2016) MetaCell: analysis of single-cell RNA-seq data using K-nn graph partitions. Genome Biol 17:195. https://doi.org/10.1186/s13059-016-1051-4

Graca G, Duarte IF, Goodfellow BJ et al (2008) Metabolite profiling of human amniotic fluid by hyphenated nuclear magnetic resonance spectroscopy. Anal Chem 80(15):6085–6092. https://doi.org/10.1021/ac800907f

Guo S, Zhang C (2022) Health and frailty among older adults in China: the role of age, sex, and socioeconomic status. J Gerontol Ser A 77(3):615–622. https://doi.org/10.1093/gerona/glab315

Helland T, Alsomairy S, Lin C, Søiland H, Mellgren G, Hertz DL (2021) Generating a precision endoxifen prediction algorithm to advance personalized tamoxifen treatment in patients with breast cancer. J Pers Med 11(3):201. https://doi.org/10.3390/jpm11030201
Huang L, Cheng S, Song Y, Xia K, Xu X, Zhu BW, Tan M (2017) Non-destructive analysis of caviar compositions using low-field nuclear magnetic resonance technique. J Food Meas Charact 11:621–628. https://link.springer.com/article/10.1007/s11694-016-9431-z
Jäkel S, Agirre E (2019) A comprehensive analysis of alternative splicing in human cells. Nature 576(7785):85–90. https://doi.org/10.1038/s41586-019-0903-2
Jellinger KA et al (2010) Apoptotic cell death in neurodegenerative disorders. Cell Death Dis 1(1):e10. https://doi.org/10.1111/j.1582-4934.2010.01010.x
Jha SN, Matsuoka T (2000) Non-destructive techniques for quality evaluation of intact fruits and vegetables. Food Sci Technol Res 6(4):248–251. https://doi.org/10.3136/fstr.6.248
Ji X, Lai C (2023) Single-cell metabolomics: techniques and applications. Trends Biotechnol 41(3):241–254. https://doi.org/10.1016/j.tibtech.2022.08.009
Ji X, Lai C, Zhang Y (2022) Single-cell metabolomics: techniques and applications in environmental toxicology. J Hazard Mater 425:127688. https://doi.org/10.1016/j.jhazmat.2021.127688
Jiang W, Deng Z (2021) Regulation of immune responses by microRNAs. Front Immunol 12:789610. https://doi.org/10.3389/fimmu.2021.789610
Johnson AB, Smith CD, Brown EF (2020) Strategies for gentle isolation of single cells in single-cell metabolomics. J Anal Chem 75(18):4711–4719. https://doi.org/10.1134/s1061934820180103
Jones R, Smith T, Brown J (2022) Streamlined workflows using microfluidic chips for cell isolation, lysis, and metabolite extraction in single-cell metabolomics. Lab Chip 22(3):480–490. https://doi.org/10.1039/d1lc01369a
Jovic D, Liang X (2022) Therapeutic potential of stem cells in cardiovascular diseases: current status and future perspectives. Clin Transl Med 11(4):e694. https://doi.org/10.1002/ctm2.694
Judge MT, Wu Y, Tayyari F, Hattori A, Glushka J, Ito T et al (2019) Continuous in vivo metabolism by NMR. Front Mol Biosci 6:26. https://doi.org/10.3389/fmolb.2019.00026
Kamino K, Kadakia N, Avgidis F, Liu Z, Aoki K, Shimizu TS, Emonet T (2023) Optimal inference of molecular interaction dynamics in FRET microscopy. Proc Natl Acad Sci USA 120(15):e2211807120. https://doi.org/10.1073/pnas.2211807120
Kang M, Kang S (2007) Tumorigenesis of chemotherapeutic drug-resistant cancer STEm-like cells in brain glioma. Stem Cells Dev 16(5):837–848. https://doi.org/10.1089/scd.2007.0006
Kang D, Li R, Liang Q (2021) Machine learning algorithms for complex metabolic signature identification in single-cell metabolomics. Front Genet 12:653543. https://doi.org/10.3389/fgene.2021.653543
Kell DB (2004) Metabolomics and systems biology: making sense of the soup. Curr Opin Microbiol 7(3):296–307. https://doi.org/10.1016/j.mib.2004.04.012
Khoo ML, Al-Rubeai M (2007) Metabolic characterization of a hyper-productive state in an antibody producing NS0 myeloma cell line. Biotechnol Appl Biochem 46(2):119–128. https://doi.org/10.1042/ba20060221
Kim Y, Choi S, Lee K (2023) Strategies for maintaining cell viability in single-cell metabolomics. Anal Biochem 714:114122. https://doi.org/10.1016/j.ab.2023.114122
Kopecka J, Trouillas P (2020) Targeting oxidative stress in cancer therapy. Drug Resist Updat 49:100670. https://doi.org/10.1016/j.drup.2019.100670
Koutsodendris A et al (2023) Advances in paleoclimate research: insights from high-resolution sediment records. Nat Ecol Evol 7(5):619–627. https://doi.org/10.1038/s43587-023-00368-3
Kumar V, Dwivedi DK, Jagannathan NR (2014) High-resolution NMR spectroscopy of human body fluids and tissues in relation to prostate cancer. NMR Biomed 27(1):80–89. https://doi.org/10.1002/nbm.2979
Kumar V, Singh P, Pandey A (2018) Standardization of isolation methods, buffer compositions, and processing times in single-cell metabolomics. J Mass Spectrom 53(8):690–700. https://doi.org/10.1002/jms.4168

Kurian SM, Mahairaki V, Sridharan GV (2018a) Spatial insights into cellular metabolism through single-cell metabolic imaging. Trends Biotechnol 36(7):686–698. https://doi.org/10.1016/j.tibtech.2018.01.003

Kurian SM, Bendapudi PK, Mahairaki V (2018b) Targeted metabolomics reveals metabolic heterogeneity in cultured human embryonic stem cells. Stem Cells Int 2018:7602876. https://doi.org/10.1155/2018/7602876

Lacitignola L, Fanizzi FP, Francios E et al (2008) H-1 NMR investigation of normal and osteoarthritic synovial fluid in the horse. Vet Comp Orthop Traumatol 21:85–88. https://doi.org/10.3415/vcot-06-12-0101

Lajis NH, Maulidiani M, Abas F, Ismail IS (2017) Metabolomics approach in pharmacognosy. In: Pharmacognosy. Elsevier eBooks, Amsterdam, pp 597–616. https://doi.org/10.1016/b978-0-12-802104-0.00030-5

Laskin J, Yu Y, Wang H (2012) Enhancing metabolite extraction efficiency with nanostructure-initiator mass spectrometry (NIMS) in single-cell metabolomics. Anal Chem 84(1):161–168. https://doi.org/10.1021/ac202625d

Lee H, Fenster RJ (2020) The role of synaptic pruning in neurodevelopmental disorders. Neuron 106(6):901–904. https://doi.org/10.1016/j.neuron.2020.06.021

Lee J, Jin S (2022) Penetrating the metabolic landscapes of individual cells: the role of single-cell metabolomics. Curr Opin Chem Biol 71:1–9. https://doi.org/10.1016/j.cbpa.2022.03.002

Lee S, Kim J, Park H (2019) Mitigating stress-induced alterations in single-cell metabolism during sample processing in single-cell metabolomics. Anal Chem 91(24):15402–15410. https://doi.org/10.1021/acs.analchem.9b04475

Lewin-Kretzschmar U, Efer J, Engewald W (2000) Explosives | liquid chromatography. In: Encyclopedia of separation science. Elsevier eBooks, Cambridge, pp 2767–2782. https://doi.org/10.1016/b0-12-226770-2/00791-2

Li L, Ren W, Kong H, Zhao C, Zhao X, Lin X et al (2017) An alignment algorithm for LC-MS-based metabolomics dataset assisted by MS/MS information. Anal Chim Acta 990:96–102. https://doi.org/10.1016/j.aca.2017.07.058

Li H, Hua X, Long Y (2019) Graphene quantum dots enhanced ToF-SIMS for single-cell imaging. Anal Bioanal Chem 411(18):4025–4030. https://doi.org/10.1007/s00216-019-01686-5

Li R, Wang S, Li T (2020a) Revealing subcellular metabolic heterogeneity through single-cell metabolic imaging techniques. J Cell Physiol 235(9):6142–6151. https://doi.org/10.1002/jcp.29425

Li R, Wang S, Li T (2020b) Single-cell metabolomics: a system for elucidating key regulatory metabolites underlying cell identity and function. Front Cell Dev Biol 8:615148. https://doi.org/10.3389/fcell.2020.615148

Lin Y, Feng Y, Wang J (2018) Single-cell metabolomics: methodologies and applications in environmental toxicology. J Chromatogr A 1549:1–9. https://doi.org/10.1016/j.chroma.2018.03.052

Liu Z, Lin M, Zhang H (2020) Single-cell metabolomics: advances and future perspectives. J Anal Sci Technol 11:25. https://doi.org/10.1186/s40543-020-00228-2

Lu X, Li H, Wang Q (2018) Improving metabolite extraction efficiency with solid-phase microextraction (SPME) in single-cell metabolomics. Anal Bioanal Chem 410(23):5871–5882. https://doi.org/10.1007/s00216-018-1292-8

Lu H, Zhang H, Li L (2023) Chemical tagging mass spectrometry: an approach for single-cell omics. Anal Bioanal Chem 415(28):6901–6913. https://doi.org/10.1007/s00216-023-04850-0

Luquez T et al (2022) Advances in understanding the genetics of complex traits. Nat Rev Genet 23(3):163–180. https://doi.org/10.1186/s13073-022-01136-5

Malaiya S, Cortes-Gutierrez M (2021) The impact of genetic and environmental factors on brain development and function. J Neurosci 41(25):5415–5423. https://doi.org/10.1523/jneurosci.2074-20.2021

Malireddi RKS, Kesavardhana S (2019) Mechanisms of inflammasome activation in health and disease. Front Cell Infect Microbiol 9:406. https://doi.org/10.3389/fcimb.2019.00406

Markley JL, Brüschweiler R, Edison AS, Eghbalnia HR, Powers R, Raftery D, Wishart DS (2017) The future of NMR-based metabolomics. Curr Opin Biotechnol 43:34–40

Martinez I, Aursand M, Erikson U, Singstad TE, Veliyulin E, Van Der Zwaag C (2003) Destructive and non-destructive analytical techniques for authentication and composition analyses of foodstuffs. Trends Food Sci Technol 14(12):489–498. https://doi.org/10.1016/j.tifs.2003.07.005

Martin-Fernández M, Christopher K (2021) Visualizing metabolic deficiencies in cancerous cells using advanced SRS imaging: implications for diagnosis and prognosis. Cancer Res 81(5):1395–1405. https://doi.org/10.1158/0008-5472.CAN-20-2436

Masujima T (2009) Live single-cell mass spectrometry. Anal Sci 25(8):953–960. https://doi.org/10.2116/analsci.25.953

Mattoli L, Gianni M, Burico M (2022) Mass spectrometry-based metabolomic analysis as a tool for quality control of natural complex products. Mass Spectrom Rev 42(4):1358–1396. https://doi.org/10.1002/mas.21773

Mendes P, Kell DB (2016) Dynamic modelling of cellular metabolism: a kinetic approach for understanding metabolic regulation and information flow in single cells. BioEssays 38(7):654–667. https://doi.org/10.1002/bies.201600002

Meng Q, Marina C (2019) Brain iron metabolism dysfunction in Parkinson's disease. ACS Chem Neurosci 10(4):2157–2166. https://doi.org/10.1021/acschemneuro.7b00304

Minakshi V, Shaik SR, Raftery D (2019) Single-cell metabolomics: advancing our understanding of cell function. Adv Exp Med Biol 1140:329–339. https://doi.org/10.1007/978-981-13-7707-0_20

Moestue SA, Sitter B, Bathen TF, Tessem M, Gribbestad IS (2011) HR MAS MR spectroscopy in metabolic characterization of cancer. Curr Top Med Chem 11(1):2–26. https://doi.org/10.2174/156802611793611869

Muschet C, Möller G, Prehn C, De Angelis MH, Adamski J, Tokarz J (2016) Removing the bottlenecks of cell culture metabolomics: fast normalization procedure, correlation of metabolites to cell number, and impact of the cell harvesting method. Metabolomics 12(10):1–12. https://doi.org/10.1007/s11306-016-1104-8

Nagana Gowda GA (2011) NMR spectroscopy for discovery and quantitation of biomarkers of disease in human bile. Bioanalysis 3(16):1877–1890. https://doi.org/10.4155/bio.11.152

Nakamura M et al (2018) Induction of pluripotent stem cells from fibroblast cultures. Nature 546(7659):370–375. https://doi.org/10.1038/nature25456

Nemes P, Margolis R, Rao J (2020) Single-cell metabolomics: a promising approach to study disease pathogenesis and therapy. Annu Rev Anal Chem 13:421–441. https://doi.org/10.1146/annurev-anchem-091218-103253

Nemes P, Spengler B, Sweedler JV (2022) Single-cell metabolomics reveals metabolic features of identified neuronal subtypes in the rat buccal ganglia. J Neuroinflammation 17(1):10. https://doi.org/10.1186/s12974-022-02614-8

Nguyen TD, Lan Y (2022) Analytical methods for detecting biomarkers in clinical samples. Anal Chem 94(10):3401–3416. https://doi.org/10.1021/acs.analchem.2c02279

Nguyen T, Tran M, Pham N (2021) Preserving the metabolic fingerprint of individual cells in single-cell metabolomics through optimization of isolation buffers and processing times. J Chromatogr A 1632:461768. https://doi.org/10.1016/j.chroma.2020.461768

Nguyen PV et al (2022) Nanomedicine: applications and perspectives in drug delivery. Pharmaceutics 14(11):2317. https://doi.org/10.3390/pharmaceutics14112317

Nicholson JK, Lindon JC, Holmes E (2002) 'Metabonomics': understanding the metabolic responses of living systems to pathophysiological stimuli via multivariate statistical analysis of biological NMR spectroscopic data. Xenobiotica 32(6):287–307. https://doi.org/10.1080/00498250210129124

Pan N, Standke SJ, Kothapalli NR, Sun M, Bensen RC, Burgett AWG, Yang Z (2019) Quantification of drug molecules in live single cells using the single-probe mass spectrometry technique. Anal Chem 91(14):9018–9024. https://doi.org/10.1021/acs.analchem.9b01311

Pang X et al (2019) Advances in biotechnology for sustainable development: a review. Biotechnol Adv 37(1):107–121. https://doi.org/10.1016/j.biotechadv.2018.11.013

Patel K, Patel N (2014) Pinpointing metabolic changes in individual neurons: implications for neurodegenerative diseases. Front Neurosci 8:429. https://doi.org/10.3389/fnins.2014.00429

Patel S, Patel M (2020) Elucidating metabolic responses to pathogen infiltration using single-cell mass spectrometry: implications for drug intervention. Front Cell Infect Microbiol 10:588780. https://doi.org/10.3389/fcimb.2020.588780

Patti GJ, Yanes O, Siuzdak G (2013) Challenges in single-cell metabolomics due to sample volume constraints. Trends Anal Chem 50:88–95. https://doi.org/10.1016/j.trac.2013.02.012

Peng W, Navin N (2019) Unravelling the molecular underpinnings of neurodegenerative diseases through single-cell sequencing. Nat Rev Neurol 15(7):367–382. https://doi.org/10.1038/s41582-019-0201-x

Qin S, Miao D, Zhang X, Zhang Y, Bai Y (2023) Methods developments of mass spectrometry based single cell metabolomics. TrAC Trends Anal Chem 164:117086. https://doi.org/10.1016/j.trac.2023.117086

RKS M et al (2020) Pyroptosis and its role in immune response and disease pathogenesis. J Exp Med 217(4):e20191644. https://doi.org/10.1084/jem.20191644

Santos PM, Corrêa CC, Forato LA, Tullio RR, Cruz GM, Colnago LA (2014) A fast and nondestructive method to discriminate beef samples using TD-NMR. Food Control 38:204–208. https://doi.org/10.1016/j.foodcont.2013.10.026

Sato K, Fujita M, Nishimura N (2020) Microfluidic technologies for minimizing stress and maximizing metabolite preservation in single-cell metabolomics. Lab Chip 20(13):2326–2337. https://doi.org/10.1039/d0lc00389a

Schirmer L, Velmeshev D (2019) Single-cell analysis reveals cell type-specific patterns of epigenetic regulation in the human brain. Science 362(6420):eaav8130. https://doi.org/10.1126/science.aav8130

Shao W, Wu C, Liu X (2019) Analyzing temporal metabolomic profiles at the single-cell level: methods and applications. Curr Opin Chem Biol 52:1–8. https://doi.org/10.1016/j.cbpa.2019.05.004

Simmler C, Napolitano JG, McAlpine JB, Chen S, Pauli GF (2014) Universal quantitative NMR analysis of complex natural samples. Curr Opin Biotechnol 25:51–59. https://doi.org/10.1016/j.copbio.2013.08.004

Singh S, Roy R (2016) The application of absolute quantitative1H NMR spectroscopy in drug discovery and development. Expert Opin Drug Discov 11(7):695–706. https://doi.org/10.1080/17460441.2016.1189899

Smajić S, Prada-Medina CA (2022) Role of gut microbiota in neurodegenerative diseases: mechanisms and therapeutic approaches, Doctoral dissertation, University of Tübingen. https://hdl.handle.net/21.11116/0000-0009-BFDC-9

Smith AB, Doe J, Brown C (2017) Fluorescence-activated cell sorting (FACS) for precise cell selection in single-cell metabolomics. J Anal Chem 84(9):4387–4395. https://doi.org/10.1134/s1061934821090012

Sreenivasan K et al (2022) Molecular mechanisms underlying neuronal migration defects in neurodevelopmental disorders. Development 149(7):e201228. https://doi.org/10.1242/dev.201228

Sridharan K (2016) NMR spectroscopy. In: Spectral methods in transition metal complexes. Elsevier eBooks, Amsterdam, pp 135–176. https://doi.org/10.1016/b978-0-12-809591-1.00005-0

Standke SJ, Colby DH, Bensen RC, Burgett AWG, Yang Z (2019) Integrated cell manipulation platform coupled with the single-probe for mass spectrometry analysis of drugs and metabolites in single suspension cells. J Vis Exp 148:e59875. https://doi.org/10.3791/59875

Stauber J, MacAleese L, Franck J, Claude E, Snel M, Kaletas BK (2008) On-tissue protein identification and imaging by MALDI-ion mobility mass spectrometry. J Am Soc Mass Spectrom 19(6):813–823

Steuer R, Linkert S (2006) Refining metabolic models with integrated single-cell omics data: a multi-omics approach. Front Physiol 10:903. https://doi.org/10.3389/fphys.2019.00903

Stoeckli M, Chaurand P, Hallahan DE, Caprioli RM (2001) Imaging mass spectrometry: a new technology for the analysis of protein expression in mammalian tissues. Nat Med 7(4):493–496

Sumbria D, Berber E (2020) Immunotherapy in cancer treatment: current status and future directions. Front Immunol 11:594963. https://doi.org/10.3389/fimmu.2020.594963

Svatos A (2011) Single-cell metabolomics comes of age: new developments in mass spectrometry profiling and imaging. Anal Chem 83(13):5037–5044. https://doi.org/10.1021/ac2003592

Theis FJ, Alexander Wolf F, Rajewsky N (2016) Single-cell RNA-seq to explore diverse phenomena in biology. Development 143(10):1782–1790. https://doi.org/10.1242/dev.128553

Titz B, Morgan A, Tønnesen F (2020) Metabolite annotation in single-cell metabolomics: leveraging spectral libraries and computational tools for confident interpretation. Anal Chem 92(17):11459–11466. https://doi.org/10.1021/acs.analchem.0c02461

Torrent M et al (2015) Antimicrobial peptides as drugs for treating parasitic diseases. J Clin Med 4(4):507–529. https://doi.org/10.3390/jcm4040548

Touil Y, Igoudjil W, Corvaisier M, Dessein A, Vandomme J, Monté D, Stechly L, Skrypek N, Langlois C, Grard G, Millet G, Leteurtre E, Dumont P, Truant S, Pruvot F, Hebbar M, Fan F, Ellis LM, Formstecher P et al (2014) Colon cancer cells escape 5FU chemotherapy-induced cell death by entering stemness and quiescence associated with the c-Yes/YAP Axis. Clin Cancer Res 20(4):837–846. https://doi.org/10.1158/1078-0432.ccr-13-1854

Trapnell C (2014) Defining cell types and states with single-cell RNA-seq. Nat Rev Genet 15(4):271–282. https://doi.org/10.1038/nrg3722

Uhlén M, Patel S (2016) Single-cell mass spectrometry: unveiling cellular variability in infectious diseases. Trends Microbiol 24(11):936–948. https://doi.org/10.1016/j.tim.2016.07.002

Usaj M et al (2021) The cell atlas of budding yeast. Cell Syst 12(5):462–470.e3. https://doi.org/10.1016/j.cels.2021.05.010

van der Velden DL, Hoes LR, Van Der Wijngaart H, van Berge Henegouwen JM, Van Werkhoven E, Roepman P, Voest EE (2019) The drug rediscovery protocol facilitates the expanded use of existing anticancer drugs. Nature 574(7776):127–131

Viola R, Tucci A, Timellini G, Fantazzini P (2006) NMR techniques: a non-destructive analysis to follow microstructural changes induced in ceramics. J Eur Ceram Soc 26(15):3343–3349. https://doi.org/10.1016/j.jeurceramsoc.2005.09.055

Vogeser M, Parhofer KG (2007) Liquid chromatography-tandem mass spectrometry (LC-MS/MS)-technique and applications in endocrinology. Exp Clin Endocrinol Diabetes 115(09):559–570. https://doi.org/10.1055/s-2007-981458

Wagner J, Wu C (2019) Unraveling intricate biological systems through single-cell metabolomics: implications for disease diagnosis, treatment, and prevention. Curr Opin Biotechnol 58:144–151. https://doi.org/10.1016/j.copbio.2019.03.009

Wang J, Zhang L, Liu Z (2019) Advancements in microfluidic platforms for high-throughput single-cell metabolite analysis in single-cell metabolomics. Lab Chip 19(15):2504–2518. https://doi.org/10.1039/c9lc00486j

Weber PK, Weber P, Yapp C (2006) Nanoscale secondary ion mass spectrometry: subcellular metabolite imaging. Anal Chem 78(13):3651–3658. https://doi.org/10.1021/ac060024x

Wei D, Xu M (2022) The roles of miR-195 in human cancers. Front Oncol 11:814085. https://doi.org/10.3389/fonc.2021.814085

Wheeler AR, Zare RN (2018) Crossing the streams in chemical analysis. Science 361(6397):640–641

Wishart DS, Lewis MJ, Morrissey JA et al (2008) The human cerebrospinal fluid metabolome. J Chromatogr B-Anal Technol Biomed Life Sci 871(2):164–173. https://doi.org/10.1016/j.jchromb.2008.05.001

Worthoff WA, Yun SD, Shah NJ (2018) Chapter 1. Introduction to magnetic resonance imaging. In: New developments in NMR. The Royal Society of Chemistry, London, pp 1–44. https://doi.org/10.1039/9781788013062-00001

Wu C, Ricard G (2021) Harnessing single-cell metabolomics for personalised medicine: gaining unprecedented insights into disease mechanisms and variability. Front Med 8:731565. https://doi.org/10.3389/fmed.2021.731565

Xue Y, Zhang L, Wang H (2019) Microfluidic devices for precise cell isolation in single-cell metabolomics. Lab Chip 19(17):2923–2936. https://doi.org/10.1039/c9lc00607j

Yi L, Dong N, Yun Y, Deng B (2017) Single-cell metabolomics: methodologies and applications. Front Cell Dev Biol 5:89. https://doi.org/10.3389/fcell.2017.00089

Yim KH, Nahm FS, Han KA, Park SY (2010) Analysis of statistical methods and errors in the articles published in the Korean Journal of Pain. Korean J Pain 23(1):35

Yin J et al (2021) The role of dopamine receptors in schizophrenia: from basic research to clinical implications. BMC Psychiatry 21(1):75. https://doi.org/10.1186/s12888-020-03021-0

Yuan M, Chen X, Zhang Q (2019) Multiple reaction monitoring (MRM) coupled with mass spectrometry (MS) for accurate metabolite quantification in single-cell metabolomics. J Mass Spectrom 54(11):1101–1112. https://doi.org/10.1002/jms.4373

Zainal Abidin SA, Yau YH, Wong YC (2020) Single-cell metabolomics: opportunities and challenges in drug discovery and development. Front Pharmacol 11:536123. https://doi.org/10.3389/fphar.2020.536123

Zeleznik OA, Pokharel Y (2018) Harnessing the potential of single-cell metabolomics: challenges and opportunities. Curr Opin Biotechnol 54:65–71. https://doi.org/10.1016/j.copbio.2018.03.013

Zhang L, Vertes A (2015) Energy charge, redox state, and metabolite turnover in single human hepatocytes revealed by capillary microsampling mass spectrometry. Anal Chem 87(20):10397–10405. https://doi.org/10.1021/acs.analchem.5b02502

Zhang Z, Yang Y, Wang L (2018) Quality control measures in single-cell metabolomics sample preparation: minimizing metabolite contamination and preserving sample integrity. J Chromatogr A 1549:1–10. https://doi.org/10.1016/j.chroma.2018.03.051

Zhang Y, Wang J, Liu Z (2019) Enhancing metabolite extraction efficiency for single-cell metabolomics: advances and challenges. Anal Chem 91(14):8707–8714. https://doi.org/10.1021/acs.analchem.9b01645

Zhang L, Xu T, Zhang J, Wong SCC, Ritchie M, Hou HW, Wang Y (2021a) Single cell metabolite detection using inertial microfluidics-assisted ion mobility mass spectrometry. Anal Chem 93(30):10462–10468. https://doi.org/10.1021/ACS.ANALCHEM.1C00106

Zhang Y, Liu Z, Wang J (2021b) Addressing intercellular variability in single-cell metabolomics: strategies for accurate quantification. Trends Anal Chem 135:116154. https://doi.org/10.1016/j.trac.2020.116154

Zhang Y, Liu Z, Wang J (2021c) Integrating single-cell metabolomic data with transcriptomic and proteomic profiles: challenges and opportunities. Front Genet 12:742162. https://doi.org/10.3389/fgene.2021.742162

Zhao Y, Xia Z (2021) Genome-wide association study of gene expression in diverse maize genotypes under high-density conditions. Front Genet 13:984273. https://doi.org/10.3389/fgene.2022.984273

Zhao Y, Xu Y (2019) Bridging experimental data and biological function: modelling approaches in single-cell metabolomics. Trends Biotechnol 37(6):595–607. https://doi.org/10.1016/j.tibtech.2018.11.006

Zhao W, Deligey F, Shekar SC, Mentink-Vigier F, Wang T (2022) Current limitations of solid-state NMR in carbohydrate and cell wall research. J Magn Reson (San Diego, Calif. 1997: Print) 341:107263. https://doi.org/10.1016/j.jmr.2022.107263

Zhu J, Zhang H, Wang L (2023a) Spatially resolved metabolomics and single-cell resolution: computational challenges and emerging frameworks. Trends Biotechnol 41(2):112–125. https://doi.org/10.1016/j.tibtech.2022.12.002

Zhu G, Zhang W, Zhao Y, Chen T, Yuan H, Liu Y, Wang X (2023b) Single-cell metabolomics-based strategy for studying the mechanisms of drug action. Anal Chem 95(10):4712–4720. https://doi.org/10.1021/acs.analchem.2c05351

Biosensors for Detection of Metabolites

5

Shreyas Patel, Malvika Shukla, Saloni Malik,
Shubhita Tripathi, and Alok Pandya

5.1 Introduction

Understanding metabolism is crucial to grasping the phenotypic behavior of all living entities (including humans) where metabolism is integral to health and appropriate functioning. Metabolites are small molecules that participate in a wide range of chemical reactions within living organisms (Altaf-Ul-Amin et al. 2019). Production of metabolites occurs naturally within the cell as a result of metabolic reactions. They serve as intermediates, substrates, and products in various metabolic pathways, which are essential for maintaining life.

Based on necessity and functions, metabolites are mainly classified into two categories: Primary metabolites and Secondary metabolites (BYJUS 2023). The cell synthesizes Primary metabolites since they are essential for their growth. They play a crucial role in growth, development, and reproduction. The most popular example includes amino acids, vitamins, alcohol, polyols, carbohydrates, proteins, lipids, etc. Secondary metabolites are also produced by living beings that are not needed for essential metabolic cycles, in spite of the fact that they can have significant ecologic and different capabilities. Steroids, essential oils, phenolics, alkaloids, etc. are secondary metabolites.

Based on origin and source, metabolites are mainly classified into two categories: Endogenous metabolites and Exogenous metabolites. Endogenous metabolites are naturally occurring compounds that are synthesized within an organism as part of its normal metabolic processes. These metabolites are integral to various physiological functions and are crucial for the organism's growth, maintenance, and overall well-being. Exogenous metabolites are metabolites that enter an organism

S. Patel · M. Shukla · S. Malik · S. Tripathi · A. Pandya (✉)
Department of Biotechnology & Bioengineering, Institute of Advanced Research, Gandhinagar, Gujarat, India
e-mail: alok.pandya@iar.ac.in

V. Singh (ed.), *Advances in Metabolomics*,
https://doi.org/10.1007/978-981-97-7459-3_5

from external sources, rather than being synthesized within the organism. These metabolites can come from the environment, food, drugs, or other external substances.

According to the Human Metabolome Database (HMDB), a total of 253,245 metabolites are there. Out of the total number, the detected and quantified metabolites are only 3444, and detected but not quantified metabolites are 20,924 (HDMB 2023).

Drug metabolites are a subset of exogenous metabolites produced when the body metabolizes pharmaceutical compounds. There are two main categories of drug metabolites: Phase I Metabolites: These are typically produced by oxidation, reduction, or hydrolysis reactions. Phase I metabolites are often more polar and can be conjugated in Phase II metabolism. Phase II Metabolites: These metabolites are formed by conjugation reactions, which typically involve the addition of small polar molecules (e.g., glucuronic acid, sulfate). Conjugation makes the drug more water-soluble, aiding in its elimination from the body (Susa et al. 2023).

5.2 The Significance of Metabolite Detection

Metabolite detection, a key component of metabolomics, holds significant importance in various fields, from biomedical research to environmental science. Metabolites are small molecules produced in living organisms during metabolic processes, and their detection provides valuable insights into the physiological state of cells, tissues, and organisms. Here, we delve into the significance of metabolite detection in various contexts:

5.2.1 Drug Development and Pharmaceuticals

- *Drug Discovery and Drug Development:* Identification and detection of metabolites give information about the functional groups and active binding sites, which should be inhibited or modified to enhance the metabolism of that drug. Identifying and quantifying the metabolites of drugs is vital during the stage of preclinical studies and animal studies. Based on these results, human trials are permitted.
- *Pharmacokinetics and Toxicology:* Understanding how drugs are metabolized in the body is crucial for drug development. Metabolite identification and detection help assess the pharmacokinetics of pharmaceutical compounds, including their distribution, metabolism, and excretion. Identification and detection of harmful metabolites produced during drug metabolism are critical for drug safety profiling.

5.2.2 Medical and Clinical Applications

- Metabolic Pathways: Metabolite detection holds profound significance in unraveling the complexities of metabolic pathways within living organisms. By identifying and quantifying metabolites, researchers gain invaluable insights into cellular functions, map out intricate metabolic pathways, and understand the dynamic interplay of molecules within cells (Altaf-Ul-Amin et al. 2019). Beyond providing a comprehensive view of metabolic flux, the detection of metabolites serves as a pivotal tool in biomarker discovery, aiding in the diagnosis and prognosis of various diseases (Lino et al. 2022).
- Disease Mechanism, Diagnosis and Monitoring: Metabolite detection holds paramount significance in unraveling disease mechanisms, diagnostics, and monitoring for several reasons. Firstly, aberrations in metabolite levels often serve as early indicators of underlying pathological conditions, enabling the identification of specific disease markers. Through metabolomics, researchers can gain comprehensive insights into the altered metabolic profiles associated with various diseases, elucidating the intricate molecular mechanisms involved. In diagnostics, the detection of disease-specific metabolites plays a pivotal role in accurate and early disease identification, facilitating timely intervention and personalized treatment strategies. Furthermore, continuous monitoring of metabolite levels provides a dynamic assessment of disease progression, treatment efficacy, and potential relapses (Lino et al. 2022; Pickup 2008). Metabolite detection thus serves as a powerful tool for precision medicine, aiding clinicians in making informed decisions based on the evolving metabolic landscape of individual patients. As technologies advance, metabolite detection is increasingly becoming integral to a holistic understanding of disease pathophysiology, offering promising avenues for early detection, targeted therapeutics, and ongoing monitoring of diverse medical conditions (Singh et al. 2020).
- Oncology—Diagnosis and Treatment Efficacy: Metabolite detection provides real-time insights into the dynamic metabolic alterations occurring during cancer progression and in response to treatments such as chemotherapy, targeted therapy, or immunotherapy. This not only enhances the precision of cancer treatment but also facilitates the identification of potential resistance mechanisms, guiding adjustments to therapeutic regimens (Pickup 2008).

5.2.3 Forensic Sciences

- *Alcohol-Related Cases:* Metabolite detection is vigorous in cases involving alcohol-related offenses, such as DUI (Driving Under the Influence). Ethanol, the main alcohol in alcoholic beverages, is metabolized into acetaldehyde and then into acetic acid. Detecting these metabolites helps determine a person's Blood Alcohol Concentration (BAC) at the time of an incident. Postmortem Analysis: Metabolite detection in postmortem samples can help determine if alcohol consumption contributed to a person's death or played a role in accidental injuries (WHO 2024).

- *Drug Abuse-Related Cases and Toxicological Examinations:* The detection of metabolites in drug abuse-related cases holds paramount significance in the realm of forensic science. It serves as a critical tool for confirming drug use, providing concrete evidence that is often contested or disputed in legal cases. Furthermore, the presence of metabolites can offer valuable insights into the timeline of drug use, aiding in the determination of when a specific drug was consumed. This temporal information is particularly valuable in criminal investigations. Additionally, metabolite detection helps in distinguishing between different drugs or drug types, which is essential for identifying the specific substances involved in drug abuse cases. Each drug produces unique metabolites, and these distinctions have legal implications, as various drugs may carry differing legal consequences and health risks (Steuer et al. 2021). The quantification of drug use is another crucial aspect of metabolite detection, as it allows forensic scientists to assess the level of impairment or potential health risks associated with drug abuse accurately. In certain cases, drug metabolites can be detected in biological samples for more extended periods than the parent drugs, extending the window of detection, which is invaluable when drug use occurred some time ago but remains relevant in a legal or investigative context. Moreover, the noninvasive nature of testing methods for drug metabolites, such as urine or hair analysis, makes them more practical for routine drug testing, facilitating their broader application in various forensic contexts. The discipline of forensic toxicology relies heavily on metabolite analysis to establish the cause of death in cases of drug overdose or poisoning, as the specific metabolites detected can shed light on whether drug toxicity played a role in a fatality (Steuer et al. 2021; Losacker et al. 2021).

5.3 Basics of Biosensors

A biosensor is a tool that measures the respective reactions by using signals proportionate to the analyte's concentration in the reactions. Biosensors are used in various field such as drug discovery, disease monitoring, and the identification of pollutants, infectious microorganisms, and disease-indicating markers in bodily fluids (blood, urine, saliva, sweat) (Zhang 2015). The components of a characteristic biosensor are shown in Fig. 5.1 and include the following.

- Analyte: It is the specific substance or material of interest that must be detected and measured in the sample. For example, in a biosensor meant to identify alcohol, alcohol (Gamella et al. 2014) and its metabolites (Helander and Beck 2005) serves as an "analyte." The identification and quantification of the analyte are crucial for the biosensor's functionality, as it is the target substance whose presence and concentration provide the necessary information for analysis and decision-making.

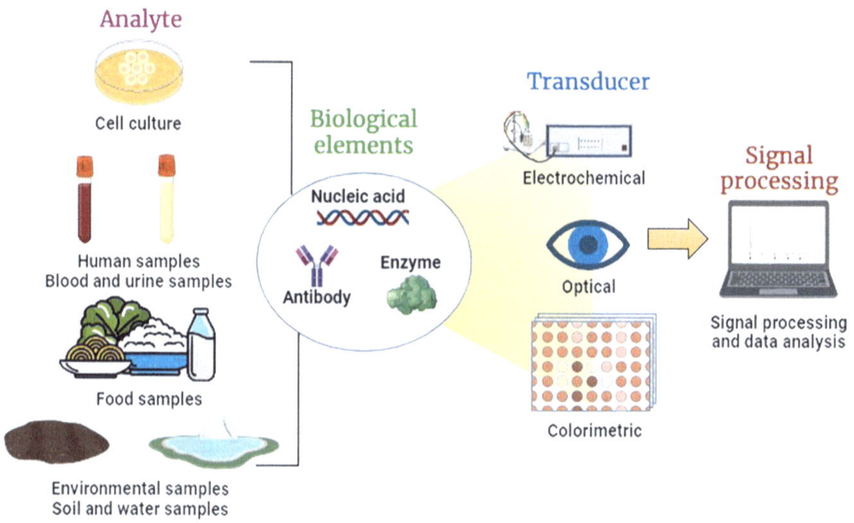

Fig. 5.1 Schematic illustration of biosensor

- Bioreceptor: A bioreceptor is a molecule or a complex of molecules capable of specifically and accurately identifying the analyte. Bioreceptors are essential components of biosensors, providing the selectivity needed for precise detection. Examples of bioreceptors include enzymes, which catalyze specific biochemical reactions (Jirakunakorn et al. 2020); cells, which can respond to specific substances; aptamers, which are short strands of DNA or RNA designed to bind specific targets (Yildirim et al. 2012); deoxyribonucleic acid (DNA), which can hybridize with complementary sequences; and antibodies, which can bind to specific antigens with high affinity (Wang et al. 2021). The choice of bioreceptor depends on the nature of the analyte and the required sensitivity and specificity of the biosensor.

- Transducer: A transducer in a biosensor is a critical component that plays a vital role in converting one form of energy into another (Zhang 2015). Specifically, in the context of biosensors, the transducer is responsible for converting a biorecognition event into a measurable signal, a process known as signalization (an energy conversion process). The interaction between the analyte and the bioreceptor typically results in optical or electrical signals, which are proportional to the magnitude of the interaction. The effectiveness of a biosensor largely depends on the sensitivity and specificity of the transducer, which ensures that the biological events are accurately reflected in the measurable signals (Administrator 2019).
- Electronics: The biosensor's electronics component is responsible for processing the signal generated by the transducer and preparing it for display. This component consists of sophisticated electronic circuitry that performs several crucial

tasks to ensure the accuracy and reliability of the biosensor's output. Signal conditioning is one of the primary functions, which involves amplifying the weak signals generated by the transducer to a level that can be easily processed. Additionally, the electronics module converts the analog signals into digital forms, a necessary step for digital signal processing and further analysis. This conversion is essential for integrating the biosensor with computer systems and other digital devices. The electronics also include filters to remove noise and enhance signal clarity, ensuring that the final signal is a true representation of the biological interaction (Yáñez-Sedeño et al. 2019).

- Display: The display unit of a biosensor serves as the user interface, where the processed data is presented in a clear and interpretable manner, either a liquid crystal display (LCD) of a computer or a direct printer that generates legible and comprehensible figures or curves. The display system typically comprises a combination of hardware and software components that work together to generate the final output. The software processes the digital signals received from the electronics module, converting them into graphical, tabular, numeric, or pictorial formats. This allows the user to quickly and accurately interpret the data. The type of display used is determined by the specific application and user requirements (Bhalla et al. 2016).

5.3.1 Working of Biosensors

A transducer and a biologically sensitive element work together to generate an electrical signal transform from biological material. The transducer's output will be either voltage or current, depending on the kind of enzyme. All is well if the output is voltage. However, if the output is current, then the next step should be to convert the current (using transimpedance amplifier, also known as Op-Amp-based current-to-voltage converter) into an equivalent voltage. Typically, a high frequency noise signal is superimposed with an output voltage signal that has a very low amplitude. Thus, a Low Pass RC Filter is used after the signal has been amplified. A signal conditioning unit, also referred to as a signal processing unit, is responsible for this signal filtering and amplification process. The signal processing unit produces an analog signal as an output that is comparable to the biological parameter being measured. Although an analogue signal can be shown directly on an LCD display, most analogue signals are passed to microcontrollers for conversion into digital signals, which are easier to process, analyze, and store (Administrator 2019).

5.3.2 Types of Biosensors

Biosensors can be categorized into two categories: those that are categorized according to the transduction technique used or the biological element that is utilized in the analysis. As previously mentioned, enzymes, DNA, antibodies, microorganisms, tissues, cellular receptors, and other elements are some of the frequently used biological elements or bio-recognition elements.

The transduction mechanism used in the biosensor or the physiochemical type that is the outcome of the sensing event, is the second and most used classification of biosensors. Additionally, there are three categories for biosensors based on the transduction mechanism. These are:

1. Mass biosensors
2. Optical biosensors
3. Electrochemical biosensors

5.3.2.1 Mass Biosensors

Mass biosensors, often referred to as mass-sensitive biosensors, are devices designed to detect and quantify mass changes occurring at the sensor surface due to biological interactions. These biosensors play a crucial role in many applications, including biotechnology, medical diagnostics, and environmental monitoring. The underlying principle relies on the fact that when a biological event, such as binding between a biomolecule (e.g., proteins, DNA, or cells) and a receptor, occurs on the sensor surface, it leads to a change in mass, which can be precisely measured.

5.3.2.2 Electrochemical Biosensors

In this type of biosensors, the biomolecules are immobilized onto a probing surface, typically using a noninterfering membrane to secure the sensing molecules in place. The sensing molecules then react with substance to be detected and produce an electrical signal corresponding to the quantity being measured (Bhalla et al. 2016). These sensing molecules interact with the target substance, generating an electrical signal proportional to the quantity of the analyte being measured. This interaction allows the conversion of chemical information into a measurable electrical signal. Various types of transducers are employed in electrochemical biosensors to achieve this, including potentiometric, amperometric, and impedimetric transducers (Yáñez-Sedeño et al. 2019). These transducers enable the precise conversion of the biochemical interactions into quantifiable electrical data, ensuring accurate detection and measurement of the analyte.

5.3.2.3 Optical Biosensors

An essential component of optical biosensors is optical fiber. Owing to the various characteristics of light, such as absorption, scattering, and fluorescence, the optical fibers enable the sensing elements to be detected (Yildirim et al. 2012; Nawrot et al. 2018; Wang et al. 2018; Nath et al. 1999). Variations in the refractive index of the interacting surface impact any of the aforementioned properties. For instance, when biological elements such as antibodies are bound to a metal layer, changes in the refractive index of the medium in contact with these biological elements will alter the layer's optical characteristics. This effect is crucial in biosensing applications, where detecting such changes can provide valuable information about the presence and concentration of analytes. Optical biosensors' nonelectrical nature is one of their key benefits. This enables them to change the light's wavelength to analyses multiple elements on a single layer (Damborský et al. 2016).

The metabolite biosensors for various applications will be discussed in further section (Fig. 5.2).

5.4 Metabolite Biosensors in Medical and Clinical Applications

Leveraging the electrical properties of metabolites, electrochemical biosensors provide a quantitative analysis of glucose and lactose. These sensors, often based on enzymatic reactions, offer rapid response times and the capability to detect metabolites at very low concentrations, providing invaluable insights into metabolic dynamics.

Lanlan Li et al. (2015) designed a flexible biosensor device capable of detecting various human metabolites, including uric acid, cholesterol, and triglycerides, utilizing hierarchically nanostructured conducting polymer hydrogels. The biosensors exhibited a wide linear range (0.07–1 mM for uric acid, 0.3–9 mM for cholesterol, and 0.2–5 mM for triglycerides) and impressive performance metrics, including high sensitivity, low detection limits, and fast response times (approximately 3 s). This is all made possible by the special properties of conducting polymer hydrogels, high permeability to bio-substrates and prompt electron transfer.

Ardalan et al. (2020) developed a wearable microfluidic patch and enhanced a smartphone with an optical filter and UV-LED lamps for the simultaneous measurement of sweat biomarkers like glucose and lactate. The patch is composed of five

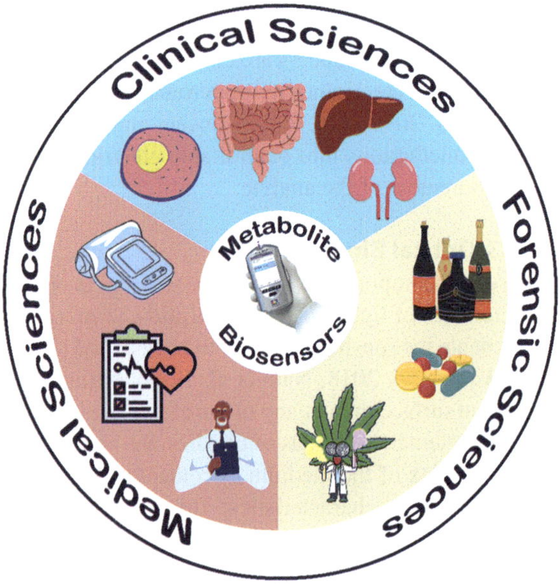

Fig. 5.2 Application of metabolite biosensors in various fields

filter papers which are coated with a fluorescent substance, along with cotton threads to collect sweat. Because each filter paper has a unique set of fluorescent probes, each biomarker can be selectively monitored and quantified. In their study, glucose is initially oxidized by glucose oxidase (GOX) to produce H_2O_2, which is then broken down by horseradish peroxidase (HRP), generating ·OH radicals that lead to the fluorescence quenching of the fluorescent substance. The detection of lactate relies on the restoration of the quenched fluorescence due to the release of a quencher (Fe^{3+}). They also designed a smartphone application to measure the fluorescence intensity with a detection algorithm, enabling the quantification of analyte concentrations (Ardalan et al. 2020; Zhang et al. 2022).

In the field of medical science, the utilization of biosensors is expanding at a rapid pace. One notable area where biosensors find extensive use is in clinical applications, particularly in the diagnosis of diabetes mellitus, where there is a critical need for accurate monitoring and control of blood glucose levels (Mehrotra 2016). Using a smartphone for signal readout, Wang et al. (2018) had developed a paper-based colorimetric sensor with enhanced color uniformity and intensity for the selective and sensitive detection of glucose and uric acid. Using chitosan, different chromogenic reagents, and horseradish peroxidase (HRP) in conjunction with specific oxidases like glucose oxidase or uricase, they modified the layers of the paper substrate to create a multilayer-modified test paper. These components were immobilized onto the paper. H_2O_2, generated by the action of the oxidases on their respective substrates, was employed to oxidize the coimmobilized chromogenic reagents. In this reaction, HRP acting as a catalyst and this resulted in the formation of colored products. To minimize the impact of external light, a straightforward approach involved placing the test paper over a LED lamp. Color pictures were captured using a smartphone camera, and subsequent quantitative analysis was performed by calculating the gray values of these images. The method that was developed exhibited a broad linear response, ranging 0.01–1.0 mM for uric acid detection and 0.02–4.0 mM for glucose detection. It achieved a low limit of detection of 0.003 and 0.014 mM for uric acid and glucose, respectively, significantly surpassing the sensitivity of formerly reported paper-based colorimetric assays (Zhang et al. 2022).

Measuring certain drugs and/or the byproducts of their breakdown (metabolites) on a frequent basis is known as therapeutic drug monitoring, and it is used to maintain a relatively constant drug concentration in the blood (Testing.com 2020). A drug's therapeutic index, which is based on its median effective and median lethal doses, indicates its margin of safety (DrugBank n.d.-a). Some drugs have a narrow "therapeutic index." In such cases, the therapeutic drug monitoring (TDM) is necessary. Irinotecan, also known as CPT-11, is a drug used to treat small cell lung cancer and colon cancer. It is sold in market under various trade names, including Camptosar (DrugBank n.d.-b). It has a low therapeutic index so the TDM is necessary. Alvau et al. developed the first biosensor for the TDM of CPT-11. The researchers constructed an enzyme-based electrochemical biosensor specifically designed for detecting and quantifying the antineoplastic prodrug CPT-11 commonly present in plasma samples of chemotherapy patients. The biosensor demonstrated the capability to measure CPT-11 within a range of 10–10,000 ng/mL, highlighting its potential utility in clinical settings for monitoring drug levels in patient samples undergoing chemotherapy treatments. The biosensor was fabricated by utilizing the enzyme couple AChE/ChOx and taking use of irinotecan's ability to block AChE. They indirectly detected and quantified the CPT-11 prodrug by detecting its metabolites (Alvau et al. 2018).

Uric acid is predominantly a byproduct of purine metabolism, which also produces the nucleic acids adenine and guanine (Jirakunakorn et al. 2020). Because urate is so insoluble, elevated levels, for example, can generate precipitation such as gout occurs in the joints and the urinary tract as stones. Hyperuricemia can be caused by a decrease in kidney excretion and an increase in nucleic acid turnover, such as in cancer (Pickup 2008). The amperometric biosensor designed by Jirakunakorn et al. aimed to detect uric acid (UA) through the immobilization of uricase on a porous cryogel (cry) platform made of graphene-incorporated chitosan (Chi) layered on a Prussian blue (PB) embedded electrode, situated atop a screen-printed carbon electrode (Uricase/Chi-Gr cry/PB/SPCE). In a flow injection system, the amperometric detection of UA catalyzed by uricase was predicated on the variation in PB's cathodic current at a potential of 0.00 V. A linear range of 0.0025–0.40 mmol/L was demonstrated by the UA biosensor, with a detection limit of 2.5 μmol/L (Jirakunakorn et al. 2020).

The most common method for diagnosing lactic acidosis is a lactic acid test. The test can also be used to assist in determining whether the body's tissues are receiving enough oxygen and to assist in the diagnosis of sepsis, a potentially fatal response to a bacterial infection (MedlinePlus n.d.). Ballesta Claver et al. developed a novel disposable biosensor based on electro-chemiluminescent detection method which detect the lactate in human saliva (Ballesta Claver et al. 2009). Their transduction mechanism uses luminol, and their identification method is based on lactate oxidase (LOx). All of the required reagents, such as luminol, LOx, BSA, electrolyte, and buffer, have been rendered immobile by a Methocel membrane that has been applied to the screen-printed electrochemical cell's working electrode.

Based on impedance spectroscopy and MIP-based synthetic receptors, a biomimetic serotonin sensor has the potential to be a quick and affordable substitute for chromatographic methods. The sensor responds specific and selectively to competitors, metabolite 5-HIAA and oxidized serotonin. The sensor features an active channel and reference channel NIP, allowing for accurate detection in nondiluted blood plasma. The concentrations obtained from the sensor exhibit minor uncertainties and correspond flawlessly with HPLC reference measurements.

One naturally occurring hormone that females naturally circulate is estradiol. It is used commercially in several hormone therapy medications that are used to treat diseases including hot flashes and vulvovaginal atrophy that are brought on by low estrogen levels. Oral tablets, injections, vaginal rings, transdermal patches, sprays, gels, and creams are a few of the forms of estradiol that are readily available. The 17β-isomer of estradiol is called 17-estradiol. It functions as a human metabolite and an estrogen (National Center for Biotechnology Information n.d.). Yildirim et al. (2012) developed a fluorescent-based biosensor to detect 17-estradiol by utilizing aptamers, which are particular short chain oligonucleotides. 17-estradiol specific DNA aptamers were employed for this aim. The analytes were stimulated by a laser beam at 635 nm and fluorescence labeled. After pumping the mixture to the photosensitive unit, the remaining free aptamers bonded to the immobilized estradiol on the sensor's surface and produced a fluorescence signal, which represents the desired estradiol concentration. The sensing procedure has a 2.1 nM detection limit and may be finished in less than 10 min (Nawrot et al. 2018).

Lactase is a disaccharide that the stomach tract's lactase enzyme breaks down into glucose, galactose, and monosaccharides. Blood lactose levels can be utilized to diagnose gastrointestinal cancer in patients (Yadav et al. 2022). Campos et al. (2014) developed a biosensor that utilizes the unique interaction between lactose and the enzyme β-galactosidase (β-Gal). β-Gal is commonly employed to catalyze the conversion of lactose into glucose and galactose, making it a crucial tool in managing patients having intolerance for lactose. The polyelectrolyte poly(ethylene-imine) (PEI) and poly(vinyl sulfonate) (PVS) were used to immobilize β-Gal in layer-by-layer films over an indium tin oxide (ITO) electrode that had been treated with a layer of Prussian Blue (PB). By adopting an amperometric detection strategy, the biosensor exhibited remarkable sensitivity (0.31 μA/mmol cm^2) and an impressively low detection limit (1.13 mMol/L). The refined ITO/PB/(PEI/PVS)1(PEI/β-Gal)30 architecture shows promise for accurate lactose determination applications, underscoring its potential impact in various analytical and diagnostic scenarios where lactose quantification is critical. This was adequate for detecting lactose in milk and for clinical tests.

S-adenosylmethionine, L-arginine, and glycine are the building blocks used by the liver, kidneys, and pancreas to generate the amino acid creatine. The creatine kinase enzyme catalyzes a reversible mechanism involving adenosine triphosphate (ATP) that mostly converts the stored creatine in muscle tissues to phosphocreatine. The body uses phosphocreatine as its main form of energy storage (Yadav et al. 2022). The final product of creatine metabolism in animals is creatinine. In clinical analysis, it is crucial to evaluate both creatine and creatinine in blood since creatinine is a useful indicator of muscle injury and creatinine is crucial for monitoring kidney function (Williamson and New 2014). Stefan et al. (2003) proposed amperometric biosensors to measure creatine and creatinine. They utilized creatinase (CI) and sarcosine oxidase (SO) for the assay of creatine and developed bienzymatic biosensor for creatine and a trienzymatic biosensor utilizing CI, SO, and creatininase (CA) for the assay of creatinine (Stefan et al. 2003).

The most significant sterol that is produced by most of the human cells, primarily in the liver, is cholesterol (Narwal et al. 2019). Monitoring cholesterol levels is essential, particularly for those who are more susceptible to heart disease. Measuring blood cholesterol levels is necessary to diagnose atherosclerosis, assess the risk of thrombosis, and determine the severity of cardiovascular disease (Yadav et al. 2022). In healthy individuals, the cholesterol level is <200 mg/dL. The cholesterol level of 200–239 mg/dL is regarded as borderline high, while 240 mg/dL and higher is regarded as a biomarker for high blood pressure, peripheral artery disease, heart attacks, strokes, and cardiovascular illnesses (Narwal et al. 2019; Taylor and Lopez 2004). Malhotra B.D. has given significant contribution toward the development of biosensor to detect the cholesterol level. By immobilizing the cholesterol oxidase (ChOx) enzyme on ITO electrodes modified with various kinds of nanomaterials and nanocomposites, he fabricated a number of cholesterol biosensors (Barik et al. 2010; Kaushik et al. 2010a, b, 2013; Matharu et al. 2012; Pradhan et al. 2022; Singh et al. 2012). Different transduction mechanism used to detect various metabolites is shown in Fig. 5.3.

5.5 Metabolite Biosensors Used in Forensic Sciences

Benzoylecgonine (BE) stands as the chief toxic metabolite following cocaine usage, and it exhibits a prolonged presence in both the human body and environment, surpassing the duration of detectability of cocaine (Wang et al. 2021). In 1999, Eldefrawi et al. developed a four-channel fiber optic based biosensor to detect the cocaine and its metabolite benzoylecgonine (BE) using a competitive fluorescence immunoassay in human urine sample (Nath et al. 1999). In this process, the binding of an anti-benzoylecgonine monoclonal antibody (mAb) with benzoylecgonine-coated optical fibers was inhibited by cocaine. The amount of bound mAb, which

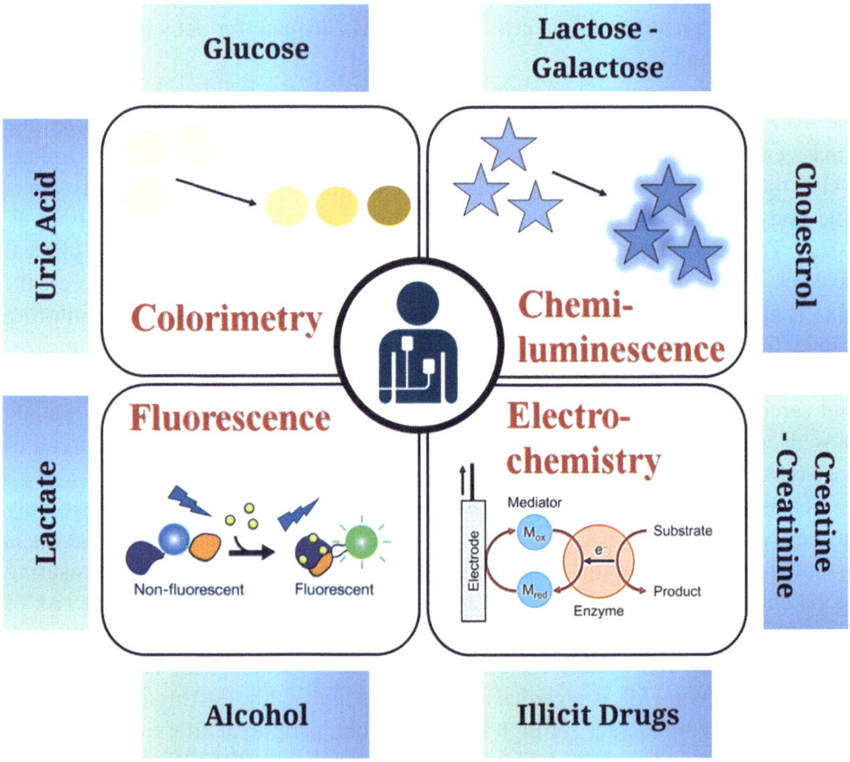

Fig. 5.3 Different transduction mechanism used to detect various metabolites

decreases with increasing cocaine concentration, was measured through the fluorescence generated when a cyanine dye-tagged antimouse antibody (CY5-Ab) is excited in the vicinity of the bound mAb. They found the effective range for benzoylecgonine (BE) inhibition of fluorescent signals is 0.75–50 ng/mL. Additionally, they mentioned that fiber optic-based biosensor had very low affinities for ecgonine and ecgonine methyl ester but comparable affinities for BE, cocaine, and cocaethylene.

Cannabis is a versatile plant known for its psychoactive and medicinal properties. It contains various compounds, with Δ-9-tetrahydrocannabinol (Δ-9-THC) and cannabidiol (CBD) being the most well-known (Ashton 2001). Of the two primary compounds found in cannabis, Δ-9-tetrahydrocannabinol (Δ-9-THC), is known to be psychoactive, meaning it produces a high or intoxicating effect. In contrast, cannabidiol (CBD) is nonpsychoactive. Because of its stimulating and euphoric properties, cannabis, often known as marijuana, is frequently used as a recreational drug. THC, the main psychoactive component in marijuana, interacts with the central nervous system through the endocannabinoid system. Stevenson H. et al. developed a biosensor using electrochemical detection method to detect the THC and its metabolite in saliva sample using anti-THC capture antibodies via impedimetric measurements (Stevenson et al. 2019).

Alcohol is a frequently consumed toxic substance by humans, and its consumption is highly linked with the human health problems and road traffic accidents (WHO 2024). As time has passed, the demand for a quick and dependable method to measure ethanol levels in biological samples has grown significantly, particularly in the fields of clinical medicine and forensic investigations (Yáñez-Sedeño et al. 2019). In alcohol biosensors, two enzymes, alcohol dehydrogenase (ADH) and alcohol oxidase (AOX), have been widely employed for the measurement of alcohol levels. This is because ADH or AOX, as individual enzyme molecules, can effectively catalyze the conversion of alcohol as a substrate, leading to an amplification effect that boosts the sensitivity of the analysis (Kuswandi and Musa 2014).

Ethyl glucuronide (EtG) and ethyl sulfate (EtS) are immediate metabolites of ethanol metabolism that can be detected in urine. The use of EtG testing is increasingly prevalent for overseeing alcohol intake, particularly in cases where individuals are mandated to refrain from drinking (Clinical Drug Testing in Primary Care n.d.). If ethanol is no longer detectable, the presence of EtG and EtS provide a clear indicator of recent drinking (Helander and Beck 2005). It is advised that EtG be used in forensic and clinical studies involving examination alcohol intake (Helander et al. 2008). Compared to standard and other traditional approaches, Kinnamon D. et al. established for the first time a more robust estimation of alcohol intake by dynamically monitoring the ethyl glucuronide (EtG). They developed a wearable biosensor gadget that could measure EtG levels in sweat sample (Lin et al. 2019). Their device consists of four subsystems (Sankhala et al. 2018). The primary analogue device, the ADuCM350 system-on-a-chip (SoC), was utilized to establish a communication channel with the other three subsystems (Bluetooth connection system, power management system and humidity measurement system) and execute an impedance measurement loop. They were patterned polymer substrates with conductive electrodes. This was achieved through a two-step shadow mask deposition process, which allowed for precise control over the deposition of materials and the formation of electrode patterns. The ZnO sensing area of every tested sensor was dispensed with a 3 μL volume of 10 mM Disodium phosphate diluted in DMSO to perform thiol-linker functionalization. After 3 h of dark incubation at room temperature, each sensor was treated for 30 min with 100 μg/mL of room temperature, diluted anti-EtG monoclonal antibodies in phosphate-buffered saline. The sensors on the benchtop platform showed that they could detect EtG in the range of 10–10,000 ng/L with a signal above noise (Lin et al. 2019).

Square-wave voltammetry (SWV) was utilized by Prasad S. et al. to electrochemically detect ethyl glucuronide (EtG) in human perspiration or sweat sample in their biosensor for chronic alcohol intake monitoring. The DSP cross-linker coupled with an EtG-specific monoclonal antibody were chemically coabsorbed onto the surface of a gold in parallel with thiolated molecule, 6-ferrocenyl-1-hexanethiol (6-FcSH), that transmits charge in a unique affinity assay. They described that one way to identify concentration-dependent EtG binding was to measure a decrease in the charge transfer of the sensor, which appears as a current response when measuring the SWV. An additional little electronic reader was built, showing sensitivity that was on par with a traditional lab instrument. Both instruments showed a linear range of 0.1–100 μg/L and an LOD of 0.1 μg/L, which correspond to the physiologically significant spectrum of human sweat EtG expression (Kinnamon et al. 2018).

A noninvasive biosensing device was recently developed by M. Gamella et al. for the in-situ analysis of sweat in real-time to detect alcohol in blood. This is because alcohol taken orally is also eliminated from the body through skin perspiration, making it simple to collect a transdermal sample (Majeed et al. 2020). They proposed model based on "bi-enzyme amperometric composite biosensors" that are delicate to the concentration of ethanol. Their prototype uses an earlier calibrated set-up which establishes a short-term correlation between the amperometric signal produced by ethanol in sweat and its concentration in blood. The features of this sensor device allow for the measurement of concentration of ethanol in both isolated as well as continuous form, providing data on a subject's blood alcohol content (8 h) at any given time or over extended periods of time (Gamella et al. 2014).

Selvam et al. demonstrated first time the wearable biosensors for the monitoring of ethanol through detecting and quantifying its metabolite ethyl glucuronide (EtG) (Panneer Selvam et al. 2016). They designed two coplanar sensors using gold and zinc oxide as sensing electrodes. They also developed a detection system utilizing LEDs to spot EtG (ethyl glucuronide) in human sweat samples. The sensor follows an affinity-based immunoassay principle, employing thiol-based chemistry used to immobilize monoclonal antibodies specific to EtG on electrodes. This configuration facilitated EtG detection in human sweat through a chemoresistive sensing method. Using an AC voltage between the two coplanar electrodes, they measured and calibrated the impedance between the sensor electrodes for physiologically relevant EtG dosages in human sweat (Yáñez-Sedeño et al. 2019).

5.6 Limitations

Biosensors are powerful tools for detecting metabolites, but they also have certain limitations. Here are some common limitations associated with biosensors in metabolite detection:

Sensitivity and Detection Limits: Biosensors may have limitations in terms of sensitivity, which can affect their ability to detect low concentrations of metabolites. The detection limits of biosensors may not be sufficient for some applications, especially when dealing with trace amounts of metabolites.

Selectivity: Selectivity denotes to the ability of the biosensor to specifically recognize and detect a target metabolite in the presence of other compounds. Cross-reactivity with similar molecules or interference from complex biological samples can compromise the selectivity of biosensors.

Stability and Shelf Life: The stability of the biological components (such as enzymes or antibodies) used in biosensors can be a limitation. These components may degrade over time, affecting the biosensor's performance. Ensuring a sufficient shelf life for biosensors is a challenge.

Calibration and Standardization: Calibration of biosensors is critical for accurate and reliable measurements. However, maintaining consistent calibration and standardization across different biosensor devices and applications can be challenging, leading to variations in results.

Matrix Effects: Biological samples, such as blood or urine, can contain various interfering substances that may affect the biosensor's performance. Matrix effects can lead to inaccuracies and reduced reliability in metabolite detection.

Response Time: The time required for biosensors to provide a measurable response can vary. Some biosensors may have relatively long response times, limiting their suitability for applications that require rapid detection of metabolites.

Miniaturization and Integration: While there have been advancements in miniaturizing biosensors, integrating them into portable and user-friendly devices can still be a challenge. Size, weight, and ease of use are important factors, especially for their applications in point-of-care diagnostics.

Cost: The cost of developing and producing biosensors, especially those incorporating specialized biological components, can be relatively high. This may limit their widespread adoption, particularly in resource-limited settings.

Long-Term Stability: Continuous and long-term use of biosensors can lead to issues related to the stability of the sensing elements. Factors such as fouling, biofouling, and denaturation of biological components over time can impact the biosensor's performance.

Limited Range of Metabolites: Some biosensors are designed for specific metabolites or classes of metabolites. Developing biosensors that can simultaneously detect a broad range of metabolites with different chemical properties remains a challenge.

5.7 Future Prospects

Despite these limitations, many of these problems are being addressed by current research and technical developments, paving the way for improved biosensor capabilities in metabolite detection. The future prospects for biosensors in metabolite detection are promising, with ongoing research and technological advancements addressing many of the current limitations. Here are some potential future developments and trends:

Advanced Biorecognition Elements: Exploration and engineering of novel biorecognition elements, such as engineered enzymes, aptamers, and molecularly imprinted polymers, may enhance the selectivity and stability of biosensors. This can lead to biosensors capable of detecting specific metabolites with high precision.

Multianalyte Detection: Future biosensors may be designed to detect multiple metabolites simultaneously. This capability would be valuable in understanding complex metabolic pathways and providing a more comprehensive analysis of biological samples.

Integration with Emerging Technologies: Integration of biosensors with other emerging technologies, like artificial intelligence (AI) and machine learning (ML), could improve data analysis and interpretation. This integration may lead to more accurate and rapid detection of metabolites, as well as improved diagnostic capabilities.

Point-of-Care and Wearable Devices: Miniaturization and the development of portable, user-friendly biosensor devices are likely to continue. Point-of-care and wearable biosensors could revolutionize healthcare by providing real-time monitoring of metabolites, enabling personalized medicine and early disease detection.

Cost Reduction and Accessibility: Efforts to optimize manufacturing processes, utilize cost-effective materials, and streamline production could contribute to a reduction in the overall cost of biosensors. This could increase accessibility and facilitate their deployment in various settings, including resource-limited environments.

Smart Materials and Surface Modifications: Advancements in smart materials and surface modifications can improve the performance of biosensors. Functionalized surfaces and responsive materials can enhance the interaction between the biosensor and the target metabolites, leading to improved sensitivity and specificity.

As technology continues to advance, biosensors are expected to play a crucial role in various fields, offering novel solutions for metabolite detection and contributing to advancements in healthcare, environmental monitoring, and personalized medicine.

Acknowledgments AP gratefully acknowledge Gujarat State Biotechnology Mission (GSBTM Project ID: FLOWG8) for financial support.

References

Administrator (2019) What are biosensors? Principle, working, types and applications [Internet]. ElectronicsHub. Available from: https://www.electronicshub.org/types-of-biosensors/

Altaf-Ul-Amin M, Kanaya S, Mohamed-Hussein ZA (2019) Investigating metabolic pathways and networks. In: Ranganathan S, Gribskov M, Nakai K, Schönbach C (eds) Encyclopedia of bioinformatics and computational biology [Internet]. Academic Press, Oxford, pp 489–503. [cited 2023 Oct 30]. Available from: https://www.sciencedirect.com/science/article/pii/B9780128096338201404

Alvau MD, Tartaggia S, Meneghello A, Casetta B, Calia G, Serra PA et al (2018) Enzyme-based electrochemical biosensor for therapeutic drug monitoring of anticancer drug irinotecan. Anal Chem 90(10):6012–6019

Ardalan S, Hosseinifard M, Vosough M, Golmohammadi H (2020) Towards smart personalized perspiration analysis: an IoT-integrated cellulose-based microfluidic wearable patch for smartphone fluorimetric multi-sensing of sweat biomarkers. Biosens Bioelectron 15(168):112450. [cited 2023 Nov 1]. Available from: https://pubmed.ncbi.nlm.nih.gov/32877780/

Ashton CH (2001) Pharmacology and effects of cannabis: a brief review. Br J Psychiatry 178:101–106

Ballesta Claver J, Valencia Mirón MC, Capitán-Vallvey LF (2009) Disposable electrochemiluminescent biosensor for lactate determination in saliva. Analyst 134(7):1423–1432

Barik A, Solanki PR, Kaushik A, Ali A, Pandey MK, Kim CG et al (2010) Polyaniline–carboxymethyl cellulose nanocomposite for cholesterol detection. J Nanosci Nanotechnol 10(10):6479–6488

Bhalla N, Jolly P, Formisano N, Estrela P (2016) Introduction to biosensors. Essays Biochem 60(1):1–8

BYJUS (2023) What are metabolites? Primary and secondary metabolites. Available from: https://byjus.com/biology/metabolites/

Campos PP, Moraes ML, Volpati D, Miranda PB, Oliveira ON Jr, Ferreira M (2014) Amperometric detection of lactose using β-galactosidase immobilized in layer-by-layer films. ACS Appl Mater Interfaces 6(14):11657–11664. Available from: https://pubmed.ncbi.nlm.nih.gov/24991705/

Clinical Drug Testing in Primary Care (n.d.) https://store.samhsa.gov/sites/default/files/sma12-4668.pdf

Damborský P, Švitel J, Katrlík J (2016) Optical biosensors. Essays Biochem 60(1):91–100. https://doi.org/10.1042/EBC20150010

DrugBank (n.d.-a) Narrow Therapeutic Index Drugs | DrugBank Online [cited 2023 Nov 30]. Available from: https://go.drugbank.com/categories/DBCAT003972

DrugBank (n.d.-b) Irinotecan: uses, interactions, mechanism of action | DrugBank Online. [cited 2023 Nov 30]. Available from: https://go.drugbank.com/drugs/DB00762

Gamella M, Campuzano S, Manso J, de Rivera GG, López-Colino F, Reviejo AJ et al (2014) A novel non-invasive electrochemical biosensing device for in situ determination of the alcohol content in blood by monitoring ethanol in sweat. Anal Chim Acta 806:1–7

HDMB (2023) Human Metabolome Database: database statistics [Internet]. Available from: https://hmdb.ca/statistics

Helander A, Beck O (2005) Ethyl sulfate: a metabolite of ethanol in humans and a potential biomarker of acute alcohol intake. J Anal Toxicol 29(5):270–274

Helander A, Bottcher M, Fehr C, Dahmen N, Beck O (2008) Detection times for urinary ethyl glucuronide and ethyl sulfate in heavy drinkers during alcohol detoxification. Alcohol Alcohol 44(1):55–61

Jirakunakorn R, Khumngern S, Choosang J, Thavarungkul P, Kanatharana P, Numnuam A (2020) Uric acid enzyme biosensor based on a screen-printed electrode coated with Prussian blue and modified with chitosan-graphene composite cryogel. Microchem J 154:104624

Kaushik A, Solanki PR, Kaneto K, Kim CG, Ahmad S, Malhotra BD (2010a) Nanostructured iron oxide platform for impedimetric cholesterol detection. Electroanalysis 22(10):1045–1055

Kaushik A, Solanki PR, Kaneto K, Kim CG, Ahmad S, Malhotra BD (2010b) Nanostructured iron oxide platform for impedimetric cholesterol detection. Electroanalysis 22(10):1045–1055. [cited 2023 Dec 12]. Available from: https://analyticalsciencejournals.onlinelibrary.wiley.com/doi/10.1002/elan.200900468

Kaushik A, Solanki PR, Chavhan PM, Bhansali S, Malhotra BD (2013) Sol-gel derived nanostructured zirconia for cholesterol detection. J Nanosci Lett 3:23. Available from: http://www.cognizure.com/abstract.aspx?p=109637280

Kinnamon D, Muthukumar S, Panneer Selvam A, Prasad S (2018) Portable chronic alcohol consumption monitor in human sweat through square-wave voltammetry. SLAS Technol 23(2):144–153

Kuswandi B, Musa A (2014) Recent progress in alcohol biosensors. OA Alcohol [Internet] 2(1):1–8. [cited 2023 Oct 30]. Available from: https://www.researchgate.net/publication/267271093_Recent_progress_in_alcohol_biosensors

Li L, Wang Y, Pan L, Shi Y, Cheng W, Shi Y, Guihua Y (2015) A nanostructured conductive hydrogels-based biosensor platform for human metabolite detection. Nano Lett [Internet] 15(2):1146–1151. [cited 2023 Dec 11]. Available from: https://pubs.acs.org/doi/abs/10.1021/nl504217p

Lin KC, Kinnamon D, Sankhala D, Muthukumar S, Prasad S (2019) AWARE: aA Wearable Awareness with Real-time Exposure, for monitoring alcohol consumption impact through ethyl glucuronide detection. Alcohol 81:93–99

Lino C, Barrias S, Chaves R, Adega F, Martins-Lopes P, Fernandes JR (2022) Biosensors as diagnostic tools in clinical applications. Biochim Biophys Acta Rev Cancer 1877(3):188726

Losacker M, Kraemer M, Philipsen A, Duecker K, Dreimueller N, Engelmann J et al (2021) Enantioselective quantification of amphetamine and metabolites in serum samples: forensic evaluation and estimation of consumption time. Metabolites 11(8):521

Majeed MI, Nawaz H, Arshad FN (2020) Detecting the presence of illicit drugs using biosensors. In: Nanobiosensors: from design to applications [Internet]. Wiley, pp 223–253. [cited 2023 Dec 11]. Available from: https://onlinelibrary.wiley.com/doi/abs/10.1002/9783527345137.ch10

Matharu Z, Solanki PR, Gupta V, Malhotra BD (2012) Mediator free cholesterol biosensor based on self-assembled monolayer platform. Analyst 137(3):747–753

MedlinePlus (n.d.) Lactic Acid Test: MedlinePlus Medical Test [Internet]. [cited 2023 Nov 30]. Available from: https://medlineplus.gov/lab-tests/lactic-acid-test/

Mehrotra P (2016) Biosensors and their applications—a review. J Oral Biol Craniofacial Res 6(2):153–159

Narwal V, Deswal R, Batra B, Kalra V, Hooda R, Sharma M et al (2019) Cholesterol biosensors: a review. Steroids 143:6–17

Nath N, Eldefrawi M, Wright J, Darwin D, Huestis M (1999) A rapid reusable fiber optic biosensor for detecting cocaine metabolites in urine. J Anal Toxicol [Internet] 23(6):460–467. [cited 2023 Oct 31]. Available from: https://academic.oup.com/jat/article/23/6/460/863318

National Center for Biotechnology Information (n.d.) PubChem. Estradiol [Internet]. [cited 2023 Dec 11]. Available from: https://pubchem.ncbi.nlm.nih.gov/compound/5757

Nawrot W, Drzozga K, Baluta S, Cabaj J, Malecha K (2018) A fluorescent biosensors for detection vital body fluids' agents. Sensors 18(8):2357. [cited 2023 Dec 11]. Available from: https://pubmed.ncbi.nlm.nih.gov/30042294/

Panneer Selvam A, Muthukumar S, Kamakoti V, Prasad S (2016) A wearable biochemical sensor for monitoring alcohol consumption lifestyle through ethyl glucuronide (EtG) detection in human sweat. Sci Rep 6(1):23111

Pickup JC (2008) Biosensors for monitoring metabolites in clinical medicine. In: Handbook of biosensors and biochips [Internet]. Wiley. [cited 2023 Oct 31]. Available from: https://onlinelibrary.wiley.com/doi/abs/10.1002/9780470061565.hbb113

Pradhan S, Nicholson BD, Albin S, Heise RL, Yadavalli VK (2022) Single-use biomimetic sensors for rapid and sensitive cortisol detection in blood. Biosens Bioelectron X 12:100280

Sankhala D, Muthukumar S, Prasad S (2018) A four-channel electrical impedance spectroscopy module for cortisol biosensing in sweat-based wearable applications. SLAS Technol 23(6):529–539

Singh K, Solanki PR, Basu T, Malhotra BD (2012) Polypyrrole/multiwalled carbon nanotubes-based biosensor for cholesterol estimation. Polym Adv Technol 23(7):1084–1091

Singh S, Kumar V, Dhanjal DS, Datta S, Prasad R, Singh J (2020) Biological biosensors for monitoring and diagnosis. In: Microbial biotechnology: basic research and applications. Springer, Singapore, pp 317–335. https://doi.org/10.1007/978-981-15-2817-0_14

Stefan RI, Bokretsion RG, van Staden JF, Aboul-Enein HY (2003) Simultaneous determination of creatine and creatinine using amperometric biosensors. Talanta 60(6):1223–1228

Steuer AE, Raeber J, Simbuerger F, Dornbierer DA, Bosch OG, Quednow BB et al (2021) Towards extending the detection window of gamma-hydroxybutyric acid—an untargeted metabolomics study in serum and urine following controlled administration in healthy men. Meta 11(3):166

Stevenson H, Bacon A, Joseph KM, Gwandaru WRW, Bhide A, Sankhala D et al (2019) A rapid response electrochemical biosensor for detecting Thc in saliva. Sci Rep 9(1):12701

Susa ST, Hussain A, Preuss CV (2023) Drug metabolism. In: StatPearls [Internet]. StatPearls Publishing, Treasure Island. [cited 2023 Oct 30]. Available from: http://www.ncbi.nlm.nih.gov/books/NBK442023/

Taylor JR, Lopez LM (2004) Cholesterol: point-of-care testing. Ann Pharmacother 38(7–8):1252–1257

Testing.com. Therapeutic drug monitoring [Internet]. 2020. [cited 2023 Nov 30]. Available from: https://www.testing.com/tests/therapeutic-drug-monitoring/

Wang J, Deng X, Wu Y, Huang Y, Hou S, Zhang Y et al (2021) Sub-lethal toxicity and elimination of the cocaine metabolite, benzoylecgonine: a narrative review. Ann Palliat Med 10(6):6936947–6936947

Wang X, Li F, Cai Z, Liu K, Li J, Zhang B et al (2018) Sensitive colorimetric assay for uric acid and glucose detection based on multilayer-modified paper with smartphone as signal readout. Anal Bioanal Chem 410(10):2647–2655

WHO. Alcohol 2024. [cited 9 Nov 2023]. Available from: https://www.who.int/news-room/fact-sheets/detail/alcohol

Williamson L, New D (2014) How the use of creatine supplements can elevate serum creatinine in the absence of underlying kidney pathology. BMJ Case Rep [Internet] 2014:bcr2014204754. [cited 2023 Dec 12]. Available from: https://casereports.bmj.com/content/2014/bcr-2014-204754

Yadav AK, Verma D, Sajwan RK, Poddar M, Yadav SK, Verma AK et al (2022) Nanomaterial-based electrochemical nanodiagnostics for human and gut metabolites diagnostics: recent advances and challenges. Biosensors 12(9):733

Yáñez-Sedeño P, Agüí L, Campuzano S, Pingarrón JM (2019) What electrochemical biosensors can do for forensic science? Unique features and applications [Internet]. Biosensors 9(4):127. [cited 2023 Oct 30]. Available from: https://www.mdpi.com/2079-6374/9/4/127

Yildirim N, Long F, Gao C, He M, Shi HC, Gu AZ (2012) Aptamer-based optical biosensor for rapid and sensitive detection of 17β-estradiol in water samples. Environ Sci Technol 46(6):3288–3294

Zhang G (2015) A brief overview of biosensors. In: Zhang G (ed) Nanoscale surface modification for enhanced biosensing: a journey toward better glucose monitoring [Internet]. Springer International Publishing, Cham, pp 1–12. [cited 2023 Nov 27]. Available from:. https://doi.org/10.1007/978-3-319-17479-2_1

Zhang M, Cui X, Li N (2022) Smartphone-based mobile biosensors for the point-of-care testing of human metabolites. Mater Today Bio 14:100254

Unveiling the Molecular Fingerprint: Mass Spectrometry in Metabolomics

6

Gholamreza Abdi, Rupal Dhariwal, Nil Patil, Bindiya Upadhyay, Mukul Jain, and Maryam Abbasi Tarighat

6.1 Introduction

6.1.1 The Intricate Dance of Metabolites: Understanding Cellular Chemistry

The cellular metabolic system consists of chemical reactions taking place within living cells, where an intricate dance of metabolites orchestrates the fundamental processes essential for life. These reactions are group into two categories: catabolic reactions, which break down nutrients to release energy, and anabolic reactions, which create larger biological molecules (Ford 2017). Metabolites are the substances involved as the reactants or products in these reactions. The genetic information flows from DNA to RNA immediately followed by protein, which is the fundamental concept of the Central Dogma of molecular biology, culminating in the formation of the metabolome (Mishra 2009). The metabolome's makeup is not constant; it varies according to the genome's expression under particular conditions. Changes in the metabolome triggered by a viral infection represent the combined effects of alterations in the host's gene expression and protein function due to the

G. Abdi (✉)
Department of Biotechnology, Persian Gulf Research Institute, Persian Gulf University, Bushehr, Iran
e-mail: abdi@pgu.ac.ir

R. Dhariwal · N. Patil · B. Upadhyay · M. Jain
Cell & Development Biology Lab, Research and Development Cell, Parul University, Vadodara, Gujarat, India

Department of Life Sciences, Parul institute of Applied Sciences, Parul University, Vadodara, Gujarat, India

M. A. Tarighat
Faculty of Nano and Bio Science and Technology, Persian Gulf University, Bushehr, Iran

V. Singh (ed.), *Advances in Metabolomics*,
https://doi.org/10.1007/978-981-97-7459-3_6

virus (Kumar et al. 2020). The temporal progression of infection, replication, and virus release is analyzed on a minutely basis through the molecular changes in these processes (Karmitros et al. 2021). Cohen proposed that chemical tools could elucidate these viral mechanisms at a molecular level. Since Cohen first coined the term "chemical virology" over 60 years ago, advancements in analytical techniques have allowed for a deeper and more precise understanding of how viruses interact with host metabolism (Cohen 2006). Metabolites can be categorized majorly into amino acids, carbohydrates, nucleotides, lipids, coenzymes, and cofactors. These categories include a vast array of molecular structures, physical and chemical properties, functions, and levels of occurrence. Given the complexity of this diversity, research often examines metabolome subgroups instead of the entire set. A fundamental subdivision made in these studies is between polar (hydrophilic) metabolites and nonpolar (hydrophobic) ones (Ryan and Robards 2006). Polar metabolites can dissolve in water and include substances like most sugars, bases that make up DNA and RNA, nucleotides and nucleosides, compounds such as acyl carnitines, various organic acids, amino acids, and phosphorylated molecules. These metabolites typically take part in processes like cellular respiration, which includes pathways like glycolysis, the TCA (tricarboxylic acid cycle), and the PPP (pentose phosphate pathway). They are also the building blocks for synthesizing DNA, RNA, proteins, and complex sugars. On the other hand, nonpolar or hydrophobic metabolites, generally termed as lipids, are involved in storing energy, making up cell membranes, and in cell signaling processes.

Given the intermediary role metabolites play in a many of cellular processes, studying them can offer invaluable insights into cell function and health. This field of study, known as metabolomics, examines the distinctive chemical signatures that are produced by cellular activities (Dunn et al. 2011). Specifically, it enables researchers to detect global changes in metabolite profiles following physiological stimulants or gene editing (Roessner et al. 2001). The comprehensive analysis of metabolites thus provides a snapshot of the physiological state of an organism at a given time point (Fiehn 2001). This can lead to the discovery of molecular indicators that signal the presence of a disease, provide a deeper understanding of disease pathology, or reveal new targets for drug therapy (Frank and Hargreaves 2003). The study of hydrophilic and hydrophobic metabolites, in particular, can illuminate how cells allocate resources between growth, maintenance, defense, and other vital functions (Moran et al. 2022). Therefore, understanding the ensemble of these small molecules is integral for advancing our knowledge of biological systems and enhancing our ability to manipulate them for therapeutic benefit (Zhou et al. 2016). Studying those metabolites are part of metabolic reactions within an organism that requires a variety of sophisticated analytical techniques due to their diversity in structure and function (Zhang et al. 2012a). Below are some key techniques used in metabolomics:

Mass Spectrometry: MS, a remarkably sensitive and adaptable analytical method, categorizes ions according to their mass-to-charge (*m*/*z*) ratio (Ren et al. 2018). It serves for both identifying and measuring metabolites. MS can be linked with chromatography techniques like liquid chromatography or gas chromatography, to separate complex mixtures of *metabolites* before detection (Theodoridis et al. 2012). LC-MS is particularly effective for analyzing a wide range of polar and nonpolar metabolites, while Mass spectrometry can be combined with chromatography methods such as liquid chromatography or gas chromatography (Drabińska et al. 2020). There are various kinds of MS available for high resolution and LOD (limit of detection), each with its advantages. Single quadrupole mass spectrometers offer simplicity and cost-effectiveness for routine analyses, while triple quadrupole mass spectrometers offer improved sensitivity and flexibility for focused quantitative analysis and tandem MS experiments in metabolomics and other fields of analytical chemistry (Gómez-Ramos et al. 2013).

Nuclear Magnetic Resonance Spectroscopy: The gentle analytical method known as nuclear magnetic resonance spectroscopy, or NMR, provides detailed information about the molecular structure of metabolites (Kumar 2016). Unlike MS, NMR does not require the separation or purification of metabolites, and it can provide quantitative data without the need for external standards (Simmler et al. 2014). NMR is highly reproducible but less sensitive than MS; thus, it is often used for targeted metabolomics where the focus is on a known set of metabolites (Bingol 2018). 1D, 2D, DOSY (diffusion-ordered spectroscopy), HR-MAS (high-resolution magic angle spinning), and qNMR (quantitative NMR) are most common types of NMR, which use in metabolomics (Gouilleux et al. 2018).

Fourier-Transform Infrared Spectroscopy: FTIR is another non-detrimental technology that detects the vibrational energy of molecular bonds within metabolites (Koczoń et al. 2023). It is useful for identifying functional groups and studying molecular interactions but is less commonly used for complex metabolite profiling due to lower sensitivity compared to MS and NMR (Coen et al. 2008).

Capillary Electrophoresis: CE is a technique of separation used in metabolomics primarily for charged metabolites. It offers high efficiency and resolution with low sample and solvent consumption. CE can be coupled with MS to enhance metabolite identification capabilities (Barbas et al. 2011).

Metabolite Imaging: Techniques like MALDI (Matrix-Assisted Laser Desorption/Ionization) and DESI (Desorption Electrospray Ionization) are used in mass spectrometry imaging to map the spatial distribution of metabolites within tissues, offering insights into metabolic processes in different cell types or regions (Gao et al. 2023).

High-Performance Liquid Chromatography: HPLC is an adaptable and sensitive technique extensively used in metabolomics for its ability to resolve a wide array of metabolites with high precision (Perez de Souza et al. 2021). The selectivity of HPLC is enhanced through various types of columns with different stationary phases that target specific metabolite classes. It is capable of analyzing both polar and nonpolar compounds by adjusting stationary and mobile phases accordingly (Alvarez-Segura et al. 2016). Coupled with sensitive detectors like mass spectrometry, HPLC facilitates not only qualitative but also quantitative analysis of metabolites, even at low concentrations (Wu et al. 2013). Sample preparation protocols for HPLC are well-developed to minimize metabolite loss from different biological matrices. The automation and high throughput capabilities of modern HPLC systems are advantageous for the large-scale sample analysis typical in metabolomics studies (Dunn et al. 2005). With its high reproducibility and precision in retention times and peak areas, HPLC is essential for targeted and untargeted metabolomics investigations. Additionally, compatibility with various online detection methods, particularly mass spectrometry, with its mass-to-charge ratio-based identification, makes HPLC an indispensable tool in the intricate analysis of the metabolome (Alseekh et al. 2021).

6.1.2 Unveiling the Metabolome: Challenges and Power of Metabolite Analysis

Metabolomics provides a magnifying glass to the subtlest changes within an organism, revealing a dynamic metabolic landscape that is sensitive to genetic and environmental influences (Van Treuren 2020). Investigating the metabolome, the totality of metabolites in a biological system, involves dissecting complex biochemical interactions and pathways (Quinones and Kaddurah-Daouk 2009). This analysis enables the tracking of disease progression, development of personalized medicine, and the understanding of plant, animal, and microbial metabolism in exquisite detail (Jain 2015). The challenges of metabolite analysis are multi-fold. First, there exists an almost bewildering diversity of metabolites, ranging from small ionic compounds to large, complex lipids. These metabolites vary greatly in polarity, stability, concentration, and solubility, making no single analytical method sufficient to detect the complete range of metabolites present in a sample (Prakash et al. 2007). Moreover, sample preparation is a crucial step that can introduce variability, requiring careful optimization to ensure consistency and minimal losses (Mire-Sluis et al. 2004). The technical aspects of metabolite analysis also present hurdles. Sensitive detection methods must be paired with selective separation technologies (Kuehnbaum and Britz-McKibbin 2013). For example, mass spectrometry, which is renowned for its sensitive and specific detection capabilities, must often be combined with liquid chromatography for the separation of complex mixtures (Di Stefano et al. 2012). Nuclear magnetic resonance spectroscopy offers unparalleled structural elucidation but requires relatively large sample quantities, limiting its sensitivity compared with MS. When considering the power of metabolite analysis, the precision and scope of current analytical techniques stand out. NMR spectroscopy not only offers insights into structural conformation but also allows quantification and determination of metabolite dynamics in living systems (Segers et al. 2019). LC coupled to MS, especially when integrated with tandem mass spectrometry (MS/MS), has revolutionized metabolite profiling, unveiling metabolic alterations linked to physio pathological conditions (Zahoor et al. 2021). Gas chromatography, often coupled with MS, excels in the analysis of substances that are volatile and semi-volatile, which is critical for the study of aroma compounds in food and plant metabolomics (Simon 2021). The arrival of high-resolution MS and the development of sophisticated software for data analysis have expanded the capacity to perform both targeted assays, focusing on a predefined set of metabolites, and untargeted metabolomic studies, which aim to capture a broad snapshot of the metabolome. Untargeted analysis, while more exploratory, can result in the identification of new biomarkers and pathways by identifying unexpected metabolic changes (Alarcon-Barrera et al. 2022). Bioinformatics plays a critical role in metabolomics by providing tools to manage, analyze, and interpret large and complex datasets generated by metabolite profiling (Johnson et al. 2015). Applying multivariate statistical methods, such as principal component analysis and orthogonal partial least squares discriminant analysis, enables the detection of patterns and associations within the data that can be correlated with specific biological conditions or phenotypes (Fonville et al. 2010). Multi-omics is an effective tool that combines metabolomics data with other "omics" data, including transcriptomics, proteomics, and genomes (Jendoubi 2021). This integration offers a more comprehensive perspective of the biological system being studied and can lead to a more comprehensive understanding of the underlying mechanisms of health and disease. Metabolomics has proven to be useful in clinical research for identifying illness biomarkers, comprehending therapeutic modes of action, and conducting toxicity assessments (Robertson et al. 2011). For example, metabolic profiling of blood or urine can reveal early biomarkers of a disease process, even before clinical symptoms appear (Mamas et al. 2011). This enables the potential for earlier interventions and better disease management. Moreover, in the field of nutrition and agriculture, metabolite analysis has provided insights into the nutritional quality of food, the metabolic processes in crops under environmental stress, and the interaction between diet and human metabolism (Hall et al. 2008). These insights can guide breeding programs to improve crop traits, inform dietary recommendations, and lead to the development of functional foods (Dwivedi et al. 2017). Despite these advances, significant challenges remain, particularly in the standardization of methods, replication of results, and functional interpretation of metabolite changes (Scalbert et al. 2009). As the field progresses, there is a pressing need for collaborative efforts to establish standardized protocols, improved analytical calibration methods, and comprehensive metabolite databases (Salek et al. 2013). The power of metabolite analysis in driving our understanding of complex biological systems holds immense promise for numerous fields such as medical science, pharmacology, environmental studies, and agriculture (Aharoni et al. 2023). As we continue to uncover the intricacies of the metabolome, the convergence of technological innovations, computational methods, and collaborative research will no doubt lead to groundbreaking discoveries and applications that will shape the future of science and medicine (Sindelar 2024).

6.1.3 Mass Spectrometry (MS): A Powerful Tool for Unveiling the Metabolite Landscape

Metabolomics, a field that intends to thoroughly investigate small molecules in biological systems, heavily relies on mass spectrometry (MS) as a primary analytical tool for identifying and quantifying metabolites more were described in Table 6.1

Table 6.1 Unveiling analytical techniques: a comparison of common methods for metabolite analysis

Sr.	Types of mass spectrometry	Advantage	Disadvantage	Reference
1	Time-of-flight mass spectrometry (TOF-MS)	• Extremely high mass range (>1 MDa) • There is no *m*/*z* scale upper limit	• Low resolution power (R = 100–2000) • Time consuming • High cost	Balcerzak (2003)
2	Quadrupole mass spectrometry	• Compact and reliable • Low cost • Determine molecular weight of multiple compounds	• Not suited for pulsed ionization method • Low accuracy (>100 ppm)	Dawson (2013)
3	Ion trap mass spectrometry	• Multi stage spectrometry • Compact mass analyzer	• Collision energy not well defined in CID MS/MS • Slow scanning speed	Nolting et al. (2019)
4	Fourier-transform ion cyclotron resonance mass spectrometry (FT-ICR-MS)	• Resolving power 1 million • Capable of detecting and quantifying metabolites over several orders of magnitude in terms of concentration	• High cost • Complex and require specialized expertise for operation, maintenance, and data analysis	Marshall et al. (1998)
5	Orbitrap mass spectrometry	• Excellent mass accuracy, typically in the sub-ppm range • Rapid data acquisition	• Require large volume of sample	Eliuk and Makarov (2015)
6	Matrix-assisted laser desorption/ ionization mass spectrometry (MALDI-MS)	• Resolution up to 10 μm • Protein detection is possible • Molar mass range up to 20 kDa	• Analytes must have low vapor pressure • Coupling MALDI with chromatography is difficult	Liébana-Martos (2018)
7	Electrospray ionization mass spectrometry (ESI-MS)	• Ability to examine biological sample with non-covalent interaction • Can be used along with LC and capillary electrophoresis	• Multiply charged ions were confusing for analysis • Not suitable for low polar compounds	Janfelt et al. (2012)
8	Gas chromatography-mass spectrometry (GC-MS)	• Fast, inexpensive, high resolution • Small sample size require • Sensitive to ppm–ppb	• Mostly volatile compound can analyze • Certain isomeric compound cannot determine by MS • Certain compound require derivatization	Krone et al. (2010)
9	Liquid chromatography-mass spectrometry (LC-MS)	• Suitable for low polar compound with low, moderate or high molecular weight compound • Derivatization is not required	• Fragmentation pattern are poorly reproducible • Time consuming • Restricted mass range	Munjanja (2017)

(Yu et al. 2020). Mass spectrometry, often combined with techniques like gas chromatography (GC-MS) or liquid chromatography (LC-MS), plays a crucial role in metabolomics research by enabling the identification of many metabolites (Li et al. 2020). Mass spectrometry-based metabolomics has been instrumental in biomarker discovery, disease diagnosis, and understanding metabolic pathways (Pinto et al. 2022). Recent advancements in mass spectrometry have led to the development of innovative approaches such as paper spray ionization-mass spectrometry (PSI-MS) for rapid and sensitive detection of metabolites, particularly in clinical settings (Groves et al. 2022). Additionally, techniques like desorption electrospray ionization mass spectrometry (DESI-MS) have emerged as valuable tools for tissue metabolomic profiling and spatial imaging (Park and Kim 2021). These developments highlight the versatility and applicability of mass spectrometry in metabolomics research. Moreover, the integration of mass spectrometry with other analytical techniques like nuclear magnetic resonance (NMR) has shown promise in providing comprehensive metabolomics data sets (Banerjee et al. 2019). This combination allows for a more holistic understanding of metabolite profiles in biological samples, aiding in The quantification of metabolites, both known and unknown crucial for various medical and biological studies (Leufken et al. 2017). When considering plant metabolomics, mass spectrometry methods like matrix-assisted laser desorption ionization (MALDI) have been utilized for high-resolution imaging to visualize metabolite distribution in plant tissues (Song et al. 2020). These applications demonstrate the significance of mass spectrometry in advancing our understanding of metabolite composition and distribution in diverse biological systems. We discuss how mass spectrometry is essential to metabolomics research in this book chapter, enabling researchers to investigate the complex domain of biological metabolites. The continuous development of mass spectrometry techniques and their integration with other analytical tools are driving innovations in metabolomics, offering valuable insights into metabolic processes, disease mechanisms, and biomarker discovery.

6.2 Unveiling the Principles of Mass Spectrometry

Identifying low-quantity substances in samples including biomarkers and single-cell metabolites, has received a lot of interest recently due to advancements in chemical and biological research. A comprehensive and methodical examination of these unique compounds is essential for an accurate understanding of related chemical and biological processes (Ma and Fernández 2024). Because of this, there is a high need for analytical tools with outstanding sensitivity and specificity, such as mass spectrometry (MS).

6.2.1 The Core Concept: Ionization and Mass Analysis—Demystifying the Molecular Fingerprint

The signal-to-noise ratio (S/N), which is derived from a given quantity of sample, quantifies the link between signal strength (S) and noise (N), and is widely used to describe the sensitivity of mass spectrometers. Either reducing noise or elevating signal strength can improve it (Li et al. 2021). Mass spectrometers measure the mass-to-charge ratio (*m*/*z*) of an ion in a molecule. Ions are created by enabling a neutral species to gain or lose charge. They are then electrostatically orientated into a mass analyzer, from where they are sorted based on mass/charge (*m*/*z*), and ultimately identified (Haag 2016). An ionization source, an ion detector, a mass analyzer, and a sample intake are the essential components for each mass spectrometer. Generally, mass analyzers produce an ion beam that appears as an electrical current when it hits a sensitive surface (Faull et al. 2008). Point detectors, such as electron and photomultipliers, are used to detect focused ion beams, whereas array detectors, such as multichannel plates, are used to detect scattered beams (Barnes and Hieftje 2004). Mass analyzers function at a vacuum, preferably at p105 Torr (1 Torr ¼ 1 mm Hg), which can be generated by two pumping stages: a rotary mechanical pump and a high vacuum pump connected in series.

- **Principle**

The molecular ion (marked M) represents the complete molecule, which serves as a template for any adduct or fragmented ions created during the ionization process. Typically, the charge on the ion is indicated. The states of the molecular ions are denoted by M^+, MH^+, depending on whether the charge results from the molecule losing an electron, adding a proton, or losing a proton (Lui and Wang 2015). The Dalton (Da), which is equivalent to one-twelfth of the mass of carbon-12, is the unit of mass measurement utilized (12C). Plotting the ion current intensity on the *y*-axis versus the *m*/*z* ratio of the ions on the *x*-axis yields a visual representation of the mass spectrum (Somogyi 2008) (Fig. 6.1).

6.2.2 Different Ionization Techniques (e.g., ESI, APCI)—Choosing the Proper Tool for the Purpose

The main source of ionization used in mass analysis before the 1980s was electron ionization (EI). In recent years, there are now a new generation of ionization methods available, including fast atom/ion bombardment (FAB), matrix-assisted laser desorption/ionization (MALDI), and electrospray ionization (ESI) (Dong et al. 2018) (Table 6.2).

6.2.2.1 Electrospray Ionization

This technique is often used for lipids, short oligonucleotides, proteins, polysaccharides, and synthetic polymers. From a liquid solution, ESI generates gaseous ionized molecules directly (Wilm 2011). The method employs applying a voltage to a capillary and driving an analyte solution through it that produces a fine haze of highly charged particles when an electric field is present. At the metal nozzle tip, which is maintained at a voltage of between 700 and 5000 V, an intense electric field sprays the sample solution out of it (Siuzdak 2004). The current-applied nozzle (or needle) scatters the solution into a thin spray of charged droplets, while the solvent is forced to evaporate from each droplet by applying either dry gas, heat, or both at atmospheric pressure (Bruins 1998). When the droplets come into contact with heat as well as opposing the bath gas flow, the solvent begins to evaporate. As the droplet size shrinks, the charge density of the droplet rises, ultimately leading to the ejection of naked ions due to charge repulsion. After emerging from the mass spectrometer with one or more proton adducts, molecules enter its high vacuum (Konermann et al. 2013). This method appears phenomenally well to create ions in the gas phase from a range of non-volatile substances. The molecular weight accuracy of ESI is remarkably good; for proteins, it is typically reported as less than ±0.01% of the mass (Schröder 2012).

6.2.2.2 Matrix-Assisted Laser Desorption/Ionization (MALDI)

Tanaka, Karas, and Hillenkamp created matrix-assisted laser desorption/ionization mass spectrometry (MALDI-MS) for the first time in 1988. A matrix-assisted laser-induced desorption phenomenon results in high ion yields of the intact analyte, and the controlled and efficient energy transfer enables the detection of molecules with sub-picomole sensitivity (Hillenkamp and Peter-Katalinic 2013). A large majority comprehend that MALDI uses laser excitation and sample matrix procedures to ionize and move samples from the gas phase into the condensed phase (Dreisewerd

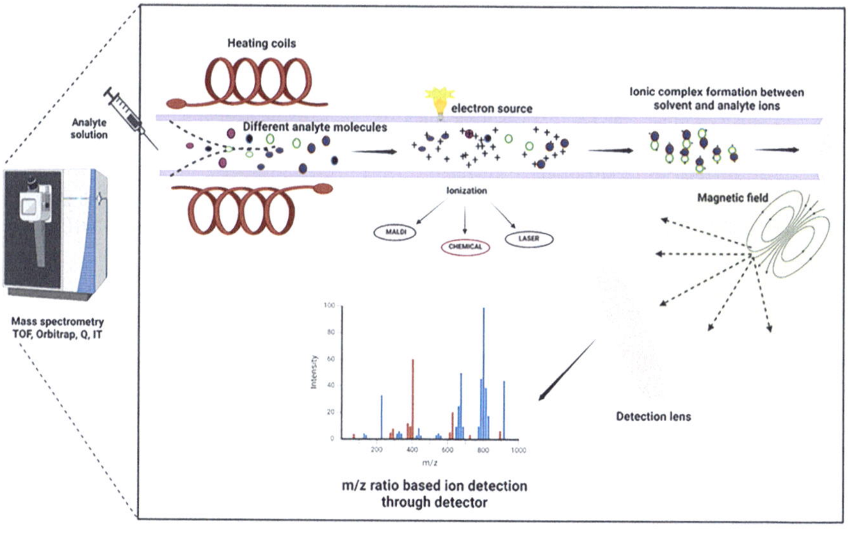

Fig. 6.1 Unveiling the MS workflow: a schematic representation of mass spectrometry analysis

Table 6.2 Unveiling the ionization toolbox: common ionization techniques in MS for metabolomics

Sr. no.	Ionization tool	Principle	Useful for sample types	Ref.
1	Electron ionization	Ions are created when atoms or molecules in the solid or gas phase interact with energetic electrons.	Gaseous and highly volatile liquid samples; tars, resins, and waxes; blood metabolites; protein	Gatchell and Zettergren (2022)
2	Inductively coupled plasma ionization	The material is ionized using an inductively coupled plasma to produce atomic and tiny polyatomic ions.	Whole blood, urine, plasma, serum, soil, water	Van Acker et al. (2023)
3	Atmospheric pressure chemical ionization	It generates primary ions on a solvent spray by using gas-phase ion–molecule reactions at atmospheric pressure (105 Pa).	Pesticides, natural substances, drugs, nonpolar lipids, and different organic molecules	Fang et al. (2020)
4	Atmospheric pressure photoionization	A vacuum ultraviolet (VUV) light source running at atmospheric pressure (105 Pa) is used to ionize molecules.	Low-polarity substances without polar functional groups include petroleum, polyatomic hydrocarbons, lipids, steroids, insecticides, and drug metabolites.	Raffaelli and Saba (2003)
5	Electrospray ionization	It uses an electrospray method to make ions, which involves applying a high voltage to a liquid to turn it into an aerosol.	Proteins, drugs, enzymes	Prabhu et al. (2023)
6	Chemical ionization	Analyte molecules in the gas phase react with the reagent gas molecules to produce analyte ions after the reagent gas molecules are ionized by electron ionization.	Organic compound	Gross (2011)
7	Matrix-assisted laser desorption/ ionization	With little fragmentation, it produces ions from big molecules using a matrix that absorbs laser energy.	Proteins, lipids, oligonucleotides, synthetic macromolecules, polymers, bacteria, fungi, parasites	Zaima et al. (2010)

2003). By absorbing the laser energy, a nonvolatile solid substance enhances the desorption and ionization process and cause the matrix and any sample placed in it to vaporize. Given that the matrix absorbs most of the incoming energy, it also helps reduce sample damage from laser radiation (Pereira et al. 2019). Before the analyte is co-crystallized in a MALDI analysis, a significantly high molarity matrix chemical is added. This chemical is typically a weak organic acid that absorbs UV light.

The analyte is carried by the matrix, which vaporizes when the analyte-matrix combination is exposed to laser light. The molecules of the co-crystallized sample likewise evaporate, but they do so without needing to directly absorb laser light (Calvano et al. 2018). As a result, direct UV laser stimulation does not harm molecules that are sensitive to laser light. Once in the gas phase, the desorbed charged molecules are driven electrostatically from the MALDI ionization source into the mass analyzer. MALDI has applications in biomolecule research since it can give intact molecules' molecular weight information.

6.2.2.3 Desorption/Ionization on Silicon (DIOS)

Combining silicon and pulsed laser desorption/ionization, DIOS is a matrix-free system that uses silicon nanowires or porous silicon as UV-absorbing semiconductors. These materials hold a significant amount of surface area, which helps to hold solvent and analyte molecules in place, while their UV absorptivity facilitates the laser energy transfer to the analyte (Venter et al. 2008). Conversely though, MALDI-LD, employed alongside a matrix, is utilized for analyzing a broad spectrum of compounds, including low-mass substances and proteins with molecular weights of more than 100 kDa (Wei et al. 1999). This ionization method tolerates inorganic salts and common biochemical additives (excluding SDS). DIOS presents distinct advantages owing to its matrix-free nature: minimal background interference within the low mass range, even distribution of aqueous samples, and simplified sample management (Li and Lipson 2013). Additionally, its chip-based design allows for automated sample handling, with rapid laser scanning between spots. These features position DIOS to expedite and streamline the high-throughput analysis of low molecular weight compounds (Lewis et al. 2003).

6.2.2.4 Fast Atom/Ion Bombardment (FAB)

It's an ionization technique that releases ions from a surface by using a liquid matrix and a continuous, high-energy particle beam (Rinehart 1982). A continuous flow of Xe neutral atoms or Cs + ions is used in fast atom bombardment, a soft ionization method, to detach the sample and matrix. For sample input, the method makes using a probe with direct insertion. As it explodes, the non-volatile liquid matrix continuously refreshes the surface with new samples. The rapid atoms or ions collapse into the matrix, causing both the matrix and analyte to transition to the gas phase (Barber et al. 1981). Charged molecules can be electromagnetically accelerated to the mass spectrometer once they are in the gas phase.

6.2.2.5 Electron Ionization (EI)

It is a crucial ionization source for analyzing small, hydrophobic, and thermally stable molecules. Operating as a "hard" ionization method, it produces abundant fragment ions (Yang et al. 2023). Samples are introduced as gas by either heating them off a probe or via gas introduction through a capillary. When a compound encounters an electron ionization source in the gas phase, electrons are excited and expelled from the molecule, causing ionization and fragmentation (Wei et al. 2019).

Beyond 400 Da molecular weight, electron ionization becomes much less useful for molecules as the following problems related to thermal desorption arise:

- In the volatility of large molecules
- Thermal breakdown
- Excessive fragmentation.

6.2.2.6 Chemical Ionization (CI)

Chemical ionization, carried out in a mass spectrometer's vacuum employs gas-phase ion–molecule reactions to generate ions from the sample molecule (Warren 2013). Initiated with reagent gases like methane or ammonia, the process begins with electron impact ionization through a narrow entrance. Ion–molecule interactions between gas ions and neutrals can be enhanced by high gas pressure inside the ionization source, promoting the formation of molecular ions with reduced fragmentation, particularly beneficial for analyte identification (Varvarezos et al. 2021). In high-pressure Chemical Ionization, magnetic, TOF, and quadrupole mass spectrometers have been used.

- **Methane**

The first CI reagent gas, still preferred for its nonselective ionization of all samples, yields dominant ions in the EI mass spectrum of methane: CH_4^+ (*m/z* 16, ~50%), CH_3^+ (*m/z* 15, ~40%), and CH_2^+ (*m/z* 14, ~10%). These ions, generated from methane, form a series (CH_5^+, $C_2H_5^+$, $C_3H_5^+$) that selectively react with the analyte (Glockler 1926). Excitation energy from CH_3^+ or CH_4^+ ions likely dissipates upon forming $C_2H_5^+$ or CH_5^+ ions, largely via methane collisions, before interacting with the analyte. Total ion current and CI sensitivity are often increased with increasing electron energy. Higher energy electrons penetrate deeper into the mass spectrometer source, enhancing ion concentration and enabling more efficient ion release (Munoz et al. 1985) Moreover, 250 eV electrons produce more ions compared to the traditional 70 eV electrons used in EI experiments.

- **Tetramethylsilane**

$(CH_3)_4Si$, the silicon equivalent of neo-pentane, is typically stored in water containers to stabilize its high vapor pressure. Its primary ion under electron ionization (EI) is $(CH_3)_3Si^+$, analogous to the *t*-butyl ion (Chizhov et al. 1991). This substance can be used either pure or combined with isobutane, nitrogen, or methane as a chemical ionization (CI) reagent. Mixing with methane helps reduce source contamination and simplifies the ion spectrum (Muñoz et al. 1986). In mass spectrometric analysis of biomedical samples, a 1/1 mixture of $(CH_3)_4Si$ and $(CD_3)_4Si$ provides equal abundances of $[M + (CH_3)_3Si]^+$ and $[M + (CD_3)_3Si]^+$ adduct ions. Tetramethylsilane CIMS is effective for characterizing alcohols, generating adduct ions like $(CH_3)_3Si]^+$ through reactions with $(CH_3)_3Si^+$.

6.2.3 Mass Analyzers (e.g., Quadrupole, TOF)—Deciphering the Fingerprint: Separating and Detecting Ions

The mass analyzer determines resolution, which is the capacity to differentiate ions with similar *m/z* values. The analyzers also determine mass accuracy, which stands for the accuracy of determining a peak's *m/z* value.

6.2.3.1 Quadrupole Mass Analyzers

Referred to as quadrupole mass filters (QMF), these instruments have been employed alongside EI sources since the 1950s and remain the predominant mass analyzers today. The quadrupole mass analyzer operates in two primary modes: The passage of ions is facilitated by the ion guide mode, while the mass filter mode enables mass analysis (Dawson 1986). Quadrupoles have three key characteristics. They are inexpensive, have a basic construction, and can withstand rather high pressures (Soler et al. 2006). Additionally, because quadrupoles have a large mass span, they can analyze ions up to 4000 *m/z*, which proves advantageous as ionization of proteins and other biomolecules by electrospray typically yields charge distributions ranging from *m/z* 1000–350 (Du and Douglas 1999). Thus, *these* quadrupole mass spectrometers are affordable instruments. These devices feature quadrupole mass analyzers linked parallel to the two components: a DC potential and an RF generator. Under a designated RF field, only ions with particular *m/z* values can traverse through the quadrupoles.

6.2.3.2 Ion Trap Mass Analyzers

Ions are trapped inside a radio frequency quadrupole field in an ion trap, as rather than passing through a quadrupole analyzer wrapped in a radio frequency field (Peng and Austin 2011). This quadrupole ion trap usually consists of two hyperbolic endcap electrodes and a ring electrode. Ions are more easily retained or released from the enclosure because of the motion that the electric field on these electrodes causes (Nolting et al. 2019). A notable attribute of ion traps is their ability to selectively isolate one ion species by expelling all others from the trap. Following their isolation, these ions may fragment due to collisional activation, and the resultant fragments may be identified (Arevalo et al. 2020). The primary advantage of quadrupole ion traps lies in their capacity to swiftly conduct multiple collision-induced dissociation experiments without necessitating multiple analyzers (Tamara et al. 2022). The major limitations of this device include:

- Ion traps prevent neutral loss scanning research and high-sensitivity scanning of precursor ions using a triple quadrupole method.
- The ratio between the lowest trapped fragment ion and the precursor *m/z* has an upper limit of around 0.3.

6.2.3.3 Time-of-Flight Mass Analyzers

The linear time-of-flight (TOF) mass analyzer stands out as the most straightforward among mass analyzers. Its principle revolves around time-of-flight analysis, wherein a set of ions is accelerated toward a detector, all receiving equal energy via an accelerating potential (Chen et al. 2021). As a consequence of their differing masses despite equal energy, lighter ions exhibit greater velocity, reaching the detector before heavier ions, which lag due to their larger mass and slower velocity (Baba et al. 2021). Consequently, the analyzer derives its name from the determination of mass based on ions' arrival times. This method boasts rapid analysis speed and a generous upper-quality limit. It proves particularly adept at handling large throughput and swiftly analyzing biological macromolecules such as peptides (Sullivan et al. 2020). Presently, the TOF reflectron, extensively employed in ESI and MALDI, enhances resolution compared to a basic TOF device by employing the reflectron to concentrate kinetic energy and lengthen the ion path.

6.2.3.4 Ion-Cyclotron Resonance (ICR) Mass Analyzer

It uses an ion confinement magnetic field and measures ions' *m/z* based on their cyclotron frequency within the magnetic field. The ICR cell's detecting plates, an amplifier, and an analog-to-digital converter make up the FT-ICR detection setup (Nagornov et al. 2022).

6.2.3.5 Tandem Mass Analyzers

Tandem Mass Analyzers involve pairing two or more mass analyzers, leveraging the strengths of each to compensate for their limitations. This approach is especially effective for analyzing samples containing intricate mixtures and significant interferences (Neagu et al. 2022).

6.2.3.6 Orbitrap Analyzers

The first entirely new mass analyzer to be released in the market after more than 20 years is the orbitrap mass analyzer. As ions revolve around an electrode, they become trapped, and the centrifugal force created by the rotation balances the electrostatic pull toward the electrode (Rakk et al. 2023).

6.3 Unveiling the Power of Applications: Utilizing MS in Metabolomics Research

A potent analytical method utilized in metabolomics research is mass spectrometry (MS), which offers several distinct advantages and significantly contributes to our understanding of cellular metabolism and its implications for health, disease, and the environment (Zhang et al. 2020; Lei et al. 2011) MS is characterized by its extreme sensitivity, enabling the detection of metabolites even at very low concentrations. This capability allows researchers to identify subtle changes in the metabolic profile of cells or organisms, which may be indicative of physiological responses, disease states, or environmental influences (Kolar et al. 2019; Zabalegui et al. 2020). By accurately quantifying these metabolites, MS facilitates the identification of metabolic biomarkers—unique signatures relevant to particular illnesses or ailments. Biomarker discovery is crucial for early disease diagnosis, prognosis, and monitoring of treatment responses. MS boasts broad coverage, meaning it can analyze a wide range of metabolites simultaneously (Monteiro et al. 2013). This capability provides researchers with a comprehensive snapshot of the metabolic landscape within a biological system. By profiling numerous metabolites in a single analysis, MS allows for a more thorough exploration of metabolic pathways and their interconnectedness. This holistic view enhances our understanding of the underlying mechanisms governing cellular metabolism and how they may be altered under various physiological or pathological conditions (Kell and Goodacre 2014) modern MS instruments are characterized by their speed and efficiency. They can rapidly analyze samples, making them well-suited for large-scale studies and high-throughput analyses. This capability accelerates research processes, allowing for the efficient screening of samples from diverse experimental conditions or patient cohorts. Additionally, the speed of MS analysis facilitates real-time monitoring of metabolic changes, providing valuable insights into dynamic metabolic processes and responses to stimuli (Miao et al. 2018).

In metabolomics research, mass spectrometry (MS) has emerged as a key technological tool, demonstrating remarkable versatility across various domains. One significant application lies in biomarker discovery, where MS enables the identification of metabolite patterns associated with specific diseases. This capability not only facilitates early disease diagnosis but also paves the way for personalized treatment strategies, as elucidated by (Jaffe et al. 2020). Moreover, MS contributes significantly to understanding disease processes by elucidating how diseases perturb metabolic pathways. Researchers, such as (Miao et al. 2018), have highlighted its role in bringing light on the mechanisms underlying disease and uncovering possible treatment approaches. Similarly, in drug development, MS plays a crucial role in evaluating the impact of drugs on cellular metabolism. Michnowicz (2002) emphasized its ability to identify metabolic alterations induced by pharmaceutical compounds, aiding in the discovery of novel therapeutic targets. Nutritional research also benefits from MS by shedding light on how dietary factors influence metabolism. Lagoumintzis and Patrinos (2023) demonstrated its role in developing personalized dietary recommendations for promoting health and preventing disease. Additionally, MS is instrumental in environmental monitoring, assessing the impact of contaminants on metabolism and informing environmental health policies and interventions. The combination of MS and metabolomics research holds immense potential in advancing our understanding of health, disease, and the environment. It serves as a powerful tool in biomedical research, providing detailed insights into the metabolic intricacies of biological systems. This, in turn, contributes to the development of innovative diagnostic, therapeutic, and preventive strategies aimed at improving human health and environmental sustainability, as highlighted. To provide detailed insights into the metabolic intricacies of biological systems, MS serves as a powerful tool in biomedical research and beyond, contributing to the development of innovative diagnostic, therapeutic, and preventive strategies for improving human health and environmental sustainability.

6.3.1 Metabolite Profiling: Mapping the Metabolic Landscape of Cells and Organisms

Metabolite profiling, facilitated by mass spectrometry (MS), serves as an invaluable tool in unraveling the complexities of cellular and organismal processes. By delving into the intricate world of metabolites, scientists can construct detailed metabolic landscapes, shedding light on the fundamental mechanisms that drive life. Metabolites stand as the end products of metabolism, the intricate web of biochemical reactions that sustain cellular function. These molecules span a diverse spectrum, encompassing essential substances like glucose and amino acids, hormones, and signaling molecules (Nicholson et al. 1999; Gavaghan et al. 2000). Each metabolite plays a crucial role in orchestrating cellular activities, influencing processes ranging from energy production to cell signaling and regulation. The advent of mass spectrometry has revolutionized the study of metabolites, offering unprecedented capabilities in their detection and characterization. Mass spectrometry enables the precise measurement of metabolite abundance, providing quantitative insights into their concentrations within biological samples. This analytical technique operates by ionizing molecules and separating them based on their mass-to-charge ratios, allowing for the identification and quantification of metabolites with remarkable accuracy and sensitivity. Metabolite profiling harnesses the potentials of mass spectrometry to generate comprehensive maps of cellular metabolomes, cataloging the diverse array of metabolites present within a biological system (Michopoulos et al. 2014). These metabolic landscapes offer a holistic view of cellular metabolism, depicting the intricate interplay between different metabolites and pathways. By analyzing changes in metabolite abundance under various conditions or perturbations, scientists can discern patterns indicative of physiological states, disease processes, or responses to external stimuli. One of the advantages of metabolite profiling lies in its ability to uncover biomarkers—specific metabolite signatures associated with particular physiological states or diseases. By identifying and quantifying these biomarkers, researchers can develop diagnostic tools for disease detection and monitoring. Moreover, metabolite profiling holds immense potential in personalized medicine, facilitating the patient's classification according to their metabolic profiles and enabling tailored treatment strategies (Gabali et al. 2014; Cannas et al. 2011) Furthermore, metabolite profiling serves as a valuable tool in systems biology, aiding in the construction of predictive models that elucidate the dynamic behavior of biological systems. Integrating metabolite data with more omics technologies, such as genomics, transcriptomics, and proteomics, enables a comprehensive understanding of cellular function at multiple levels (Anal.Chem. et al. 2009). This systems-level approach offers insights into the intricate networks and interactions underlying biological processes, paving the way for advancements in fields such as drug discovery, biotechnology, and agricultural science. In addition to its applications in basic research and clinical diagnostics, metabolite profiling holds promise in fields such as environmental science and food technology (Tischfield et al. 2011). By analyzing metabolites in environmental samples or food products, scientists can assess the impact of pollutants, toxins, or processing techniques on human health and ecosystems. In conclusion, metabolite profiling powered by mass spectrometry offers a powerful means of elucidating the intricacies of cellular metabolism and organismal physiology (Aharoni et al. 2023).

6.3.2 Biomarker Discovery: Identifying Signatures of Disease in the Metabolome

In the field of biological research, the identification, confirmation, and measurement of novel biomarkers, also referred to as "biological markers," are crucial pillars. Metabolites are one such target for biomarker identification. Metabolomics is the study of metabolic patterns in a biological system and is the last step in the omics cascade that starts with activated genes (genome), gene transcripts (transcriptome), and proteins (proteome) (Dettmer et al. 2007; Jain 2010; Dunn et al. 2011; Roberts et al. 2012). Many diseases naturally result in changes at the metabolite level, and these changes frequently manifest in biological fluids before the onset of clinical symptoms Biomarker Discovery, an expanding field propelled by mass spectrometry (MS), offers a compelling avenue for unraveling disease signatures embedded within an organism's metabolome. This sophisticated technique holds promise in identifying distinctive molecular patterns associated with various pathological conditions, revolutionizing disease diagnosis and treatment strategies (Van et al. 2011). At the heart of Deep Biomarker Discovery lies metabolomics analysis, a process wherein biological samples such as blood, urine, or tissue extracts undergo scrutiny by MS (Borovecki et al. 2005; Hu et al. 2007; Hu et al. 2007). Leveraging the principle of the mass-to-charge ratio, MS meticulously separates and characterizes the myriad metabolites present within these samples, furnishing researchers with a comprehensive snapshot of the organism's metabolic landscape. The ensuing data acquisition phase yields a wealth of information, encapsulating the identities and relative abundances of diverse metabolites (Gowda et al. 2008; Zhang et al. 2012b).

6.4 The Biomarker Discovery Workflow: A Step-by-Step Guide

(a) **Sample Collection:** Biological samples relevant to the disease under investigation are collected from both diseased and healthy individuals. Blood, urine, and tissue samples are commonly used (Want et al. 2013).
(b) **Metabolite Extraction:** Efficient methods are employed to extract metabolites from the samples, considering their diverse chemical properties. This step often involves quenching metabolic activity, followed by solvent extraction and purification.
(c) **LC-MS Analysis:** The extracted metabolites are separated by LC using techniques like reversed-phase chromatography. The separated metabolites are then ionized and analyzed by MS, generating metabolite profiles.

(d) **Data Processing:** The raw MS data undergoes extensive processing to remove noise, correct for background signals, and align peaks across different samples. Normalization techniques are applied to account for changes in the size of the sample or the instrument's response.
(e) **Statistical Analysis:** Advanced statistical techniques are employed to find metabolites that show substantial differences between groups with and without illness. To illustrate group separation and determine the most influential metabolites, multivariate statistical techniques such as Principal Component Analysis (PCA) and Partial Least Squares-Discriminant Analysis (PLS-DA) are frequently used.
(f) **Biomarker Validation:** Putative biomarkers identified through statistical analysis undergo rigorous validation in independent studies with larger sample sizes. This step ensures the robustness and generalizability of the identified biomarkers for real-world applications (Dunn et al. 2011; Roberts et al. 2012).

This massive dataset serves as the foundation for subsequent statistical analyses, facilitated by advanced software tools. By juxtaposing metabolite profiles from healthy and diseased cohorts, researchers unearth significant alterations in metabolite levels that hold potential as disease biomarkers. Candidate metabolites exhibiting perturbed levels are subjected to meticulous scrutiny, often involving the interrogation of metabolic pathways or consultation of comprehensive databases to unravel their biological significance. This meticulous curation process culminates in the selection of promising biomarker candidates warranting further validation. The adoption of MS-based biomarker discovery confers several distinct advantages. Firstly, its high-throughput nature enables the simultaneous analysis of multiple metabolites, facilitating comprehensive profiling of the metabolome (Jain 2010; Dunn et al. 2011) Moreover, MS exhibits remarkable sensitivity, capable of detecting subtle fluctuations in metabolite concentrations, potentially enabling early disease detection and intervention. Additionally, by offering a panoramic view of the metabolic landscape, MS illuminates the intricate interplay between metabolic pathways, fostering a deeper understanding of disease mechanisms. However, the pursuit of Deep Biomarker Discovery via MS is not without its challenges (Mayerle et al. 2018, Ottosson et al. 2019). The inherent complexity of MS data necessitates the utilization of sophisticated bioinformatics tools for data processing, analysis, and interpretation. Meticulous attention to sample preparation protocols is imperative to ensure data integrity and metabolite stability throughout the analytical workflow (Nho et al. 2019; Deng et al. 2019). Furthermore, while MS-based biomarker candidates hold promise, their clinical utility hinges on rigorous validation in large-scale clinical studies to ascertain their diagnostic or prognostic efficacy. In conclusion, Deep Biomarker Discovery empowered by mass spectrometry represents a paradigm-shifting approach to disease characterization and diagnosis.

6.4.1 Metabolic Pathway Elucidation: Understanding the Flow of Cellular Chemistry

Metabolic pathways can be likened to a complex system of conveyor belts within a cell, where raw materials undergo a series of chemical reactions to produce end products. Understanding these pathways is essential for various reasons. Firstly, disruptions in these pathways are often associated with diseases such as diabetes and cancer, making them crucial targets for drug development. Additionally, comprehending these pathways aids in biotechnology, allowing scientists to engineer organisms for more efficient production of certain compounds (Kwan et al. 2023). Moreover, unraveling metabolic pathways enhances our basic biological knowledge, providing insights into the molecular mechanisms of living organisms Scientists use enzyme assays to gauge enzyme activity, providing clues about the reactions occurring in the pathway (Kumar et al. 2020). They also utilize radioactive or stable isotope tracers to track labeled molecules and reveal the sequence of reactions. Advanced analytical techniques like mass spectrometry and nuclear magnetic resonance spectroscopy aid in identifying and quantifying metabolites within the pathway. Despite the challenges posed by the intricate nature of cellular reactions, advancements in technology such as high-throughput screening and computational modeling are accelerating the elucidation process. In essence, metabolic pathway elucidation is a captivating field that sheds light on the intricate workings of life at its most fundamental level (Rajesh et al. 2023).

6.5 Conclusion

In metabolomics research, mass spectrometry (MS) has become an essential instrument, offering unparalleled insights into the intricate world of metabolites within biological systems. The capabilities of MS have transformed the study of metabolomics, which explores the thorough examination of all metabolites. Through the precise measurement of mass-to-charge ratios of ions, MS enables the molecular identification and quantification in complex mixtures, providing a snapshot of the relative levels of compounds present. This analytical technique, coupled with chromatography and nuclear magnetic resonance, has significantly advanced biomarker discovery, metabolic pathway elucidation, and disease diagnostics. The development of high-resolution accurate mass (HRAM) MS workflows has enhanced the quantitative capabilities of metabolomics, allowing for targeted tandem MS approaches that ensure data reliability and reduce errors. Isotope-labeled internal standards further enable the absolute quantification of metabolites, although challenges such as cost and availability persist. Moreover, the advent of ambient ionization methods in MS has bypassed the need for chromatography separation, offering rapid and sensitive analysis of biological samples. Not only has MS-based metabolomics produced significant understandings of metabolic fluxes, pathways, and disease biomarkers, but has also paved the way for powerful isotope labeling and tracing methods. The field continues to evolve, with new techniques like MS imaging enabling the visualization of spatial distributions of chemical compositions without the need for complex sample preparation. Despite the remarkable progress, challenges such as biomarker validation and inter laboratory reproducibility remain, underscoring the need for continued advancements in MS-based metabolomics to unlock its full potential in understanding phenotypic changes influenced by genetic, environmental, and nutritional factors. In conclusion, mass spectrometry stands as a beacon in metabolomics research, driving innovation, and offering a deeper understanding of the intricate metabolic landscape within living organisms. Its role in unraveling the complexities of metabolites and their roles in health, disease, and beyond solidifies its position as a cornerstone in modern biochemistry and systems biology.

References

Aharoni A, Goodacre R, Fernie AR (2023) Plant and microbial sciences as key drivers in the development of metabolomics research. Proc Natl Acad Sci 120(12):e2217383120

Alarcon-Barrera JC, Kostidis S, Ondo-Mendez A, Giera M (2022) Recent advances in metabolomics analysis for early drug development. Drug Discov Today 27(6):1763–1773

Alseekh S, Aharoni A, Brotman Y, Contrepois K, D'Auria J, Ewald J, Fernie AR (2021) Mass spectrometry-based metabolomics: a guide for annotation, quantification and best reporting practices. Nat Methods 18(7):747–756

Alvarez-Segura T, Torres-Lapasió JR, Ortiz-Bolsico C, García-Alvarez-Coque MC (2016) Stationary phase modulation in liquid chromatography through the serial coupling of columns: a review. Anal Chim Acta 923:1–23

Arevalo R, Ni Z, Danell RM (2020) Mass spectrometry and planetary exploration: a brief review and future projection. J Mass Spectrom 55(1):e4454. https://doi.org/10.1002/jms.4454

Baba T, Ryumin P, Duchoslav E, Chen K, Chelur A, Loyd B, Chernushevich I (2021) Dissociation of biomolecules by an intense low-energy electron beam in a high sensitivity time-of-flight mass spectrometer. J Am Soc Mass Spectrom 32(8):1964–1975. https://doi.org/10.1021/jasms.0c00425

Balcerzak M (2003) An overview of analytical applications of time of flight-mass spectrometric (TOF-MS) analyzers and an inductively coupled plasma-TOF-MS technique. Anal Sci 19(7):979–989

Banerjee S, Wong A, Yan X, Wu B, Zhao H, Tibshirani R, Brooks J (2019) Early detection of unilateral ureteral obstruction by desorption electrospray ionization mass spectrometry. Sci Rep 9(1):11007. https://doi.org/10.1038/s41598-019-47396-x

Barbas C, Moraes EP, Villasenor A (2011) Capillary electrophoresis as a metabolomics tool for non-targeted fingerprinting of biological samples. J Pharm Biomed Anal 55(4):823–831

Barber M, Bordoli RS, Sedgwick RD, Tyler AN (1981) Fast atom bombardment of solids as an ion source in mass spectrometry. Nature 293(5830):270–275. https://doi.org/10.1038/293270a0

Barnes JH, Hieftje GM (2004) Recent advances in detector-array technology for mass spectrometry. Int J Mass Spectrom 238(1):33–46. https://doi.org/10.1016/j.ijms.2004.08.004

Bingol K (2018) Recent advances in targeted and untargeted metabolomics by NMR and MS/NMR methods. High Throughput 7(2):9

Borovecki F, Lovrecic L, Zhou J, Jeong H, Then F, Rosas H, Jensen R (2005) Genome-wide expression profiling of human blood reveals biomarkers for Huntington's disease. Proc Natl Acad Sci USA 102:11023–11028

Bruins AP (1998) Mechanistic aspects of electrospray ionization. J Chromatogr A 794(1–2):345–357. https://doi.org/10.1016/S0021-9673(97)01110-2

Calvano CD, Monopoli A, Cataldi TRI, Palmisano F (2018) MALDI matrices for low molecular weight compounds: an endless story? Anal Bioanal Chem 410(17):4015–4038. https://doi.org/10.1007/s00216-018-1014-x

Cannas S, Molicotti P, Bua A, Usai D, Sechi LA, Scanu AM, Zanetti S (2011) Interaction between Mycobacterium tuberculosis, Mycobacterium bovis, Mycobacterium avium subspecies paratuberculosis with the enteric glia and microglial cells. Gut Pathog 3:19. https://doi.org/10.1186/1757-4749-3-19

Chen X-F, Hou X, Xiao M, Zhang L, Cheng J-W, Zhou M-L, Huang J-J, Zhang J-J, Xu Y-C, Hsueh P-R (2021) Matrix-Assisted Laser Desorption/Ionization Time of Flight Mass Spectrometry (MALDI-TOF MS) analysis for the identification of pathogenic microorganisms: a review. Microorganisms 9(7):1536. https://doi.org/10.3390/microorganisms9071536

Chizhov OS, Kadentsev VI, Stomakhin AA (1991) Chemical ionization mass spectrometry of halogenoalkanes with tetramethylsilane as reagent gas. Org Mass Spectrom 26(9):757–760. https://doi.org/10.1002/oms.1210260909

Coen M, Holmes E, Lindon JC, Nicholson JK (2008) NMR-based metabolic profiling and metabonomic approaches to problems in molecular toxicology. Chem Res Toxicol 21(1):9–27

Cohen JI (2006) Virology and molecular biology of Epstein–Barr virus. In: Epstein-Barr virus. CRC Press, Boca Raton, pp 43–60

Dawson PH (1986) Quadrupole mass analyzers: performance, design and some recent applications. Mass Spectrom Rev 5(1):1–37. https://doi.org/10.1002/mas.1280050102

Dawson PH (ed) (2013) Quadrupole mass spectrometry and its applications. Elsevier

Deng L, Ismond KP, Liu Z, Constable J, Wang H, Alatise OI, Chang D (2019) Urinary metabolomics to identify a unique biomarker panel for detecting. Cancer Epidemiol Biomarkers Prev 28(8):1283–1291

Dettmer K, Aronov PA, Hammock BD (2007) Mass spectrometry-based metabolomics. Mass Spectrom Rev 26:51–78

Di Stefano V, Avellone G, Bongiorno D, Cunsolo V, Muccilli V, Sforza S, Vékey K (2012) Applications of liquid chromatography–mass spectrometry for food analysis. J Chromatogr A 1259:74–85

Dong Y, Liu J, Guo T (2018) Introduction of mass spectrometry and ambient ionization techniques. In: Dong Y (ed) Direct analysis in real time mass spectrometry, 1st edn. Wiley, pp 1–42. https://doi.org/10.1002/9783527803705.ch1

Drabińska N, Młynarz P, de Lacy Costello B, Jones P, Mielko K, Mielnik J, Ratcliffe NM (2020) An optimization of liquid–liquid extraction of urinary volatile and semi-volatile compounds and its application for gas chromatography-mass spectrometry and proton nuclear magnetic resonance spectroscopy. Molecules 25(16):3651

Dreisewerd K (2003) The desorption process in MALDI. Chem Rev 103(2):395–426. https://doi.org/10.1021/cr010375i

Du Z, Douglas DJ (1999) A novel tandem quadrupole mass analyzer. J Am Soc Mass Spectrom 10(11):1053–1066. https://doi.org/10.1016/S1044-0305(99)00094-X

Dunn WB, Bailey NJ, Johnson HE (2005) Measuring the metabolome: current analytical technologies. Analyst 130(5):606–625

Dunn WB, Broadhurst DI, Atherton HJ, Goodacre R, Griffin JL (2011) Systems level studies of mammalian metabolomes: the roles of mass spectrometry and nuclear magnetic resonance spectroscopy. Chem Soc Rev 40(1):387–426

Dwivedi SL, Scheben A, Edwards D, Spillane C, Ortiz R (2017) Assessing and exploiting functional diversity in germplasm pools to enhance abiotic stress adaptation and yield in cereals and food legumes. Front Plant Sci 8:275463

Eliuk S, Makarov A (2015) Evolution of orbitrap mass spectrometry instrumentation. Annu Rev Anal Chem 8:61–80

Fang J, Zhao H, Zhang Y, Lu M, Cai Z (2020) Atmospheric pressure chemical ionization in gas chromatography-mass spectrometry for the analysis of persistent organic pollutants. Trends Environ Anal Chem 25:e00076. https://doi.org/10.1016/j.teac.2019.e00076

Faull KF, Dooley AN, Halgand F, Shoemaker LD, Norris AJ, Ryan CM, Laganowsky A, Johnson JV, Katz JE (2008) Chapter 1 An introduction to the basic principles and concepts of mass spectrometry. In: Comprehensive analytical chemistry, vol 52. Elsevier, pp 1–46. https://doi.org/10.1016/S0166-526X(08)00201-8

Fiehn O (2001) Combining genomics, metabolome analysis, and biochemical modelling to understand metabolic networks. Comp Funct Genomics 2(3):155–168

Fonville JM, Richards SE, Barton RH, Boulange CL, Ebbels TM, Nicholson JK, Dumas ME (2010) The evolution of partial least squares models and related chemometric approaches in metabonomics and metabolic phenotyping. J Chemom 24(11–12):636–649

Ford BJ (2017) Cellular intelligence: microphenomenology and the realities of being. Prog Biophys Mol Biol 131:273–287

Frank R, Hargreaves R (2003) Clinical biomarkers in drug discovery and development. Nat Rev Drug Discov 2(7):566–580

Gabali AM, Jazaerly T, Chang CC, Cleveland R, Kass L (2014) Simultaneous hepatosplenic T-cell lymphoma and myelofibrosis. Avicenna J Med 4(2):34–36. https://doi.org/10.4103/2231-0770.130343

Gao SQ, Zhao JH, Guan Y, Tang YS, Li Y, Liu LY (2023) Mass spectrometry imaging technology in metabolomics: a systematic review. Biomed Chromatogr 37(7):e5494

Gatchell M, Zettergren H (2022) Open questions on the interaction dynamics of molecules and clusters in the gas phase. Commun Chem 5(1):28. https://doi.org/10.1038/s42004-022-00646-y

Gavaghan CL, Holmes E, Lenz E et al (2000) An NMR-based metabonomic approach to investigate the biochemical consequences of genetic strain differences: application to the C57BL10J and Alpk:ApfCD mouse. FEBS Lett 484:169–174

Glockler G (1926) The ionization potential of methane. J Am Chem Soc 48(8):2021–2026. https://doi.org/10.1021/ja01419a002

Gómez-Ramos MM, Ferrer C, Malato O, Agüera A, Fernández-Alba AR (2013) Liquid chromatography-high-resolution mass spectrometry for pesticide residue analysis in fruit and vegetables: screening and quantitative studies. J Chromatogr A 1287:24–37

Gouilleux B, Rouger L, Giraudeau P (2018) Ultrafast 2D NMR: methods and applications. Annu Rep NMR Spectrosc 93:75–144

Gowda GN, Zhang S, Gu H, Asiago V, Shanaiah N, Raftery D (2008) Metabolomics-based methods for early disease diagnostics. Expert Rev Mol Diagn 8:617–633

Gross JH (2011) Chemical ionization. In: Gross JH (ed) Mass spectrometry. Springer, Berlin/Heidelberg, pp 351–380. https://doi.org/10.1007/978-3-642-10711-5_7

Groves R, Chan C, Wildman S, Gregson D, Rydzak T, Lewis I (2022) Rapid LC–MS assay for targeted metabolite quantification by serial injection into isocratic gradients. Anal Bioanal Chem 415(2):269–276. https://doi.org/10.1007/s00216-022-04384-x

Haag AM (2016) Mass analyzers and mass spectrometers. In: Mirzaei H, Carrasco M (eds) Modern proteomics—sample preparation, analysis and practical applications, vol 919. Springer International Publishing, pp 157–169. https://doi.org/10.1007/978-3-319-41448-5_7

Hall RD, Brouwer ID, Fitzgerald MA (2008) Plant metabolomics and its potential application for human nutrition. Physiol Plant 132(2):162–175

Hillenkamp F, Peter-Katalinic J (eds) (2013) MALDI MS: a practical guide to instrumentation, methods, and applications, 1st edn. Wiley. https://doi.org/10.1002/9783527335961

Hu S, Wang J, Meijer J, Ieong S, Xie Y, Yu T et al (2007) Salivary proteomic and genomic biomarkers for primary Sjögren's syndrome. Arthritis Rheum 56:3588–3600

Jaffe K, ter Horst E, Gunn LH, Zambrano JD, Molina G (2020) A network analysis of research productivity by country, discipline, and wealth. PLoS One 15(5):e0232458. https://doi.org/10.1371/journal.pone.0232458

Jain K (2010) The handbook of biomarkers. Springer, New York, pp 10–11

Jain KK (2015) Textbook of personalized medicine. Springer Nature

Jendoubi T (2021) Approaches to integrating metabolomics and multi-omics data: a primer. Metabolites 11(3):184

Johnson CH, Ivanisevic J, Benton HP, Siuzdak G (2015) Bioinformatics: the next frontier of metabolomics. Anal Chem 87(1):147–156
Karamitros T, Pogka V, Papadopoulou G, Tsitsilonis O, Evangelidou M, Sympardi S, Mentis A (2021) Dual Rna-Seq enables full-genome assembly of measles virus and characterization of host–pathogen interactions. Microorganisms 9(7):1538
Kell DB, Goodacre R (2014) Metabolomics and systems pharmacology: why and how to model the human metabolic network for drug discovery. Drug Discov Today 19:171–182. https://doi.org/10.1016/j.drudis.2013.07.014
Koczoń P, Hołaj-Krzak JT, Palani BK, Bolewski T, Dąbrowski J, Bartyzel BJ, Gruczyńska-Sękowska E (2023) The analytical possibilities of FT-IR spectroscopy powered by vibrating molecules. Int J Mol Sci 24(2):1013
Kolar P, Grbić MS, Hrabar S (2019) Sensitivity enhancement of NMR spectroscopy receiving chain used in condensed matter physics. Sensors 19(14):3064. https://doi.org/10.3390/s19143064
Konermann L, Ahadi E, Rodriguez AD, Vahidi S (2013) Unraveling the mechanism of electrospray ionization. Anal Chem 85(1):2–9. https://doi.org/10.1021/ac302789c
Krone N, Hughes BA, Lavery GG, Stewart PM, Arlt W, Shackleton CH (2010) Gas chromatography/mass spectrometry (GC/MS) remains a pre-eminent discovery tool in clinical steroid investigations even in the era of fast liquid chromatography tandem mass spectrometry (LC/MS/MS). J Steroid Biochem Mol Biol 121(3–5):496–504
Kuehnbaum NL, Britz-McKibbin P (2013) New advances in separation science for metabolomics: resolving chemical diversity in a post-genomic era. Chem Rev 113(4):2437–2468
Kumar D (2016) Nuclear magnetic resonance (NMR) spectroscopy for metabolic profiling of medicinal plants and their products. Crit Rev Anal Chem 46(5):400–412
Kumar R, Ghosh M, Prasad M (2020) Single cell metabolomics: a future tool to unmask cellular heterogeneity and virus-host interaction in context of emerging viral diseases. Front Microbiol 11:530147
Kwan BD et al (2023) Leveraging synthetic biology and metabolic engineering to overcome obstacles in plant pathway elucidation. Curr Opin Plant Biol 71:102330. https://doi.org/10.1016/j.pbi.2022.102330
Lagoumintzis G, Patrinos GP (2023) Triangulating nutrigenomics, metabolomics and microbiomics toward personalized nutrition and healthy living. Hum Genomics 17(1):109. https://doi.org/10.1186/s40246-023-00561-w
Lei Z, Huhman DV, Sumner LW (2011) Mass spectrometry strategies in metabolomics. J Biol Chem 286(29):25435–25442. https://doi.org/10.1074/jbc.R111.238691
Leufken J, Niehues A, Sarin L, Wessel F, Hippler M, Leidel S et al (2017) Pyqms enables universal and accurate quantification of mass spectrometry data. Mol Cell Proteomics 16(10):1736–1745. https://doi.org/10.1074/mcp.m117.068007
Lewis WG, Shen Z, Finn MG, Siuzdak G (2003) Desorption/ionization on silicon (DIOS) mass spectrometry: background and applications. Int J Mass Spectrom 226(1):107–116. https://doi.org/10.1016/S1387-3806(02)00973-9
Li J, Lipson RH (2013) Insights into Desorption Ionization on Silicon (DIOS). J Phys Chem C 117(51):27114–27119. https://doi.org/10.1021/jp4074653
Li Y, Bouza M, Wu C, Guo H, Huang D, Doron G et al (2020) Sub-nanoliter metabolomics via mass spectrometry to characterize volume-limited samples. Nat Commun 11(1):5625. https://doi.org/10.1038/s41467-020-19444-y
Li C, Chu S, Tan S, Yin X, Jiang Y, Dai X, Gong X, Fang X, Tian D (2021) Towards higher sensitivity of mass spectrometry: a perspective from the mass analyzers. Front Chem 9:813359. https://doi.org/10.3389/fchem.2021.813359
Liébana-Martos C (2018) Indications, interpretation of results, advantages, disadvantages, and limitations of MALDI-TOF. In: The use of mass spectrometry technology (MALDI-TOF) in clinical microbiology. Academic Press, pp 75–86
Liu S, Wang Y (2015) Mass spectrometry for the assessment of the occurrence and biological consequences of DNA adducts. Chem Soc Rev 44(21):7829–7854. https://doi.org/10.1039/C5CS00316D

Ma X, Fernández FM (2024) Advances in mass spectrometry imaging for spatial cancer metabolomics. Mass Spectrom Rev 43(2):235–268. https://doi.org/10.1002/mas.21804
Mamas M, Dunn WB, Neyses L, Goodacre R (2011) The role of metabolites and metabolomics in clinically applicable biomarkers of disease. Arch Toxicol 85:5–17
Marshall AG, Hendrickson CL, Jackson GS (1998) Fourier transform ion cyclotron resonance mass spectrometry: a primer. Mass Spectrom Rev 17(1):1–35
Mayerle J, Kalthoff H, Reszka R, Kamlage B, Peter E, Schniewind B et al (2018) Metabolic biomarker signature to differentiate pancreatic ductal adenocarcinoma from chronic pancreatitis. Gut 67:128–137
Miao B, Li FW, Zhang SW, Wang H, Qi WJ, Wang C (2018) Efficacy and safety of Tongfu powder in acute pancreatitis patients with gastrointestinal dysfunction: a clinical trial. Drug Des Devel Ther 12:3665–3673. https://doi.org/10.2147/DDDT.S163645
Michnowicz J (2002) Mass spectrometry in drug discovery and development. Nat Rev Drug Discov 1:651. https://doi.org/10.1038/nrd886
Michopoulos F, Whalley N, Theodoridis G et al (2014) Targeted profiling of polar intracellular metabolites using ion-pair-high performance liquid chromatography and -ultra high performance liquid chromatography coupled to tandem mass spectrometry: applications to serum, urine and tissue extracts. J Chromatogr A 1349:60–68
Mire-Sluis AR, Barrett YC, Devanarayan V, Koren E, Liu H, Maia M et al (2004) Recommendations for the design and optimization of immunoassays used in the detection of host antibodies against biotechnology products. J Immunol Methods 289(1–2):1–16
Mishra B (2009) Intelligently deciphering unintelligible designs: algorithmic algebraic model checking in systems biology. J R Soc Interface 6(36):575–597
Monteiro MS, Carvalho M, Bastos ML, Guedes de Pinho P (2013) Metabolomics analysis for biomarker discovery: advances and challenges. Curr Med Chem 20(2):257–271. https://doi.org/10.2174/0929867311320020006
Moran MA, Kujawinski EB, Schroer WF, Amin SA, Bates NR, Bertrand EM et al (2022) Microbial metabolites in the marine carbon cycle. Nat Microbiol 7(4):508–523
Munjanja BK (2017) Liquid Chromatography Mass Spectrometry (LC-MS). In: Multiresidue methods for the analysis of pesticide residues in food. CRC Press, pp 197–234
Munoz RC, Cumming JB, Holroyd RA (1985) Ionization of tetramethylsilane by alpha particles. Chem Phys Lett 115(6):477–480. https://doi.org/10.1016/0009-2614(85)85174-5
Muñoz RC, Cumming JB, Holroyd RA (1986) Ionization of liquid hydrocarbons and tetramethylsilane by 241Am alpha particles. J Chem Phys 85(2):1104–1115. https://doi.org/10.1063/1.451306
Nagornov KO, Tsybin OY, Nicol E, Kozhinov AN, Tsybin YO (2022) Fourier transform ion cyclotron resonance mass spectrometry at the true cyclotron frequency. Mass Spectrom Rev 41(2):314–337. https://doi.org/10.1002/mas.21681
Neagu A-N, Jayathirtha M, Baxter E, Donnelly M, Petre BA, Darie CC (2022) Applications of tandem mass spectrometry (MS/MS) in protein analysis for biomedical research. Molecules 27(8):2411. https://doi.org/10.3390/molecules27082411
Nho K, Kueider-Paisley A, Ahmad S, Mahmoudiandehkordi S, Arnold M, Risacher SL et al (2019) Association of altered liver enzymes with Alzheimer disease diagnosis, cognition, neuroimaging measures, and cerebrospinal fluid biomarkers. JAMA Netw Open 2:e197978
Nicholson JK, Lindon JC, Holmes E (1999) "Metabonomics": understanding the metabolic responses of living systems to pathophysiological stimuli via multivariate statistical analysis of biological NMR spectroscopic data. Xenobiotica 29:1181–1189
Nolting D, Malek R, Makarov A (2019) Ion traps in modern mass spectrometry. Mass Spectrom Rev 38(2):150–168. https://doi.org/10.1002/mas.21549
Ottosson F, Smith E, Gallo W, Fernandez C, Melander O (2019) Purine metabolites and carnitine biosynthesis intermediates are biomarkers for incident type 2 diabetes. J Clin Endocrinol Metab 104:4921–4930

Park M, Kim Y (2021) Mass spectrometry based metabolomics approach on the elucidation of volatile metabolites formation in fermented foods: a mini review. Food Sci Biotechnol 30(7):881–890. https://doi.org/10.1007/s10068-021-00917-9
Peng Y, Austin DE (2011) New approaches to miniaturizing ion trap mass analyzers. TrAC Trends Anal Chem 30(10):1560–1567. https://doi.org/10.1016/j.trac.2011.07.003
Pereira I, De Aguiar DVA, Vasconselos G, Vaz BG (2019) Fourier transform mass spectrometry applied to petroleomics. In: Fundamentals and applications of Fourier transform mass spectrometry. Elsevier, pp 509–528. https://doi.org/10.1016/B978-0-12-814013-0.00016-8
Perez de Souza L, Alseekh S, Scossa F, Fernie AR (2021) Ultra-high-performance liquid chromatography high-resolution mass spectrometry variants for metabolomics research. Nat Methods 18(7):733–746
Pinto F, Mahmud I, Rubio V, Máquina A, Durans A, Neto W et al (2022) Data-driven soft independent modeling of class analogy in paper spray ionization mass spectrometry-based metabolomics for rapid detection of prostate cancer. Anal Chem 94(4):1925–1931. https://doi.org/10.1021/acs.analchem.1c04004
Prabhu GRD, Williams ER, Wilm M, Urban PL (2023) Mass spectrometry using electrospray ionization. Nat Rev Methods Primers 3(1):23. https://doi.org/10.1038/s43586-023-00203-4
Prakash C, Shaffer CL, Nedderman A (2007) Analytical strategies for identifying drug metabolites. Mass Spectrom Rev 26(3):340–369
Quinones MP, Kaddurah-Daouk R (2009) Metabolomics tools for identifying biomarkers for neuropsychiatric diseases. Neurobiol Dis 35(2):165–176
Raffaelli A, Saba A (2003) Atmospheric pressure photoionization mass spectrometry. Mass Spectrom Rev 22(5):318–331. https://doi.org/10.1002/mas.10060
Rajesh R et al (2023) Dysregulation of metabolic pathways in pulmonary fibrosis. Pharmacol Ther 246:108436. https://doi.org/10.1016/j.pharmthera.2023.108436
Rakk D, Kukolya J, Škrbić BD, Vágvölgyi C, Varga M, Szekeres A (2023) Advantages of multiplexing ability of the orbitrap mass analyzer in the multi-mycotoxin analysis. Toxins 15(2):134. https://doi.org/10.3390/toxins15020134
Ren JL, Zhang AH, Kong L, Wang XJ (2018) Advances in mass spectrometry-based metabolomics for investigation of metabolites. RSC Adv 8(40):22335–22350
Rinehart KL (1982) Fast atom bombardment mass spectrometry. Science 218(4569):254–260. https://doi.org/10.1126/science.218.4569.254
Roberts LD, Souza AL, Gerszten RE, Clish CB (2012) Targeted metabolomics. Curr Protoc Mol Biol 30(32):31–30.32.24
Robertson DG, Watkins PB, Reily MD (2011) Metabolomics in toxicology: preclinical and clinical applications. Toxicol Sci 120(suppl_1):S146–S170
Roessner U, Luedemann A, Brust D, Fiehn O, Linke T, Willmitzer L, Fernie AR (2001) Metabolic profiling allows comprehensive phenotyping of genetically or environmentally modified plant systems. Plant Cell 13(1):11–29
Ryan D, Robards K (2006) Analytical chemistry considerations in plant metabolomics. Sep Purif Rev 35(4):319–356
Salek RM, Steinbeck C, Viant MR, Goodacre R, Dunn WB (2013) The role of reporting standards for metabolite annotation and identification in metabolomic studies. Gigascience 2(1):2047–217X
Scalbert A, Brennan L, Fiehn O, Hankemeier T, Kristal BS, van Ommen B et al (2009) Mass-spectrometry-based metabolomics: limitations and recommendations for future progress with particular focus on nutrition research. Metabolomics 5:435–458
Schröder D (2012) Applications of electrospray ionization mass spectrometry in mechanistic studies and catalysis research. Acc Chem Res 45(9):1521–1532. https://doi.org/10.1021/ar3000426
Segers K, Declerck S, Mangelings D, Heyden YV, Eeckhaut AV (2019) Analytical techniques for metabolomic studies: a review. Bioanalysis 11(24):2297–2318
Simmler C, Napolitano JG, McAlpine JB, Chen SN, Pauli GF (2014) Universal quantitative NMR analysis of complex natural samples. Curr Opin Biotechnol 25:51–59

Simon CD (2021). Volatile characterization: optimizing GC-MS methods to reveal the chemistry and aroma of savoury food ingredients. Doctoral dissertation, Wageningen University and Research

Sindelar RD (2024) Genomics, other "OMIC" technologies, precision medicine, and additional biotechnology-related techniques. In: Pharmaceutical biotechnology: fundamentals and applications. Springer, Cham, pp 209–254

Siuzdak G (2004) An introduction to mass spectrometry ionization. An excerpt from the expanding role of mass spectrometry in biotechnology, 2nd ed.; MCC Press: San Diego, 2005. JALA. Journal of the Association for Laboratory Automation 9(2):50–63. https://doi.org/10.1016/j.jala.2004.01.004

Soler C, Hamilton B, Furey A, James KJ, Mañes J, Picó Y (2006) Optimization of LC–MS/MS using triple quadrupole mass analyzer for the simultaneous analysis of carbosulfan and its main metabolites in oranges. Anal Chim Acta 571(1):1–11. https://doi.org/10.1016/j.aca.2006.04.033

Somogyi Á (2008) Mass spectrometry instrumentation and techniques. In: Medical applications of mass spectrometry. Elsevier, pp 93–140. https://doi.org/10.1016/B978-044451980-1.50008-2

Song X, Yang X, Narayanan R, Shankar V, Ethiraj S, Wang X et al (2020) Oral squamous cell carcinoma diagnosed from saliva metabolic profiling. Proc Natl Acad Sci 117(28):16167–16173. https://doi.org/10.1073/pnas.2001395117

Sullivan GL, Gallardo JD, Jones EW, Hollliman PJ, Watson TM, Sarp S (2020) Detection of trace sub-micron (nano) plastics in water samples using pyrolysis-gas chromatography time of flight mass spectrometry (PY-GCToF). Chemosphere 249:126179. https://doi.org/10.1016/j.chemosphere.2020.126179

Tamara S, Den Boer MA, Heck AJR (2022) High-resolution native mass spectrometry. Chem Rev 122(8):7269–7326. https://doi.org/10.1021/acs.chemrev.1c00212

Theodoridis GA, Gika HG, Want EJ, Wilson ID (2012) Liquid chromatography–mass spectrometry based global metabolite profiling: a review. Anal Chim Acta 711:7–16

Tischfield MA, Cederquist GY, Gupta ML Jr, Engle EC (2011) Phenotypic spectrum of the tubulin-related disorders and functional implications of disease-causing mutations. Curr Opin Genet Dev 21(3):286–294. https://doi.org/10.1016/j.gde.2011.01.003

Van Acker T, Theiner S, Bolea-Fernandez E, Vanhaecke F, Koellensperger G (2023) Inductively coupled plasma mass spectrometry. Nat Rev Methods Primers 3(1):52. https://doi.org/10.1038/s43586-023-00235-w

Van Treuren W (2020) Advancing microbial metrology to reveal microbe-microbe, microbe-host, and microbe-environment interactions. Stanford University, California

Van QN, Veenstra TD, Issaq HJ (2011) Metabolic profiling for the detection of bladder cancer. Curr Urol Rep 12:34–40

Varvarezos L, Costello JT, Long C, Achner AJ, Wagner R, Meyer M, Grychtol P (2021) Ionization—dissociation of methane in ultrashort 400 nm and 800 nm laser fields. Chem Phys Lett 775:138687. https://doi.org/10.1016/j.cplett.2021.138687

Venter A, Nefliu M, Graham Cooks R (2008) Ambient desorption ionization mass spectrometry. TrAC Trends Anal Chem 27(4):284–290. https://doi.org/10.1016/j.trac.2008.01.010

Want EJ, Masson P, Michopoulos F, Wilson ID, Theodoridis G, Plumb RS et al (2013) Global metabolic profiling of animal and human tissues via UPLC-MS. Nat Protoc 8(1):17–32

Warren CR (2013) Use of chemical ionization for GC–MS metabolite profiling. Metabolomics 9(S1):110–120. https://doi.org/10.1007/s11306-011-0346-8

Wei J, Buriak JM, Siuzdak G (1999) Desorption–ionization mass spectrometry on porous silicon. Nature 399(6733):243–246. https://doi.org/10.1038/20400

Wei JN, Belanger D, Adams RP, Sculley D (2019) Rapid prediction of electron–ionization mass spectrometry using neural networks. ACS Cent Sci 5(4):700–708. https://doi.org/10.1021/acscentsci.9b00085

Wilm M (2011) Principles of electrospray ionization. Mol Cell Proteomics 10(7):M111.009407. https://doi.org/10.1074/mcp.M111.009407

Wu H, Guo J, Chen S, Liu X, Zhou Y, Zhang X, Xu X (2013) Recent developments in qualitative and quantitative analysis of phytochemical constituents and their metabolites using liquid chromatography–mass spectrometry. J Pharm Biomed Anal 72:267–291

Yang Q, Ji H, Xu Z, Li Y, Wang P, Sun J, Fan X, Zhang H, Lu H, Zhang Z (2023) Ultra-fast and accurate electron ionization mass spectrum matching for compound identification with million-scale in-silico library. Nat Commun 14(1):3722. https://doi.org/10.1038/s41467-023-39279-7

Yu H, Xing S, Nierves L, Lange P, Huan T (2020) Fold-change compression: an unexplored but correctable quantitative bias caused by nonlinear electrospray ionization responses in untargeted metabolomics. Anal Chem 92(10):7011–7019. https://doi.org/10.1021/acs.analchem.0c00246

Zabalegui N, Manzi M, Depoorter A, Hayeck N, Roveretto M, Li C et al (2020) Atmos Chem Phys 20(10):6243. https://doi.org/10.5194/acp-20-6243-2020

Zahoor I, Rui B, Khan J, Datta I, Giri S (2021) An emerging potential of metabolomics in multiple sclerosis: a comprehensive overview. Cell Mol Life Sci 78:3181–3203

Zaima N, Hayasaka T, Goto-Inoue N, Setou M (2010) Matrix-assisted laser desorption/ionization imaging mass spectrometry. Int J Mol Sci 11(12):5040–5055. https://doi.org/10.3390/ijms11125040

Zhang A, Sun H, Wang P, Han Y, Wang X (2012a) Modern analytical techniques in metabolomics analysis. Analyst 137(2):293–300

Zhang A, Sun H, Wu X, Wang X (2012b) Urine metabolomics. Clin Chim Acta 414:65–69

Zhang XW, Li QH, Xu ZD, Dou JJ (2020) Mass spectrometry-based metabolomics in health and medical science: a systematic review. RSC Adv 10(6):3092–3104. https://doi.org/10.1039/c9ra08985c

Zhou W, Wang Y, Lu A, Zhang G (2016) Systems pharmacology in small molecular drug discovery. Int J Mol Sci 17(2):246

7 Utilizing HPLC for Efficient Metabolite Purification

Maryam Abbasi Tarighat, Abhishek Barnwal, Mohd Tariq, Prachi Mishra, Shipra Choudhary, Mukul Jain, and Gholamreza Abdi

M. A. Tarighat (✉)
Faculty of Nano and Bio Science and Technology, Persian Gulf University, Bushehr, Iran
e-mail: matarighat@pgu.ac.ir

A. Barnwal · P. Mishra
Department of Life Sciences, Parul Institute of Applied Sciences, Parul University, Vadodara, Gujarat, India

M. Tariq
Department of Life Sciences, Parul Institute of Applied Sciences, Parul University, Vadodara, Gujarat, India

Department of Biotechnology, Graphic Era (Deemed to be University), Uttarakhand, Dehradun, India

S. Choudhary
Department of Biotechnology and Microbiology, Meerut Institute of Engineering and Technology, Meerut, India

M. Jain
Department of Life Sciences, Parul Institute of Applied Sciences, Parul University, Vadodara, Gujarat, India

Cell & Developmental Biology Lab, Centre of Research for Development, Parul University, Vadodara, Gujarat, India

G. Abdi
Department of Biotechnology, Persian Gulf Research Institute, Persian Gulf University, Bushehr, Iran
e-mail: abdi@pgu.ac.ir

V. Singh (ed.), *Advances in Metabolomics*,
https://doi.org/10.1007/978-981-97-7459-3_7

7.1 Introduction

7.1.1 The Significance of Metabolites: Understanding Cellular Chemistry

Metabolites are small molecules that play crucial roles in the intricate biochemical networks that sustain life (Patti et al. 2012). This sentence emphasizes the significance of metabolites, low-molecular-weight products of cellular metabolic processes (Bouatra et al. 2013). These small molecules serve three critical functions within cells: providing the building blocks for more complex structures, fueling cellular activities, and acting as signaling molecules that coordinate cellular responses (Bouatra et al. 2013). Unveiling the identities and concentrations of metabolites within a cell or organism, a process known as metabolite profiling, equips researchers with powerful tools to gain deep understanding of an organism's physiological state (Dettmer et al. 2007). This information sheds light on the underlying mechanisms of diseases and offers promising targets for therapeutic interventions (Dettmer et al. 2007).

7.1.1.1 The Metabolome: A Biochemical Snapshot

The term "metabolome" refers to the entire collection of small molecule metabolites present within a biological system at a particular moment and under defined environmental conditions (Fiehn 2002). This dynamic landscape is constantly in flux, influenced by a variety of factors including genetics, epigenetic modifications, and the surrounding environment (Patti et al. 2012). By analyzing the metabolome, researchers gain valuable insights into the functional state of the cell. The metabolome essentially acts as a functional readout, integrating the contributions of an organism's genetic makeup (genome), its gene expression patterns (transcriptome), and the proteins it produces (proteome) (Dettmer et al. 2007). Unlike the genome, which remains relatively static throughout an organism's lifetime, the metabolome is highly sensitive to perturbations, making it an invaluable tool for studying biological processes and disease states (Patti et al. 2012). By analyzing the metabolic profiles of cells, tissues, or biofluids, researchers can elucidate the interplay between genetic and environmental factors, unraveling the underlying mechanisms that govern cellular metabolism (Dettmer et al. 2007).

7.1.1.2 Metabolite Profiling: Unraveling Cellular Biochemistry

Metabolite profiling, also known as metabolomics, has emerged as a powerful analytical tool for comprehensively characterizing the complete set of metabolites present within a biological system (Dettmer et al. 2007). This field is rapidly evolving, and researchers leverage sophisticated analytical techniques like high-performance liquid chromatography (HPLC), mass spectrometry (MS), and nuclear magnetic resonance (NMR) spectroscopy to generate high-resolution metabolic profiles, providing a detailed picture of a cell's metabolite landscape (Patti et al. 2012).

The process begins by extracting the metabolites of interest from a biological sample, such as cells, tissues, or even biofluids (Stavru et al. 2021). This extraction is akin to collecting all the residents from the cellular neighborhood. Once extracted, the metabolites are introduced into the HPLC system. Here, a specialized column acts like a sophisticated sorting station (Stavru et al. 2021). The metabolites are separated based on their unique physicochemical properties, such as polarity, electrical charge, and how readily they interact with water. Imagine marbles of different sizes and colors rolling down an inclined plane; those with similar properties will group together during separation.

Following separation, the individual metabolites are detected and quantified using various methods (Dettmer et al. 2007). Mass spectrometry (MS) is a popular choice, acting like a high-precision scale that weighs each metabolite, revealing its abundance. Other techniques like ultraviolet-visible (UV-Vis) spectroscopy or fluorescence detectors can also be employed for detection.

The beauty of HPLC-based metabolite profiling lies in its ability to provide a detailed picture of the cellular metabolome. By separating, identifying, and quantifying the metabolites present, researchers gain valuable insights into various cellular processes. The specific types and abundance of metabolites can reveal information about energy production, waste removal, and signaling pathways within the cell. A healthy cell will have a distinct metabolic profile compared to a diseased cell, where disruptions in these processes may lead to altered metabolite levels. Therefore, HPLC-based metabolite profiling holds immense potential for disease diagnosis, as well as the development of targeted therapeutic strategies.

When it comes to metabolite profiling, HPLC-MS stands out as a particularly powerful technique due to its exceptional sensitivity, selectivity, and its capacity to elucidate the structures of the identified metabolites (Stavru et al. 2021). This approach strategically combines the separation capabilities of HPLC with the high-resolution power of mass spectrometry (MS). This marriage of techniques allows researchers to not only identify a wide range of metabolites but also quantify their abundance within a sample (Zhou et al. 2012a, b). The identification process relies on two key parameters: retention times during HPLC separation and the mass-to-charge ratio (m/z) of each metabolite as measured by MS (Zhou et al. 2012a, b).

HPLC-MS doesn't stop at just identification and quantification. The incorporation of tandem mass spectrometry (MS/MS) techniques, such as collision-induced dissociation (CID) or higher-energy collisional dissociation (HCD), empowers researchers to delve deeper. These MS/MS techniques fragment the identified metabolites, generating unique patterns that provide clues about their structures (Zhou et al. 2012a, b). Imagine a detective analyzing fingerprints at a crime scene. HPLC-MS is like identifying the individual fingerprints, while MS/MS is akin to analyzing the fingerprint patterns to reveal details about the individual who left them.

The applications of HPLC-based metabolomics extend far beyond the realm of basic research (Stavru et al. 2021). This powerful approach has found widespread use in various fields, including the discovery of biomarkers for disease diagnosis, the development of new drugs, environmental monitoring, and research on plants and microbes (Stavru et al. 2021). By providing comprehensive metabolite profiles, HPLC-MS has significantly contributed to advancements in several areas. Researchers can utilize this technique to elucidate metabolic pathways, identify potential biomarkers for diseases, and characterize how organisms respond to environmental stressors or genetic modifications (Dettmer et al. 2007). In essence,

HPLC-MS acts as a powerful tool for unraveling the intricate language of metabolites, offering valuable insights into cellular function, disease development, and environmental impact. Mass spectrometry-based metabolomics platforms offer unparalleled sensitivity and resolution, enabling the detection and identification of a wide range of metabolites (Zhou et al. 2012a, b). By coupling MS with various separation techniques, such as liquid chromatography (LC) or gas chromatography (GC), researchers can resolve complex metabolite mixtures and quantify individual metabolites with high accuracy (Dettmer et al. 2007).

NMR spectroscopy, on the other hand, provides a complementary approach to metabolite profiling, offering a non-destructive and highly reproducible method for simultaneously detecting and quantifying a broad range of metabolites in complex biological samples (Markley et al. 2017). NMR-based metabolomics has proven particularly valuable in studying metabolic fluxes, elucidating reaction mechanisms, and investigating metabolite-protein interactions (Markley et al. 2017).

By integrating data from multiple analytical platforms, researchers can obtain a comprehensive view of the metabolome, capturing a wide range of metabolites with diverse physicochemical properties (Patti et al. 2012). This multi-platform approach enhances the coverage and reliability of metabolite identification and quantification, enabling a deeper understanding of cellular biochemistry (Zhou et al. 2012a, b).

7.1.1.3 Metabolic Pathways: The Intricate Network of Life

Metabolic pathways are the intricate networks of biochemical reactions that govern the flow of matter and energy within cells (Rohrig and Viksna 2022). These interconnected series of enzymatic reactions transform substrates into products, enabling the synthesis of essential biomolecules, the generation of energy, and the elimination of waste products (Rohrig and Viksna 2022).

The study of metabolic pathways is crucial for understanding cellular metabolism and its role in maintaining homeostasis (Karp et al. 2021). By mapping the metabolites onto known metabolic pathways, researchers can elucidate the biochemical mechanisms underlying various biological processes, such as energy production, biosynthesis, and signal transduction (Karp et al. 2021).

Metabolic pathway analysis can reveal insights into the regulatory mechanisms that govern metabolic fluxes, identifying key enzymes, cofactors, and rate-limiting steps (Karp et al. 2021). This knowledge is essential for developing therapeutic interventions targeting specific metabolic pathways implicated in diseases, such as cancer, metabolic disorders, and neurological conditions (Karp et al. 2021).

The analysis of metabolic pathways plays a crucial role in systems biology, a field dedicated to unraveling the complex interplay between various cellular components and their collective behavior (Rohrig and Viksna 2022). By strategically merging metabolomics data with information from other omics fields, such as transcriptomics (gene expression) and proteomics (protein analysis), researchers can construct comprehensive models of cellular metabolism (Karp et al. 2021). These models offer a deeper understanding of the intricate regulatory networks that govern cellular processes, providing valuable insights into how these networks influence health and disease.

7.1.1.4 Metabolites as Biomarkers: Unveiling Disease Mechanisms

Insights into disease processes and prospective treatment targets can be gained via the use of metabolites, which have become strong biomarkers for a variety of disorders (Beger et al. 2022). Before clinical symptoms appear, dysregulated cellular processes might be reflected by altered metabolite levels, which can act as early indications of illness development (Beger et al. 2022). In cancer research, metabolomics has revealed distinct metabolic signatures associated with various tumor types, reflecting the reprogrammed metabolism that supports rapid cell growth and proliferation (Patel and Ahmed 2015). For instance, elevated levels of lactate, a byproduct of aerobic glycolysis (the Warburg effect), have been observed in many cancers, indicating a shift toward glycolytic metabolism (Patel and Ahmed 2015).

The pathophysiology of metabolic diseases as obesity, diabetes, and nonalcoholic fatty liver disease (NAFLD) has also been clarified using metabolomics (Serlie and Kalkhoven 2021). Variations in the amounts of metabolites, such as carbs, lipids, and amino acids, can indicate changes in insulin signaling, lipid homeostasis, and energy metabolism, offering important information about the course of the illness and possible targets for treatment (Serlie and Kalkhoven 2021).Metabolomics has identified dysregulated metabolic pathways including energy generation, oxidative stress, and amino acid metabolism in neurodegenerative illnesses like Parkinson's and Alzheimer's (Wilkins and Preshad 2021). According to Wilkins and Preshad (2021), these metabolic disruptions could be a factor in neuronal malfunction and death, providing prospective biomarkers for early diagnosis and tracking the course of the disease. Furthermore, metabolomics has proven invaluable in identifying metabolic signatures associated with environmental exposures, nutritional status, and gut microbiome composition (Bedia 2022; Deng et al. 2019). These findings have implications for understanding the complex interplay between environmental factors, metabolism, and health outcomes, paving the way for personalized interventions and precision medicine (Bedia 2022; Deng et al. 2019).

7.1.1.5 Metabolite Interactions: Navigating the Complexity of Cellular Networks

Metabolites do not function in isolation; rather, they engage in intricate interactions with other biomolecules, such as proteins, lipids, and nucleic acids (Pervushin et al. 2022). These interactions play crucial roles in regulating cellular processes, modulating enzyme activities, and orchestrating signaling cascades (Pervushin et al. 2022).

Metabolite-protein interactions are particularly important in cellular signaling pathways, where metabolites can act as allosteric regulators, binding to specific protein domains and inducing conformational changes that modulate enzyme activity or protein-protein interactions (Pervushin et al. 2022). For example, the metabolite acetyl-CoA can directly bind to and activate the histone acetyltransferase enzyme, influencing gene expression and chromatin remodeling (Pervushin et al. 2022).

Metabolite-lipid interactions are also vital for maintaining membrane integrity, energy homeostasis, and signal transduction (Pervushin et al. 2022). Lipids can serve as reservoirs for metabolic intermediates, modulating their bioavailability and distribution within cells (Pervushin et al. 2022). Additionally, metabolites can influence lipid metabolism, with implications for various pathological conditions, such as metabolic disorders and neurodegeneration (Pervushin et al. 2022).

Metabolite-nucleic acid interactions play crucial roles in epigenetic regulation and gene expression (Pervushin et al. 2022). Certain metabolites, such as acetyl-CoA, *S*-adenosylmethionine (SAM), and α-ketoglutarate, serve as cofactors for enzymes involved in DNA and histone modifications, modulating chromatin structure and transcriptional activity (Pervushin et al. 2022). These interactions provide fresh insights into the control of gene expression and the emergence of illnesses linked to epigenetic dysregulation by connecting cellular metabolism with epigenetic processes (Pervushin et al. 2022).

Understanding the intricacy of cellular processes and their consequences for health and illness requires deciphering the complex web of metabolite interactions (Pervushin et al. 2022). Researchers may trace the interactions between metabolites and other biomolecules and decipher the complex regulatory processes that control cellular activity by combining metabolomics data with other omics techniques, such as proteomics and interactomics (Pervushin et al. 2022).

7.1.1.6 Metabolic Engineering: Harnessing Cellular Chemistry

The goal of the quickly expanding discipline of metabolic engineering is to alter and optimize cellular metabolism for a variety of uses, such as the creation of useful chemicals and biofuels and the creation of medicinal treatments (Lee et al. 2012). By leveraging our understanding of metabolic pathways and metabolite interactions, researchers can rationally design and manipulate cellular systems to achieve desired outcomes (Lee et al. 2012).

Producing high-value molecules, such as medications, nutraceuticals, and fine chemicals, is one important use of metabolic engineering (Lee et al. 2012). By introducing or modifying specific enzymes or metabolic pathways in microbial hosts, researchers can redirect cellular metabolism toward the efficient synthesis of target molecules (Lee et al. 2012). The synthesis of several substances, such as vitamins, biofuels, and antibiotics, has been accomplished with success using this method (Lee et al. 2012). Metabolic engineering also holds promise in the development of cell-based therapies for treating metabolic disorders and other diseases (Dwivedee et al. 2021). By modulating metabolic pathways implicated in disease pathogenesis, researchers can develop genetically engineered cells or tissues that exhibit enhanced therapeutic properties or improved tolerance to metabolic stress (Dwivedee et al. 2021).

Furthermore, metabolic engineering principles can be applied to enhance crop productivity and improve stress tolerance in plants (Nouri and Komatsu 2022). By manipulating metabolic pathways involved in nutrient assimilation, biomass production, and stress responses, researchers can develop crop varieties with improved yields, enhanced nutritional value, and increased resilience to environmental stressors (Nouri and Komatsu 2022).

The triumphs of metabolic engineering hinge on a comprehensive understanding of a cell's metabolic machinery, the intricate pathways metabolites traverse, and how these metabolites interact with each other (Lee et al. 2012). Researchers are increasingly leveraging the power of omics, the study of large sets of biological molecules, to achieve this goal. By strategically integrating metabolomics data (focused on metabolites) with information from other omics fields and employing computational modeling techniques, researchers can gain a deeper understanding of the complex regulatory networks that govern cellular metabolism (Lee et al. 2012). This newfound knowledge empowers them to design more precise and efficient metabolic engineering strategies.

7.1.1.7 Conclusion

Within the intricate choreography of cellular biochemistry, metabolites reign supreme as the fundamental currency of life (Patti et al. 2012). These ubiquitous small molecules act as the building blocks for more complex structures, fuel cellular processes through energy transduction, and function as signaling molecules to orchestrate cellular homeostasis (Dettmer et al. 2007). Deciphering the metabolic landscape, encompassing the complete repertoire of metabolites (the metabolome) and their intricate pathways (metabolic pathways), empowers researchers to elucidate the mechanisms underlying diverse biological phenomena. This knowledge sheds light on both healthy cellular function and the dysregulated pathways associated with disease states. Furthermore, it offers promising avenues for the identification of novel therapeutic targets (Dettmer et al. 2007).

Advanced analytical techniques, such as mass spectrometry and NMR spectroscopy, have revolutionized the field of metabolomics, granting researchers unparalleled resolution to comprehensively characterize the cellular metabolome (Patti et al. 2012; Markley et al. 2017). This newfound ability to map the intricate metabolic tapestry has paved the way for a more holistic understanding of cellular function through the integration of metabolomics with other omics disciplines like genomics, transcriptomics, and proteomics (Karp et al. 2021). By fostering a systems biology approach, researchers can now examine the intricate interplay between all cellular components, from the blueprint encoded in genes to the functional proteins and the small molecule metabolites that orchestrate cellular life. This comprehensive understanding holds immense promise for advancements in disease diagnosis, drug discovery, and the development of personalized medicine strategies. Moreover, the study of metabolite interactions with other biomolecules, such as proteins, lipids, and nucleic acids, has unveiled the intricate mechanisms by which metabolites modulate cellular processes, including enzyme activity, signal transduction, and epigenetic regulation (Pervushin et al. 2022). These discoveries have profound implications for our understanding of health and disease, as well as the development of targeted therapeutic interventions.

Metabolic engineering, a rapidly evolving field, harnesses our knowledge of cellular metabolism and metabolite interactions to modify and optimize cellular systems for various applications, ranging from the production of valuable compounds to the development of cell-based therapies and improved crop varieties (Lee et al. 2012; Dwivedee et al. 2021; Nouri and Komatsu 2022).

As our understanding of cellular chemistry continues to deepen, the significance of metabolites will only become more apparent. By unraveling the intricate tapestry of metabolic pathways and metabolite interactions, we can unlock the secrets of life itself, paving the way for groundbreaking discoveries in medicine, biotechnology, and our understanding of the natural world.

7.1.2 Unveiling the Metabolome: Challenges in Metabolite Purification

The whole collection of tiny molecules found in a biological system, or metabolome, is a treasure trove of data that provides priceless insights into the complex biochemical processes that keep life possible (Patti et al. 2012). Researchers are now able to clarify disease processes, find biomarkers, and discover possible treatment targets by using metabolomics, also known as metabolite profiling, to sort through the complexity of the metabolome (Dettmer et al. 2007). But there are several obstacles in the way of fully understanding the range of metabolites, especially when it comes to metabolite purification (Tian et al. 2021).

7.1.2.1 The Complexity of Biological Matrices

Biological samples are intrinsically complex, comprising a wide range of metabolites with different physicochemical characteristics. These samples include cells, tissues, biofluids, and environmental matrices (Dettmer et al. 2007; Tian et al. 2021). These metabolites are found in complex mixes with other biomolecules such proteins, lipids, and nucleic acids, and their quantities vary widely, covering many orders of magnitude (Chaleckis et al. 2019). The presence of these interfering components can pose significant challenges in metabolite purification, as they can co-extract or co-elute with the metabolites of interest, leading to ion suppression, matrix effects, and compromised analytical performance (Chaleckis et al. 2019). Furthermore, the diverse chemical nature of metabolites, including their varying polarity, solubility, and stability, necessitates the development of tailored purification strategies to ensure comprehensive metabolome coverage (Tian et al. 2021).

7.1.2.2 Sample Preparation: The Cornerstone of Metabolite Purification

Effective sample preparation is crucial for successful metabolite purification and subsequent analysis (Tian et al. 2021). The quality and dependability of the metabolomic data can be greatly impacted by each of the various processes that are usually included in this process, which include sample collection, extraction, and pretreatment (Patti et al. 2012).

7.1.2.3 Gathering and Preserving Samples

To protect the integrity of the metabolome and reduce degradation or changes in metabolite profiles, appropriate sample collection and storage procedures are crucial (Dettmer et al. 2007). The stability of metabolites can be influenced by several factors, including sample type, handling techniques, and storage conditions. Therefore, it is necessary to design standardized methods that are customized to particular sample matrices (Chaleckis et al. 2019).

For instance, biofluids like blood or urine require careful handling to prevent enzymatic degradation or microbial contamination, while tissue samples may require rapid quenching or snap-freezing to halt metabolic processes (Chaleckis et al. 2019). Additionally, long-term storage conditions, such as temperature and storage duration, can impact metabolite stability, necessitating appropriate measures to minimize degradation (Dettmer et al. 2007).

7.1.2.4 Metabolite Extraction

The extraction of metabolites from biological matrices is a critical step in metabolite purification, as it determines the efficiency and selectivity with which metabolites are isolated from the sample matrix (Tian et al. 2021). Various extraction techniques have been developed, including pressurized liquid extraction (PLE), solid-phase extraction (SPE), and liquid–liquid extraction (LLE) each with its unique strengths and limitations (Bedia et al. 2022).

A popular method known as "liquid–liquid extraction" (LLE) depends on the metabolites being divided between two immiscible solvents, such as water, and organic solvents, such as methanol or chloroform (Bedia et al. 2022). This approach can effectively separate metabolites based on their polarity and solubility, enabling the selective extraction of targeted metabolite classes (Bedia et al. 2022).

Solid-phase extraction (SPE) is another popular technique that utilizes solid sorbent materials to selectively retain and concentrate metabolites from complex matrices (Bedia et al. 2022). SPE is appropriate for high-throughput applications because to its many benefits, which include great selectivity, the capacity to eliminate interfering substances, and the possibility of automation (Bedia et al. 2022).

In order to improve the extraction efficiency of metabolites from solid materials, such as plant tissues or environmental matrices, pressurized liquid extraction (PLE), often referred to as accelerated solvent extraction (ASE), uses higher temperature and pressure (Bedia et al. 2022). This technique can improve extraction yields, reduce solvent consumption, and enable the extraction of thermally labile or tightly bound metabolites (Bedia et al. 2022).

The sample matrix, the target metabolite classes, and the analytical platforms to be used all influence the extraction process selection (Tian et al. 2021). Furthermore, optimizing extraction parameters is essential for increasing recovery of metabolites and reducing matrix effects. These factors include solvent composition, pH, temperature, and extraction duration (Bedia et al. 2022).

7.1.2.5 Sample Pretreatment and Cleanup

Following metabolite extraction, sample pretreatment and cleanup steps may be necessary to remove interfering compounds, concentrate analytes, and optimize sample compatibility with downstream analytical platforms (Tian et al. 2021). These steps can include techniques such as protein precipitation, evaporation and reconstitution, filtration, and solid-phase extraction (SPE) (Chaleckis et al. 2019).

Protein precipitation is a common pretreatment step for biofluids like plasma or serum, where proteins are selectively removed to prevent ion suppression and fouling of analytical instruments (Chaleckis et al. 2019). This can be achieved through the addition of organic solvents, such as acetonitrile or methanol, or the use of precipitating agents like trichloroacetic acid (TCA) or perchloric acid (PCA) (Chaleckis et al. 2019).

Evaporation and reconstitution techniques are often employed to concentrate dilute samples or exchange solvents for better compatibility with analytical platforms (Tian et al. 2021). This process involves evaporating the solvent from the extracted sample, typically under vacuum or nitrogen flow, and reconstituting the residue in a suitable solvent or buffer (Chaleckis et al. 2019).

Filtration methods, such as syringe filters or centrifugal filters, can be used to remove particulates or high-molecular-weight contaminants from samples, improving analytical performance and preventing instrument fouling (Chaleckis et al. 2019).

Solid-phase extraction (SPE) can also serve as a pretreatment step, allowing for selective enrichment of metabolites, removal of interfering compounds, and desalting of samples (Tian et al. 2021). By carefully selecting the appropriate sorbent materials and elution solvents, SPE can enhance metabolite purification and improve analytical sensitivity (Bedia et al. 2022).

7.1.2.6 Analytical Platforms for Metabolite Purification

To enable thorough metabolome coverage and precise metabolite identification, the purification of metabolites is frequently combined with cutting-edge analytical methods (Patti et al. 2012). The sample matrix, the particular goals of the investigation, and the physicochemical characteristics of the target metabolites all influence the analytical platform selection (Tian et al. 2021).

7.1.2.7 Unveiling the Cellular Orchestra: A Deep Dive into LC-MS for Metabolite Profiling

Unraveling the intricate tapestry of cellular metabolites is a cornerstone of understanding health and disease. Liquid Chromatography-Mass Spectrometry (LC-MS) stands as a powerful analytical maestro, conducting a multi-step symphony to achieve this goal (Dettmer et al. 2007).

7.1.2.8 The Separation Act: A Masterful Conductor

The first act of the LC-MS symphony involves meticulous separation. Imagine a complex orchestra where all the instruments are jumbled together. LC, the "liquid chromatography" component, acts as a meticulous conductor, separating the metabolites based on their unique physicochemical properties (Stavru et al. 2021). These properties, like polarity (think oil vs. water) and charge, determine how strongly a metabolite interacts with the stationary phase within the LC column. This interaction dictates how long each metabolite spends traveling through the column, resulting in their separation (Dettmer et al. 2007). Different chromatographic modes, akin to using different musical scores, are employed for optimal separation (Stavru et al. 2021). Reversed-phase (RP) chromatography excels at separating non-polar metabolites like lipids, while hydrophilic interaction (HILIC) caters to highly polar ones like amino acids (Stavru et al. 2021).

7.1.2.9 The Mass Spectrometry Stage: Unveiling Identities

Following this separation act, the individual metabolites enter the mass spectrometry (MS) stage. Here, they are ionized, becoming electrically charged particles.

Imagine the MS as a high-precision scale that weighs these charged metabolites with exceptional sensitivity and selectivity (Dettmer et al. 2007). This precise measurement of the mass-to-charge ratio (*m/z*) acts as a unique fingerprint for each metabolite, allowing researchers to identify them (Stavru et al. 2021).

7.1.2.10 The Encore: Unveiling Structures with Tandem MS

But the symphony does not end there. Tandem MS (MS/MS) techniques take center stage for an encore performance (Zhou et al. 2012a, b). These techniques fragment the ionized metabolites, generating a characteristic pattern of smaller fragments. Think of this fragmentation as breaking down a complex musical chord into its individual notes. By analyzing these fragmentation patterns, researchers can gain insights into the structure of the original metabolite, offering valuable information beyond simple identification (Zhou et al. 2012a, b).

7.1.2.11 The Power of the Symphony: Unveiling the Cellular Metabolome

In essence, LC-MS acts as a powerful tool for untangling the complex web of metabolites within a cell. Through its masterful separation techniques and high-resolution mass spectrometry, LC-MS provides an unparalleled view of the cellular metabolome (Dettmer et al. 2007). This knowledge is instrumental in various fields, from deciphering the mechanisms of disease to developing targeted therapeutic strategies. By understanding the intricate dance of metabolites within the cellular orchestra, researchers can compose new melodies for promoting health and combating diseases.

7.1.2.12 Gas Chromatography–Mass Spectrometry (GC-MS)

Another popular analytical tool for metabolite profiling is gas chromatography combined with mass spectrometry (GC-MS), which is especially useful for the study of volatile and thermally stable metabolites (Dettmer et al. 2007). According to their boiling points and interactions with the GC column's stationary phase, metabolites are separated in GC-MS (Dettmer et al. 2007). GC-MS excels in the analysis of metabolites such as fatty acids, amino acids, and organic acids, offering high chromatographic resolution and efficient separation of isomeric compounds (Dettmer et al. 2007). However, to facilitate GC-MS analysis, metabolites often require derivatization to improve their volatility and thermal stability, which can introduce additional challenges in sample preparation and data interpretation (Dettmer et al. 2007).

7.1.2.13 Nuclear Magnetic Resonance (NMR) Spectroscopy

Unlike some techniques, nuclear magnetic resonance (NMR) spectroscopy stands out as a non-destructive and reproducible workhorse in metabolite profiling (Markley et al. 2017). Not only can it analyze a broad range of metabolites simultaneously, but it also offers the unique advantage of revealing their detailed structures without the need for extensive sample preparation or chemical modifications often required by other methods. This makes NMR a powerful tool for comprehensively characterizing the cellular metabolome. NMR spectroscopy is particularly valuable for studying metabolic fluxes, elucidating reaction mechanisms, and investigating metabolite-protein interactions (Markley et al. 2017). However, NMR-based metabolomics is generally less sensitive than MS-based techniques, limiting its application in the analysis of low-abundance metabolites or highly complex samples (Markley et al. 2017).

7.1.2.14 Mass Spectrometry Using Capillary Electrophoresis (CE-MS)

A new analytical tool for metabolite profiling is capillary electrophoresis combined with mass spectrometry (CE-MS), which is especially useful for the study of highly charged and polar metabolites (Ramautar et al. 2018). For the investigation of metabolites such amino acids, nucleotides, and carbohydrates, CE-MS is a useful method because it has excellent separation efficiency, minimal sample consumption, and the capacity to resolve isomeric and isobaric molecules (Ramautar et al. 2018).

With the use of CE-MS, metabolites may be separated according to their charge, size, and hydrophobicity using a variety of modes, such as capillary zone electrophoresis (CZE), micellar electrokinetic chromatography (MEKC), and capillary electrochromatography (CEC) (Ramautar et al. 2018). Nevertheless, in comparison to LC-MS and GC-MS, CE-MS is often less reliable and more vulnerable to matrix effects, requiring meticulous sample preparation and separation condition adjustment (Ramautar et al. 2018).

7.1.2.15 Multi-platform Approaches and Data Integration

Given the diverse physicochemical properties of metabolites and the inherent limitations of individual analytical platforms, the integration of data from multiple platforms has become increasingly important for achieving comprehensive metabolome coverage (Patti et al. 2012; Tian et al. 2021). Researchers may identify and quantify metabolites more reliably and capture a larger variety of metabolites by combining the strengths of several methods, such as LC-MS, GC-MS, NMR, and CE-MS (Zhou et al. 2012a, b).

Data integration and fusion from multiple analytical platforms pose significant challenges in terms of data processing, alignment, and normalization (Patti et al. 2012). However, advances in bioinformatics and computational approaches have facilitated the development of powerful data integration tools and workflows, enabling researchers to leverage the complementary information provided by different analytical platforms (Patti et al. 2012; Tian et al. 2021).

7.1.2.16 Computational Approaches for Metabolite Identification and Annotation

In order to evaluate metabolic profiles and clarify biological findings, it is essential for metabolomics research to identify and annotate metabolites (Domingo-Almenara et al. 2022). However, metabolite identification and annotation are severely hampered by the great chemical variety of metabolites and the incomplete coverage of databases of metabolites that are made accessible to the public (Domingo-Almenara et al. 2022).

7.1.2.17 Metabolite Databases and Spectral Libraries

The availability of comprehensive metabolite databases and spectral libraries is essential for accurate metabolite identification and annotation (Domingo-Almenara et al. 2022). These resources provide reference information, such as chemical structures, molecular formulas, and spectral data (e.g., mass spectra, retention times, NMR spectra), enabling the comparison and matching of experimental data with known metabolites (Domingo-Almenara et al. 2022).

Popular metabolite databases include the Human Metabolome Database (HMDB), the Kyoto Encyclopedia of Genes and Genomes (KEGG), and the Lipid Maps Structure Database (LMSD), among others (Domingo-Almenara et al. 2022). Additionally, spectral libraries, such as the NIST Mass Spectral Library and the Global Natural Products Social Molecular Networking (GNPS) library, provide valuable resources for metabolite identification based on mass spectral data (Domingo-Almenara et al. 2022).

7.1.2.18 *In Silico* Metabolite Annotation and Dereplication

With the increasing complexity of metabolomics data and the vast number of unknown metabolites, computational approaches have become indispensable for metabolite annotation and dereplication (Domingo-Almenara et al. 2022). *In silico* metabolite annotation involves the prediction of metabolite structures and identities based on experimental data, such as accurate mass measurements, isotope patterns, and fragmentation spectra (Domingo-Almenara et al. 2022).

Various software tools and algorithms have been developed for *in silico* metabolite annotation, including molecular formula generators, structure generators, and spectral matching algorithms (Domingo-Almenara et al. 2022). These tools leverage computational methods, such as combinatorial chemistry, machine learning, and pattern recognition, to propose candidate structures and identities for unknown metabolites (Domingo-Almenara et al. 2022).

Dereplication is another computational approach used to rapidly identify known metabolites in complex mixtures, reducing the number of unknowns that require further characterization (Domingo-Almenara et al. 2022). Dereplication algorithms compare experimental data, such as accurate mass, fragmentation patterns, retention time against databases of known metabolites, enabling the rapid identification of previously reported compounds (Domingo-Almenara et al. 2022).

7.1.2.19 Molecular Networking and Substructure Recognition

To view and annotate metabolites in complicated mixtures, molecular networking is a potent computational technique that makes use of the ideas of molecular similarity and structural relatedness (Wang et al. 2016). Using this method, molecular networks are created, with nodes standing in for particular metabolites and edges for their structural similarity (Wang et al. 2016). By analyzing the patterns and clusters within these molecular networks, researchers can identify metabolite families, recognize substructures and chemical motifs, and facilitate the annotation of unknown metabolites based on their structural relationships with known compounds (Wang et al. 2016). Molecular networking has proven particularly valuable in the analysis of natural product mixtures, enabling the rapid dereplication and prioritization of novel metabolites for further characterization (Wang et al. 2016). Substructure recognition algorithms complement molecular networking by identifying specific structural motifs or substructures within metabolites (Domingo-Almenara et al. 2022). These algorithms can facilitate the annotation of metabolites by recognizing characteristic fragmentation patterns, functional groups, or substructural features, providing valuable insights into the chemical nature and potential biological activities of the metabolites (Domingo-Almenara et al. 2022).

7.1.2.20 Machine Learning and Artificial Intelligence for Metabolite Annotation

There are now more options for metabolite annotation and identification thanks to the quick development of machine learning and artificial intelligence (AI) techniques (Domingo-Almenara et al. 2022). In order to predict and annotate novel metabolites, these computational methods make use of massive datasets and sophisticated algorithms to identify patterns and links between experimental data and known metabolite structures (Domingo-Almenara et al. 2022). Metabolite annotation challenges have demonstrated significant potential for deep learning architectures, including convolutional neural networks (CNNs) and recurrent neural networks (RNNs) (Domingo-Almenara et al. 2022). These models can learn to recognize patterns in mass spectral data, retention times, and other experimental features, enabling the prediction of metabolite identities and the discovery of novel metabolites (Domingo-Almenara et al. 2022).

Additionally, natural language processing (NLP) techniques have been applied to metabolite annotation, leveraging the wealth of information contained in scientific literature and text-based databases (Domingo-Almenara et al. 2022). NLP models can assist in the annotation of metabolites, make it easier to find new metabolites, and offer insights into the possible biological functions and activities of metabolites by extracting and mining pertinent information from textual sources (Domingo-Almenara et al. 2022).

The integration of AI techniques and Machine Learning with traditional computational approaches, such as molecular networking and substructure recognition, holds great promise for enhancing metabolite annotation and identification capabilities (Domingo-Almenara et al. 2022). But the availability of solid validation methodologies, high-quality training data, and the capacity to decipher and explain the models' predictions are all necessary for the effective use of these approaches (Domingo-Almenara et al. 2022).

7.1.2.21 Challenges and Future Perspectives

Despite the significant advances in metabolite purification and analytical techniques, several challenges remain in the quest to unveil the full complexity of the metabolome (Tian et al. 2021). These challenges include the vast chemical diversity of metabolites, the presence of isomeric and isobaric compounds, the wide dynamic range of metabolite concentrations, and the impact of matrix effects on analytical performance (Tian et al. 2021).

One of the major challenges in metabolite purification is the efficient separation and isolation of isomeric and isobaric metabolites (Tian et al. 2021). These compounds share identical molecular formulas and often exhibit similar physicochemical properties, making their resolution and identification particularly challenging (Tian et al. 2021). Advanced separation techniques, such as ion mobility spectrometry (IMS) and multidimensional chromatography, offer promising solutions for resolving isomeric and isobaric metabolites, but their implementation and data analysis remain complex (Tian et al. 2021).

Another significant challenge is the wide dynamic range of metabolite concentrations in biological samples, spanning several orders of magnitude (Tian et al. 2021). This dynamic range can exceed the linear range of most analytical instruments, necessitating the development of strategies for metabolite enrichment, depletion of highly abundant species, or the use of multiple analytical platforms with complementary detection capabilities (Tian et al. 2021).

Matrix effects, caused by the co-elution of interfering compounds from complex biological matrices, can also significantly impact analytical performance and quantitative accuracy (Tian et al. 2021). Careful sample preparation, cleanup procedures, and the use of appropriate internal standards or matrix-matched calibration strategies are essential to mitigate these effects and ensure reliable quantitative data (Tian et al. 2021).

Future perspectives in metabolite purification and metabolomics will likely involve the integration of advanced separation techniques, such as multidimensional chromatography, ion mobility spectrometry, and microfluidic devices, with high-resolution mass spectrometry and complementary analytical platforms (Tian et al. 2021). Additionally, the development of automated sample preparation workflows, miniaturized analytical systems, and high-throughput data acquisition strategies will be crucial for enhancing metabolome coverage and enabling large-scale metabolomics studies (Tian et al. 2021).

The future of metabolite profiling hinges on the powerful marriage of cutting-edge computational tools and a holistic omics approach (Patti et al. 2012; Tian et al. 2021; Domingo-Almenara et al. 2022). Machine learning, artificial intelligence, and molecular networking techniques will revolutionize metabolite identification, annotation, and even the discovery of entirely new metabolites. Furthermore, integrating metabolomics data with information from genomics, transcriptomics, and proteomics will paint a more comprehensive picture of biological systems, allowing researchers to unravel the intricate regulatory networks that govern cellular function. This combined approach holds immense promise for advancements in various fields. Collaborations and interdisciplinary efforts among analytical chemists, biologists, bioinformaticians, and computational scientists will be essential to address the challenges in metabolite purification and metabolomics, ultimately unveiling the full complexity of the metabolome and its implications for biological research and healthcare applications (Tian et al. 2021).

7.1.3 HPLC as a Powerful Ally: Advantages for Metabolite Purification

The pursuit of unveiling the intricate tapestry of the metabolome has been a driving force in the field of metabolomics, fueling the development of powerful analytical techniques and purification strategies (Patti et al. 2012). Among these, high-performance liquid chromatography (HPLC) has emerged as a versatile and indispensable tool, offering numerous advantages for the purification and analysis of a wide range of metabolites (Stavru et al. 2021). This chapter delves into the strengths and capabilities of HPLC, exploring its pivotal role in metabolite purification and its contributions to advancing our understanding of cellular biochemistry.

7.1.3.1 HPLC: A Versatile Separation Technique

High-performance liquid chromatography (HPLC) stands out as a powerful tool for separating the intricate world of cellular metabolites (Stavru et al. 2021; Zhou et al. 2012a, b). It works by exploiting the unique properties of each metabolite, such as polarity and charge. Imagine a moving stream (mobile phase) carrying a mixture of metabolites, which interact differently with a stationary phase within the column. Based on these interactions, some metabolites travel through the column faster than others, achieving a clean separation of the complex mixture. The beauty of HPLC lies in its versatility. By employing different separation modes, researchers can tailor this technique to effectively isolate a wide range of metabolites present in biological samples.

7.1.3.2 Reversed-Phase Liquid Chromatography (RPLC)

Within the vast toolbox of metabolite profiling techniques, reversed-phase liquid chromatography (RPLC) emerges as a dominant force, particularly adept at separating non-polar and moderately polar metabolites (Stavru et al. 2021). This technique acts like a maestro, meticulously orchestrating the separation of these metabolites based on their unique interactions with the chromatographic environment (Zhou et al. 2012a, b).

The underlying principle of RPLC revolves around the concept of "like dissolves like." The stationary phase, the heart of the separation process, is meticulously crafted to be non-polar. Imagine tiny silica beads coated with long, greasy chains (e.g., C18 or C8)—akin to a reversed oil slick (Zhou et al. 2012a, b). The mobile phase, on the other hand, is a dynamic duo of polar solvents like water and methanol, often blended with non-polar solvents like hexane or isopropanol (Zhou et al. 2012a, b).

The magic of RPLC unfolds as the mobile phase, carrying the complex mixture of metabolites, navigates through the stationary phase. Metabolites with a higher affinity for grease (more hydrophobic) experience a stronger attraction to the non-polar stationary phase. Think of these metabolites as eagerly clinging to the oil slick. Consequently, they spend a longer time interacting with the stationary phase, resulting in a slower elution (emergence) from the column (Zhou et al. 2012a, b). Conversely, metabolites with a greater affinity for the polar mobile phase (more hydrophilic) experience a weaker attraction to the stationary phase. Imagine them zipping past the oil slick with minimal interaction. This translates to a faster elution time (Stavru et al. 2021).

By carefully manipulating the composition of the mobile phase (the polarity of the solvent mixture), researchers can effectively control the strength of these interactions, fine-tuning the separation process (Zhou et al. 2012a, b). This tailored approach makes RPLC a powerful tool for analyzing a specific range of metabolites within the cellular orchestra. It excels at separating lipids, steroids, and other lipophilic (fat-loving) compounds that play crucial roles in cellular function (Stavru et al. 2021).

In essence, RPLC offers a versatile and selective separation technique for non-polar and moderately polar metabolites. By exploiting the concept of differential partitioning based on hydrophobicity, RPLC allows researchers to isolate and identify these crucial cellular components, paving the way for a deeper understanding of cellular metabolism and its role in health and disease.

7.1.3.3 Hydrophilic Interaction Liquid Chromatography (HILIC)

While RPLC reigns supreme for non-polar metabolites, the cellular orchestra also harbors a diverse range of highly polar and ionizable players. Here, Hydrophilic Interaction Liquid Chromatography (HILIC) emerges as a complementary technique, specifically tailored for their separation (Stavru et al. 2021). In HILIC, the stationary phase becomes the hydrophilic counterpart, akin to a water-loving surface. This is achieved by attaching hydrophilic groups like diols, amides, or zwitterions to silica particles (Zhou et al. 2012a, b). The mobile phase, on the other hand, takes a dramatic shift, becoming predominantly organic solvents like acetonitrile or methanol, with a minimal amount of water buffer (Zhou et al. 2012a, b).

The magic of HILIC lies in the contrasting polarities of the stationary and mobile phases. Highly polar and ionizable metabolites, such as amino acids, carbohydrates, and nucleosides, establish stronger interactions with the hydrophilic stationary phase through mechanisms like hydrogen bonding (Stavru et al. 2021). Think of these metabolites readily sticking to the water-loving surface. Consequently, they spend a longer time interacting with the stationary phase, leading to a slower elution from the column (Zhou et al. 2012a, b). Conversely, less polar metabolites experience weaker interactions and flow through the column more rapidly.

By manipulating the organic solvent content in the mobile phase, researchers can fine-tune the strength of these interactions, achieving optimal separation for the desired polar metabolites (Zhou et al. 2012a, b). This targeted approach makes HILIC an invaluable tool for analyzing this specific class of cellular components. It plays a crucial role in deciphering the intricate metabolic pathways involving amino acids, sugars, and signaling molecules, all vital for cellular function (Stavru et al. 2021).

7.1.3.4 Ion-Exchange Chromatography (IEC)

Another HPLC mode that is particularly helpful for the separation of charged metabolites, such as organic acids, amino acids, and nucleotides, is ion-exchange chromatography (IEC) (Stavru et al. 2021). Depending on the charge of the analytes, anion-exchange chromatography (AEC) and cation-exchange chromatography (CEC) are two more subtypes of IEC (Zhou et al. 2012a, b).

According to Zhou et al. (2012a, b), the stationary phase in AEC is positively charged and usually made up of silica particles functionalized with quaternary ammonium groups. An aqueous buffer with variable pH and ionic strength makes up the mobile phase (Stavru et al. 2021). Negatively charged metabolites, such as organic acids, interact with the positively charged stationary phase through electrostatic interactions, leading to their separation based on charge density and molecular size (Zhou et al. 2012a, b).

According to Zhou et al. (2012a, b), the stationary phase in CEC is negatively charged and usually consists of silica particles functionalized with carboxylic or sulfonic acid groups. The mobile phase and separation principles are similar to those of AEC, but in this case, positively charged metabolites, such as amino acids and nucleotides, interact with the negatively charged stationary phase (Stavru et al. 2021).

7.1.3.5 Multidimensional HPLC Approaches

To enhance metabolite coverage and resolution, multidimensional HPLC approaches have been developed, combining different separation modes in a single analytical run (Stavru et al. 2021). These approaches include two-dimensional HPLC (2D-HPLC) and online heart-cutting techniques (Stavru et al. 2021).

In 2D-HPLC, two different HPLC columns with orthogonal separation mechanisms are coupled in series (Stavru et al. 2021). For example, a combination of RPLC and HILIC can be used to separate metabolites based on their hydrophobicity and polarity, respectively (Zhou et al. 2012a, b). This approach provides enhanced peak capacity and resolving power, enabling the separation of metabolites with diverse physicochemical properties (Stavru et al. 2021).

Online heart-cutting techniques involve the selective transfer of target analytes from the first HPLC column to a second column with different separation characteristics (Stavru et al. 2021). This approach allows for the focused analysis of specific metabolites or metabolite classes, improving sensitivity and resolution while reducing analysis time (Zhou et al. 2012a, b).

7.1.3.6 HPLC Coupled with Mass Spectrometry: A Powerful Combination

The integration of mass spectrometry (MS) detection with high-performance liquid chromatography (HPLC) has transformed the area of metabolomics by offering an effective analytical framework for thorough metabolite identification and profiling (Zhou et al. 2012a, b). High sensitivity, selectivity, and the capacity to collect structural information using tandem mass spectrometry (MS/MS) are just a few benefits of HPLC-MS (Dettmer et al. 2007).

7.1.3.7 High-Resolution Mass Spectrometry (HRMS)

The capabilities of HPLC-MS in metabolite analysis have been greatly improved with the introduction of high-resolution mass spectrometry (HRMS) (Dettmer et al. 2007). HRMS analyzers provide precise mass measurements with high resolving power and mass accuracy. Examples of these analyzers are time-of-flight (TOF) or Orbitrap mass spectrometers (Zhou et al. 2012a, b). This enables the precise determination of molecular formulae and the differentiation of isomeric and isobaric metabolites, which is crucial for reliable metabolite identification (Dettmer et al. 2007).

7.1.3.8 Tandem Mass Spectrometry (MS/MS)

In addition to accurate mass measurements, HPLC-MS/MS provides valuable structural information through fragmentation experiments (Dettmer et al. 2007). In "tandem mass spectrometry," precursor ions are isolated and fragmented to produce product ion spectra that are useful for metabolite identification and structural clarification (Zhou et al. 2012a, b).

Various fragmentation techniques, such as, higher-energy collisional dissociation (HCD), collision-induced dissociation (CID), and electron transfer dissociation (ETD), can be employed to induce targeted fragmentation and generate informative product ion spectra (Zhou et al. 2012a, b). These spectra can be searched against spectral libraries or used in combination with *in silico* fragmentation tools for metabolite annotation and identification (Domingo-Almenara et al. 2022).

7.1.3.9 Ion Mobility Spectrometry (IMS)

The ever-evolving landscape of metabolite analysis welcomes a powerful alliance: the integration of ion mobility spectrometry (IMS) with HPLC-MS (Zhou et al. 2012a, b). This innovative approach goes beyond traditional separation techniques, offering an extra layer of metabolite discrimination. Imagine HPLC-MS as a highly skilled duo separating a complex mixture of metabolites. Here, IMS acts as a third collaborator, introducing a new dimension of separation based on how quickly ions travel through a drift gas (Lanucara et al. 2014). This orthogonal separation adds another level of specificity, allowing researchers to distinguish between structurally similar metabolites that might otherwise appear identical in HPLC-MS analysis alone. By combining these techniques, scientists gain a more refined understanding of the cellular metabolite landscape, paving the way for advancements in metabolite identification and characterization. IMS-MS offers several advantages, including improved peak capacity, enhanced separation of isomeric and isobaric compounds, and the ability to measure collision cross-sections, which provide valuable structural information for metabolite identification (Lanucara et al. 2014). By coupling IMS with HPLC-MS, researchers can achieve better metabolite resolution, reduce ion suppression effects, and improve the overall confidence in metabolite assignments (Zhou et al. 2012a, b).

7.1.3.10 Advantages of HPLC for Metabolite Purification

HPLC offers numerous advantages for metabolite purification, contributing to its widespread adoption in the field of metabolomics (Stavru et al. 2021). These advantages include:

High Resolution and Selectivity

High resolution and selectivity in metabolite separation are provided by HPLC, making it possible to efficiently purify complex biological mixtures (Stavru et al. 2021). The versatility of HPLC modes, such as RPLC, HILIC, and IEC, allows for the tailored separation of metabolites based on their physicochemical properties, improving metabolite coverage and reducing matrix effects (Zhou et al. 2012a, b).

Compatibility with Various Sample Matrices

Numerous sample matrices may be used using HPLC, such as tissue extracts, environmental samples, and biofluids (such as plasma, urine, and cerebrospinal fluid) (Stavru et al. 2021). Because of its adaptability, HPLC may be used in a variety of contexts, including environmental monitoring, food and plant analysis, and biological research (Zhou et al. 2012a, b).

Quantitative Capabilities

HPLC offers accurate and reproducible quantification of metabolites, making it suitable for quantitative metabolomics studies (Stavru et al. 2021). By employing appropriate internal standards and calibration strategies, HPLC can provide reliable quantitative data for metabolites of interest, enabling the investigation of metabolic pathways and the identification of potential biomarkers (Zhou et al. 2012a, b).

Scalability and Automation

HPLC systems can be easily scaled and automated, facilitating high-throughput analysis and improving analytical efficiency (Stavru et al. 2021). Automated sample handling, injection systems, and data acquisition workflows enable the processing of large sample sets, which is essential for large-scale metabolomics studies and clinical applications (Zhou et al. 2012a, b).

Robustness and Reproducibility

Well-established HPLC methods and protocols offer robustness and reproducibility, ensuring consistent and reliable results across different laboratories and studies (Stavru et al. 2021). This is particularly important for metabolomics research, where data reproducibility and standardization are crucial for meaningful comparisons and data integration (Zhou et al. 2012a, b).

Complementarity with Other Analytical Techniques

HPLC can be coupled with various detection methods, such as Fluorescence, UV-Vis, and mass spectrometry, providing complementary information for metabolite identification and characterization (Stavru et al. 2021). This versatility enables the integration of HPLC with other analytical platforms, such as NMR spectroscopy and capillary electrophoresis, further enhancing metabolome coverage and confidence in metabolite assignments (Zhou et al. 2012a, b).

7.1.3.11 Applications of HPLC in Metabolite Purification

The advantages of HPLC have contributed to its widespread application in various fields of metabolomics research, including biomarker discovery, metabolic pathway elucidation, and the characterization of natural products and environmental contaminants (Zhou et al. 2012a, b; Stavru et al. 2021).

Biomarker Discovery

The identification of putative biomarkers for a range of illnesses and physiological states has been made possible using HPLC-based metabolomics (Zhou et al. 2012a, b). HPLC has made it easier to identify metabolic patterns linked to illnesses including cancer, metabolic disorders, and neurological diseases by enabling the thorough profiling and quantification of metabolites in biofluids and tissue samples (Patti et al. 2012). These metabolic indicators have the potential to monitor therapy responses, help with early diagnosis, provide light on disease causes, and eventually support the advancement of customized medicine techniques (Patti et al. 2012; Zhou et al. 2012a, b).

7.1.3.12 Metabolic Pathway Elucidation

HPLC plays a pivotal role in elucidating metabolic pathways by enabling the comprehensive profiling and quantification of metabolites involved in various biochemical processes (Stavru et al. 2021). By mapping metabolite levels and fluxes onto known metabolic pathways, researchers can gain insights into the regulatory mechanisms governing cellular metabolism, identify rate-limiting steps, and uncover potential targets for therapeutic intervention (Patti et al. 2012).

HPLC-based metabolomics has been particularly valuable in studying metabolic pathways in various biological systems, including microorganisms, plants, and mammalian cells, contributing to our understanding of energy metabolism, secondary metabolite biosynthesis, and the interplay between metabolism and cellular signaling (Zhou et al. 2012a, b).

7.1.3.13 Natural Product and Environmental Metabolite Characterization

Purification and identification of bioactive chemicals from complicated mixtures are made possible by high-performance liquid chromatography (HPLC), which is essential for the characterization of natural products and environmental metabolites (Stavru et al. 2021). HPLC has made it easier to isolate and determine the structural makeup of several metabolites that have potential medical uses, including antibacterial, anti-inflammatory, and anticancer drugs (Zhou et al. 2012a, b). Similarly, in environmental metabolomics, HPLC has been employed to characterize and monitor metabolites derived from environmental contaminants, enabling the assessment of their impact on human health and ecosystems (Stavru et al. 2021). Additionally, HPLC-based metabolomics has contributed to the study of xenobiotic metabolism, providing insights into the biotransformation pathways and metabolic fates of environmental pollutants and pharmaceuticals (Zhou et al. 2012a, b).

7.1.3.14 Quality Control and Authenticity Assessment

Food, beverage, and pharmaceutical sectors, among others, have found HPLC to be a beneficial technique for authenticity evaluation and quality control (Stavru et al. 2021). By enabling the comprehensive profiling of metabolites and the identification of characteristic markers, HPLC can be used to verify the authenticity of products, detect adulteration or contamination, and ensure compliance with regulatory standards (Zhou et al. 2012a, b).

In the food and beverage industry, HPLC-based metabolomics has been employed to authenticate the origin and quality of products, such as wines, olive oils, and honey, by analyzing their metabolite profiles and comparing them to established reference databases (Stavru et al. 2021). Similarly, in the pharmaceutical industry, HPLC is used to monitor the quality and purity of drug substances, ensuring product safety and efficacy (Zhou et al. 2012a, b).

7.1.3.15 Challenges and Future Perspectives

Despite the numerous advantages and widespread applications of HPLC in metabolite purification, several challenges remain that warrant further research and development (Zhou et al. 2012a, b; Stavru et al. 2021).

Matrix Effects and Ion Suppression

One of the major challenges in HPLC-based metabolomics is the occurrence of matrix effects and ion suppression, which can compromise quantitative accuracy and sensitivity (Stavru et al. 2021). These effects can be caused by the co-elution of interfering components from complex biological matrices, leading to signal suppression or enhancement during ionization in mass spectrometry detection (Zhou et al. 2012a, b).

To mitigate these effects, strategies such as sample cleanup, derivatization, and the use of internal standards or matrix-matched calibration curves are often employed (Stavru et al. 2021). However, further research is needed to develop more robust and efficient methods for reducing matrix effects, ensuring reliable quantitative data across diverse sample matrices (Zhou et al. 2012a, b).

Separation of Isomeric and Isobaric Metabolites

The separation and identification of isomeric and isobaric metabolites pose significant challenges in HPLC-based metabolomics (Stavru et al. 2021). These metabolites share identical molecular formulas and often exhibit similar physicochemical properties, making their resolution and identification particularly challenging (Zhou et al. 2012a, b).

While multidimensional HPLC approaches and the integration of IMS have shown promise in improving the separation of isomeric and isobaric metabolites, further advancements in separation techniques and data analysis strategies are needed to fully resolve these challenges (Lanucara et al. 2014).

Metabolite Identification and Annotation

In investigations utilizing HPLC for metabolomics, the identification and annotation of metabolites continue to be a major obstacle (Domingo-Almenara et al. 2022). The enormous chemical array of metabolites and the incomplete coverage of publically accessible databases provide major obstacles, even with the availability of spectrum libraries and computational tools (Stavru et al. 2021). Continuous efforts are needed to expand and curate metabolite databases, develop advanced computational tools for *in silico* metabolite annotation, and integrate complementary analytical techniques, such as NMR spectroscopy and ion mobility spectrometry, to enhance metabolite identification capabilities (Domingo-Almenara et al. 2022).

Throughput and Automation

While HPLC systems can be automated and scaled for high-throughput analysis, further improvements in automation workflows, data acquisition, and data processing are needed to meet the increasing demands of large-scale metabolomics studies (Stavru et al. 2021). The development of miniaturized and integrated analytical platforms, coupled with advanced data handling and processing capabilities, could significantly enhance throughput and efficiency (Zhou et al. 2012a, b).

Integration with Other Omics Technologies

A thorough understanding of biological systems requires the integration of metabolomics data derived by HPLC with other omics technologies, including transcriptomics, proteomics, and genomes (Patti et al. 2012). However, this integration poses challenges in data processing, normalization, and the development of computational frameworks for multi-omics data analysis (Zhou et al. 2012a, b).

To enable the smooth integration of multi-omics data, interdisciplinary collaborations and the creation of cutting-edge bioinformatics tools are required. This will allow for the clarification of intricate regulatory networks and the discovery of fresh biological insights (Patti et al. 2012; Domingo-Almenara et al. 2022).

Standardization and Data Sharing

The lack of standardized protocols and data reporting formats in metabolomics research can hinder data reproducibility, comparability, and sharing (Stavru et al. 2021). Efforts toward the development of community-accepted standards for experimental design, data acquisition, and data reporting are crucial for enabling effective data sharing and meta-analysis across different laboratories and studies (Zhou et al. 2012a, b).

Initiatives such as the Metabolomics Standards Initiative (MSI) and the establishment of public metabolomics repositories are important steps toward achieving standardization and promoting open data sharing in the field (Domingo-Almenara et al. 2022).

Despite these challenges, the field of HPLC-based metabolomics continues to evolve rapidly, driven by technological advancements and interdisciplinary collaborations (Stavru et al. 2021). The integration of cutting-edge analytical techniques, such as multidimensional chromatography, ion mobility spectrometry, and high-resolution mass spectrometry, coupled with advanced computational approaches and data analysis strategies, holds immense promise for enhancing metabolite purification, identification, and quantification (Zhou et al. 2012a, b; Domingo-Almenara et al. 2022).

As our understanding of cellular biochemistry deepens, the role of HPLC in metabolite purification will continue to be pivotal, contributing to the elucidation of metabolic pathways, the discovery of biomarkers, and the development of novel therapeutic interventions across various domains of life sciences research (Patti et al. 2012).

7.2 Unveiling the Principles of HPLC

High-performance liquid chromatography (HPLC) leverages the exquisite principles of chromatographic separation for meticulous analysis of complex mixtures. This technique meticulously orchestrates the movement of a dissolved sample (analyte mixture) within a mobile phase through a column packed with a stationary phase. As the mobile phase traverses the column, the individual analytes exhibit varying degrees of interaction with the stationary phase, governed by their unique physicochemical properties (e.g., polarity, size, charge, and affinity). This differential interaction dictates the separation of the analytes, with those exhibiting stronger affinity for the stationary phase eluting (emerging from the column) later. The separated analytes emerge from the column at distinct time intervals, registering as well-defined peaks on a detector's output, enabling both qualitative identification and quantitative analysis of the original mixture's components. HPLC's versatility and remarkable control over the separation process, achieved by meticulously tailoring mobile phase composition, flow rate, and column temperature, solidify its position as a dominant force in analytical chemistry.

7.2.1 A Journey Through the System: Core Components of an HPLC Setup

Exploring the HPLC system reveals a carefully planned and executed process of analytical competence. The procedure starts with putting a sample into the system. Once inside, the sample comes into contact with the mobile stage, which is a solvent that the pump facilitates moving through the system precisely. The sample enters the column, an essential part of the separation process, while the mobile phase travels. Within the column, the sample molecules interact with the stationary phase, with each compound crossing this small-scale environment at a speed determined by its chemical characteristics. The detector quickly sees their emergence, ready to collect their signals when they depart the column. The procedure provides useful information on the concentration and composition of the sample and is driven by variables like the mobile phase's composition and flow rate. It takes place during real-time data collecting. At the end of the crossing, the HPLC system produces distinct peaks on chromatograms, with each peak serving as evidence of the amazing potential of separation technology and the analytical accuracy of HPLC. The core components of an HPLC setup typically includes:

1. The **pump** is tasked with providing the mobile phase (solvent) continuously through the system, maintaining a steady flow rate and pressure. This guarantees a uniform and repeatable division.
2. An **injector** is used to introduce the sample into the stream of the mobile phase. The device must have the ability to deliver accurate and consistent sample quantities.
3. The **column** serves as the central component of the HPLC system, responsible for the separation of different components. The column contains a stationary phase that interacts with the sample components in distinct ways, resulting in their separation according to diverse physicochemical parameters.
4. The **detector** observes the eluent exiting the column and generates a signal that is directly proportional to the concentration of analytes present. Typical detectors used in this context include UV-Vis detectors, fluorescence detectors, refractive index detectors, and mass spectrometers.
5. A **column oven/heater** is used for some HPLC analyses to provide accurate temperature control of the column, which is crucial for optimizing separation efficiency and repeatability. A column oven or heater ensures a consistent temperature throughout the analysis.
6. **Data Acquisition System**: This system captures and analyzes the signals from the detector, transforming them into data that can be effectively utilized. It can include software designed for the purpose of analyzing and interpreting data.
7. **Mobile Phase Reservoirs** are containers designed to hold the solvents that are utilized as the mobile phase. These reservoirs frequently incorporate solvent degassing systems to eliminate dissolved gasses that may disrupt the chromatographic process.
8. **Degasser:** A device that eliminates dissolved gases from the mobile phase in order to prevent the production of bubbles, which may disrupt flow rates and compromise the accuracy of the analysis.
9. **A gradient controller** is utilized in gradient HPLC to regulate the amounts of solvents entering the system as the composition of the mobile phase changes over time.
10. **The trash container** is responsible for gathering the waste generated during the chromatographic process, which includes both the mobile phase and sample waste.
11. **The pressure regulator and safety features** of HPLC systems are designed to maintain and control the high pressures required for efficient separations. Pressure regulators and safety features are crucial for preserving the integrity of the system and avoiding mishaps.

7.2.2 The Separation Game: Exploring Different Modes of HPLC

It explores the wide range of chromatographic modes that can be used in high-performance liquid chromatography (HPLC). Reverse-phase chromatography utilizes a non-polar stationary phase and a polar mobile phase, while normal-phase chromatography uses a polar stationary phase and a nonpolar mobile phase. These two modes of chromatography provide distinct methods for separation. Size exclusion chromatography separates chemicals by their size, whereas ion-exchange chromatography separates compounds using ionic interactions. Affinity chromatography is a technique that captures molecules based on unique ligand–receptor interactions, while chiral chromatography is used to distinguish between enantiomers, which is important in the field of medicines. Analysts can effectively utilize a wide set of tools by traveling between different modes. Various instruments are specifically designed to address the various requirements of analytical issues, hence improving their capacity to accurately and effectively analyze complex combinations.

7.2.2.1 Reversed-Phase HPLC: A Popular Choice

Reversed-phase high-performance liquid chromatography (RP-HPLC) has gained prominence in the field of current analytical chemistry due to its versatility and extensive use. Using this method, a polar mobile phase (typically a mixture of water and organic solvents like methanol or acetonitrile) and a non-polar stationary phase (typically composed of hydrophobic groups like C18 bound silica) are used. This polarity contrast causes analytes to elute from the column based on their hydrophobicity. Compounds that are more hydrophobic are held in the column for a longer time, which helps in separating them. RP-HPLC is widely used in diverse domains such as pharmaceutical analysis, environmental monitoring, and food and beverage testing due to its capacity to effectively separate complicated mixtures with great sensitivity and consistency. The popularity of RP-HPLC is due to its strong performance, ability to work with various analytes, and simplicity in developing methods. This makes it a trustworthy and effective alternative for chromatographers looking to achieve accurate separations in their analytical work.

One of the most used methods for separating proteins and peptides is RP-HPLC. Its brilliant resolving power, versatility, stability, reproducibility, and convenience make it one of the most prominent chromatography in laboratory practices. It has been extensively utilized for nano, micro, and analytical scaling in large industrial setup. It has great compatibility with mass spectrometry, making of perfect and crucial tool in proteomic research (Josic and Kovac 2010).

Reversed-phase high-performance liquid chromatography (RP-HPLC) reigns supreme for separating peptides and proteins (Žuvela et al. 2019). This workhorse technique relies on an alkyl silica-based sorbent as the stationary phase. Imagine a microscopic maze with walls coated in long hydrocarbon chains (alkyl groups). Peptides and proteins are then eluted (washed out) using gradients of increasing organic solvent concentration, typically acetonitrile (Žuvela et al. 2019). Think of this gradient as a gradually intensifying current that strategically pushes the components through the maze. To further enhance separation, an ionic modifier like trifluoroacetic acid (TFA) is often incorporated (Žuvela et al. 2019).

The beauty of modern RP-HPLC lies in its remarkable sensitivity. This sophisticated instrumentation allows separation and detection of incredibly small quantities—even picomolar amounts of chemical molecules (Žuvela et al. 2019). That is like finding a single grain of sand on a vast beach! The vast array of commercially available RP-HPLC columns (over 600!) caters to diverse needs, offering a spectrum of selectivities and production qualities (Žuvela et al. 2019). These columns essentially act as specialized versions of the maze, each potentially offering a unique path for optimal separation.

The power of RP-HPLC does not stop there. Researchers can fine-tune the separation process by adjusting various parameters. For instance, modifying the gradient slope, temperature, ionic modifier concentration, or the organic solvent composition can significantly impact how peptides and proteins navigate the maze (Žuvela et al. 2019). This level of control allows for the separation of complex mixtures containing both enzymatically derived peptides (produced by enzymes) and synthetically generated ones (Žuvela et al. 2019).

In essence, RP-HPLC stands as a versatile and powerful tool for separating a wide range of peptides and proteins. Through its meticulous design, exceptional sensitivity, and tunable parameters, RP-HPLC continues to be a cornerstone technique in various research fields.

Conditions for Reverse-Phase Chromatography

Stationary Phase: In RP-HPLC, stationary phase consists of a non-polar hydrophobic material, for instance, such as silica C-18 polymer.

Mobile Phase: The mobile phase is a hydrophilic polar solvent that can include methanol or acetonitrile in addition to water. Polar molecules in the sample are eluted using this polar mobile phase. Now the combination of analytes that makes up the sample is introduced into the column.

Segregation: Non-polar molecules in the sample have a stronger affinity for the nonpolar stationary phase and travel more slowly through the column when the mobile component moves through it, whereas polar molecules have a weaker connection and move more quickly.

Detection and Collection of Data

As the separated molecules elute from the column, they pass through a detectorare detected and quantified based on their characteristic properties. The separated components in the sample can be further analyzed through various sequencing techniques like automated Edman sequencing or electrospray mass spectroscopy. When purifying synthetic peptide molecules, the complexity of the crude synthetic product is evaluated after separation on an analytical scale. Large-scale product collection and purification are then carried out, and an aliquot of the purified sample is subjected to additional chromatography under various HPLC conditions to achieve high-resolution separation. Important factors that play a crucial role in the purification and analysis of proteins and peptides include the influence on protein conformation, subunit assembly, and degree of microheterogeneity. RP-HPLC has been widely utilized in separation carotene. The analysis is done usually through utilizing C18 and C30 columns. RP-HPLC has successfully been able to differentiate between α-carotene and β-carotenes (Panfili et al. 2004).

7.2.2.2 Normal-Phase HPLC: An Alternative Approach

While reversed-phase HPLC reigns supreme for many separations, normal-phase HPLC (NP-HPLC) offers a valuable alternative, leveraging contrasting principles (Žuvela et al. 2019). This technique hinges on the concept of polarity—the "like attracts like" principle in the realm of chemistry. Imagine a microscopic landscape where the stationary phase is a polar terrain, akin to a wet, water-loving surface. This is often achieved using materials like silica gel, alumina, or cyano-bonded silica (Žuvela et al. 2019). The mobile phase, on the other hand, takes on a less polar character, typically consisting of nonpolar solvents like hexane, dichloromethane, or ethyl acetate (Žuvela et al. 2019). Think of this mobile phase as a gentle stream flowing through the polar landscape.

The magic of NP-HPLC unfolds through these contrasting polarities. Analytes, particularly those with more polar characteristics, establish stronger interactions with the polar stationary phase. Imagine these polar analytes readily sticking to the wet surface. Consequently, they spend a longer time traversing the column, leading to a slower elution (emergence) (Žuvela et al. 2019). Conversely, less polar analytes experience weaker interactions with the stationary phase and travel through the column more rapidly. By manipulating the mobile phase composition, researchers can fine-tune the separation process, making NP-HPLC a valuable tool for analyzing a specific range of analytes based on their polarity (Žuvela et al. 2019). The analytes, which differ in polarity, interact differentially with the stationary and mobile phases, resulting in their separation based on relative affinities**.** The separation process involves injecting a sample into an HPLC system and passing it through a stationary phase-containing column. The nonpolar analytes interact weakly with the polar stationary phase and are thus eluted first because they have a higher affinity for the less polar mobile phase. Polar analytes, on the other hand, interact strongly with the stationary phase and elute later because they are maintained in the column for longer periods of time. Normal-phase HPLC is widely used for separating and analyzing nonpolar or moderately polar substances. It is widely employed in the examination of natural products such lipids, steroids, and fat-soluble vitamins. Additionally, it is used in the pharmaceutical industry to analyze medicinal molecules and their metabolites. Normal-phase HPLC is frequently used in environmental studies to separate and quantify pollutants and other substances. A comprehensive representation highlighting the key features and applications of NP-HPLC are presented in Table 7.1.

Substances like lipids, sterols, and certain natural products can be separated using NP-HPLC, especially in situations where reversed-phase techniques don't provide the optimum resolution. NP-HPLC is an essential tool in analytical chemistry because of its versatility and suitability for a wide range of sample kinds. It

Table 7.1 NP-HPLC and its key features

Aspect	Description
Stationary phase	Polar stationary phase (e.g., silica)
Mobile phase	Nonpolar mobile phase (e.g., hexane, dichloromethane)
Separation mechanism	Based on polarity differences between compounds
Retention time	More polar compounds retained longer; less polar compounds elute earlier
Applications	– Separation of polar compounds such as lipids, sterols, and natural products – Analysis of highly polar compounds where reversed-phase methods may not be suitable – Purification of polar compounds for further analysis or characterization
Advantages	– High selectivity for polar compounds – Compatibility with a wide range of sample types – Versatile in separating compounds based on polarity differences
Limitations	– Limited retention of nonpolar compounds – Requires careful optimization of mobile phase composition and column conditions

offers researchers a different approach to accomplish accurate and trustworthy separations.

7.2.3 Optimizing Separation: Key Factors Affecting Retention Time and Resolution

Retention time and resolution in high-performance liquid chromatography (HPLC) are influenced by several key factors:

1. The **mobile phase composition** affects retention time and resolution significantly. This includes the kind and quantity of solvents and additives. Analytes' interactions with the stationary phase can be affected by changes in the solvent's polarity, pH, or buffer concentration, which can thus affect the analytes' retention and separation.
2. Retention and resolution depend heavily on the characteristics of the stationary phase used in the column. Features including functional groups, surface chemistry, pore size, and particle size affect the separation process's efficiency and selectivity.
3. **The temperature of the column** has an impact on the interactions between the substances being analyzed and the stationary phase. Raising the temperature often reduces retention durations by decreasing the contacts between the analytes and the stationary phase. Nevertheless, changes in temperature can also impact resolution, as the most favorable separation circumstances may differ depending on the temperature.

4. **The flow rate** at which the mobile phase passes through the column affects both the amount of time a substance is retained and the degree of separation between different substances. Increased flow rates often lead to decreased retention durations, but this might potentially impact resolution by reducing the amount of time analytes have to interact with the stationary phase.
5. The parameters of the HPLC column, specifically its **length and diameter**, have an effect on both retention and resolution. Increased column length typically results in higher resolution but can also result in longer retention times. In contrast, columns that are shorter in length result in shorter retention durations, but there may be a trade-off in terms of resolution.
6. **Gradient Conditions:** In gradient elution HPLC, the alteration of the mobile phase's composition over time impacts the retention periods and resolution through the gradient slope and duration. It is crucial to carefully choose the most effective gradient conditions in order to get the desired outcomes of separation.
7. **Analyte Properties**: The physical and chemical characteristics of analytes, including their size, polarity, charge, and attraction to the stationary phase, have a substantial impact on their retention and separation. Comprehending these characteristics enables chromatographers to customize HPLC settings for the most effective separation.

7.3 Crafting the Perfect Strategy: HPLC-Based Metabolite Purification Workflow

Developing an ideal approach for purifying metabolites using HPLC involves a complex procedure that necessitates a thorough comprehension of the fundamental scientific principles and the most recent technology breakthroughs. HPLC is a very adaptable and potent method that enables researchers to effectively isolate, detect, and cleanse metabolites from intricate biological materials (Sobolewska and Biesaga 2024). Through the utilization of HPLC's exceptional high-resolution capabilities, scientists are able to precisely separate and identify individual metabolites (Wang et al. 2023). This enables groundbreaking research to be conducted in several domains including pharmacology, metabolomics, and natural product discovery. The workflow commences with meticulous selection and preparation of the biological sample, which is subsequently followed by customized extraction techniques to enhance the sample with specific metabolites. Subsequently, the sample is put into the HPLC system, where separation takes place as a result of both the well-controlled mobile phase and the analyte-stationary phase interaction (Shinde and Rajurkar 2024). The method may perform high-resolution separation by utilizing gradient elution techniques and sophisticated column technologies, enabling the isolation of molecules that are closely related. In addition, advanced techniques like mass spectrometry (MS) can be included to provide immediate monitoring and identification of metabolites, hence assisting in the collection of pure fractions (Mattoli et al. 2023). Researchers can improve the efficiency, precision, and reproducibility of metabolite purification in the bio sciences by using a strategic strategy that combines advanced technologies with careful planning and analysis. This method promotes innovation and discovery (Patra et al. 2023).

7.3.1 Sample Preparation: Laying the Foundation for Success

A critical step in the analytical process that influences the precision and dependability of the findings is sample preparation (Matuszewski et al. 2003). It is the most critical stage in extracting, isolating, and concentrating the analyte of interest from complicated matrices (Tang et al. 2016). Recent developments in sample preparation include miniaturization, automation, high-throughput performance, online connectivity with analytical equipment, decreased solvent use, and operating time (Maciel et al. 2021). Many effective methods for preparing samples for HPLC have been developed and employed, including solid-phase extraction, liquid-phase microextraction, and related microextraction procedures (Li et al. 2024). To reduce time and produce a more accurate and robust HPLC analysis, it is helpful to have a fundamental understanding of the concepts of these sample preparation processes. Then, a systematic approach should be used to choose, optimize, and validate a protocol. Sample preparation necessitates precise handling and manipulation of samples from when they are brought into the laboratory until they are positioned on the tray. It is important to think about which stages of the sample handling procedure may allow unnoticed errors to enter the technique. Sample preparation involves using several methods, including conventional techniques such as weighing, dilution, and filtration, as well as more recent advancements like Quenchers (Pathak et al. 2024). The older approaches are significantly more prevalent. The conversion of a raw sample into a format suitable for injection into an instrument can include multiple sequential procedures, each of which has the potential to introduce variability to the overall technique. Gravimetric steps often have the lowest error rates, provided the balances are adequately maintained and calibrated, and correct operating protocols are followed (Zhang 2024). Dilutions, however, are frequently overlooked as a significant cause of variability. The process of preparing samples is frequently the primary cause of variation in a system, making it the most crucial component to manage and design effectively. It is frequently overlooked because of its lower cost and less exciting nature compared to other components of a strategy. Relying solely on the column cannot always substitute for a sample preparation that lacks appropriate selectivity. Similarly, even a mass spectrometer with the utmost resolving capacity remains vulnerable to ion suppression caused by a sample matrix that has not been adequately cleansed (Henderson et al. 2024). For optimal performance in any type of separation, it is essential to use clean samples, since this reduces the need for frequent instrument maintenance. Utilizing well-established and standardized methodologies can effectively reduce variance, whether it is due to daily fluctuations, individual differences, or variations between laboratories. Liquid–liquid extraction (LLE), solid-phase extraction (SPE), solid-phase microextraction (SPME), quenchers, and other solid-phase techniques are a few sample preparation techniques that can be regarded as independent separation processes.

Considering them as a possibly independent dimension from a main separation is a helpful approach to simplify an otherwise intricate sample. When it comes to separations, it is crucial to consider both the equilibrium and kinetics to ensure the sensitivity and repeatability of a procedure.

7.3.2 Choosing the Right Column: Matching the Column to the Metabolite

One useful analytical technique for separating, identifying, and quantifying ingredients in a mixture is HPLC, including metabolites (Ahmed et al. 2023; Sobolewska and Biesaga 2024; Pantiya et al. 2024). The selection of the proper column for a certain application is a critical decision in HPLC (Calderón and Lämmerhofer 2023). The selection of the column has a direct influence on the effectiveness, clarity, and consistency of the analysis (Brinco et al. 2023). To achieve optimal separation, numerous criteria must be taken into account when matching the column to the metabolite. Initially, analyze the chemical characteristics of the metabolite, including its polarity, molecular weight, and chemical structure. These characteristics will assist you in choosing a stationary phase that will offer the most optimal interaction with the metabolite for efficient separation. Columns designed for hydrophilic interaction liquid chromatography (HILIC) can be used to efficiently separate polar metabolites; however, non-polar metabolites might require the use of a reversed-phase column (Jandera and Hájek 2020). Furthermore, the size of the particles, the size of the pores, and the length of the column all have a significant impact on the effectiveness of the separation process (Žuvela et al. 2019). Reducing the size of particles often improves resolution, but the size of the pores should be compatible with the metabolite size to enhance interaction with the stationary phase. The pH and composition of the mobile phase are crucial factors to consider, as they might affect the ionization state of the metabolite and consequently its retention on the column (Gritti 2021). Finally, it is crucial to take into account the sample matrix and any interference to select a column that can effectively manage intricate combinations while preserving durability and longevity. By accurately aligning the column with the metabolite, one can obtain precise, efficient, and consistent analytical outcomes. A range of columns are utilized in HPLC to separate and analyze a wide variety of metabolites (Perez de Souza et al. 2021). The chemical composition of the metabolites and the type of interaction necessary for separation dictate the column selection (Girel et al. 2023). Table 7.2 provides a summary of prominent HPLC column types and the corresponding metabolite applications.

7.3.3 Mastering the Mobile Phase: Solvent Selection and Gradient Elution

High-performance liquid chromatography (HPLC) relies on a powerful duo: the mobile phase and the stationary phase, to separate complex mixtures. The mobile

Table 7.2 Different types of columns used in HPLC based on their properties and type of metabolites

Column	Stationary phase	Metabolite type	Uses
Reverse phase	Nonpolar (e.g., C18, C8)	Lipids, hydrophobic	Most popular kind, adaptable, and effective for a variety of metabolites
Normal phase	Polar (e.g., silica)	Polar metabolites, vitamins	Useful for the separation of isomers and polar compounds
Ion exchange	Ion-exchange resin	Amino acids, peptides, proteins, organic acids	Effective for separating charged metabolites and proteins
Size exclusion	Porous polymer or silica gel	Proteins, peptides, polymers, large molecules	Divides molecules by to their molecular weight, or size
Chiral	Chiral stationary phase	Enantiomers, chiral drugs	Used for separating chiral compounds
Hydrophilic interaction (HILIC)	Hydrophilic phase (e.g., silica with polar coating)	Small polar metabolites	Compounds, such as sugars and amino acids

phase, a solvent or solvent mixture, plays a critical role, with its strength dictating how quickly it removes analytes from the column. Think of strong solvents like organic modifiers in reversed-phase HPLC acting like a forceful current, rapidly pulling analytes through the column, while weaker solvents like water hold onto analytes for a longer separation. The key is the concentration of the strongest solvent in the mobile phase—the higher the concentration, the faster the elution. Understanding solvent strength and how it varies across common HPLC modes like reversed-phase, normal-phase, and HILIC is crucial for researchers to optimize separations and effectively identify and quantify their analytes of interest.

7.3.3.1 Selecting the Right Solvents

The selection of solvents and the use of gradient elution are two crucial factors in high-performance liquid chromatography (HPLC; Lozano-Sánchez et al. 2018), significantly influencing the effectiveness and selectivity of the separation process. The process of selecting a solvent is crucial in HPLC as it directly influences the effectiveness of chemical separation and analysis (Žuvela et al. 2019). To ensure optimal separation, the selected solvents, often referred to as mobile phases, must be compatible with both the stationary phase and the analytes. The primary solvents utilized in HPLC are water and organic solvents, specifically acetonitrile and methanol (Joshi and Adhikari 2019). HPLC solvent selection revolves around a balancing act between analyte polarity and stationary phase. In popular reverse-phase HPLC, a non-polar stationary phase (like C18) attracts non-polar analytes more strongly, causing them to elute (emerge from the column) later. Conversely, polar analytes travel faster through the column with a polar mobile phase (often water and organic solvent mix). This strategic selection, along with adjustments to mobile phase pH using buffers or acid/base modifiers (Djajić et al. 2022), allows researchers to fine-tune separations for optimal analysis of their target analytes. For instance, analytes with high acidity may necessitate the use of a mobile phase that is buffered to maintain a neutral state and achieve successful separation. Furthermore, it is important to take into account the viscosity of the solvent, as it might have an impact on the effectiveness and pressure of the HPLC system (Yabré et al. 2018).

7.3.3.2 Utilizing Gradient Elution for Enhanced Separation

Gradient elution is a method employed to enhance the differentiation of intricate mixtures and analytes with a broad spectrum of polarity (Pirok et al. 2018). Throughout the separation process, this approach gradually modifies the composition of the mobile phase. This is accomplished by combining two or more solvents in varying quantities using high-pressure mixing pumps. A commonly used technique in gradient elution includes altering the proportion of an aqueous phase (such as water with or without a buffer) to an organic phase (such as acetonitrile or methanol) during the study.

The gradient profile can exhibit either a linear or nonlinear behavior, contingent upon the specific requirements for separation. A linear gradient refers to a consistent and gradual shift in the composition of the solvent over time, while a nonlinear gradient can involve abrupt changes or more intricate variations in composition (Navarro-Huerta et al. 2019). This method can enhance the optimization of the separation of compounds with different polarity and retention durations. For instance, commencing with a greater ratio of water and progressively augmenting the ratio of organic solvent can induce a potent separation of polar chemicals succeeded by nonpolar compounds.

Gradient elution has various benefits, including greater resolution, reduced running time, and improved selectivity (Snyder and Dolan 2021). Additionally, it can improve the form of the peak and reduce the chances of co-elution in intricate samples. Moreover, gradient elution is especially advantageous for assessing materials with uncertain or diverse compositions since it offers a wide range of solvent conditions to accommodate various types of analytes.

The use of gradient elution is a crucial factor in achieving successful HPLC analysis (Gritti 2021). It is necessary to customize the selection of solvents and gradient methods based on the unique properties of the sample and the intended separation. Analysts can attain optimal separation performance, repeatability, and accuracy in their HPLC analyses by carefully addressing these factors.

The pH of the mobile phase can be changed by buffers or acid/base modifiers (Alvarez-Segura et al. 2019). This changes the charge and solubility of the analytes, which impacts the separation process. For example, using a buffered mobile phase to maintain a neutral pH and accomplish successful separation can be necessary if the analytes are acidic. Moreover, the solvent's viscosity needs to be considered since it might affect the HPLC system's efficiency and pressure (De Vos et al. 2021).

7.3.4 Unveiling the Essence: Detection Methods in HPLC

The selection of a detection method in HPLC reigns supreme, acting as the gatekeeper to analyte analysis (Žuvela et al. 2019). An unsuitable detector renders valuable chromatographic information invisible, highlighting the importance of a well-considered choice. Four crucial parameters guide this selection:

- **Chemical Composition of Analytes:** Understanding the analytes' chemical makeup is paramount.
- **Interference Potential:** Identifying potential interferences within the sample matrix is crucial.
- **Limits of Detection (LOD) and Quantification (LOQ):** Sensitivity is key—determining the minimum detectable and quantifiable amounts of the analyte is vital.
- **Linearity Range and Cost-Effectiveness:** The detector's ability to maintain a proportional response across a concentration range, alongside its affordability, is an important consideration.

For trace analysis, where minute quantities are of interest, fluorescence, electrochemical, or mass spectrometry detectors are often the preferred choices (Žuvela et al. 2019). In preparative HPLC, where the focus lies on isolating larger quantities, the refractive index (RI) detector reigns supreme due to its tolerance for high analyte concentrations without saturation (Žuvela et al. 2019).

7.3.4.1 Shining a Light: Ultraviolet (UV) Detectors

Ultraviolet (UV) detectors, specifically designed for ultraviolet/visible absorbance (UV/Vis), are the workhorses of HPLC detection (Turner et al. 2020). Their popularity stems from several factors: durability, affordability, and adaptability. Most substances exhibit some degree of light absorption, particularly at lower UV wavelengths, making them amenable to UV detection (Turner et al. 2020). For enhanced versatility, diode array detectors (DADs) allow for simultaneous monitoring across a spectrum of wavelengths.

However, a key limitation of UV detectors is their lack of inherent analyte specificity (Zhang et al. 2019). They rely on the analyte absorbing light more significantly than the sample matrix at the chosen wavelength. To optimize both sensitivity and specificity, careful selection of the detection wavelength is crucial. It's important to consider that mobile phase solvents and buffer components can cause slight deviations in the analyte's UV maximum compared to reference values (Attimarad et al. 2019). Therefore, verifying the analyte's absorbance within the actual mobile phase used is recommended. Operating at wavelengths well above the UV cut-off of the mobile phase and buffer components is essential to avoid diminished sensitivity and increased system noise, resulting in an unstable baseline. Steering clear of wavelengths below 200 nm is advisable due to the inherent rise in detector noise at these lower wavelengths. For enhanced selectivity, employing higher wavelengths is generally advantageous.

7.3.4.2 Mass Spectrometry (MS): Unveiling the Molecular Fingerprint

Mass spectrometry (MS) has established itself as a cornerstone detection technology, often hyphenated with various separation methods including liquid chromatography (LC) (Alvarez-Rivera et al. 2019). Unlike simpler detectors like UV, RI, and FL, MS offers a deeper layer of information—the ability to decipher molecule masses and glean specific structural properties (Ventouri et al. 2022). This power allows for the differentiation of co-eluting peaks using selected ion monitoring mode. While this mode reduces the need for exceptional chromatographic resolution prior to detection, achieving well-separated peaks is still preferable to avoid potential ion suppression or enhancement effects.

7.3.4.3 Delving into the MS Realm: The Quadrupole Mass Analyzer

One prevalent type of mass analyzer is the quadrupole. This workhorse features four parallel rods subjected to a combination of fixed direct current (DC) and alternating radio frequency (RF) potentials. The HPLC system, operating at atmospheric pressure (760 Torr), delivers the dissolved analytes to the MS. Within the MS, a dramatic shift in pressure occurs as the ionized samples transition from a gaseous form (HPLC) to a high vacuum environment (10^{-5}–10^{-6} Torr) for detection. An interface bridges the gap between the LC and MS, gently transforming the analyte solution into an aerosol, followed by ionization and solvent removal. These transformed ions are then directed through the center of the quadrupole rods. Their trajectory is influenced by the electric fields within the rods, allowing only ions with a specific mass-to-charge ratio (m/z) to follow a stable path toward the detector. By adjusting the radio frequency (RF), the instrument focuses ions with different m/z values onto the detector, ultimately generating a mass spectrum. The type of interface employed depends on the analyte's physical characteristics and molecular mass. These interfaces differ in their ionization techniques and the pressure applied throughout the process. Currently, popular ionization techniques like electrospray ionization (ESI) and atmospheric pressure chemical ionization (APCI) operate at atmospheric pressure.

References

Ahmed F, Eticha T, Hymete A, Ashenef A (2023) Principles and applications of ultra-high-performance liquid chromatography. In: High performance liquid chromatography-recent advances and applications. IntechOpen, London

Alvarez-Rivera G, Ballesteros-Vivas D, Parada-Alfonso F, Ibañez E, Cifuentes A (2019) Recent applications of high resolution mass spectrometry for the characterization of plant natural products. TrAC Trends Anal Chem 112:87–101

Alvarez-Segura T, Subirats X, Rosés M (2019) Retention-pH profiles of acids and bases in hydrophilic interaction liquid chromatography. Anal Chim Acta 1050:176–184

Attimarad M, Shahzad Chohan M, Ahmed Balgoname A (2019) Simultaneous determination of moxifloxacin and flavoxate by RP-HPLC and ecofriendly derivative spectrophotometry methods in formulations. Int J Environ Res Public Health 16(7):1196

Bedia C (2022) Metabolomics in environmental toxicology: applications and challenges. Trends Environ Anal Chem 34:e00161

Bedia C et al (2022) Plant metabolite extraction procedures and associated challenges. Metabolomics 18(3):26
Beger RD et al (2022) Metabolite biomarkers in metabolomics: a review with emphasis on diagnostic performance. Meta 12(3):217
Bouatra S et al (2013) The human urine metabolome. PLoS One 8(9):e73076
Brinco J, Guedes P, da Silva MG, Mateus EP, Ribeiro AB (2023) Analysis of pesticide residues in soil: a review and comparison of methodologies. Microchem J 195:109465
Calderón C, Lämmerhofer M (2023) Basic principles for the selection of liquid chromatographic modes for specific applications. In: Liquid chromatography. Elsevier, Amsterdam, pp 81–157
Chaleckis R et al (2019) Challenges for setting up a robust validation strategy for quantitative clinical metabolomics studies with mass spectrometry. Meta 9(9):198
De Vos J, Stoll D, Buckenmaier S, Eeltink S, Grinias JP (2021) Advances in ultra-high-pressure and multi-dimensional liquid chromatography instrumentation and workflows. Anal Sci Adv 2(3–4):171–192
Deng P, Li X, Petriello MC, Wang C, Morris AJ, Hennig B (2019) Application of metabolomics to characterize environmental pollutant toxicity and disease risks. Rev Environ Health 34(3):251–259
Dettmer K et al (2007) Mass spectrometry-based metabolomics. Mass Spectrom Rev 26(1):51–78
Djajić N, Krmar J, Rmandić M, Rašević M, Otašević B, Zečević M et al (2022) Modified aqueous mobile phases: a way to improve retention behavior of active pharmaceutical compounds and their impurities in liquid chromatography. J Chromatogr Open 2:100023
Domingo-Almenara X et al (2022) Computational approaches for metabolite annotation in metabolomics data. Curr Opin Biotechnol 74:43–52
Dwivedee BP et al (2021) Metabolic engineering: a promising approach for therapeutic cell development. Mol Biotechnol 63(6):433–445
Fiehn O (2002) Metabolomics—the link between genotypes and phenotypes. Plant Mol Biol 48(1–2):155–171
Girel S, Guillarme D, Fekete S, Rudaz S, González-Ruiz V (2023) Investigation of several chromatographic approaches for untargeted profiling of central carbon metabolism. J Chromatogr A 1697:463994
Gritti F (2021) Perspective on the future approaches to predict retention in liquid chromatography. Anal Chem 93(14):5653–5664
Henderson A, Heaney LM, Rankin-Turner S (2024) Ambient ionisation mass spectrometry for drug and toxin analysis: a review of the recent literature. Drug Test Anal. https://doi.org/10.1002/dta.3644. Epub ahead of print
Jandera P, Hájek T (2020) Dual-mode hydrophilic interaction normal phase and reversed phase liquid chromatography of polar compounds on a single column. J Sep Sci 43(1):70–86
Joshi DR, Adhikari N (2019) An overview on common organic solvents and their toxicity. J Pharm Res Int 28(3):1–18
Josic D, Kovac S (2010) Reversed-phase high performance liquid chromatography of proteins. Curr Protoc Protein Sci 61:8–7. https://doi.org/10.1002/0471140864.ps0807s61
Karp PD et al (2021) Metabolic pathway databases for plant and microbial metabolic engineering. Metab Eng 64:102–118
Lanucara F et al (2014) The power of ion mobility-mass spectrometry for structural elucidation and the mapping of molecules onto molecular landscapes. Nat Rev Chem 2:571–584
Lee JW et al (2012) Systems metabolic engineering of microorganisms for natural and non-natural chemicals. Nat Chem Biol 8(6):536–546
Li N, Zhang Z, Li G (2024) Recent advance on microextraction sampling technologies for bioanalysis. J Chromatogr A 1720:464775
Lozano-Sánchez J, Borrás-Linares I, Sass-Kiss A, Segura-Carretero A (2018) Chromatographic technique: high-performance liquid chromatography (HPLC). In: Modern techniques for food authentication. Academic, Cambridge, pp 459–526
Maciel EVS, Medina DAV, Borsatto JVB, Lanças FM (2021) Towards a universal automated and miniaturized sample preparation approach. Sustain Chem Pharm 21:100427

Markley JL et al (2017) The future of NMR-based metabolomics. Curr Opin Biotechnol 43:34–40
Mattoli L, Gianni M, Burico M (2023) Mass spectrometry-based metabolomic analysis as a tool for quality control of natural complex products. Mass Spectrom Rev 42(4):1358–1396
Matuszewski BK, Constanzer ML, Chavez-Eng CM (2003) Strategies for the assessment of matrix effect in quantitative bioanalytical methods based on HPLC– MS/MS. Anal Chem 75(13):3019–3030
Navarro-Huerta JA, Gisbert-Alonso A, Torres-Lapasió JR, García-Alvarez-Coque MC (2019) Benefits of solvent concentration pulses in retention time modelling of liquid chromatography. J Chromatogr A 1597:76–88
Nouri MZ, Komatsu S (2022) Metabolic engineering of plants for stress tolerance. Plants 11(5):608
Panfili G, Fratianni A, Irano M (2004) Improved normal-phase high-performance liquid chromatography procedure for the determination of carotenoids in cereals. J Agric Food Chem 52(21):6373–6377. https://doi.org/10.1021/jf0402025
Pantiya P, Guiard BP, Gotti G (2024) Sensitive and fast detection of monoamines and their metabolites by high-performance liquid chromatography coupled with an electrochemical detector (HPLC-ECD) under isocratic conditions: application to intracerebral microdialysis in mice treated by fluoxetine and atomoxetine. Chromatographia 87:1–11
Patel S, Ahmed S (2015) Emerging field of metabolomics: big promise for cancer biomarker identification and drug discovery. J Pharm Biomed Anal 107:63–74
Pathak A, Verma N, Tripathi S, Mishra A, Poluri KM (2024) Nanosensor based approaches for quantitative detection of heparin. Talanta 273:125873
Patra P, Disha BR, Kundu P, Das M, Ghosh A (2023) Recent advances in machine learning applications in metabolic engineering. Biotechnol Adv 62:108069
Patti GJ et al (2012) Metabolomics: the apogee of the omics trilogy. Nat Rev Mol Cell Biol 13(4):263–269
Perez de Souza L, Alseekh S, Scossa F, Fernie AR (2021) Ultra-high-performance liquid chromatography high-resolution mass spectrometry variants for metabolomics research. Nat Methods 18(7):733–746
Pervushin K et al (2022) Mapping metabolite interactions with proteins and nucleic acids by high-resolution NMR approaches. Biochim Biophys Acta Proteins Proteom 1870(9):140867
Pirok BW, Gargano AF, Schoenmakers PJ (2018) Optimizing separations in online comprehensive two-dimensional liquid chromatography. J Sep Sci 41(1):68–98
Ramautar R et al (2018) CE-MS for metabolomics: developments and applications in the period 2016–2018. Electrophoresis 39(1):111–125
Rohrig U, Viksna J (2022) Metabolic pathway applications for understanding disease and strain engineering. J Mol Biol 434(7):167449
Serlie M, Kalkhoven E (2021) Metabolic dysregulation in disease: the NAFLD-obesity-diabetes connection uncovered. Int J Mol Sci 22(8):4328
Shinde S, Rajurkar V (2024) Advances in high-performance liquid chromatography (HPLC) method development and validation: a comprehensive review. J Drug Deliv Biother 1(03):16–26
Snyder LR, Dolan JW (2021) The linear-solvent-strength model of gradient elution. In: Advances in chromatography. CRC Press, Boca Raton, pp 115–187
Sobolewska E, Biesaga M (2024) High-performance liquid chromatography methods for determining the purity of drugs with weak UV chromophores–a review. Crit Rev Anal Chem:1–15. https://doi.org/10.1080/10408347.2023.2291815
Stavru VM, Stavru AG, Gheorghe A (2021) Recent advances in HPLC techniques for metabolomics. Curr Anal Chem 17(3):239–249. https://doi.org/10.2174/1573411017666210317132411
Tang S, Zhang H, Lee HK (2016) Advances in sample extraction. Anal Chem 88(1):228–249
Tian R et al (2021) Challenges in metabolite purification and metabolome coverage in biological sample analysis. TrAC Trends Anal Chem 143:116377
Turner J, Igoe D, Parisi AV, McGonigle AJ, Amar A, Wainwright L (2020) A review on the ability of smartphones to detect ultraviolet (UV) radiation and their potential to be used in UV research and for public education purposes. Sci Total Environ 706:135873

Ventouri IK, Loeber S, Somsen GW, Schoenmakers PJ, Astefanei A (2022) Field-flow fractionation for molecular-interaction studies of labile and complex systems: a critical review. Anal Chim Acta 1193:339396

Wang M et al (2016) Sharing and community curation of mass spectrometry data with Global Natural Products Social Molecular Networking. Nat Biotechnol 34(8):828–837

Wang Z, Zhu H, Xiong W (2023) Advances in mass spectrometry-based multi-scale metabolomic methodologies and their applications in biological and clinical investigations. Sci Bull 68:2268–2284

Wilkins HM, Preshad S (2021) Metabolomics in neurological disorders. Int J Mol Sci 22(17):9236

Yabré M, Ferey L, Somé IT, Gaudin K (2018) Greening reversed-phase liquid chromatography methods using alternative solvents for pharmaceutical analysis. Molecules 23(5):1065

Zhang C (2024) Fundamentals of environmental sampling and analysis. Wiley, Hoboken

Zhang K, Kurita KL, Venkatramani C, Russell D (2019) Seeking universal detectors for analytical characterizations. J Pharm Biomed Anal 162:192–204

Zhou B, Xiao JF, Tuli L, Ressom HW (2012a) LC-MS-based metabolomics. Mol BioSyst 8(2):470–481. https://doi.org/10.1039/C1MB05350G

Zhou B et al (2012b) Comprehensive metabolic profiling reveals the distinct metabolic phenotypes that different resources partition in saccharomyces cerevisiae. BMC Genomics 13(1):1–15

Žuvela P, Skoczylas M, Jay Liu J, Baczek T, Kaliszan R, Wong MW, Buszewski B (2019) Column characterization and selection systems in reversed-phase high-performance liquid chromatography. Chem Rev 119(6):3674–3729

NMR Spectroscopy for Characterization of Metabolite

8

Maryam Abbasi Tarighat, Dhwani Upadhyay, Prasad Andhare, Harshit Kumar, Mohd Tariq, Mukul Jain, and Gholamreza Abdi

M. A. Tarighat (✉)
Faculty of Nano and Bio Science and Technology, Persian Gulf University, Bushehr, Iran
e-mail: matarighat@pgu.ac.ir

H. Kumar
Department of Life Sciences, Parul Institute of Applied Sciences, Parul University, Waghodia, Gujarat, India

M. Tariq
Department of Life Sciences, Parul Institute of Applied Sciences, Parul University, Waghodia, Gujarat, India

Department of Biotechnology, Graphic Era (Deemed to be University), Uttarakhand, Dehradun, India

D. Upadhyay
Department of Life Sciences, School of Sciences, GSFC University, Gujarat, Vadodara, India
e-mail: dhwani.upadhyay@gsfcuniversity.ac.in

P. Andhare
Biological Sciences, PDPIAS, Charotar University of Science and Technology, Changa, Gujarat, India

M. Jain
Department of Life Sciences, Parul Institute of Applied Sciences, Parul University, Waghodia, Gujarat, India

Cell & Developmental Biology Lab, Research and Development Cell, Parul University, Vadodara, Gujarat, India

G. Abdi
Department of Biotechnology, Persian Gulf Research Institute, Persian Gulf University, Bushehr, Iran
e-mail: abdi@pgu.ac.ir

V. Singh (ed.), *Advances in Metabolomics*,
https://doi.org/10.1007/978-981-97-7459-3_8

8.1 Introduction

In 1936, a Dutch physicist named Gorter predicted the existence of nuclear magnetic resonance. He attempted to demonstrate this phenomenon in lithium fluoride but was unsuccessful. In 1939, Rabi made a significant advancement in beam methods. He carried a stream of hydrogen molecules via two different magnetic fields that weren't uniform and a homogeneous magnetic field, exposing them to radiofrequency (RF) electromagnetic energy. The molecules absorbed energy with a highly defined frequency, resulting in a modest but observable deflection of the beam. Rabi received the Nobel Prize in 1944 for his pioneering work in making the first NMR observations. These investigations, however, were restricted to tiny molecule nuclei in a molecular beam operating at a very high vacuum. The resonance was detected by measuring the deflection of the beam. In 1937, it was discovered that nuclear magnetic moments contribute to bulk magnetic susceptibility for hydrogen at low temperatures. Bloch, Hansen, and Packard attempted the experiment in 1945 with a water sample. Purcell, Torrey, and Pound were able to quantify the minute absorption of radiofrequency energy by proton magnetic moments in a paraffin block the same year (Becker 1993). For their contributions to the area of NMR spectroscopy, Bloch and Purcell received the Prize in 1952. The first useful NMR experiments were developed as a result of their contributions. Over time, NMR spectroscopy has evolved from a physical experiment to a useful tool for chemists in structure elucidation. The discovery of the "chemical shift" helped to solidify its usefulness. Initially, NMR spectrometers were continuous wave (CW) equipment with electromagnets, but their usefulness ended in the 1970s with the introduction of superconducting magnets. However, it wasn't until Ernst created the fundamentals of the Fourier transformation (FT) approach that modern NMR spectroscopy methods were established. In 1991, Ernst became the first chemist to win the Nobel Prize since NMR spectroscopy was previously limited to physicists. A decade later, Wüthrich became the second chemist to be awarded the prize. In 2002, he won the prize for explaining the three-dimensional structures of macromolecules. It is not surprising that the NMR technique was honored again the following year, with the Nobel Prizes being awarded to Peter Lauterbur and Peter Mansfield for their contributions to magnetic resonance imaging (Emsley and Feeney 1995).

Technological developments in NMR data capture and interpretation have demonstrated advantages in analyzing extremely complex biological mixtures. Despite their broad spectrum of uses, most metabolomics applications have been directed toward the comprehension, prevention, diagnosis, and treatment of human diseases (Nagana Gowda and Raftery 2022). Since sample preparation and separation procedures greatly increase analytical variability, NMR makes it possible to analyze whole tissues and biofluids without the requirement for the processing or separation of samples. Being non-destructive, NMR allows the material to be preserved for future NMR reanalysis or use with other techniques like mass spectrometry. Using a range of stable isotope-labeled precursors, NMR makes it possible to monitor metabolic fluxes and follow metabolic pathways. NMR may identify metabolites using one or more atomic nuclei, such as ^{31}P, ^{1}H, ^{15}N, or ^{13}C (Trammell and Brenner 2013; Kumar et al. 2013; Purwaha et al. 2014; Nagana Gowda et al. 2015).

8.1.1 The Core Concept: Nuclear Magnetic Resonance—A Dance of Nuclei and Radiowaves

The foundation of NMR spectroscopy is the existence of spin (I) and magnetic moment (μ) in specific atomic nuclei. Angular momentum is quantized, as per quantum theory. An energy spectrum emerges when the magnetic moment of such nuclei interacts with an external magnetic field (B_0). For a nucleus with spin I, the $2I + 1$ stationary or energetic eigenstates correspond to the direction of the angular momentum with respect to any arbitrary axis z. In the simplest situation of nuclei with spin ½, there are two spin states that correspond to the magnetic quantum numbers mI = ½ and −½. The eigenfunctions $\alpha >$ (mI = ½) and $\beta >$ (mI = −½) represent these two spin states (Fig. 8.1).

Both energy levels are equal when there is no magnetic field (B_0). This, however, alters when B_0 is present. The two states' combined energy is equal to $-1/2\gamma B_0$. The nucleus under study has a unique property known as the gyromagnetic ratio ($\gamma = \mu/I(h/2\pi)$). Between the two levels, the energy separation (ΔE) is equal to γB_0 and is proportional to B_0. The distribution of spin populations between the two states in a sample at a specific temperature T is described by the Boltzmann relation for a large number of such nuclei:

$$N\beta / N\alpha = \exp(-\Delta E / \mathrm{kT}) = \exp(-\mathrm{hv0} / \mathrm{kT}) \approx 1 - \mathrm{hv0} / 2\pi kT \qquad (8.1)$$

The number of nuclei in their ground and excited states is represented by $N\alpha$ and $N\beta$, respectively. ΔE stands for the difference in energy between these states. k refers to the Boltzmann constant, and T represents the absolute temperature, also known as the spin temperature (Ts). As the energy difference ΔE is relatively low, the equilibrium state will have a slightly higher number of nuclei in the ground state ($N\alpha$) as compared to the excited state ($N\beta$). At a magnetic field strength of 2.35 T and ambient temperature, the energy difference (ΔE) for protons is approximately 0.04 J mol^{-1}. The number of protons in the lower energy state affects the probability of a transition, which in turn affects the sensitivity of the experiment. This number is very small, only about 0.002% or roughly 20 nuclei in 10^6. Therefore, NMR

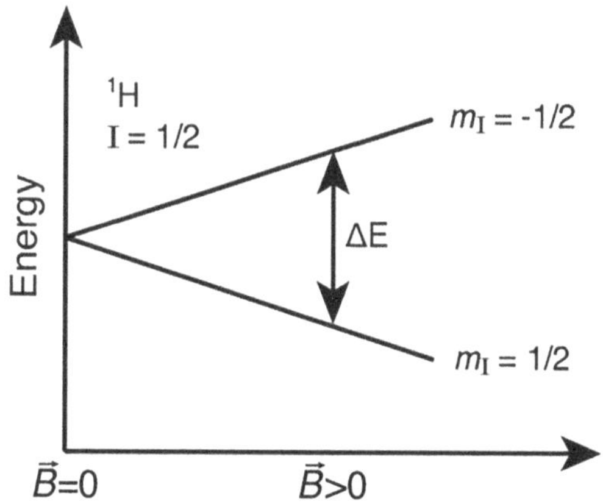

Fig. 8.1 Splitting of nuclei spin energies in an external magnetic field (Chary and Govil 2008)

spectroscopy must be able to detect very weak signals. Equation (8.1) shows that the sensitivity of NMR experiments can be improved by increasing the magnetic field strength and decreasing the temperature.

The three most critical components of an NMR spectrometer are the magnet, RF source, and detector. During the investigation process, the substance being studied is placed within a sample tube, which is essentially a glass tube that is around 15 cm long and has a diameter of 5 or 10 mm. The sample tube is placed within the external magnetic field B_0. An RF coil, known as the transmitter, generates RF radiation, which then stimulates a signal that is detected by either the same coil or a different coil, known as the receiver. Once the signal is amplified and transmitted to an x, y plotter or computer, the spectrum and resonance frequencies can be monitored. The NMR spectra can be measured using CW (continuous wave) and FT (Fourier transform) NMR spectroscopy.

Atomic nuclei do not all have magnetic moments. In fact, spin quantum values larger than 1/2 are possible for nuclei larger than the proton. The magnetic quantum numbers mI = $I, I - 1, I - 2, ..., - I$, define the spin states of such nuclei. For even-even nuclei with even mass and even atomic number, $I = 0$. For all other nuclei, I is greater than or equal to 1/2. Even-odd nuclei have I as an integral multiple of 1, whereas odd-even and odd-odd nuclei have I as an integral multiple of 1/2. Therefore, magnetic resonance is applicable in nuclei with an odd number of either protons or neutrons.

The NMR spectroscopy of proton is well-known. However, the spectra of other nuclei like carbon-13 (^{13}C), nitrogen-15 (^{15}N), fluorine-19 (^{19}F), silicon-29 (^{29}Si), and phosphorus-31 (^{31}P) have also been extensively studied. Moreover, many metal nuclei are currently available for NMR investigations (Chary and Govil 2008; Günther 2013).

8.1.1.1 Nuclear Relaxation

Within the domain of nuclear magnetic resonance (NMR), "relaxation" describes the mechanism through which magnetic atomic nuclei achieve thermal equilibrium with their disordered molecular surroundings. The duration of this process varies based on the sample and surrounding circumstances, ranging from a few seconds to several minutes. Relaxation is beneficial for NMR spectroscopists because it generates the nuclear magnetization that forms the basis of all NMR signals. However, relaxation may also be detrimental because it limits the duration of any different types of spin order. For a long time, it was believed that relaxation was an inherent property of the sample and could not be easily controlled or avoided. Recent investigations have uncovered situations where nuclear spin systems can exist in specific configurations known as long-lived states, which are relatively immune to relaxation. Two types of relaxation processes can be characterized:

1. Spin–lattice relaxation (longitudinal relaxation) T_1
2. Spin–spin relaxation (transverse relaxation) T_2

Spin–lattice relaxation (T_1) occurs when energy is transferred from the spins of magnetic nuclei to the surrounding environment, resulting in heat. The mobility of molecules containing magnetic nuclei causes fluctuating magnetic fields, which can be phased and oriented to match the processional frequency. This allows an upper spin nucleus to relax to the ground state by transferring its excess energy to the lattice as rotational and translational energy. This process is called nuclear-dipolar interaction and does not alter the system's total energy. The strength of the local fields and their rate of fluctuation determine how effective the relaxation mechanism is. A high T_1 number indicates that the relaxation process is ineffective and the state of excitement lasts longer. The spontaneous relaxation rate of nuclear spin orientations is almost negligible. T_1 relaxation occurs due to temporary magnetic fields (often caused by molecular motion) at the Larmor frequency. T_1 relaxation is most effective when the molecule reorientation rate in space is on average equal to the Larmor precession frequency. On the other hand, typical molecule rotation rates in mobile liquids at normal temperature are several orders of magnitude larger than the Larmor frequency, so only a very small fraction of the movements are at the appropriate frequency, leading to very inefficient relaxation (long T_1). Molecular processes slowdown in larger molecules and more viscous fluids, resulting in more efficient relaxation. However, at some time, the average molecular movements become slower than ν_0, causing T_1 to increase again (Mlynárik 2017).

T_2 relaxation, also known as spin–spin relaxation, occurs when nearby atomic nuclei exchange energy without releasing any energy to the surrounding environment. This results in a loss of magnetization in the perpendicular plane. After the radiofrequency (RF) excitation, all of the proton magnetic moments align in the $x - y$ plane and rotate together with a phase difference of 0°. This alignment produces the transverse magnetization that is detected by the voltage it generates while rotating. When the excitation pulse is stopped, the synchronous rotation is gradually lost, causing some moments to speed up and others to slow down, ultimately resulting in a 360° distribution around the z-axis. When protons spin, they produce weak magnetic fields around them. These magnetic fields interact with each other, causing slight fluctuations in the speed of their precession. This interaction is called dephasing, which leads to a loss of coherence between the spins. When nearby spins exchange energy, this decay is called spin–spin relaxation. As a result of this relaxation, there is a loss of phase coherence. Consequently, the induced voltage created by the net magnetic vector cutting over B_0 decreases gradually over time, leading to the corresponding MR signal gradually diminishing and ultimately disappearing. The time constant for this decay process is T_2, and it is determined by the transfer of energy between proton spins. Typical T_2 relaxation periods range from 100 to 300 ms. Although T_2 relaxation is fully independent of T_1 relaxation, it occurs concurrently. T_2 relaxation, like T_1 relaxation, is an exponential function and is defined as the duration for a drop in transverse magnetization by 63% (Valori and Nicot 2019) (Fig. 8.2).

8.1.1.2 Sample Preparation and Standardization

Preparing samples for NMR spectroscopy is a simple process. To conduct ^1H NMR, you need solvents without protons, but this is not a big issue as deuterated substances are widely available. Deuterated solvents are used, including $CDCl_3$ (deuterochloroform), which is the most common solvent for organic molecules. Other deuterated solvents, such as dimethyl sulfoxide, acetone, acetonitrile, benzene and

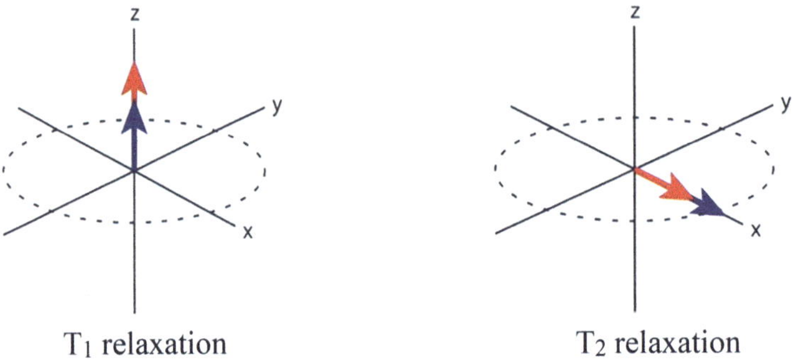

Fig. 8.2 T_1 relaxation and T_2 relation (Valori and Nicot 2019)

heavy water, are used for weakly soluble materials and other purposes. The lock channel uses deuteron resonance to stabilize the spectrometer. In certain situations, the choice of solvent for NMR spectroscopy is influenced by the fact that the compound being studied is formed in the NMR tube by reacting the solvent or another component with a starting material. For example, protonated ketones can be obtained by this method using trifluoroacetic acid. Simple carbonium ions were first observed by combining aliphatic fluorides with antimony(v) fluoride in sulfur dioxide, ultimately leading to the creation of hexafluoroantimonate. Similarly, a mixture of fluorosulfonic acid and antimony(v) fluoride produces similar effects.

Small amounts of partially deuterated solvent molecules can produce low-intensity signals in the spectrum, typically under 0.5%. Dimethyl sulfoxide and acetone show spin–spin splitting due to the presence of CHD2 groups. The proton interacts with two deuterons, resulting in spin values of +2, +1, 0, −1, and −2 in the statistical ratio of 1:2:3:2:1. The proton resonance signals in partially deuterated solvents display the same δ values as non-deuterated substances, except for a slight high-field shift caused by a deuterium-induced isotope effect on the 1H chemical shift.

To prepare a sample for measurement, take around 5–10 mg or 10 μL of the material and insert it into the sample cell, also known as the NMR tube. The NMR tube is a 17-cm-long, cylindrical glass tube with a 5-mm outside diameter. Then, dissolve it with approximately 0.5 mL of solvent. Finally, add a standard or reference substance, usually tetramethylsilane (TMS). However, handling TMS in tiny amounts can be challenging due to its low boiling point and high vapor pressure. Hence, it is recommended to keep 5% (v/v) TMS solutions in the commonly used solvents available. Alternatively, the solvent signal can be utilized as a reference, and the frequency difference to TMS can be incorporated into the plotter's control software. For certain nuclei, such as 15N, larger sample tubes are helpful if an appropriate probe-head is available. When only small amounts of a compound are available, special microcells and tubes with diameters ranging from 1 to 3 mm, known as Shigemi tubes, have been developed for measurements. The choice of standard, like the choice of solvent, depends on the substance being investigated. For example, tetramethylsilane, cyclopropane derivatives, and silyl compounds can produce a superposition of resonance and reference signals. In some cases, cyclohexane, methylene chloride, or benzene can serve as references. In general, any substance whose protons produce a sufficiently strong signal can be used as a reference (Akitt and Mann 2017).

Metabolomics research involves the utilization of biological samples, such as human tissue, animal models, urine, and blood plasma or serum. The identification of metabolites by a mix of 1D and 2D NMR techniques, database searching, spiking with genuine substances, and lastly measuring the found metabolites with just one internal benchmark are crucial phases in the process. The next step involves using metabolite concentrations to differentiate between disease and control groups. This is typically accomplished through univariate or multivariate statistical analysis, the creation and validation of classification models, and the evaluation of the models' sensitivity and specificity using the area under the receiver operating characteristic (ROC) curve. Furthermore, the utilization of metabolite concentrations in the identification of changed metabolic pathways contributes to the mechanistic knowledge of cellular function and offers insights into potential pharmacological targets for therapeutic development as well as translational options for disease prevention or treatment.

The interference from an enormous number of serum/plasma proteins (6–8 g/dL) presents a significant difficulty for metabolite profiling of intact serum/plasma. NMR experiments that uses a T_2 (transverse relaxation) filter, like the CPMG sequence (Beckonert et al. 2007), are commonly used to suppress the undesired macromolecular signals from proteins. Metabolites are retained by the CPMG sequence because they typically have longer T_2 relaxation times than macromolecules. Thus, intact blood serum and plasma metabolomics have long benefited from the CPMG-based NMR investigation.

The urine analysis does not require sample pretreatment because macromolecules are not present in urine and so the intact urine samples are typically utilized. The pH range of typical human urine, which is roughly between 5 and 8, is significant (Hernandez et al. 2001; Rylander et al. 2006; Welch et al. 2008). Because many peaks in the NMR spectra are sensitive to pH variations between samples, urine analysis poses a serious challenge. As a result, urine samples are usually combined in a 1:1 (v/v) ratio (at pH = 7.4) with a phosphate buffer solution in D_2O. Beckonert et al. (2007) released a comprehensive study that provides a full protocol for urine analysis by NMR. Detection process of metabolic disease by NMR indicated in Fig. 8.3.

Metabolites in biological specimens, including tissue and cells, can be examined by two distinct techniques. One technique allows for examining whole cells or tissue without the need to process the sample. This approach is appealing, but it has two main problems:

1. Low spectrum resolution because of the sample's non-uniform magnetic susceptibility and unwanted intra- and intermolecular interactions.
2. Changed metabolite profiles because of enzyme activity. The biological specimens are placed in sample tubes that are rotated at a magic angle (54.7° with respect to the magnetic field) during NMR data collecting in order to address the

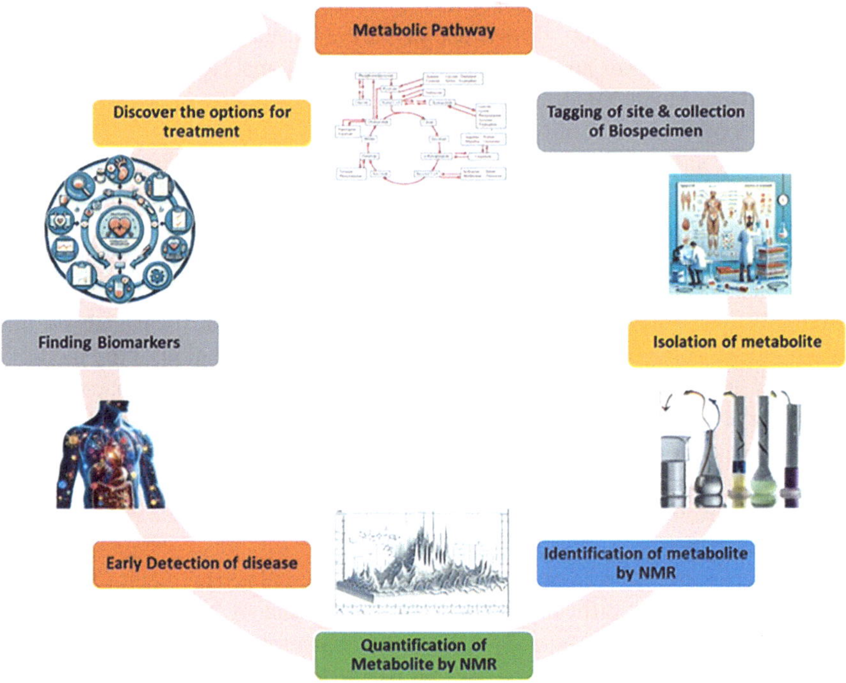

Fig. 8.3 Detection process of metabolic disease by NMR

first problem. To tackle the second problem, the sample is usually maintained at a low temperature (4 °C) throughout the investigation.

Pre-analyzing cells and tissues by extracting their metabolites is an alternative method. This method removes one of the main obstacles to the examination of intact materials. To extract aqueous metabolites, cells or tissue samples are usually homogenized in a solvent mixture of methanol and cold water. Chloroform is added to the water-methanol combination, allowing for the simultaneous extraction of lipid and aqueous metabolites.

Two phases are formed by this mixture of three solvents: Water- and methanol-containing upper phase dissolves aqueous metabolites, while Lipid metabolites are dissolved by the bottom phase, which comprises methanol and chloroform. By drying, the two phases are divided and the solvents are eliminated. A solution of deuterated chloroform ($CDCl_3$), deuterated methanol (CD_3OD), and water (D_2O) is used to dissolve the residue that contains lipids, usually in a 16:7:1 (v/v/v) ratio (which does not phase separate), for NMR analysis. D_2O is then added to the dry residue to dissolve the aqueous metabolites present in it.

8.1.2 Decoding the Spectral Score: Chemical Shifts and Coupling Constants

When atoms or molecules are placed in a magnetic field, the observed frequency of the nuclei's NMR differs from their theoretical value. This difference is known as the chemical shift or shielding effect. It is caused by the magnetic field that is formed by the electrons as they move around the nucleus, which acts against the applied field. The nucleus detects a lower field than the external magnetic field as a result, and the difference between the two is equivalent to the induced magnetic field.

$$H_{eff} = H_{applied} - H_{induced}$$

Chemical shift described the dependence of nuclear magnetic energy levels on the electronic environment in a molecule

$$\delta = \frac{\text{Difference in expression frequency between two nucleus}}{\text{Operating frequency of the magnet}}$$
$$= \frac{300\,\text{Hz}}{1\times10^{6}\,\text{Hz}} = 1\times10^{-6}\,\text{PPM}$$

The chemical shift is the difference in the position of a hydrogen atom's spectral signal in different chemical environments compared to an arbitrarily determined standard. The difference is commonly represented as the chemical shift (δ), described as

$$\delta = \frac{H_{sample} - H_{ref} \times 10^{-6}}{H_{ref}}$$

Electrons that orbit around the nucleus are affected by a magnetic field, which creates a magnetic field of their own. This is different from the magnetic field around the nucleus of a proton in a C–H bond. In Fig. 8.4, the difference between the signal and TMS is called the chemical shift (Nerli et al. 2018).

Spin–spin coupling: The actual field surrounding a proton or nucleus is not determined solely by the surrounding electrons, but also by the impact of magnetically active nuclei. This impact is due to spin–spin interaction. A proton's spin can be either α (MI = +1/2) or β (MI = −1/2). Consider proton A and proton B present in the surroundings. The field detected by proton A will depend on the spin of proton B.

Protons are typically surrounded by an electrical charge caused by nearby atoms and bonds. When exposed to a magnetic field (Bo), electrons start circulating, generating an induced field (Bi). Some spectra feature undistorted multiplets, where the multiplicity follows the (n + 1) rule. The splitting patterns can be easily identified as there are only two sets of non-equivalent protons in each situation. However, in some cases, when a proton has chemically comparable protons close by, the signal is divided into (n_a + 1), (n_b + 1), (n_c + 1), and so on. This leads to more complex multiplet patterns that are helpful in determining the structure (Krivdin 2018).

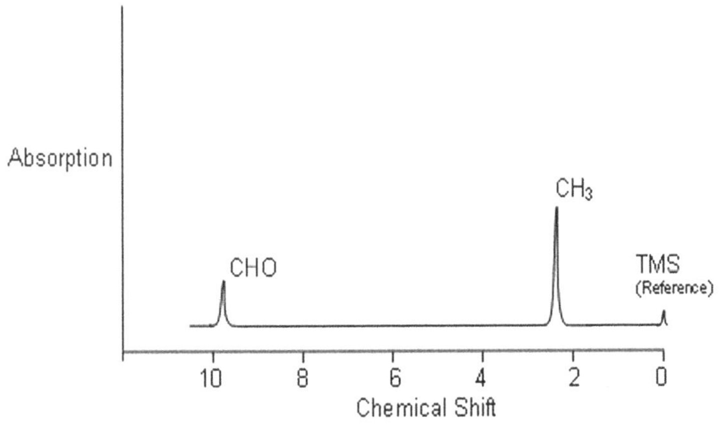

Fig. 8.4 NMR spectrum of ethanal (Nerli et al. 2018)

8.1.2.1 Factors Influencing Chemical Shift

1. **Inductive Effect (Electronegativity)**: A proton found in an organic compound is connected through a covalent sigma bond to carbon, nitrogen, oxygen, or other atoms. When exposed to a strong magnetic field, this substance, the electrons of the sigma bond circulate and create a small magnetic field that resists the applied field. The chemical shift of halomethane protons is directly proportional to the electronegativity of the halogen atom that is present. Fluorine (F) is particularly effective in removing electron density from the methyl group, which makes it more deshielded than other halogens (Dračínský 2017).
2. **Hydrogen Bonding**: When a proton forms a hydrogen bond, it causes a shift in its magnetic field, which is known as a downfield shift, compared to when it is not bonded. In phenol and carboxylic acids, hydrogen bonding causes changes in magnetic susceptibility at a chemical shift value greater than 10. Other factors such as the deshielding impact created by the anisotropy of the carbonyl group and the van der Waals deshielding effect can also impact the chemical shift (Hansen and Spanget-Larsen 2017).
3. **Magnetic Anisotropy:** It refers to the variation in magnetic characteristics at distinct sites in space caused by the application of a magnetic field. Protons connected to C=C in alkenes are less protected than in alkanes, with maxima at higher δ values. Alkyne (C≡C–H) has a modest δ value (1.5–3.5), but is expected to have a higher value. The movement of π electrons in >C≡C< produces an induced magnetic field that is parallel to the applied field. The alkyne molecule is linear and has a symmetrical distribution of π electrons. This is why the alkyne protons are shielded. When a proton experiences shielding effects caused by electronic circulations originating in other parts of the molecule, it means that the electron cloud around the proton is distorted in a way that makes it less sensitive to external magnetic fields. However, such effects resulting from electronic currents in other parts of the molecule do not contribute to the shielding of a proton because they are averaged to zero by rapid thermal motions. In a relatively rigid molecule, these currents can affect either shielding or deshielding of

a proton, depending on the orientation of the proton relative to the induced magnetic currents. These effects are called anisotropy effects (Nardelli et al. 2020).

8.1.2.2 Coupling Constant

The separation between the centers of two neighboring peaks in a multiplet is known as the spin–spin coupling constant (J) or coupling constant. The value of coupling constant is generally measured in Hz instead of δ (ppm) and is constant and independent of the magnetic field. As a result, the distance between two peaks in a multiplet stays stable. On the other hand, if this distance varies, it means that they are either independent signals or do not belong to the same peak or multiplet. In two connected multiplets, the peak separations (J values) are precisely equal. For example, 1,1,2-trichloroethane's 1H NMR spectra shows two multiplets, one doublet, and one triplet. The J values of 6.1 Hz are the same for the doublet and triplet.

The coupling constant J is used to quantify the magnetic interaction between the Ha and Hb hydrogen sets. This is of equal magnitude in both directions, implying that Ha affects Hb as much as Hb affects Ha. When studying multifaceted NMR spectra, the reciprocal coupling constants are a useful concept in determining the coupling relationships between different proton sets.

8.1.2.3 Factors Affecting Coupling Constant

The efficiency of spin–spin coupling is measured by the coupling constant. Its value, denoted by J, depends on the type, number, and arrangement of bonds between linked nuclei. When two nuclei are separated by one bond, the coupling is stronger than when separated by two bonds, and so on. The coupling constant is independent of the applied field strength, in contrast to chemical shift, which is measured in Hz and changes with the field strength. This is because the magnetic moment of a neighboring proton, which causes the spin–spin coupling, is not affected by the applied field strength. The coupling constant, denoted by J_{ab}, measures the strength of interaction between two atomic nuclei. The superscripts ^1J, ^2J, and ^3J signify one, two, or three bonds engaged in the coupling interaction, respectively, and indicate the number of bonds via which this interaction happens. However, subscript (ab) indicates nuclei (H_a and H_b) involved in the connection. Several factors influence the magnitude of coupling constants in various types of coupling (Krivdin 2020).

8.2 Geminal Coupling

Geminal protons are protons that are linked to the same carbon atom and are separated by two bonds. Coupling only occurs when these protons are not chemically equivalent. Geminal coupling constants (^{2}JHH) are typically negative or close to zero, but for simplicity, we will ignore the negative sign in this discussion. The value of the geminal coupling constant is influenced by factors such as the hybridization of the carbon atom, bond angles, and substituents like electronegative atoms. The overall geminal coupling constant value falls and turns positive as the s-character grows. Furthermore, the geminal coupling constant likewise drops and turns positive if an electronegative group is joined to the linked protons.

Variations in the HCH bond angle may have an impact on the ^{2}JHH value. The ^{2}J coupling constant tends to rise in magnitude when the bond angle between the coupled protons decreases. This is due to the fact that when the angle drops, the two atoms go closer to one another, strengthening the contact between their electron spins. The geminal coupling constant helps determine the ring size, with the absolute value of the constant increasing as the ring size increases. An increase in the germinal coupling constant results from a complementary HCH bond angle decreasing from 3 to 6 members as the CCC bond angle in the ring increases (Krivdin 2018).

8.3 Vicinal Coupling

Vicinal protons are protons that are connected to carbon atoms located nearby and separated by three bonds. The dihedral angle (the angle at which HCCH rotates), the length of the carbon–carbon bond, and the behaviors of electronegative atoms all contribute to the vicinal coupling constant, which is always positive. The dihedral angle ranges from 0 to 180°, with the angle at 90° resulting in the lowest (zero) vicinal coupling constant. When the two C–H bond orbitals are parallel, the side-by-side overlap of the orbitals achieves its greatest at 0 °C, and when they are perpendicular, it reaches its minimum at 90 °C. The orbitals' back lobes overlap at 180 °C. When an atom connected to the system becomes more electro-negative, the vicinal coupling constant falls (Burns et al. 2019).

8.4 Long Range Coupling

The strength of J reduces considerably as the space between atoms increases when there are more than three covalent bonds, the coupling is usually negligible, although in certain unusual circumstances, it can still be seen. Long-range coupling is the term used to describe this type of connection that occurs beyond three bonds. Olefins, acetylenes, aromatics, heteroaromatics, and strained ring systems (small or bridged rings) are the most common examples of this coupling (Charisiadis et al. 2014).

8.5 Allylic Coupling

Alkenes show a slight interaction between protons on the other end of the double bond and those replaced on carbons α, in addition to germinal and vicinal coupling. Allylic coupling is the coupling that occurs between four bonds. At roughly 0–3 Hz, the allylic coupling constant is quite low. The connection between the concerned nuclei seems to be facilitated by the π-electrons present in the double bond. The degree of overlap between the π bond and the carbon–hydrogen σ bond determines the size of this connection. In acetylenes, this kind of coupling also takes place.

8.6 Homoallylic Systems

There is coupling between the C–H σ-bonds on both sides of the double bond in certain alkenes. Because homoallylic coupling takes place over a greater distance (five links) than allylic coupling, it is weaker than allylic coupling. Although this form of coupling (5 *J*) is typically insignificant or even absent in most molecules, it may occasionally appear in the NMR spectrum.

8.7 Aromatic Coupling

Aromatic systems exhibit coupling that extends beyond the carbon atoms that are located nearby (more than three bonds away). While interactions can occur between protons at ortho, meta, and para positions, orthocoupling is the strongest, followed by meta and para. Usually, paracoupling is not encountered (Achar et al. 2019).

8.8 Rigid Systems

Although it is less frequent, long-range coupling can nevertheless happen in compounds without π-systems. This fascinating phenomenon is observed in a saturated system by a prearrangement of atoms that resembles the letter "W," with hydrogens located at each end. These inflexible strained arrangements possess the right shape for the involved overlaps, even though the coupling constant is usually quite modest, with the exception of severely strained ring systems. Utmost W-type couplings result in peak broadening instead of apparent splitting (Sengupta et al. 2013).

8.8.1 Unveiling the Nuclei: Different NMR Experiments

8.8.1.1 1H NMR Spectroscopy

By focusing on hydrogen-1 nuclei, proton nuclear magnetic resonance (proton NMR), sometimes referred to as hydrogen-1 NMR or 1H NMR, is a scientific method that uses nuclear magnetic resonance (NMR) spectroscopy to determine a substance's molecular structure. In samples that contain natural hydrogen (H), almost all of the hydrogen is made up of the isotope ^{1}H. The term "protium" refers to a full ^{1}H atom. For NMR, deuterated solvents are preferred, such as deuterated water, acetone, methanol, dimethyl sulfoxide, and chloroform. On the other hand, hydrogen-free solvents like carbon disulfide or tetrachloride can also be utilized. The proton NMR spectra of most organic compounds show chemical shifts between +14 and −4 ppm. Furthermore, protons interact with one other's spins. Each proton's integration curve shows how abundant certain protons are. Simple spectra are found in simple substances. For instance, a quartet at 3.5 ppm and a triplet at 1.5 ppm make up the 3:2 ratio of ethyl chloride's spectrum. In contrast, the diamagnetic ring current causes the benzene spectra to have a single peak at 7.2 ppm (Krivdin 2019).

Typically, 50–100 metabolites can be identified and quantified at once from a biofluid or tissue extract using its ^{1}H NMR spectrum (Holmes et al. 2000; Lindon et al. 2000). Several references hundreds of known metabolites' ^{1}H NMR spectra have been collected and kept up to date in a number of public databases, which substantially facilitates this identification process. Furthermore, a plethora of software tools are available to process and interpret 1D ^{1}H NMR spectra (offline) automatically or semi-automatically in a matter of minutes or seconds after the spectra are obtained (Hao et al. 2012; Ravanbakhsh et al. 2015; Rohnisch et al. 2018). These software programs are capable of quantifying (Rohnisch et al. 2018; Tardivel et al. 2017) and identifying (Cañueto et al. 2018) metabolites. Nowadays, ^{1}H NMR metabolomics is the preferred technique for conducting extensive population studies for examining thousands of owing to its benefits in speed and repeatability. Comparing simple 1D ^{1}H NMR spectra to all other forms of NMR spectroscopy, they are very useful for quantifying metabolites. This is because they are typically obtained without the use of polarization transfer procedures and in completely relaxed settings. In this method, ^{1}H NMR signals obtained in a complex mixture offer an accurate depiction of the distribution of proton nuclei inside the molecules and the various concentration levels of the respective metabolites.

8.8.1.2 ^{13}C NMR Spectroscopy

The utmost common form of carbon isotope is C-12, which is NMR inactive due to its nuclear spin of $I = 0$. On the other hand, C-13, or ^{13}C, has a nuclear spin of $I = 1/2$, much like a proton, and its natural abundance is 1.1%. The nuclei C-13 is hence NMR active. Inopportunely, because of C-13's lower magnetogyric ratio (g), which is only one-fourth of that of protons (^{1}H), and low natural abundance, it is more difficult to notice the resonances of C-13 nuclei than those of protons. Consequently, the sensitivity of C-13 nuclei is significantly lower, with an overall sensitivity of around 1/5700th that of the proton. The inferred information about a molecule's carbon skeleton that proton NMR offers is one of the primary benefits of ^{13}C NMR over ^{1}H NMR (Merchak et al. 2016).

8.9 Chemical Shift

The ^{13}C signals of organic compounds typically range from 0 to 220 ppm, with the exception of CH4 and CH3I, which exhibit signals of −2.4 and −24.0 ppm, respectively. Nonetheless, the heavy atom effect—spin–orbit coupling is present in this instance—causes the CI4 signal to appear at −293.0 ppm. These values in transition metal complexes can vary greatly, reaching values greater than +400 ppm.

Furthermore, M-CH3 carbon is often significantly upfield displaced, appearing at a lower frequency below TMS, which is 0 ppm.

8.10 Spin–Spin Splitting in ^{13}C NMR

It is rare to find ^{13}C nuclei in nature. Therefore, it is highly unlikely to discover two neighboring ^{13}C atoms in the same molecule. As a result, there is no spin–spin splitting between ^{13}C and ^{13}C nuclei. However, spin–spin splitting occurs when hydrogen is bonded to a certain ^{13}C atom. This type of spectrum is called the proton-linked ^{13}C NMR spectrum. The splitting of the ^{13}C signal follows the $n + 1$ rule. Because of their high coupling constant ($J = 100–320$ Hz), the proton-coupled 13C NMR spectra are difficult to interpret. When there is no coupling between ^{1}H and ^{13}C nuclei, the ^{13}C spectrum is known as a proton-decoupled ^{13}C NMR spectrum or $^{13}C\{^{1}H\}$. This reduces the spectrum to just singlets. Nevertheless, the information regarding the hydrogens that were attached was lost. Proton decoupled spectra comprise the majority of ^{13}C NMR spectra (Balci 2005).

8.11 Nuclear Overhauser Effect (NOE)

In comparison to a proton-coupled experiment, the intensity of some carbon resonances increases dramatically when we acquire a $^{13}C\{1H\}$ NMR spectrum. The greatest enhancement occurs when carbon atoms are directly bonded to hydrogen atoms, and this enhancement rises (albeit not always linearly) with the hydrogen atom count linked to a given carbon. This phenomenon is called the Nuclear Overhauser Effect (NOE), which refers to both the magnitude of the effect and the effect itself. NOE is essentially the change in a nucleus's signal strength caused by radiation from a nearby nucleus, and this effect occurs across space. This interaction can have either a negative or positive effect, depending on the type of atom involved. When the hydrogens are irradiated, the interaction between ^{13}C and ^{1}H has a positive effect on the carbon signals' intensities (Neuhaus 2007).

8.12 Distortionless Enhancement by Polarization Transfer (DEPT)

This method is employed to determine the protons numbers that are straight linked to each ^{13}C nucleus. As a result, we are able to decipher CH_3, CH_2, CH, and C signals. By using the DEPT approach, the sample is subjected to an intricate series of pulses. As a result, carbon atoms that are bonded with different numbers of hydrogen exhibit distinct phases. The duration of these complex pulses causes each type of carbon to behave slightly differently. It is possible to detect discrepancies and present the spectra from each experiment. The plotting four distinct sub spectra is a standard procedure for representing the results of a DEPT test. Every sub spectrum has distinct information that can be used to identify signals for every carbon atom in the molecule (Zheng et al. 2015).

^{13}C enrichment can be used in conjunction with clever spin-manipulation (as demonstrated in the DEPT experiment) to improve ^{13}C NMR signals. In plant metabolomic research, for example, ^{13}C-enriched CO_2 is mostly utilized to label the metabolites, while ^{13}C-enriched glucose is utilized to label the metabolites in the microbial metabolomics investigations (Wushensky et al. 2018) and mammalian cell line studies (Buescher et al. 2015). However, in most cases, ^{13}C enrichment is not practical for experiments involving humans or mammals. Shanaiah et al. used a chemical tagging technique or derivatization technique to concentrate unlabeled metabolites with ^{13}C in order to circumvent this issue (Gowda et al. 2012). In instance, this is possible to tag ^{13}C-specific classes of metabolites (like amino acids) through the ^{13}C acetylation of amines by incorporating ^{13}C-acetic anhydride into serum or urine samples. Through selective labeling, this tagging makes it possible to quickly collect ^{13}C NMR spectra while also greatly simplifying NMR spectrum of complicated biofluids, including urine. Urine samples from patients with inborn metabolic defects were effectively tagged with chemo-selective tagging, allowing for the comparison and identification of amino acids and associated metabolites (Shanaiah et al. 2007). We think there are numerous further uses for this elegant and straightforward method of boosting ^{13}C NMR signals in human or animal metabolomics.

8.12.1 2-D NMR Spectroscopy

A standard NMR spectrum is a one-dimensional plot of intensity versus frequency which does not provide complete information about the coupling between nuclei such as H–H, C–H, etc. However, it is possible to separate interactions between carbon and hydrogen atoms of sample compounds using a two-dimensional NMR (2D-NMR) plot, which involves two axes specified for frequency. This aids in determining the relationship between coupling nuclei, such as pairing of protons with carbons or protons. In 2D-NMR, both the x and y axes have chemical shift scales, and the spectra are presented on a grid, similar to a map. You can obtain information by analyzing the peaks in the grid and matching them to the axes. However, this method cannot be utilized to obtain coupling constant values. Nonetheless, if two resonances are coupled, it can still provide useful information for interpreting the spectrum.

Correlation spectroscopy is a key and extensively researched 2D-NMR method. This technique can be either homonuclear or heteronuclear. An example of homonuclear proton-proton is known as COSY (as shown in Fig. 8.5), while the heteronuclear carbon–proton correlation case is called HETCOR (as shown in Fig. 8.6).

Homonuclear COSY, which is proton-proton correlation spectroscopy, provides the proton NMR spectra of any molecule along the x-axis and repeats it along the y-axis. The signals are repeated and the contour of an off-diagonal cross-peak indicates when a proton connects with another proton. For example, In the COSY of

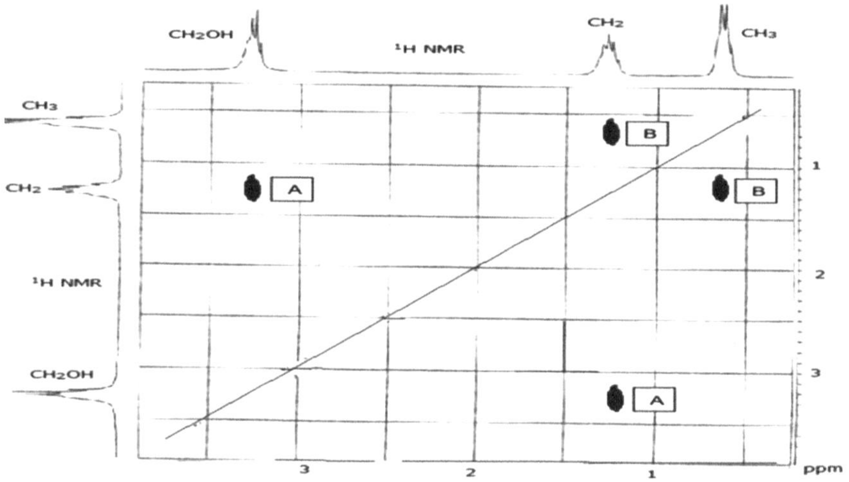

Fig. 8.5 COSY of 1-propanol (Nilsson and Morris 2007)

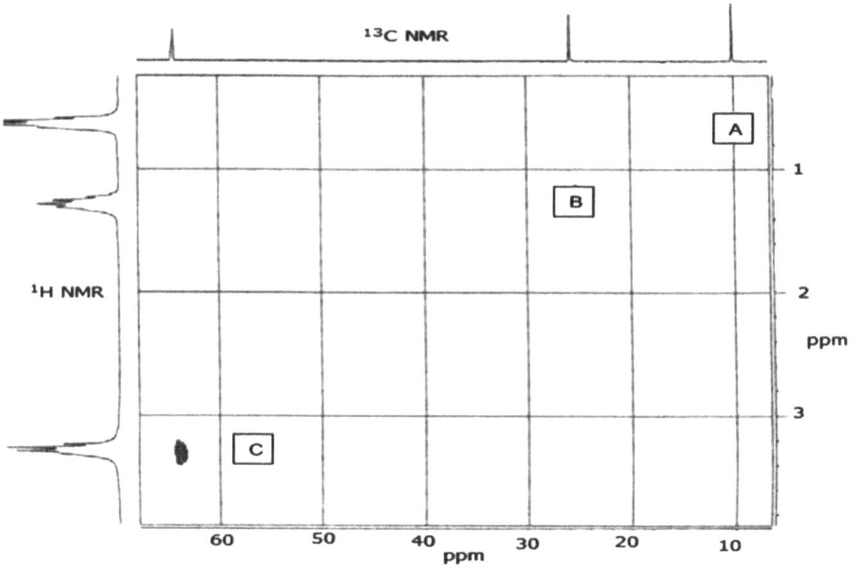

Fig. 8.6 HETCOR of 1-propanol (Nilsson and Morris 2007)

1-propanol ($CH_3CH_2CH_2OH$), there are two labeled peaks, A and B. Peak A shows a coupling interaction between the H at 1.5 ppm and the H at 3.6 ppm, which indicates a connection between the CH_2 group and the neighboring H on the CH_2OH group. Peak B, on the other hand, represents a coupling interaction between the H at 0.7 and 1.3 ppm, indicating the connection of the CH_2 and CH_3 in 1-propanol.

There is another set of equivalent peaks labeled A and B on the opposite side of the diagonal (Nilsson and Morris 2007) (Fig. 8.5).

Carbon–proton correlation spectroscopy, also known as heteronuclear HETCOR, is a technique used to generate the ^{13}C NMR spectrum of any molecule along the *x*-axis and the proton NMR (^{1}H NMR) spectrum along the *y*-axis. The contour in this technique is used to describe when a carbon atom bonds with a proton. For instance, in the case of 1-propanol's HETCOR ($CH_3CH_2CH_2OH$), the A peak indicates a coupling contact between the H at 0.7 ppm and the C at 10 ppm. This relates to group CH_3. The peak labeled B signifies a coupling interaction between the H atom at 1.3 ppm and the C atom at 25 ppm, which corresponds to the CH_2 group. On the other hand, the peak labeled C represents a coupling interaction between the H atom at 3.6 ppm and the C atom at 65 ppm, which relates to the CH_2OH group (Yan et al. 2022).

When dealing with overlapping resonances in metabolomics, most often, 2D NMR is used to extend the peaks into a second dimension by taking into account another "orthogonal" physical attribute of the atom or atoms of interest (e.g., a covalently linked neighbor, a relaxation period, a coupling constant, etc.). It may be possible to identify and detect additional metabolites by 2D NMR than with 1D NMR due to the increased resolution that 2D NMR offers. For a long time, metabolomics studies have routinely used heteronuclear ^{13}C, ^{1}H, single quantum coherence (^{1}H-^{13}C-HSQC) and heteronuclear multiple bond correlation (HMBC) experiments in addition to various homonuclear 2D ^{1}H-^{1}H-NMR experiments, including numerous variations by total correlation spectroscopy (TOCSY) (Sandusky and Raftery 2005), correlation spectroscopy (COSY) (Le Guennec et al. 2014), and nuclear Overhauser effect (NOESY) experiments (Kumar et al. 1980).

8.12.2 NMR and Metabolomics

8.12.2.1 Advantages of Metabolomics

As shown in Fig. 8.7, in the study of metabolomics, the NMR provides the consecutive reproducible data that is highly reliable for the separation of various metabolic compounds from different sources. Moreover, metabolic flow measurements in vivo and in vitro are made possible by NMR. The accurate measurement of precursors and products is made possible by its intrinsic quantitative nature. Using NMR makes it relatively simple to map the locations of stable isotopes and incorporate points into molecules. Also, the method enables high-resolution magic-angle sample spinning (HRMAS) NMR to detect metabolites in tissues at trace levels. Typically, less than 200 metabolites are clearly identified and detected in a single measurement, depending on the spectral resolution (Fig. 8.7).

Since NMR is non-destructive, a sample can be subjected to multiple analyses on the same one. Furthermore, the sample has a lengthy recovery and storage period. The sensitivity is inherently poor; however, it can be increased by using cryo-cooled microprobes, stronger magnet field strength, numerous scans (duration), and hyperpolarization techniques (Nagana Gowda et al. 2019).

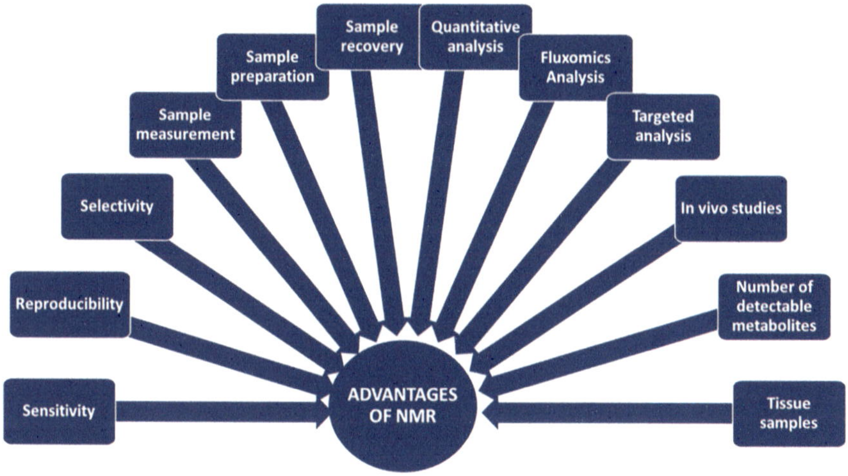

Fig. 8.7 Advantages of NMR in Metabolomics study

8.12.2.2 Identification and Quantification of Metabolomics by Databases

The most important approach is the NMR identification of metabolites; therefore, the metabolomics is associated by the several databases. The reference chemical shifts or, for pure chemicals, reference spectra are necessary for the NMR identification of metabolites. By comparing the observed spectra or observed chemical shifts, peak intensities, and/or coupling patterns to reference spectra, NMR spectroscopists can identify and even quantify compounds from measured NMR spectra. The majority of chemical compounds' NMR databases were assembled and stored in books. But today's NMR spectral databases have to be online due to their sheer size and complexity, particularly the ones necessary for metabolomics. Numerous excellent web-based NMR spectrum libraries that contain reference NMR spectra collected across a wide range of spectrometer frequencies for hundreds of compounds and for a variety of nuclei are now available, thanks to the work of various metabolomics labs worldwide (Sokolenko et al. 2013). The details of the important databases for the study of biological and chemical samples are mentioned in Table 8.1.

The Human Metabolome Database (HMDB), established and hosted by the University of Alberta, is a comprehensive, publicly available resource that provides complete data on over 114,000 metabolites. These entries are sourced from an extensive array of literature, including thousands of journal articles, books, and electronic databases. Individual entry in the HMDB is rich in data, detailing metabolite reactions, associated enzymes, transport mechanisms, chemical structures, physical and chemical properties, and links to related diseases and literature references. Additionally, the database encompasses a vast array of experimental data from numerous metabolomic studies performed on blood, urine, and cerebrospinal

Table 8.1 The NMR spectral databases for the study of metabolomics

Sr. no.	Name of database	Study	Developing unit	Link of database	References
1	Human Metabolome Database	Study of human or mammalian metabolomics	University of Alberta, Edmonton, Alberta,	HMDB, www.hmdb.ca	Wishart et al. (2012)
2	NMRshiftdb2 Database	Study of organic compounds	Stefan Kuhn and Christoph Steinbeck	http://nmrshiftdb.nmr.uni-koeln.de/	Farag et al. (2016)
3	National Institute of Advanced Industrial Science and Technology (AIST), Japan	Study of organic compounds	AIST, Japan	SDBS, http://sdbs.db.aist.go.jp/sdbs/cgi-bin/cre_index.cgi	Emwas et al. (2019)
4	Madison-Qingdao Metabolomics Consortium Database	Study of metabolomics of biological samples	National Magnetic Resonance Facility at the University of Wisconsin (Madison)	http://mmcd.nmrfam.wisc.edu/	Cui et al. (2008)
5	Biological Magnetic Resonance Data Bank	Study of metabolomics of biological samples	National Institute of General Medical Sciences	https://bmrb.io/	Ulrich et al. (2007)

fluid by nuclear magnetic resonance (NMR) and mass spectrometry (MS) techniques, enhancing its utility in research.

Another essential resource is the Madison-Qingdao Metabolomics Consortium Database (MMCD), which is maintained by the National Magnetic Resonance Facility at the University of Wisconsin-Madison. It supports high-throughput quantification and identification of metabolites through NMR and MS. The MMCD contains detailed information on over 20,000 substances, including a broad spectrum of NMR spectral data, both experimental and literature-derived.

The NMRshiftdb2 database, specifically designed for organic chemists, offers a repository of NMR spectra for over 40,000 substances, mostly non-biological. Despite this, the database is increasingly including metabolite spectra to support growing interest in metabolomics.

Similarly, the Spectral Database for Organic Compounds (SDBS) preserved by Japan's National Institute of Advanced Industrial Science and Technology provides multiple types of spectral data for a large number of organic compounds. While the majority are non-metabolites, this resource is invaluable for chemical analysis.

Innovations in metabolomics continue to advance, such as new methods for the accurate automated quantification and identification of urine metabolites using synthetic samples. This approach significantly enhances the predictability and identification of metabolites from NMR spectra.

Furthermore, software developments, such as the COLMAR algorithm by the Bruschweiler group and MetaboMiner, facilitate more effective metabolite identification and assessment from complex 2D NMR spectra, though these tools are primarily focused on metabolite identification due to the intrinsic challenges of extracting quantitative information from 2D NMR data.

8.12.2.3 Advances in NMR Equipments

NMR techniques created by physicists or chemists are always evolving to get beyond some of the intrinsic instrumental restrictions of NMR. New NMR technologies that boost resolution, quicken spectrum acquisition, or increase sensitivity are also being developed by NMR engineers. Exciting advancements in NMR technology, such as enhanced probes and powerful high field magnets, are currently in development or on the horizon.

The responsiveness and resolution of NMR are intricately linked to the amplitude of the magnetic field. Specifically, sensitivity or signal-to-noise (S/N) improves by ~Bo3/2 with increasing field strength. Furthermore, for the resolution of 1D NMR increases in a proportional way with the intensity of the magnetic field; for 2D NMR, it increases by the square of the field. Consequently, Considerable efforts are underway to build NMR spectrometers with extremely high magnetic fields. In particular, a brand-new superconducting substance that increases the NMR magnets' magnetic field intensity while lowering the expense of helium pumping and refrigeration. The development of ultra-high-field spectrometers involves the utilization of 2.2 thousand inner chambers. As of right now, 950 MHz and 1 GHz are offered for sale, and 1.2 GHz magnets should be produced in around 2020. In comparison to magnets with lower magnetic intensity, these high-field magnets enable the detection of a wider range of compounds in one testing session due to their increased sensitivity and improved resolution. The advancement of cryostats that require no maintenance (for instant the systems of Bruker Aeon), which eliminate the need for a liquid N2 shield and autonomously recycle liquid helium, is another result of recent advancements in magnet technology (Tadanki et al. 2012).

The NMR probe fitted with technological devices, both receiving and transmitting coils, and extra machinery such as shimming, securing, and tuning infrastructure, in order to send and receive electromagnetic waves into an NMR specimen. The advancements in probe design and development have made it possible to conduct new tests with ever-increasing sensitivity and a larger variety of nuclei detected. In the last decade, several advancements in probe technology have greatly enhanced the field of NMR-based metabolomics. Micro-coil probes, one of the most recent developments in micro-probes, not only increase sensitivity but also minimize the sample size needed to a few microliters. Cryogenically cooled probes, also known as cryoprobes (Grimes and O'Connell 2011), for instance, can boost signal sensitivity by a ratio of three to four. In fact, metabolomics investigations of urine and serum have effectively employed micro coil NMR probes, which have shown considerable gains in mass sensitivity over conventional 5 mm probes.

8.12.3 NMR Limitations in Metabolomics

Firstly, even with the improvements brought about by stronger magnetic fields, digital signal and cryo-probes processing, many significant, low-abundance metabolites remain undetectable with current NMR technology (Wishart et al. 2012).

Additionally, NMR spectroscopy has always been limited primarily by its low sensitivity. Although the number of acquisition scans can be increased to achieve a theoretically infinite sensitivity for NMR, the practical detection limits remain in the low-μmolar to high-nanomolar limits.

Next, the overlapping resonances are a persistent problem in the quantification and identification of metabolites, particularly in the case of proton NMR spectra of biofluids. The majority of NMR-based metabolomics investigations employ ^{1}D ^{1}H NMR spectroscopy, which exhibits a restricted chemical shift dispersion and a small range of resonances between 1 and 4 ppm. Peak assignment and compound identification remain quite challenging as a result, particularly at lower magnetic field strengths.

Finally, in comparison to mass spectrometers and other standard analytical instruments, NMR spectrometers are generally highly expensive. Furthermore, NMR equipment needs a large amount of laboratory area which should have non-vibrational floor and separation from radiofrequency and magnetic interference, with highly qualified operators.

8.12.4 Conclusion

This comprehensive review has underscored the pivotal role that NMR spectroscopy is a versatile and robust technique that is used in the field of metabolomics to analyze complicated biological mixtures. Through advancements in technology and methodology, NMR spectroscopy has proven indispensable for identifying and quantifying metabolites, aiding in the understanding of metabolic pathways, and contributing significantly to disease diagnosis and monitoring. The non-destructive nature of NMR allows for repeated analyses of the same sample, which is crucial for validating scientific findings and improving experimental reproducibility. Despite challenges such as its relatively low sensitivity compared to other analytical techniques and the high costs associated with NMR equipment, the continuous improvements in magnetic field strength and probe design promise to enhance resolution and sensitivity. As we look toward the future, the integration of NMR with other omics technologies and the development of more sophisticated data analysis tools will likely open new avenues for research and clinical diagnostics, cementing NMR's role as a cornerstone of molecular biology and biochemistry.

References

Achar TK, Zhang X, Mondal R, Shanavas MS, Maiti S, Maity S, Maiti D (2019) Palladium-catalyzed directed meta-selective C– H allylation of arenes: unactivated internal olefins as allyl surrogates. Angew Chem 131(30):10461–10468

Akitt JW, Mann BE (2017) NMR and chemistry: an introduction to modern NMR spectroscopy. CRC Press, Boca Raton

Balci M (2005) Spin-spin splitting in 1H-NMR spectra. In: Basic 1H-and 13C-NMR spectroscopy. Elsevier Science, Amsterdam

Becker ED (1993) A brief history of nuclear magnetic resonance. Anal Chem 65(6):295A–302A

Beckonert O, Keun HC, Ebbels TM, Bundy J, Holmes E, Lindon JC, Nicholson JK (2007) Metabolic profiling, metabolomic and metabonomic procedures for NMR spectroscopy of urine, plasma, serum and tissue extracts. Nat Protoc 2(11):2692

Buescher JM, Antoniewicz MR, Boros LG, Burgess SC, Brunengraber H, Clish CB, Fendt SM (2015) A roadmap for interpreting 13C metabolite labeling patterns from cells. Curr Opin Biotechnol 34:189–201

Burns DC, Mazzola EP, Reynolds WF (2019) The role of computer-assisted structure elucidation (CASE) programs in the structure elucidation of complex natural products. Nat Prod Rep 36(6):919–933

Cañueto D, Gómez J, Salek RM, Correig X, Cañellas N (2018) rDolphin: a GUI R package for proficient automatic profiling of 1D 1 H-NMR spectra of study datasets. Metabolomics 14:1–5

Charisiadis P, Kontogianni VG, Tsiafoulis CG, Tzakos AG, Siskos M, Gerothanassis IP (2014) 1H-NMR as a structural and analytical tool of intra-and intermolecular hydrogen bonds of phenol-containing natural products and model compounds. Molecules 19(9):13643–13682

Chary KVR, Govil G (2008) Basic concepts in NMR spectroscopy. In: NMR in biological systems: from molecules to humans. Springer, pp 1–41

Cui Q, Lewis IA, Hegeman AD, Anderson ME, Li J, Schulte CF et al (2008) Metabolite identification via the Madison metabolomics consortium database. Nat Biotechnol 26(2):162–164

Dračínský M (2017) The chemical bond: the perspective of NMR spectroscopy. Annu Rep NMR Spectrosc 90:1–40

Emsley JW, Feeney J (1995) Milestones in the first fifty years of NMR. Prog Nucl Magn Reson Spectrosc 28(1):1–9

Emwas AH, Roy R, McKay RT, Tenori L, Saccenti E, Gowda GN, Wishart DS (2019) NMR spectroscopy for metabolomics research. Metabolites 9(7):123

Farag MA, Porzel A, Al-Hammady MA, Hegazy MEF, Meyer A, Mohamed TA, Wessjohann LA (2016) Soft corals biodiversity in the Egyptian Red Sea: a comparative MS and NMR metabolomics approach of wild and aquarium grown species. J Proteome Res 15(4):1274–1287

Gowda GAN, Shanaiah N, Raftery D (2012) Isotope enhanced approaches in metabolomics. In: Atreya HS (ed) Isotope labeling in biomolecular NMR, vol 992. Springer, Berlin/Heidelberg, pp 147–164

Grimes JH, O'Connell TM (2011) The application of micro-coil NMR probe technology to metabolomics of urine and serum. J Biomol NMR 49:297–305

Guennec AL, Giraudeau P, Caldarelli S (2014) Evaluation of fast 2D NMR for metabolomics. Anal Chem 86(12):5946–5954

Günther H (2013) NMR spectroscopy: basic principles, concepts and applications in chemistry. Wiley, Hoboken

Hansen PE, Spanget-Larsen J (2017) NMR and IR investigations of strong intramolecular hydrogen bonds. Molecules 22(4):552

Hao J, Astle W, De Iorio M, Ebbels TM (2012) BATMAN—an R package for the automated quantification of metabolites from nuclear magnetic resonance spectra using a Bayesian model. Bioinformatics 28(15):2088–2090

Hernandez ME, Lopez AC, Calatayud AG et al (2001) Vesical uric acid lithiasis in a child with renal hypouricemia. An Esp Pediatr 55(3):273–276

Holmes E, Nicholls AW, Lindon JC, Connor SC, Connelly JC, Haselden JN, Nicholson JK (2000) Chemometric models for toxicity classification based on NMR spectra of biofluids. Chem Res Toxicol 13(6):471–478
Krivdin LB (2018) Carbon-carbon spin-spin coupling constants: practical applications of theoretical calculations. Prog Nucl Magn Reson Spectrosc 105:54–99
Krivdin LB (2019) Computational 1H NMR: part 1 theoretical background. Magn Reson Chem 57(11):897–914
Krivdin LB (2020) Recent advances in computational 31P NMR: part 2. Spin–spin coupling constants. Magn Reson Chem 58(6):500–511
Kumar A, Ernst RR, Wüthrich K (1980) A 2D NOE experiment for the elucidation of complete proton-proton cross-relaxation networks in biological macromolecules. Biochem Biophys Res Commun 95:1–6
Kumar M, Chatterjee A, Khedkar AP, Kusumanchi M, Adhikary L (2013) Mass spectrometric distinction of in-source and in-solution pyroglutamate and succinimide in proteins: a case study on rhG-CSF. J Am Soc Mass Spectrom 24(2):202–212
Lindon JC, Nicholson JK, Holmes E, Everett JR (2000) Metabonomics: metabolic processes studied by NMR spectroscopy of biofluids. Concepts Magn Reson An Educ J 12(5):289–320
Merchak N, Silvestre V, Rouger L, Giraudeau P, Rizk T, Bejjani J, Akoka S (2016) Precise and rapid isotopomic analysis by 1H–13C 2D NMR: application to triacylglycerol matrices. Talanta 156:239–244
Mlynárik V (2017) Introduction to nuclear magnetic resonance. Anal Biochem 529:4–9
Nagana Gowda GA, Raftery D (2022) Quantitative NMR methods in metabolomics. In: Metabolomics and its impact on health and diseases. Springer International Publishing, Cham, pp 143–164
Nagana Gowda GA, Gowda YN, Raftery D (2015) Expanding the limits of human blood metabolite quantitation using NMR spectroscopy. Anal Chem 87(1):706–715
Nagana Gowda GA, Abell L, Tian R (2019) Extending the scope of 1H NMR spectroscopy for the analysis of cellular coenzyme A and acetyl coenzyme A. Anal Chem 91(3):2464–2471
Nardelli F, Borsacchi S, Calucci L, Carignani E, Martini F, Geppi M (2020) Anisotropy and NMR spectroscopy. Rend Lincei Sci Fis e Nat 31(4):999–1010
Nerli S, McShan AC, Sgourakis NG (2018) Chemical shift-based methods in NMR structure determination. Prog Nucl Magn Reson Spectrosc 106:1–25
Neuhaus D (2007) Nuclear overhauser effect. In: eMagRes. Wiley, Hoboken
Nilsson M, Morris GA (2007) Pure shift proton DOSY: diffusion-ordered 1 H spectra without multiplet structure. Chem Commun 9:933–935
Purwaha P, Silva LP, Hawke DH, Weinstein JN, Lorenzi PL (2014) An artifact in LC-MS/MS measurement of glutamine and glutamic acid: in-source cyclization to pyroglutamic acid. Anal Chem 86(12):5633–5637
Ravanbakhsh S, Liu P, Bjordahl TC, Mandal R, Grant JR, Wilson M, Wishart DS (2015) Accurate, fully-automated NMR spectral profiling for metabolomics. PLoS One 10(5):e0124219
Röhnisch HE, Eriksson J, Müllner E, Agback P, Sandström C, Moazzami AA (2018) AQuA: an automated quantification algorithm for high-throughput NMR-based metabolomics and its application in human plasma. Anal Chem 90(3):2095–2102
Rylander R, Remer T, Berkemeyer S, Vormann J (2006) Acid-base status affects renal magnesium losses in healthy, elderly persons. J Nutr 136(9):2374–2377
Sandusky P, Raftery D (2005) Use of selective TOCSY NMR experiments for quantifying minor components in complex mixtures: application to the metabonomics of amino acids in honey. Anal Chem 77(8):2455–2463
Sengupta I, Nadaud PS, Jaroniec CP (2013) Protein structure determination with paramagnetic solid-state NMR spectroscopy. Acc Chem Res 46(9):2117–2126
Shanaiah N, Desilva MA, Nagana Gowda GA, Raftery MA, Hainline BE, Raftery D (2007) Class selection of amino acid metabolites in body fluids using chemical derivatization and their enhanced 13C NMR. Proc Natl Acad Sci 104(28):11540–11544

Sokolenko S, McKay R, Blondeel EJ, Lewis MJ, Chang D, George B, Aucoin MG (2013) Understanding the variability of compound quantification from targeted profiling metabolomics of 1D-1 H-NMR spectra in synthetic mixtures and urine with additional insights on choice of pulse sequences and robotic sampling. Metabolomics 9:887–903

Tadanki S, Colon RD, Moore J, Waddell KW (2012) Double tuning a single input probe for heteronuclear NMR spectroscopy at low field. J Magn Reson 223:64–67

Tardivel PJ, Canlet C, Lefort G, Tremblay-Franco M, Debrauwer L, Concordet D, Servien R (2017) ASICS: an automatic method for identification and quantification of metabolites in complex 1D 1 H NMR spectra. Metabolomics 13:1–9

Trammell SA, Brenner C (2013) Targeted, LCMS-based metabolomics for quantitative measurement of NAD+ metabolites. Comput Struct Biotechnol J 4(5):e201301012

Ulrich EL, Akutsu H, Doreleijers JF, Harano Y, Ioannidis YE, Lin J et al (2007) BioMagResBank. Nucleic Acids Res 36(suppl_1):D402–D408

Valori A, Nicot B (2019) A review of 60 years of NMR wettability. Petrophysics 60(02):255–263

Welch AA, Mulligan A, Bingham SA, Khaw KT (2008) Urine pH is an indicator of dietary acid–base load, fruit and vegetables and meat intakes: results from the European Prospective Investigation into Cancer and Nutrition (EPIC)-Norfolk population study. Br J Nutr 99(6):1335–1343

Wishart DS, Jewison T, Guo AC, Wilson M, Knox C, Liu Y, Scalbert A (2012) HMDB 3.0—the human metabolome database in 2013. Nucleic Acids Res 41(D1):D801–D807

Wushensky JA, Youngster T, Mendonca CM, Aristilde L (2018) Flux connections between gluconate pathway, glycolysis, and pentose–phosphate pathway during carbohydrate metabolism in Bacillus megaterium QM B1551. Front Microbiol 9:425123

Yan Z, Ye YQ, Zhang R (2022) 2D HETCOR solid-state NMR spectroscopy for multiphase materials with mobility contrast. J Phys Chem C 126(31):13311–13318

Zheng C, Zhu M, Zhang D (2015) Characterisation of asphaltenes extracted from an Indonesian oil sand using NMR, DEPT and MALDI-TOF. Energy Procedia 75:847–852

Transcriptional Control for Metabolite Production

9

Ajay Kumar, Juveriya Israr, Suruchi Gaur,
and Indrajeet Singh

Abbreviations

ABI3	Abscisic acid-insensitive 3
BAN	Banyuls
bHLH	Basic-Helix-Loop-Helix
CKII	Casein kinase II
COI1	Coronatin-insensitive 1
DFR	Dihydroflavonol-4-reductase
Em	Embryonic maturation
JA	Jasmonic acid
LAR	Leucoanthocyanidin reductase
MeJA	Methyl jasmonate
Pti4	Pto-interactor 4
TIA	Terpenoid indole alkaloids
vp1	Maize viviparous 1

A. Kumar (✉) · I. Singh
Department of Biotechnology, Faculty of Engineering and Technology, Rama University, Kanpur, India
e-mail: drajay.fet@ramauniversity.ac.in

J. Israr
Institute of Biosciences and Technology, Shri Ramswaroop Memorial University, Barabanki, Uttar Pradesh, India

S. Gaur
Biochemistry and Molecular Biology Lab, Department of Zoology, Banaras Hindu University, Varanasi, India

V. Singh (ed.), *Advances in Metabolomics*,
https://doi.org/10.1007/978-981-97-7459-3_9

9.1 Introduction

An important process that regulates the amount of secondary metabolites created in plant cells is the coordinated transcriptional regulation of biosynthetic genes. Metabolic processes are impacted by certain transcription factors. A class of proteins known as transcription factors attaches itself to particular DNA sequences in order to interact with gene promoters. Their main job is to control the rate at which RNA polymerase II initiates the synthesis of messenger RNA (mRNA). These proteins respond to a wide range of physiological and environmental cues, including hormones produced by bacteria and plants, UV light, and the particular type of tissue being studied, in order to regulate the transcription of genes. Events outside of the body can set in motion the process of signal creation. When exposed to the hormone jasmonic acid (JA), which is produced by elicitors, certain plant species benefit from the accumulation of secondary metabolites, according to Memelink et al. (2001). Two metabolic pathways in plants, flavonoid synthesis and terpenoid indole alkaloids (TIA) synthesis, have their transcription factors identified and are now under investigation. Upon completion of the flavonoid production process, the end result is anthocyanin pigments. Some plant species can regulate the expression of genes that code for anthocyanins, according to scientific research. There are two separate categories of transcription factors that impact this regulation. According to Mol et al. (1998), the c-MYB proto-oncogene in vertebrates produces a protein that resembles the Basic-Helix-Loop-Helix (bHLH) protein. The c-MYC proto-oncogene encodes it. Methyl jasmonate (MeJA) enhances the synthesis of terpenoid indole alkaloids (TIA) in *Catharanthus roseus*, commonly known as Madagascar periwinkle. When exposed to JA, the Octadecanoid-Responsive Catharanthus AP2/ERF-domain transcription factors (ORCA transcription factors) activate many TIA-producing genes (van der Fits and Memelink 2000; Memelink et al. 2001). Memelink et al. (2001) and Mol et al. (1998) have conducted studies on the role of these transcription factors in secondary metabolism. Therefore, this review will solely offer a concise summary of their function and portrayal. These regulators have the ability to adapt their levels and functions in response to both internal and external stimuli, enabling them to respond appropriately to growth or stress. Researching and speculating on ways to influence genetic and protein processes is the primary goal of this undertaking.

9.2 The Function of MYB and bHLH Proteins in the Process of Flavonoid Metabolism

The majority of plant MYB proteins have two helix-turn-helix motifs, R2 and R3, that are located in close proximity to one another. It is possible to bind specific regions of DNA using these patterns. Huge gene families encode the 125 R2R3-MYB genes in *Arabidopsis thaliana*. Several members of the MYB family collaborate with specific members of the bHLH protein family to regulate the flavonoid pathway. It has been suggested that MYB and bHLH proteins may physically interact (Mol et al. 1998). Researchers have primarily studied petunias, snapdragons, and maize to understand the MYB/bHLH proteins, which regulate the production of anthocyanins. The Arabidopsis system for controlling anthocyanin levels has recently been the subject of much study. It is believed that the MYB gene C1 and the bHLH protein-encoding gene R regulate most genes involved in anthocyanin biosynthesis in the aleurone layer of the kernels. Mol et al. (1998) found that various members of the gene family have the responsibility of regulating these genes in different areas of the plant. When the R and C1 genes were excessively expressed in colorless maize cells cultivated in a controlled laboratory environment, it led to the synthesis and accumulation of anthocyanins. This occurred because the bulk of the genes involved for anthocyanin synthesis were activated at the same time, as demonstrated in Fig. 9.1 (Grotewold et al. 1998). The production of 3-deoxy flavonoids in maize is dependent on MYB-type transcriptional regulator P, which functions independently from R and C1 in the transcriptional process. These flavonoids have antifungal and insecticidal properties. A specific set of biosynthetic genes were simultaneously activated in maize cells when an excess of P was detected. According to Grotewold et al. (1998) and Bruce et al. (2000), this resulted to an accumulation of flavonoids that were not regulated by C1/R. An increase in genes involved in flavonoid production was seen in Arabidopsis plants that had the MYB transcription factor overexpressed, also known as production of anthocyanin pigment 1 (PAP1). Most parts of the plant were clearly purple in color (Borevitz et al. 2000). Because of their shared sequence with C1, PAP1, and PAP2, a closely related gene, are likely to be It is possible that ectopically generated PAP1 can function independently of bHLH transcription factors or, more probable, requires always-present bHLH proteins, since the PAP1-mediated phenotype is not dependent on the simultaneous synthesis of a bHLH protein partner. Damage to the transparent testa2 (TT2) gene, which encodes the R2R3 MYB protein, causes the accumulation of concentrated tannins, also known as proanthocyanidins. The seed's color is changed directly as a result of this damage, which reduces the amount of brown tannins in the inner layer. In addition, it significantly downregulated the expression of many structural genes. The primary enzyme responsible for producing proanthocyanidin, leucoanthocyanidin reductase (LAR), is believed to be encoded by the gene banyuls (BAN), which was aberrantly activated by TT2 in gain-of-function experiments. A number of genes, including dihydroflavonol-4-reductase (DFR) and BAN, are uniquely induced by TT2. The production of flavonoids is regulated by these genes. The presence of TT8, a bHLH protein, is necessary for the activation of this mechanism (Nesi et al. 2001). By interacting with one another, the TT2 and TT8 proteins can regulate target genes. If TT2 overexpression causes BAN gene abnormalities, then TT2 must be the key gene that determines BAN expression. Research into the TT2's temporal control is currently of the utmost importance. During seed development, the process of gene expression is set up. There seems to be a function for the MYB and bHLH transcriptional activators in gene regulation. Research into what variables activate genes that produce flavonoids in reaction to UV light, illnesses, or stress has yielded no clear results. When the AmMYB308 and AmMYB330 genes were introduced into tobacco from snapdragon, they prevented the accumulation of

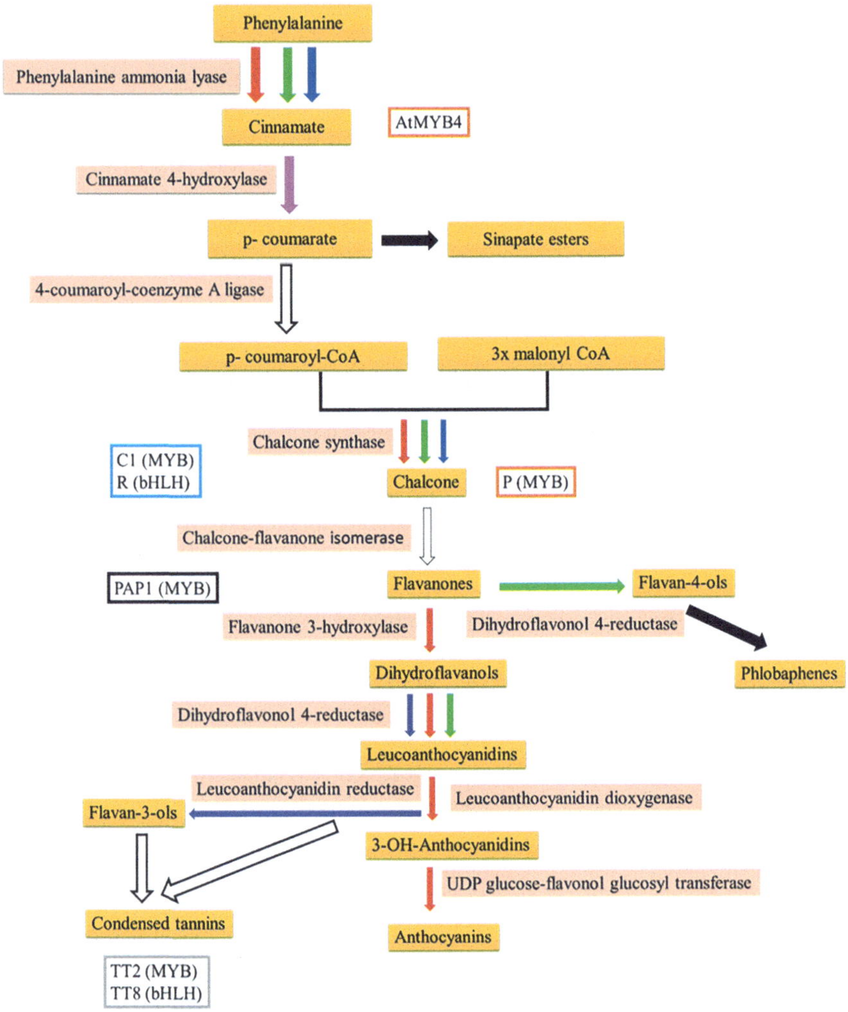

Fig. 9.1 Flavonoid biosynthetic pathway

hydroxycinnamic acid and monolignol, demonstrating the potential of MYB transcription factors to restrict gene activity. These compounds were shown to be produced due to the downregulation of gene expression (Tamagnone et al. 1998). Weak activators/repressors encoded by AmMYB308 and AmMYB330 compete for control of target gene expression with stronger activators associated with MYB, which could explain their behavior. In order to suppress its normal activity, the AtMYB4 gene was the target of a genetic mutation that was induced in the Arabidopsis plant. Because the mutant accumulated more sinapate ester, it was better able to resist exposure to UV-B light (Jin et al. 2000). The Arabidopsis gene is known as AtMYB4. Overexpression of the red strawberry-derived transcription factor FaMYB1 in tobacco resulted in a decrease in anthocyanin and flavonol production (Aharoni et al. 2001). The conserved motif pdLNLD/ELxiG/S is located in the C-terminal region of all receptor MYB factors. This motif aids in halting transcription, as revealed by an analysis of AtMYB4 mutations and deletions (Jin et al. 2000). There is a striking similarity between this motif and another well-established repressor motif, which is known as ERF-associated amphiphilic repression as well. Additionally, the EAR motif can be found in transcriptional repressors of class II AP2/ERF as well as zinc finger proteins that perform this function (Ohta et al. 2001). A complex network of different MYB transcription factors controls many aspects of phenylpropanoid metabolism; repressors containing the EAR-like domain may be active repressors or use a passive strategy dependent on competition. Their bHLH protein partners determine the function of some of these components (Fig. 9.1).

9.3 ORCA Proteins Are Involved in the Metabolic Processes of Alkaloids

The production of terpenoid indole alkaloids occurs via a secondary metabolic pathway that has undergone thorough investigation (Fig. 9.2). Methyl jasmonate (MeJA) enhances the synthesis of TIA by activating genes associated with biosynthesis at the same time (van der Fits and Memelink 2000; Menke et al. 1999a; Collu et al. 2001). Researchers have found that a particular segment of the TIA biosynthetic gene, known as JERE, controls the activation of genes in reaction to elicitors and jasmonate. Using JERE as bait in yeast one-hybrid screening, Menke et al. (1999b) made the discovery that the AP2/ERF (APETALA2/ethylene responsive factor)-domain class of transcription factors was discovered. The study conducted by Memelink et al. (2001) states that the AP2/ERF transcription factors are unique compared to other transcription factors due to the presence of the AP2/ERF DNA-binding domain. These characteristics are only exhibited by plants. Following the administration of an elicitor into MeJA, there were transient elevations in the expression of the ORCA2 gene. Another finding from Menke et al. (1999b) reveals that the ORCA2 protein boosts the activity of the STR promoter by binding to the JERE region. van der Fits and Memelink (2000) presented a comprehensive explanation of the activation of T-DNA in *C. roseus* by ORCA3, a transcription factor belonging to the AP2/ERF-domain family. ORCA3 and ORCA2 exhibit similarities in the AP2/ERF DNA-binding domain, but they differ substantially in the amino acid composition of the other regions. Concurrently with the initiation of ORCA2 gene expression, the ORCA3 gene was stimulated by MeJA. Memelink and van der Fits observed that the ORCA3 protein establishes interactions with the promoter regions of the TIA biosynthesis genes CPR, TDC, and STR. Through a brief experiment, it was observed that there was an increase in gene transcription activity (van der Fits and Memelink 2000). The initiation of STR promoter activation was facilitated by a specific interaction mediated by ORCA3, as explained by van der Fits and Memelink (2001). A substantial enhancement in the synthesis of TIA was observed when the

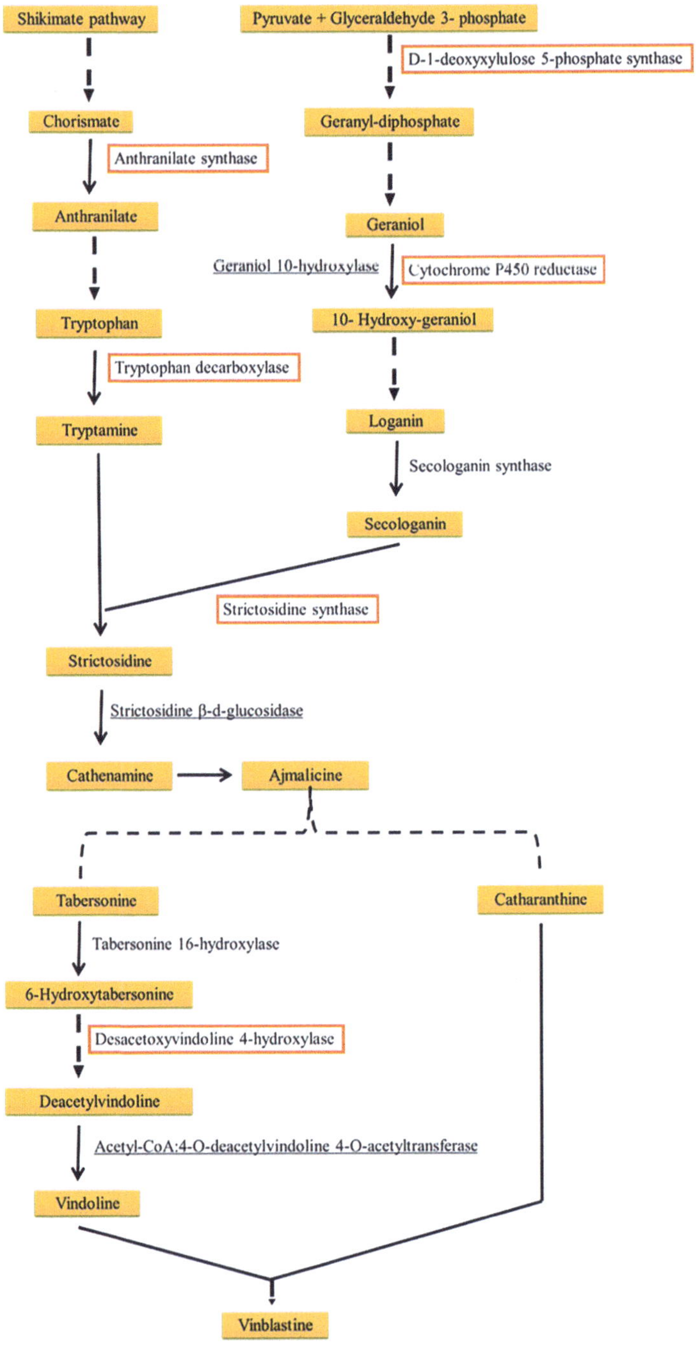

Fig. 9.2 Periwinkle's biosynthetic pathway for terpenoid indole alkaloids

expression of ORCA3 was artificially increased in transgenic *C. roseus* cell suspension cultures. This was accomplished by stimulating multiple genes that participate in both primary and secondary metabolisms. Several genes were activated, including CPR, STR, and TDC. When given a dietary supply of the precursor loganin, the overexpression of ORCA3 resulted in an increased synthesis of TIA. Van der Fits and Memelink (2001) argue that JA has the capacity to influence the expression of TIA biosynthesis genes by activating ORCA proteins. The information in question originates from. Presently, there is considerable focus on investigating the mechanisms by which jasmonates stimulate the activation of genes and proteins that possess AP2/ERF domains. A protein with an AP2/ERF-domain controls multiple genes that are involved in the metabolism responsive to jasmonic acid (JA). Other stress-responsive metabolic pathways in plants may also have a regulatory mechanism that our discovery reveals. Plants and plant cell cultures may find AP2/ERF-domain proteins useful for increasing the production of taxol and related taxoids, as well as other alkaloids, and other JA-inducible secondary metabolites, given the important role that jasmonates play in this process.

9.4 Control of mRNA Levels for Transcription Factors

There are various ways in which transcriptional regulators might be impacted. An effective strategy for controlling a component's concentration is to modify mRNA synthesis. An example that is rich in literature is the MYB protein C1. Research conducted by Mol et al. (1998) indicates that C1 plays a regulatory role in the synthesis of anthocyanin in developing maize seeds. An issue observed in maize viviparous 1 (vp1) mutant seeds is the absence of C1 gene expression during their development. As a result, the seeds lack anthocyanins. The reporter gene is regulated by the C1 promoter, which can be activated by the VP1 protein. The development of dicotyledonous plant seeds is thought to be aided by the B3 domain of VP1, which is situated within the abscisic acid-insensitive 3 (ABI3) factor. This region collaborates with other molecules to form a strong and specific bond with DNA. The protein–DNA interaction is specifically concentrated in regions that have been proven to be crucial in the stimulation of the C1 gene by maize cell VP1 (Suzuki et al. 1997). VP1 directly binds to the promoter region to control the expression of the MYB protein C1 in actively growing seeds. VP1 regulates the transcription of genes that play a role in the later stages of seed development. There is a high probability that the regulation of genes that are involved in the production of secondary metabolites in plants is comparable to the regulation of genes that typically respond to pathogens (PR). Because of this, these genes are susceptible to being influenced by a wide variety of stress conditions, including those that are induced by signal molecules that are released by microbes. The promoter of the PR gene was analyzed by researchers, and they discovered a number of cis-acting motifs that were similar to one another, as well as the binding proteins that were associated with them. The GCC box can be found in the promoter region of a number of different PR genes. Stepanova and Ecker (2000) demonstrated that the cis-acting element functions as a responsive element to ethylene. Tobacco possesses Ethylene-Responsive Element Binding Proteins (EREBPs) while Arabidopsis possesses Ethylene Response Factors (ERFs), along with other proteins. The ethylene-sensitive GCC-box is the site of DNA attachment (Stepanova and Ecker 2000). AP2/ERF domain transcription factors are the family of transcription factors that they belong to. A protein called EIN3 is essential for starting the transcription of ERF1; it binds to specific DNA regions. Stepanova and Ecker (2000) found that this process triggers the transcription of ethylene-regulated genes. So, when ethylene gas is present, a cascade of transcriptional regulators is set in motion by the EIN3 and ERF1 transcription factors. A gene's expression level can be controlled via auto-regulation rather than by relying on a transcriptional cascade. This gene encodes a transcription factor. Proteins like Parsley's WRKY1 are part of a class of proteins called zinc-finger type proteins that control genes that are involved in protecting plants from harmful pathogens. Parsley's WRKY1 protein has been found to bind most favorably with PR1 gene promoter regions containing TGAC sequences, according to in vitro research. Research into transient gene expression in parsley protoplasts has revealed that the WRKY1 gene requires and is adequately activated by a certain configuration of Wbox elements in its promoter region when exposed to elicitors. Further evidence that WRKY1 interacts with these components directly points to a role for WRKY1 in responding to stimuli by initiating its own expression. MeJA causes a short-lived but noticeable spike in ORCA mRNA levels, with the peak of ORCA expression happening just before the spike peaks. The identification of specific target genes, such as TDC and STR, provides evidence of the presence of a transcriptional cascade. The study conducted by Van der Fits and Memelink (2001) demonstrated that the protein synthesis inhibitor cycloheximide did not have any effect on the expression of TDC and STR induced by MeJA. This indicates that the generation of initial transcription factors is not required. MeJA has a broader impact than just enhancing TIA gene expression by controlling ORCA or increasing the quantity of ORCA protein through a transcriptional cascade. However, it seems to stimulate the already present ORCA protein (Fig. 9.3).

9.5 Control of Transcription Factor Protein Functionality

Cellular regulation of transcription factors is possible via posttranslational modifications, interactions with other proteins, and transcriptional or posttranscriptional regulation. Proteins can undergo various posttranslational modifications, such as phosphorylation, acetylation, hydroxylation, nitrosylation, glutathionylation, the formation of S-S bridges between cysteines within the same protein or between different proteins, myristoylation, farnesylation, ubiquitination, or glycosylation. Modifying transcription factors can change the shape of proteins, enabling them to interact with other regulatory proteins or affect where they are located inside a cell. These alterations can subsequently impact the affinity for DNA binding, the capacity for activation, the localization within the nucleus, and/or the stability of the protein. One well-known way that mammals control the activity of their

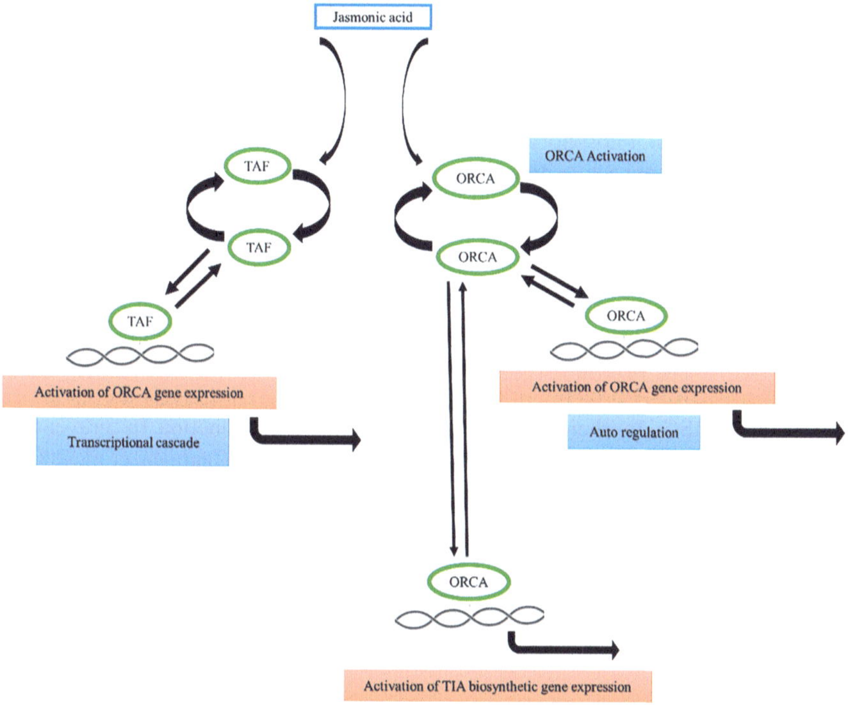

Fig. 9.3 Proposed framework for the activation of gene expression by jasmonic acid (JA) through the involvement of ORCA transcription factors

transcription factors is by phosphorylation and dephosphorylation. Phosphorylation can also control the function of transcription factors in plants, according to a number of lines of evidence. As an illustration of a potential phosphorylation of a plant transcription factor, take into consideration the AP2/ERF protein known as Pto-interactor 4 (Pti4). The receptor kinase Pto, which aids in tomato plant defense against diseases, was the initial target of Pti4's discovery. To facilitate Pti4's attachment to the GCC box, Pto phosphorylated it in a controlled laboratory setting (Gu et al. 2000). Based on these data, it can be inferred that phosphorylation is likely to improve the capacity of Pti4 to stimulate the production of PR genes that have GCC boxes in tomatoes. There is currently no proof to support the idea that mechanisms involving protein modification control the activity of plant MYB proteins in living things. In a laboratory environment, it was found that phosphorylation negatively impacts the DNA binding ability of the AmMYB340 protein derived from snapdragon. This protein has the function of controlling the metabolism of flavonoids (Moyano et al. 1996). It is possible that the phosphorylation of AmMYB340 in living organisms will make it even more effective at controlling transcription. This will be accomplished by working with the various ways in which it interacts with other members of the MYB family and working for them. Due to the fact that cycloheximide did not inhibit the production of MeJA-induced target genes, it is highly probable that ORCA is activated through post-translational modification and/or protein–protein interaction. JA-responsive gene products, such as TDC and STR, are altered in their expression when protein kinase inhibitors are present during the process. Menke et al. (1999a) state that this proves phosphorylation is involved in TIA gene activation. Phosphorylation may activate ORCAs, or it may be necessary for ORCAs to become active. Interactions between proteins can influence transcription factor function. An earlier illustration of this was the interaction that occurred between the transcription factors MYB and bHLH. By interacting with one another, transcription factors can improve their binding to particular promoters and, in turn, speed up transcription. Evidence suggests that 14-3-3 is one of several regulatory proteins that can interact with transcription factors. The RSG transcriptional activator, which inhibits shoot growth in tobacco plants and yeast two-hybrid systems, interacts with 14-3-3 proteins in both organisms. Gibberellin levels are naturally regulated by RSG. Further, 14-3-3 proteins inhibit RSG function by preventing RSG entry to the nucleus, as supported by good evidence (Igarashi et al. 2001). Specifically, phosphoserine residues in conserved amino acid sequences are the target of 14-3-3 proteins. The presence of an amino acid known as serine is necessary for the interaction between RSG and 14-3-3. The binding motif known as 14-3-3 is well conserved and includes serine. Altering this specific serine increases RSG transcriptional activity in tobacco protoplasts, according to research (Igarashi et al. 2001). Cell localization of RSG is controlled by phosphorylation of its interaction with 14-3-3 proteins. One way 14-3-3 proteins regulate the nucleus is by being a part of transcriptional complexes. Researchers used anti-14-3-3 antibodies to identify and learn more about 14-3-3 proteins in nuclear extracts of rice. As part of their complex formation, these proteins can bind to a specific region of the Em (embryonic maturation) gene promoter that is involved in the gene's response to abscisic acid and VP1 (Schultz et al. 1998). In a controlled laboratory environment, the rice 14-3-3 protein GF14b interacted with the Em1a-binding bZIP proteins VP1 and EmBP1. It is reasonable to assume that 14-3-3 proteins structurally link VP1 and Em1a-binding proteins because these proteins are likewise unable to bind DNA and a vp1 mutant that is unable to bind DNA can still activate Em transcription. This connection forms a compound with the DNA that is located in the nucleus. Among the few genes that have been found to regulate flavonoid biosynthesis gene expression, the MYB and bHLH transcription factor genes stand out. The petunia AN11 protein (de Vetten et al. 1997) and the Arabidopsis transparent testa glabra1 (TTG1; Walker et al. 1999) are examples of regulators found in this group. The AN11 and TTG1 proteins are very similar to one another. Over four WD repeat motifs, the two proteins are identical. Motifs are collections of amino acid sequences that typically terminate in Trp-Asp. Similar patterns are also found in the b-subunits of heterotrimeric G proteins and a number of other regulatory proteins. By synthesizing MYB protein AN2, it is feasible to restore DFR gene expression in Petunia an11 mutants. The level of operation of AN11 is higher than that of MYB. According to De Vetten et al. (1997), the AN11 protein controls the activity of MYB proteins. This is supported by the finding that the an11 mutant has no effect on the level of AN2 mRNA. Among the many pleiotropic defects seen in the Arabidopsis ttg1 mutant are trichomes and anthocyanin synthesis. Although Arabidopsis bHLH protein glabra3 (GL3) only partially restores the ttg1 mutant's phenotype, maize bHLH protein R can entirely restore it. According to these findings, TTG1 performs better than at least one bHLH protein (Payne et al. 2000). Overexpression of GL3 prevents the production of anthocyanin, a hallmark of the ttg1-1 mutant phenotype. However, restoring a typical amount of trichomes is not achieved. However, ttg1-1 mutant consequences are not mitigated by upregulating the MYB transcription factor GL1. Research conducted by Payne et al. (2000) found that F1 hybrid plants with increased GL1 and GL3 expression on a ttg1-1 background exhibited remarkably similar trichomes number and arrangement to wild type plants. This discovery lends credence to the idea that, even in the absence of TTG1, GL1 and GL3 can work together in Arabidopsis plants to maximize the effects of both. The fact that there were slightly fewer trichomes than in the wild type, however, indicates that TTG1 is necessary for complete GL3: GL1 involvement. In addition, Payne et al. (2000) demonstrated through their yeast two-hybrid assay that GL3 had a physical connection with both GL1 and TTG1. It is likely that the activation, modification, stability, or some other beneficial way in which TTG1 enhances the GL3 protein's capacity to aid gene transcription because of the fact that the two proteins are connected. The function of MYB-bHLH complexes is likely regulated by AN11 and TTG1, two proteins that share comparable functions, according to researchers.

9.6 Control of the Amount of Transcription Factor Proteins

The abundance of transcription factors can be managed by adjusting the rates at which proteins are degraded, in addition to controlling cellular levels through transcriptional regulation.

Common methods for controlling protein stability include interactions with other proteins as well as covalent modifications such as phosphorylation and ubiquitination. Using MeJA, Arabidopsis significantly increases the production of indole glucosinolate. In order for the body to produce these secondary metabolites when stimulated, the coronatin-insensitive 1 (COI1) gene must be functioning properly (Brader et al. 2001). The COI1 gene was found to be essential for jasmonate-dependent Arabidopsis defense and fertilization, according to research conducted by Xie et al. (1998). An F-box protein is encoded by COI1, according to its sequence analysis. All three of these proteins—F-box, SKP1, and Cullin—are essential parts of the SCF complexes. These molecules keep an eye on the target proteins going through the ubiquitin-mediated proteolysis pathway and destroy them when it's time. According to the findings of Gray et al. (2001), the Arabidopsis F-box protein transport inhibitor response 1 (TIR1) is responsible for regulating the AUX/IAA repressors through the process of ubiquitin-mediated proteolysis. One possible explanation is that the auxin-response pathway is blocked because AUX/IAA proteins negatively regulate auxin response transcription factors (ARF). One possible explanation is that auxin makes SCFTIR1 ubiquitin ligase and AUX/IAA proteins work together better. Once ubiquitinated, these proteins can be broken down by the proteasome. The auxin response pathway is kept open and the ARF transcription factors are activated by degradation (Gray et al. 2001). Previous research suggests that COI1 may degrade a JA-responsive repressor, which in turn inhibits AP2/ERF-domain transcription factors. The Arabidopsis photo morphogenesis-promoting bZIP transcription factor HY5 is an example of a regulator that uses targeted protein phosphorylation and degradation to control gene expression. Through the process of binding to a conserved G box sequence in the promoters of light-responsive genes, such as chalcone synthase (CHS), the HY5 protein exhibits its ability to interact with these genes. By increasing anthocyanin accumulation and activating CHS gene expression, it acts as a growth regulator, according to Hardtke and Deng (2000). The concentration of HY5 protein in seedlings increases dramatically, approximately 20 times higher than in the dark. However, the ratio of HY5 mRNA levels in dark and light environments is barely two to three times different. This provides more evidence that the stability of the protein regulates the concentration of HY5. HY5 is able to inhibit photo morphogenesis through its interaction with COP1, which is a RING-finger protein that contains WD repeats. Furthermore, according to Hardtke and Deng (2000), the ability of RING-finger proteins to perform the function of E3 ubiquitin-protein ligases suggests that these proteins have the ability to designate proteins for degradation. There are two distinct phosphorylation states for HY5, but its nuclear localization remains the same as it is in both states. When light isn't present, the nuclear protein COP1 accumulates and phosphorylate-specifically breaks down the unphosphorylated form of HY5. In their research conducted in the year 2000, Hardtke and Deng recorded the aforementioned process. Similarly, casein kinase II (CKII) contributes to the cellular upkeep of a subset of phosphorylated HY5. This HY5 variant has a lower physiological activity and is more resistant to degradation. While HY5-related CKII activity decreases in response to light, COP1 is ejected from the nucleus (Hardtke and Deng 2000). Because of its greater physiological activity, unphosphorylated HY5 is more abundant in light-exposed environments. Genes that are controlled by light are activated when this happens. Controlling nuclear COP1 and phosphorylation levels at the same time is necessary to increase HY5 activity. The end result is HY5 that is both more reliable and precise.

9.7 Conclusions

When it comes to precisely directing the production of secondary metabolites in plant cells, transcription factors play a crucial role. They do this by coordinating the control of biosynthetic genes throughout the cell. During this session, authors examined multiple examples of how the regulation of these entities can be controlled. A successful technique entails controlling the abundance of transcription factors during the gene expression process. In the ethylene response, the EIN3/ERF1 proteins do this, and the VP1/C1 proteins do the same thing during seed maturation. The WRKY1 transcription factor gene allows the plant to self-regulate during pathogen defense, which is a defense mechanism against this. Many transcription factors are influenced by protein interactions and changes. Protein phosphorylation studies in plants have been quite extensive. One example of how protein phosphorylation states can alter protein interactions is the RSG/14-3-3 interaction that occurs during gibberellin synthesis. Controlling the amounts of proteins by selectively breaking down certain regulators, as seen in photomorphogenesis and the auxin response, appears to be a significant process. The phosphorylation status of the HY5 protein has the capacity to impact its degradation. The regulation of tissue-specific secondary metabolism is presumably achieved by managing the levels of mRNA encoding transcription factors, potentially in conjunction with tissue-specific regulation of protein levels. In signal transduction pathways that necessitate quick gene activation, cells can effectively control a transcription factor's activity by interacting with and modifying proteins. Still, it has been seen that in plants, the same signal quickly activates genes that contain transcription factors linked to quick response pathways. These genes include the WRKYs and the ORCAs. Looking into how proteins interact with each other and how transcription factors change covalently will help us understand secondary metabolism better and make genetic engineering tools for making secondary metabolites better.

References

Aharoni A, De Vos CH, Wein M, Sun Z, Greco R, Kroon A, Mol JN, O'Connell AP (2001) The strawberry FaMYB1 transcription factor suppresses anthocyanin and flavonol accumulation in transgenic tobacco. Plant J 28:319–332

Borevitz JO, Xia Y, Blount J, Dixon RA, Lamb C (2000) Activation tagging identifies a conserved MYB regulator of phenylpropanoid biosynthesis. Plant Cell 12:2383–2394

Brader G, Tas E, Palva ET (2001) Jasmonate-dependent induction of indole glucosinolates in Arabidopsis by culture filtrates of the nonspecific pathogen Erwinia carotovora. Plant Physiol 126:849–860

Bruce W, Folkerts O, Garnaat C, Crasta O, Roth B, Bowen B (2000) Expression profiling of the maize flavonoid pathway genes controlled by estradiol-inducible transcription factors CRC and P. Plant Cell 12:65–79

Collu G, Unver N, Peltenburg-Looman AM, van der Heijden R, Verpoorte R, Memelink J (2001) Geraniol 10-hydroxylase, a cytochrome P450 enzyme involved in terpenoid indole alkaloid biosynthesis. FEBS Lett 508:215–220

de Vetten N, Quattrocchio F, Mol J, Koes R (1997) The an11 locus controlling flower pigmentation in petunia encodes a novel WD-repeat protein conserved in yeast, plants, and animals. Genes Dev 11:1422–1434

Gray WM, Kepinski S, Rouse D, Leyser O, Estelle M (2001) Auxin regulates SCF(TIR1)-dependent degradation of AUX/IAA proteins. Nature 414:271–276

Grotewold E, Chamberlin M, Snook M, Siame B, Butler L, Swenson J, Maddock S, St Clair G, Bowen B (1998) Engineering secondary metabolism in maize cells by ectopic expression of transcription factors. Plant Cell 10:721–740

Gu YQ, Yang C, Thara VK, Zhou J, Martin GB (2000) Pti4 is induced by ethylene and salicylic acid, and its product is phosphorylated by the Pto kinase. Plant Cell 12:771–786

Hardtke CS, Deng X-W (2000) The cell biology of the COP/DET/ FUS proteins. Regulating proteolysis in photomorphogenesis and beyond? Plant Physiol 124:1548–1557

Igarashi D, Ishida S, Fukazawa J, Takahashi Y (2001) 14–3-3 proteins regulate intracellular localization of the bZIP transcriptional activator RSG. Plant Cell 13:2483–2497

Jin H, Cominelli E, Bailey P, Parr A, Mehrtens F, Jones J, Tonelli C, Weisshaar B, Martin C (2000) Transcriptional repression by AtMYB4 controls production of UV-protecting sunscreens in Arabidopsis. EMBO J 19:6150–6161
Memelink J, Verpoorte R, Kijne JW (2001) ORCAnisation of jasmonate-responsive gene expression in alkaloid metabolism. Trends Plant Sci 6:212–219
Menke FLH, Parchmann S, Mueller MJ, Kijne JW, Memelink J (1999a) Involvement of the octadecanoid pathway and protein phosphorylation in fungal elicitor-induced expression of terpenoid indole alkaloid biosynthetic genes in Catharanthus roseus. Plant Physiol 119:1289–1296
Menke FLH, Champion A, Kijne JW, Memelink J (1999b) A novel jasmonate- and elicitor-responsive element in the periwinkle secondary metabolite biosynthetic gene Str interacts with a jasmonate- and elicitor-inducible AP2-domain transcription factor, ORCA2. EMBO J 18:4455–4463
Mol J, Grotewold E, Koes R (1998) How genes paint flowers and seeds. Trends Plant Sci 3:212–217
Moyano E, Martinez-Garcia JF, Martin C (1996) Apparent redundancy in myb gene function provides gearing for the control of flavonoid biosynthesis in antirrhinum flowers. Plant Cell 8:1519–1532
Nesi N, Jond C, Debeaujon I, Caboche M, Lepiniec L (2001) The Arabidopsis TT2 gene encodes an R2R3 MYB domain protein that acts as a key determinant for proanthocyanidin accumulation in developing seed. Plant Cell 13:2099–2114
Ohta M, Matusui K, Hiratsu K, Shinshi H, Ohme-Takagi M (2001) Repression domains of class II ERF transcriptional repressors share an essential motif for active repression. Plant Cell 13:1959–1968
Payne CT, Zhang F, Lloyd AM (2000) GL3 encodes a bHLH protein that regulates trichome development in Arabidopsis through interaction with GL1 and TTG1. Genetics 156:1349–1362
Schultz TF, Medina J, Hill A, Quatrano RS (1998) 14–3-3 proteins are part of an abscisic acid-VIVIPAROUS1 (VP1) response complex in the Em promoter and interact with VP1 and EmBP1. Plant Cell 10:837–847
Stepanova AN, Ecker JR (2000) Ethylene signaling: from mutants to molecules. Curr Opin Plant Biol 3:353–360
Suzuki M, Kao CY, McCarty DR (1997) The conserved B3 domain of VIVIPAROUS1 has a cooperative DNA binding activity. Plant Cell 9:799–807
Tamagnone L, Merida A, Parr A, Mackay S, Culianez-Macia FA, Roberts K, Martin C (1998) The AmMYB308 and AmMYB330 transcription factors from antirrhinum regulate phenylpropanoid and lignin biosynthesis in transgenic tobacco. Plant Cell 10:135–154
van der Fits L, Memelink J (2000) ORCA3, a jasmonate-responsive transcriptional regulator of plant primary and secondary metabolism. Science 289:295–297
van der Fits L, Memelink J (2001) The jasmonate-inducible AP2/ ERF-domain transcription factor ORCA3 activates gene expression via interaction with a jasmonate-responsive promoter element. Plant J 25:43–53
Walker AR, Davison PA, Bolognesi-Winfield AC, James CM, Srinivasan N, Blundell TL, Esch JJ, Marks MD, Gray JC (1999) The transparent testa glabra1 locus, which regulates trichome differentiation and anthocyanin biosynthesis in Arabidopsis, encodes a WD40 repeat protein. Plant Cell 11:1337–1350
Xie D-X, Feys BF, James S, Nieto-Rostro M, Turner JG (1998) COI1: an Arabidopsis gene required for jasmonate-regulated defense and fertility. Science 280:1091–1094

Microfluids in Metabolites Detection, Production, and Optimization

10

Gholamreza Abdi, Neha Chhabra, Jueeli Shiriskar, Ganga Srinivasan, Rupal Dhariwal, and Mukul Jain

10.1 Introduction

Metabolites, which are complex chemicals that coordinate essential activities in biological systems, serve as the foundation for cellular activity and regulation. Gaining insight into the functions and interactions of living creatures is crucial for comprehending their intricacies. This introduction explores the importance of metabolites in biological systems, the ongoing difficulties faced in conventional metabolite detection methods, and the promise of microfluidics to revolutionize the analysis, manufacturing, and optimization of metabolites.

10.1.1 The Significance of Metabolites in Biological Systems

Metabolites are essential for life as they are the primary unit of measurement and are involved in various metabolic pathways. They consist of substances like carbohydrates, amino acids, lipids, and nucleotides, each with specific functions. Metabolites connect cellular processes, allowing cells to adjust to varying

G. Abdi (✉)
Department of Biotechnology, Persian Gulf Research Institute, Persian Gulf University, Bushehr, Iran
e-mail: abdi@pgu.ac.ir

N. Chhabra · J. Shiriskar · G. Srinivasan
Vivekanand Education Society's College of Pharmacy, Chembur, India

R. Dhariwal · M. Jain
Cell and Development Biology Lab, Research and Development Cell, Parul University, Vadodara, Gujarat, India

Department of Life Sciences, Parul Institute of Applied Sciences, Parul University, Vadodara, Gujarat, India

V. Singh (ed.), *Advances in Metabolomics*,
https://doi.org/10.1007/978-981-97-7459-3_10

environmental conditions and metabolic requirements. They also function as biomarkers, providing information about organisms' physiological condition and health, sickness, and potential therapeutic effects. The complex interactions of metabolites highlight their importance in biological systems, necessitating thorough analytical methods for characterizing and measuring them (Table 10.1).

10.1.2 Obstacles in Conventional Metabolite Detection Methods

Despite the significant importance of metabolites, conventional methods of detection have always faced intrinsic constraints, impeding our capacity to comprehensively capture the complete range of metabolomic dynamics. Conventional methodologies, such as nuclear magnetic resonance (NMR) spectroscopy and mass spectrometry (MS), often face obstacles associated with diminished sensitivity, limited manufacturing capacity, and cumbersome sample preparation procedures. Furthermore, the utilization of these methodologies may necessitate significant quantities of samples, hence presenting difficulties in the examination of metabolites in scarce or valuable biological material. The study conducted by Kell and Adla (2021) highlights the intricate nature of metabolite detection, encompassing the

Table 10.1 Comparison of traditional and microfluidic methods for metabolite detection

Aspect	Traditional methods	Microfluidic methods (Johnson et al. 2019b)
Sample volume	Requires relatively large sample volumes (microliters)	Requires small sample volumes (nanoliters to picoliters)
Analytical time	Longer analysis time (hours to days)	Shorter analysis time (minutes to hours)
Equipment size	Typically bulky instrumentation	Compact and portable devices
Throughput	Lower throughput due to sequential analysis	Higher throughput due to parallel analysis
Automation	Limited automation capabilities	Greater automation potential
Reagent consumption	Higher reagent consumption	Lower reagent consumption
Sensitivity	Moderate sensitivity	High sensitivity
Multiplexing capability	Limited multiplexing capability	Greater multiplexing capability
Cost	Higher cost due to equipment and reagents	Lower cost due to miniaturization and reduced reagent consumption
Complexity	Often requires skilled operators and extensive sample preparation	Simplified operation and reduced sample preparation
Integration with analysis techniques	Limited integration with other analysis techniques (Garcia et al. 2018)	Integration with various analysis techniques (e.g., mass spectrometry, spectroscopy) (Wang et al. 2017b)

extensive chemical variety of metabolites, fluctuating concentration ranges, and the influence of matrix effects. These factors can complicate precise measurement and interpretation. Therefore, there is an urgent want for novel analytical approaches that can effectively address these obstacles and offer full understanding of the metabolome.

10.1.3 Introduction to the Field of Microfluidics

The utilization of microfluidic platforms provides a wide range of capabilities that are specifically designed to address the distinct requirements of metabolite analysis, synthesis, and optimization (Fig. 10.1). These devices contain a range of capabilities that surpass the constraints of standard analytical methods, including capillary electrophoresis (CE) microchips for high-resolution separation of metabolites and microfluidic biosensors for real-time detection. In addition, microfluidic methodologies have brought about a significant transformation in the field of metabolite production and optimization. These techniques have allowed for meticulous regulation of reaction parameters and have greatly facilitated the efficient screening of metabolic pathways at a large scale. The review conducted by Gong et al. (2023) highlights the significance of combining microfluidics with computer modeling and simulation in order to gain a comprehensive understanding of intricate metabolic networks, forecast metabolite yields, and enhance production parameters. The field of metabolomics is positioned to explore novel opportunities in personalized medicine, biomarker identification, and biotechnological breakthroughs, thanks to the combined progress made in microfluidic technology (Schmidt et al. 2018).

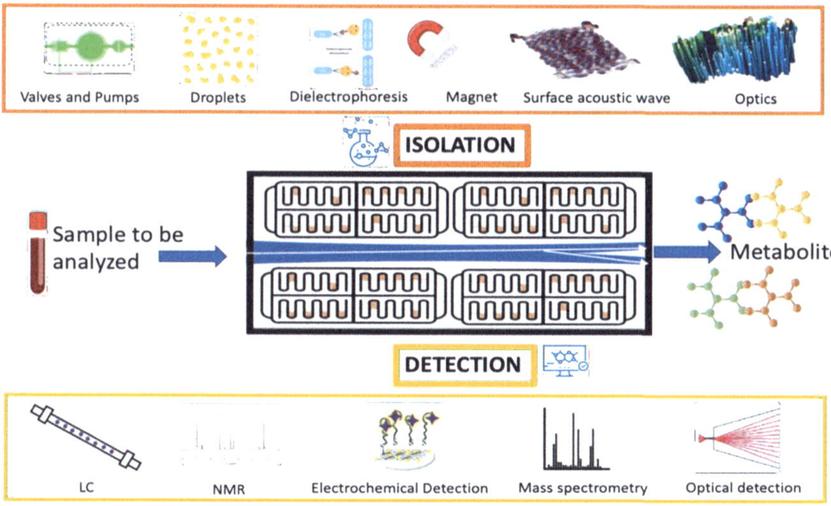

Fig. 10.1 Microfluidic platform for high-throughput screening of metabolite

10.2 Microfluidic Devices for Metabolite Detection

10.2.1 Capillary Electrophoresis (CE) Microchips

10.2.1.1 Principles of CE

A channel containing an electrolyte solution and a variety of chemical species is located inside the capillary electrophoresis equipment. Analyte migration is facilitated by electroosmotic flow (EOF), which arises as the driving force under the influence of an electric potential. When a voltage is applied, the solution obeys and moves through the channel, starting the careful compound separation process. This demonstrates the effectiveness of capillary electrophoresis and attests to the accuracy and sophistication of contemporary analytical methods. Analytes are injected at one end of the capillary and then propelled toward the detector at a speed that depends on their charge/mass ratio and EOF. The analyte ions; varying electrophoretic mobilities are then used to determine separation. In contrast to the parabolic flow observed in liquid chromatography due to mechanical pumping, the EOF generates a flow with a greater velocity. When flow is observed at a high velocity, sometimes referred to as "plug flow"; the phenomenon of band broadening is diminished, resulting in an augmentation of plate numbers.

CE is the optimal approach for analyzing many types of molecules, considering parameters such as charge, size, stereospecificity, and hydrophobicity. There are several advantages associated with it. Key features encompass rapid analysis, exceptional separation efficacy, a diverse array of applications, and negligible waste generation. The technology possesses the ability to analyze intricate substances, such as biological macromolecules encompassing peptides, proteins, and DNA. The detection of various sorts of molecules can be facilitated through the utilization of diverse detectors (Table 10.2, Fig. 10.2).

MCE enables high-throughput analysis with small sample volumes (nanoliter scale or smaller). The several microchips utilized in MCE provide quicker analysis, reduced usage of reagents and samples, increased throughputs additionally less time invested. Additionally, improved portability is crucial to enabling analysis outside of labs. This is one of great significance in a variety of domains, including therapeutic, forensic, environmental, and space. The microfluidic devices market encompasses several sectors, including pharmaceuticals, medical devices, and in vitro diagnosis. Among these, the in vitro diagnosis sector dominated with over 60% market share in 2020. Furthermore, the medical devices sector is projected to experience the most substantial annual growth rate in the foreseeable future.

Miniaturized CE, also known as microfluidic CE or microchip CE (MCE), is considered a significant advancement over regular CE, providing numerous advantages. Capillary electrophoresis (CE) is the most suitable separation technology for microchips as it does not require pumps for fluid movement. MCE facilitates efficient analysis with small sample volumes, often at the nanoliter scale or smaller. The utilization of several microchips in MCE contributes to expedited analysis, decreased reliance on reagents and specimens, enhanced throughputs, and reduced time requirements. Moreover, increased portability is essential for facilitating

Table 10.2 Applications of microfluidic devices for metabolite production

Application	Description	Reference
High-throughput screening	Rapid screening of enzyme or microbial libraries for metabolite production using microscale reaction chambers	Smith et al. (2018)
Metabolic engineering	Precise control over microenvironment for optimizing metabolic pathways and enhancing metabolite production	Johnson et al. (2019a)
Biosynthesis pathway analysis	Real-time monitoring of biosynthetic pathways to identify key metabolite intermediates and optimize production	Garcia et al. (2020)
Single-cell analysis	Study of metabolite production at the single-cell level, enabling characterization of cell-to-cell variability	Wang et al. (2017a)
Continuous production	Continuous flow reactors for steady-state metabolite production, minimizing product inhibition and increasing yields	Example et al. (2016a, b)
Protein engineering	Directed evolution of enzymes for improved catalytic activity and specificity in metabolite synthesis	Chen et al. (2019)
In situ monitoring	Real-time monitoring of metabolite production within cells or tissues to study dynamic metabolic processes	Li et al. (2018)
Organ-on-a-chip	Mimicking organ microenvironments to study organ-specific metabolism and drug metabolism	Zhang et al. (2020)
Environmental monitoring	Detection of metabolites in environmental samples for monitoring microbial activity and pollutant degradation	Gupta et al. (2017)
Personalized medicine	Microfluidic platforms for point-of-care metabolite analysis, enabling personalized treatment strategies	Lee et al. (2019)

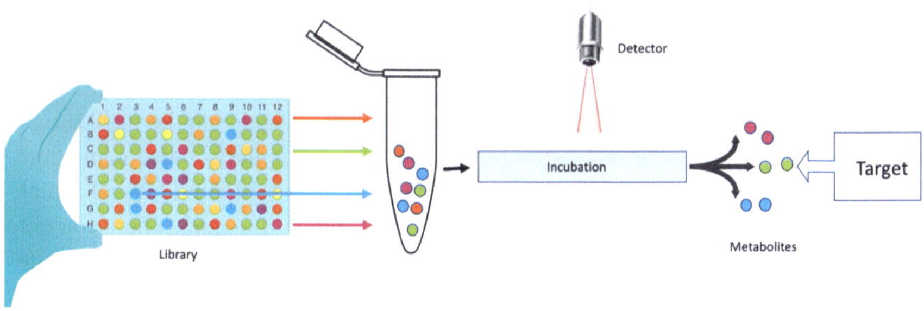

Fig. 10.2 Schematic of a microfluidic chip for metabolite detection

analysis beyond laboratory settings. This phenomenon holds considerable importance across multiple domains, encompassing therapeutic, forensic, environmental, and spatial settings. The microfluidic devices market can be segmented into different sectors such as pharmaceuticals, medical instruments, and in vitro diagnostics. Among these sectors, the medical devices sector maintains the largest market share, surpassing 60% in 2020. Furthermore, it is anticipated that the medical devices sector will experience the highest annual growth rate in the foreseeable future.

An established analytical method for quickly and effectively separating charged analytes in small sample volumes is capillary electrophoresis (CE). The ability to create small, portable systems due to the modest volumes may greatly simplify the employment of bio/chemical handling and analytical systems. CE is a method of separation that uses an electric potential difference across a microchip channel or narrow-bore capillary that is filled with an electrolyte solution, usually aqueous. A somewhat thin zone is inserted into the capillary containing a specimen mixture of several chemicals in solution. When an electric potential is applied across the capillary, electroosmotic flow (EOF) induces the movement of solution in bulk.

10.2.1.2 Design and Fabrication of CE Microchips for Metabolite Separation

Capillary electrophoresis (CE) is a method of separation that uses an electric potential difference across a microchip channel or narrow-bore capillary filled with an electrolyte solution, usually aqueous. When an electric potential is applied across the capillary, electroosmotic flow (EOF) induces the movement of solution in bulk. This method is effective for analyzing many types of molecules, including biological macromolecules like peptides, proteins, and DNA.

Miniaturized CE, also known as microfluidic CE or microchip CE (MCE), is considered a significant advancement over regular CE, providing numerous advantages. MCE facilitates high-throughput analysis with small sample volumes, often at the nanoliter scale or smaller. The use of several microchips in MCE contributes to expedited analysis, decreased reliance on reagents and specimens, enhanced throughputs, and reduced time requirements. Additionally, improved portability is crucial for enabling analysis beyond laboratory settings.

The microfluidic devices market can be segmented into different sectors such as pharmaceuticals, medical instruments, and in vitro diagnosis. The in vitro diagnosis sector dominated with over 60% market share in 2020, and it is expected to experience the highest annual growth rate in the foreseeable future.

The growing interest in MCE research over the past decade can be attributed to recent breakthroughs in micromachining techniques, particularly those using commonly used semiconductor processing techniques. These technological breakthroughs enable the integration of sampling, sample processing, and separation onto a single microchip device, allowing for high-throughput bioanalytical procedures without necessitating additional connectors.

The initial introduction of electrophoresis to flat planar microchips occurred approximately two decades ago, with capillary channels incorporated into planar glass substrates. Electrophoresis and micromachining procedures were used to establish separation in the channels, and in optimal conditions, it is possible to produce more than 100,000 theoretical plates. A key benefit of employing microchips for analyte separation via electrophoresis is the marked improvement in heat dissipation compared to capillaries of equivalent dimensions, allowing for stronger electrical fields in narrower separation channels, leading to considerably faster separations.

Initially, glass and quartz substrates served as the primary materials for microchips in CE research, owing to their exceptional electroosmotic flow properties, comparable chemistry to fused silica capillaries utilized in traditional CE, and optical transparency (Harrison et al. 1993). Nevertheless, there has been a shift in focus toward polymer-based chips as a result of the labor-intensive and time-consuming manufacturing procedure, along with the requirement for costly equipment and sterile environments (Shang et al. 2012). These options are characterized by their lower cost and require only a minimal set of fundamental technologies and fabrication equipment. MCE applications have utilized a diverse range of polymers, each possessing distinct advantages and disadvantages. The composite materials encompassed in this category consist of polyester, cyclo-olefin polymer (COP), poly(dimethylsiloxane) (PDMS), poly(methyl methacrylate) (PMMA), polycarbonate (PC), polyethylene terephthalate (PET), and SU-8. PDMS is widely employed due of its remarkable flexibility and optical transparency. However, a notable drawback of this approach is its limited separation efficiency (Kitagawa and Otsuka 2011). Hybrid PDMS/glass microchips are frequently used to address this challenge. In recent times, there has been increasing attention toward employing paper and polyester-toner (PT) substrates as economical, adaptable, lightweight, and readily available bases for manufacturing disposable medical devices (Coltro et al. 2010; Gabriel et al. 2013). Although uncoated channels are commonly utilized, they can be coated to prevent peptides and proteins from adhering to their surface, thereby ensuring consistent results.

Glass (Lithography)

Microfluidic devices (MCE) are crucial for high-throughput analysis with small sample volumes, offering improved portability and reduced usage of reagents and samples. The market includes sectors like pharmaceuticals, medical devices, and in vitro diagnosis, with the medical devices sector dominating the market. Conventional photolithographic techniques are used for fabricating glass microfluidic chips, which are then subjected to etching using hydrofluoric acid solution. After cleaning, the chip is prepared for bonding at elevated temperatures.

Polymer materials have gained interest due to their inherent advantages, such as reduced costs, simplified design of complex microchannel architectures, and lower manufacturing requirements. Various microfabrication techniques are available for creating polymer chips, including injection molding, laser ablation, hot embossing, casting, and imprinting (Nikcevic et al. 2007; Yap et al. 2013). The Subsequent Modification and Replication by Thermoforming (SMART) approach and optofluidic maskless lithography (OFML) are promising methods for producing microstructures like microparticles within fluidic channels.

Hybrid glass/quartz-polymer microchips have seen significant progress in recent years. However, leakage and obstruction are primary challenges encountered during the development of MCE. To achieve optimal bonding between diverse materials, clean room environments and fabrication processes like oxygen plasma bonding are necessary (Beh et al. 2012; Shameli et al. 2011). Laminar flow is a prominent feature of fluid flow in microchannels, making it essential for efficient analysis outside labs. The medical devices sector is expected to exhibit the highest annual growth rate in the near future.

10.2.1.3 Detection Techniques for CE Microchips (e.g., UV-Vis, Fluorescence)

Microfluidic devices (MCE) are a rapidly growing field that offers high-throughput analysis with small sample volumes. The use of microchips in MCE allows for quicker analysis, reduced usage of reagents and samples, increased throughputs, and less time invested. This is particularly important in various domains such as therapeutic, forensic, environmental, and space analysis. The microfluidic devices market includes sectors like pharmaceuticals, medical devices, and in vitro diagnosis.

Isoelectric focusing (IEF) is a valuable technique for precise separation of amphoteric molecules, such as peptides, proteins, and components in complex biological mixtures. Capillary-based isoelectric focusing (CIEF) is a compact technique that uses fused silica for IEF (Shen et al. 2000; Tang et al. 1996). However, single point detection methods often require chemical, hydrodynamic, or electroosmotic mobilization techniques, which can increase the analysis duration and decrease accuracy. Column imaging detection approach (WCID) is an alternative that circumvents these challenges.

Two commonly used detection methods for IEF applications are ultraviolet (UV) absorbance detection and laser-induced fluorescence detection (LIF). LIF is widely used due to its exceptional selectivity and sensitivity, but it requires expensive and time-consuming derivatization of proteins using a fluorescent marker. Labeling reactions can also lead to decreased separation efficiency. UV absorbance detection has been limited due to its low detection sensitivity, which is directly proportional to the length of its light route. There is a growing interest in developing methods to fully capitalize on UV absorbance detection in microchip-based separations.

10.2.2 Microfluidic Chromatography Devices

10.2.2.1 Principles of Microfluidic Chromatography

Microfluidic devices (MCE) are enabling high-throughput analysis with small sample volumes, reducing reagent usage and sample time. This technology is crucial in various domains, including therapeutic, forensic, environmental, and space analysis. The microfluidic devices market includes sectors like pharmaceuticals, medical devices, and in vitro diagnosis. The in vitro diagnosis sector dominated in 2020, with the medical devices sector expected to experience significant growth in the future. Micrototal analysis systems (μTAS) offer benefits like reduced sample usage, swift analysis time, and portability (Feng et al. 2020; Sarma et al. 2019). On-chip electrophoresis and liquid chromatography (LC) separation performance has improved, leading to increased interest in this field. Microfluidics offers comprehensive platforms for online manipulation and examination of biological samples, reducing sample loss. On-chip LC systems can lyse cells, extract biomolecules, and be coupled with detection systems (Oliva and Llabrés 2019; Song et al. 2020; Wali et al. 2020).

Affinity-Based Systems

Affinity-based separation techniques involve specific interactions between ligands and biological molecules, such as antibodies or metal oxides, for chromatography. For example, titanium dioxide (TiO_2) is used for phosphopeptide enrichment due to its strong affinity for phosphor groups. Miniaturized metal oxide affinity chromatography systems have been developed for analyzing limited clinical samples, using microchips with TiO_2 coatings. These devices offer high recovery rates and can handle small sample volumes (Zhao et al. 2016a). Antibody-based microfluidics investigates the precise interplay between antibodies and antigens. Cell affinity chromatography, introduced in the 1970s, involves cell separation using cell surface receptors and immobilized ligands. Aptamers, oligonucleotides or peptides, have advantages over antibodies due to their in vitro production via SELEX. Aptamer-based microchip LC has been used for biomolecule separation, such as in the capture of cancer cells and DNA amplifying for gene mutation analysis (Reinholt and Craighead 2018).

Adsorption-Based Separation

Adsorption-based separation relies on partition coefficients during adsorption and desorption processes. Reversed-phase chromatography is widely used for biomolecule analysis. On-chip adsorption separation has been developed, with optimization ongoing. Liu et al. (2013) developed a nanoLC chip-MS/MS assay for toxin quantification in food, achieving high reproducibility and low detection limits. Detobel et al. developed porous-shell pillar array columns using UV photolithography and sol-gel procedures, achieving minimal plate heights and optimized pillar design, resulting in one million plate numbers and decreased elution time compared to traditional columns (Table 10.1).

Size Exclusion and Hydrodynamic-Based Separation

Size exclusion chromatography and hydrodynamic chromatography are two column types used for protein purification and polymer weight distribution analysis. Size exclusion chromatography uses porous bead-packed columns, while hydrodynamic chromatography uses open tubular or non-porous columns. Millitte et al. introduced a microfluidic system in 2015 that combined size exclusion, ion exchange, and affinity interaction chromatography. The system, featuring 3D fluidic bridges, successfully enriched green fluorescent protein (eGFP) from *E. coli* samples, demonstrating the effectiveness of these methods (Millet et al. 2015).

Ion Exchange

Ion exchange chromatography, a chromatographic technique with the ability to separate various ionic species (either positively charged cations or negatively charged anions), is widely utilized in environmental analysis. It is particularly valuable for monitoring the levels of different ions (such as nitrate (NO_3−) and phosphate

(PO_4^{3-})) in wastewater. Additionally, ion chromatography has been increasingly used in the analysis of biomolecules such as proteins and metabolites. However, despite its significance, on-chip ion chromatography devices are relatively scarce compared to affinity or adsorption-based separation techniques. Nevertheless, ion chromatography remains a valuable complementary method for separating unique samples.

The creation of on-chip microfluidics ion exchange LC devices was pioneered by Murrihy et al. The feasibility of employing the conventional latex nanoparticle-coating technique for the production of on-chip ion chromatography was proven in 2001. The microfluidic device employed in their study exhibited narrow microchannels with a range of 0.5 to 10 μm, which effectively promoted improved interaction between the stationary phase and analytes. The apparatus effectively achieved the separation of nitrite, nitrate, and iodide ions by employing KCl as the mobile phase. The utilization of a UV detector, in conjunction with the optimized device and separation conditions, facilitated the efficient separation of the three ions on the chip within a time frame of only 3 min. According to Murrihy et al. (2001), the linearity of $NO2^{2-}$ and NO^{3-} detection was seen within the concentration range of 5 μM to 1 mM, with a limit of detection of 0.5 μM.

Other Separation Techniques

Novel separation mechanisms for on-chip separation have been investigated by researchers, encompassing expanded nanospace channels and T-nanomixers. The expanded nanospace channels (10–100 nm) were constructed by Kato et al. through the use of electron-beam lithography, with the electric double layer effect employed for separation purposes. The researchers were able to obtain quick separation of fluorescein and sulforhodamine B within 30 seconds by adjusting the length of the electric double layer and the size of the channel. This was possible even with sample volumes as little as 280 aL (Kato et al. 2009). Ishibashi et al. (2012a, b) have devised an automated injection system that effectively manages small sample volumes. This system facilitates normal phase separation mode, utilizing toluene as the mobile phase. The system successfully achieves the separation of Pyrromethene 597 and Coumarin 460 within a time frame of 5 s. In their study, Shimizu et al. (2016) demonstrated an enhancement in separation efficiency through the elongation of the nano channel, resulting in a plate number of up to 1.4 × 104. In a study conducted by Smirnova et al. (2016), a T-nanomixer was devised for the purpose of on-chip gradient separation. This experimental setup successfully facilitated the separation of a mixture of 17 amino acids within a time frame of 50 s. Notably, the majority of chromatographic peaks seen exhibited a plate height of less than 1 μm.

10.2.2.2 Stationary Phases for Microfluidic Chromatography of Metabolites

LC chips, or stationary phases for LC systems, have been developed to miniaturize to a few square centimeters. These microchips are used in chromatographic separations, often conducted in open channels with high surface-to-volume ratios. However, monolithic columns have drawbacks such as potential wall effects and limited reproducibility and characterization of commercial columns. Particle-based stationary phases have the potential to deliver significantly improved sample loading capacity and increased separation efficiency.

Documented progress in LC chip technology has been limited compared to microfluidic electrophoresis devices. The primary challenge encountered in miniaturizing traditional particle-based LC systems was the inherent difficulty in consistently fabricating compact chromatographic micropackings within tiny channel systems that can withstand elevated pressure levels during operation. Additionally, the consistent creation of nanoliter sample plugs and their responsive identification are more manageable.

LC chips that are readily available for purchase can only be utilized in specific devices, particularly if they are hyphenated with MS. The main utilization of these commercially available LC chip devices has been in proteomic research. However, the development of LC chip technology is ongoing, with a significant demand for extensive study in various areas such as micropacking preparation methods, sample injection performance, detection techniques, parallelization, and hyphenation.

Particle retention in microfluidic channels can be achieved using various devices, including frits, Teflon membranes, and glass wool. Fritless techniques and frit-like devices are used to create dense and homogenous packings for efficient chromatographic separations. The keystone effect is a method used to retain C18 particles in chromatographic separations, which can be achieved by using tapered outlet geometries. However, the resulting chromatographic packing was only suitable for basic separations involving mixtures of two or three components.

Huft et al. introduced a method for implementing parallel chromatographic columns on PDMS microchips using multilayer soft lithography (MSL) and a low-pressure packing approach. However, there is a lack of published research on this system due to the sophisticated approach employed in the preparation and implementation of these packings (Huft et al. 2013).

Microfluidics has been used in the segregation of diverse substances, including chromatographic separation structures. Fritless packing systems have drawbacks, such as relying solely on a tapering or bottleneck to sustain the front end of the packing, resulting in unidirectional liquid flow. Self-assembly colloidal silica particles in microchips can enhance separation efficiency and mitigate backpressure in hydraulic flow.

One potential approach for incorporating chromatographic particles onto microchips involves selective retention of particles solely on the surface of the channels, enabling the development of open-channel chromatographic devices. Magnetic beads can be attracted to the surface by magnetic forces, allowing them to be easily controlled within the channels using externally placed magnets. The immobilization of functionalized gold nanoparticles on the surface of microfluidic channels can enhance the wall's capacity as a stationary phase.

Microcontact printing (mCP) facilitates the formation of a self-assembled bead layer, allowing for the generation of patterns with diverse surface functionalities. The inclusion of self-assembled beads in the chip channels enhances their stability amid liquid flow. One notable benefit of microfluidics is its capacity to accommodate many analytical systems on a solitary microchip, enabling the concurrent execution of multiple chromatographic separations.

Sample concentration is crucial for the advancement of multifunctional integrated microfluidic devices, as it facilitates the identification of minute or scarce species, which is crucial in clinical diagnostics, forensics, environmental monitoring, and biodefense applications. Preconcentration also enhances detection sensitivity and reliability by increasing the signal-to-noise ratio (Vlčková et al. 2008).

The smallest volume of sample dispensed into a microdevice is constrained to between 0.5 and a few microliters, primarily due to practical constraints such as syringe and interconnect sizes. Microchip assays typically analyze volumes ranging from 100 pL to a few nanoliters, allowing for easy application of preconcentration steps that increase the concentration of analytes without the need for additional samples. Methods for sample preconcentration include surface-binding methods, electrokinetic equilibrium techniques, and nanoporous membrane/nanochannel approaches.

10.2.3 Microfluidic Biosensors for Metabolite Detection

10.2.3.1 Types of Biosensors (e.g., Enzymatic, Aptamer-Based)

Electrochemical biosensors are widely used in electrochemical methodologies for measuring glucose levels. These biosensors typically fall into two main categories: enzyme-based and non-enzymatic biosensors. Enzyme-based microfluidic biosensors require immobilization to ensure their ongoing functionality during the detection process. Techniques for enzyme immobilization include physical adsorption, sol-gel, polymeric materials, covalent binding, or cross-linking.

Studies have demonstrated the linear detection range of various enzyme-based microfluidic biosensors, such as Ferreira et al.'s (2013) study, Kurita et al.'s (2002) study, Huang et al.'s (2007) study, Pu et al.'s (2016) study, and Picher et al.'s (2013) application.

Current studies have focused on the integration of cell culturing and continuous glucose monitoring within microfluidic chips, with a specific focus on the study of individual cells. Weltin et al.'s (2013) chip demonstrated the ability to detect cancer cells and glucose simultaneously using electrodes immobilized with glucose oxidase (GOx). Obeidat et al.'s multi-sensor setup aimed at monitoring metabolic indicators during embryonic development showed a dynamic range from 0 to 9 mM with a limit of detection (LOD) of 0.5 mM.

The incorporation of cell culturing and detection presents benefits such as integrability, operability, and cost but also presents challenges such as potential impact of electrical factors on cells, electrode corrosion, and stability. Mitigating these concerns can be achieved by separating detection zones from the cell culture within the chip. Examples of this include Vonau et al.'s use of a nano-porous membrane to isolate cells from electrodes, Rodrigues et al. positioning electrodes at both the entrance and exit of a cell culture chamber, and Gu et al. presenting a droplet-based microfluidic sensor that uses Pt-black microelectrodes to rapidly identify the production of H_2O_2 resulting from glucose breakdown.

Metabolite identification during cell culture plays a critical role in drug screening. Weltin et al.'s microsystem demonstrated the effectiveness of their microsystem in drug screening through the observation of metabolic alterations following drug administration. Prill et al. developed a method to evaluate the harmful effects of drugs by manipulating the concentration of rotenone and tracking glucose and lactate amounts, providing valuable data on liver damage.

Despite their high linearity, easy integration, and low cost, challenges like enzyme activity decline and electrode fouling restrict their application in this domain.

Detecting and measuring metabolite concentrations are crucial for individual health, disease prevention, and management. Changes in these biomarkers can provide diagnostic clues about a person's health status, especially in areas with limited resources or non-clinical environments. Continuous, real-time monitoring of biomarkers in bio-fluids is essential for understanding dynamic biochemical processes. Laboratory-based techniques, such as chromatography, spectrophotometry, and electrochemical analysis, are expensive, require specialized apparatus, skilled staff, and involve laborious sample preparation, leading to delayed diagnoses and potentially severe outcomes.

Point-of-care testing (POCT) presents a straightforward, economical, convenient, expeditious, and portable substitute for laboratory-based testing, requiring fewer samples. The use of point-of-care testing in conjunction with biosensors with notable selectivity and specificity has the potential to greatly enhance the diagnosis and treatment of pathological conditions, especially in regions with limited resources. Biosensors generally comprise a bioreceptor responsible for identifying targets and a transducer that transforms biological signals into quantifiable outputs.

Smartphones are cost-effective, efficient, and portable tools for point-of-care testing (POCT) and mobile diagnostics, enabling real-time monitoring in non-laboratory settings. Integrating cutting-edge technologies like cloud computing, artificial intelligence, machine learning, paper-based sensors, and microfluidic chips will enhance the capabilities of smartphone-based POCT applications. Recent developments in smartphone-based platforms designed for the detection of human metabolites include integrated printed microfluidic biosensors, which enable efficient production turnaround times and convenient customization (Wilhelm 2018).

Sensors can be categorized into different groups depending on their specific components or substances they are meant to identify. Active sensors require an external power source to operate, while passive sensors generate signals without needing external power. Contactless sensors function without physical contact, while contact-based sensors require direct interaction with stimuli. Absolute sensors respond to stimuli with significant magnitude, while relative sensors detect changes in relation to a fixed or variable reference. Analog sensors produce a signal based on a dynamic physical property, while digital sensors provide output in Boolean format.

Sensors can be grouped into four categories based on signal detection type: chemical, thermal, physical, and biological. Continuous, real-time monitoring of biomarkers in bio-fluids is essential for understanding dynamic biochemical processes and improving healthcare outcomes.

10.2.3.2 Microfluidic Platforms for Biosensing Applications

In the realm of biosensing, the marriage of microfluidics and cutting-edge technologies has paved the way for groundbreaking advancements. Microfluidic platforms, with their ability to manipulate minuscule volumes of fluids in precise and controlled manners, have catalyzed a revolution in biosensing applications.

At the heart of microfluidic platforms lies the intricate manipulation of fluids at the microscale. These platforms, typically constructed from materials like polymers, glass, and silicon, enable the creation of microchannels, valves, and pumps that govern fluid flow with unparalleled precision. This meticulous control over fluid dynamics forms the foundation upon which a myriad of biosensing applications is built.

The synergy between microfluidic platforms and biosensors is the cornerstone of their success in biosensing applications. Biosensors, capable of detecting specific biological molecules or analytes, are seamlessly integrated into microfluidic devices to enable real-time, high-throughput analysis. These biosensors come in various forms, including optical, electrochemical, and surface plasmon resonance (SPR) sensors, each offering unique advantages in terms of sensitivity and specificity.

Microfluidic platforms find applications across a broad spectrum of biosensing domains, ranging from healthcare to environmental monitoring and beyond. In healthcare, microfluidic biosensors facilitate rapid and accurate diagnosis of diseases, enabling personalized medicine and point-of-care testing. Environmental monitoring benefits from the portability and sensitivity of microfluidic platforms, allowing for on-site detection of pollutants and contaminants (Zhuang et al. 2022).

10.2.3.3 Signal Transduction and Detection Methods

Biosensors play a crucial role in the detection of viruses, pathogens, and illnesses. These devices consist of an analyte, biological substance, transducer, electronic module, and display unit. When introduced into an electrolytic solution, they can be identified as containing substances like alcohol, glucose, lactose, and ammonia. The process of biorecognition initiates signal generation through interactions between analytes and bioreceptors, converting this signal into a measurable electrical representation. These signals are then communicated to cloud servers for data storage and analysis.

Calorimetric biosensors quantify the reactions or structural dynamics of dissolved biomolecules. Microelectromechanical (MEMS) materials are commonly used in metabolic applications for monitoring metabolic processes. Acoustic biosensors use piezoelectric materials to alter the physical properties of an acoustic wave based on changes in analyte absorption. Two types of mass balance acoustic transducers are used: bulk acoustic wave (BAW) and surface acoustic wave (SAW) devices.

Field-effect transistors (FETs) are widely used in biological applications due to their three-terminal mechanism for regulating electric current flow via an electric field. Transistors-based sensing platforms like metal-oxide semiconductor field-effect transistors (MOSFETs) and ion-sensitive field-effect transistors (ISFETs) are commonly used.

The transduction mechanisms of electrochemical biosensors, such as potentiometric, amperometric, impedimetric, conductometric, and voltammetric techniques, are used to classify them. Conductometric biosensors evaluate the conductance between electrodes, voltammetric biosensors monitor current by controlling voltage variations to identify analytes, and potentiometric biosensors detect charge buildup at the working electrode.

Quartz crystal microbalance (QCM) is a biosensor platform that uses mass sensing to track changes in mass or thickness of layers on a quartz crystal surface. SAW biosensors generate electromagnetic impulse signals for real-time protein detection without labels.

10.3 Microfluidic Techniques for Metabolite Production

10.3.1 Cell Culture and Microfluidic Bioreactors

The application of microfluidic bioreactor systems for the cell culture has great potential for several applications, including tissue engineering, drug development, bioprocessing optimization, and cell-based evaluating investigations (Pasirayi et al. 2011; Steinhaus et al. 2007). Microfluidic bioreactor systems, with their length scales corresponding to cell dimensions, are ideal for studying cell behavior and organization, offering superior biological function, quality data, reagent volume, integration, and cost-effectiveness (Baudoin et al. 2011; Martin et al. 2004).

10.3.1.1 Advantages of Microfluidics for Culturing Metabolically Active Cells

Microfluidic systems improve the comprehension of single-cell dynamics, population heterogeneity, and chemical gradients, impacting disciplines like cellular phenotyping, drug assessment, toxicological evaluation, and disease investigation by creating stable gradients (Allard et al. 2022; Luke et al. 2016). Further advancements in the comprehension of cellular processes and the advancement of therapeutic uses are promised by the forthcoming prospects of microfluidics in single-cell cultivation.

1. Precise Control of Microenvironment: Shear stress, pH, oxygen levels, and levels of nutrients can all be precisely controlled with the use of microfluidic systems. Microfluidics makes it easier to cultivate cells in environments that closely mimic physiological situations by more accurately simulating the in vivo microenvironment than is possible with conventional culture methods (Rothbauer et al. 2018).
2. Improved Spatial and Temporal Resolution: Real-time observation of cellular responses to stimuli is made possible by the high spatial and temporal resolution

provided by microfluidicsystems. This capability is crucial for researching dynamic processes like metabolism, migration, and cell signaling because it sheds light on the complex processes regulating cellular behavior (Yeh and Hsu 2019).

3. Lower Sample and Reagent Consumption: Compared to conventional culture techniques, the miniaturization of culture volumes in microfluidic technology greatly lowers sample and reagent usage. This reduces expenses while enabling high-throughput screening as well as evaluation of cell-mediated responses to different stimuli, which facilitates applications in personalized medicine and drug discovery (Falconnet et al. 2011).
4. Integration with Analytical Techniques: Thorough assessment of cellular processes is made possible by the seamless integration of microfluidic equipment with analytical methods like mass spectrometry, spectroscopy, and microscopy. By improving the knowledge of cellular metabolism, growth, and differentiation, this integration creates new opportunities for fundamental study and practical applications (García Alonso et al. 2019).

10.3.1.2 Design Considerations for Microfluidic Bioreactors (e.g., Nutrient Delivery, Waste Removal)

Stabilizing tissue-engineered structures in a bioreactor framework is difficult than just giving the cells food and oxygen. Only when cellular structures mature in conditions that correspond to their individual in vivo environments will they be able to fulfill their full operational capacity (Martin et al. 2004; Pörtner et al. 2005). As a result, different constructs need different kinds of stimulatory specifications. These specifications could include adding mechanical, electrical, or chemical stress-producing processes to a culture environment that is hermetically sealed and free of contaminants (Bilodeau and Mantovani 2006; Massai et al. 2016). Therefore, when designing a tissue-specific bioreactor framework, particular engineering parameters like mass transfer, mechanical stimulation, and electrical stimulation should be taken into account.

When building bioreactors for tissue engineering applications, choosing materials plays a crucial role. Physical and chemical specifications like transparency, stability, and compatibility with manufacturing processes are important factors to take into account when designing (Lei and Ferdous 2016; Salehi-Nik et al. 2013; Zhao et al. 2016b). Ideally, materials should be conveniently sterilized, non-leachable, and non-degradable. The FDA has to permit the materials used to build bioreactors in order to ensure the safety of clinical products. Mass transfer is crucial in bioreactor processes, supplying oxygen and nutrients, eliminating metabolic waste, and improving cellular microarchitecture and differentiation. Effective mass transfer should be given the highest priority in bioreactor design, and mechanisms for regulating the physical and physiological environment should be included to cater to the unique requirements of target tissues (Ginner et al. 2020; Martin et al. 2008). Electrical fields in bioreactors promote cell growth, tissue creation, and upregulation of contractile proteins. Sensors monitor temperature, pH, glucose, and lactate concentrations, while automation decreases contamination and enhances efficiency with minimal human involvement (Kirsch et al. 2010; Viens et al. 2011). The development of functional tissue alternatives with potential clinical utility is aided by these technologies.

10.3.1.3 Examples of Microfluidic Bioreactors for Metabolite Production

Microfluidic systems offer potential for studying live microscopic cultures of cells, tissue specimens, and small creatures. Incorporating hyperpolarization into high-resolution NMR spectroscopy can help explore metabolism with volume-limited technologies. Sylwia Barker et al. used a microfluidic chip to make and identify a hyperpolarized metabolite simultaneously, showing that it can be useful in technologies with limited volume (Barker et al. 2022).

Finkbeiner et al. introduced a microfluidic bioreactor for plant-cell culture, replicating cell interactions in plant tissues. The chip, made of polycarbonate housing and ultrasonic welding, was transparent and biocompatible. The researchers demonstrated three proof-of-concept functions, showing that the growth factor generated by multiplying cells regulates tobacco BY-2 cell division, vindoline can be produced by merging two species of *Catharanthus roseus* cells, and the fungus *Neofusicoccum parvum's* phytotoxin production. B. O. Sullivan et al. demonstrated the potential use of "fed-batch" substrate feeding techniques in continuous-flow microreactor technology. They used the TK-catalyzed condensation of HPA and GA as a test reaction and developed a flexible microfluidic reactor structure. The design was simple to manufacture using laser ablation and can be modified to create customized reactions for various feeding schemes and biocatalysts. The modularity of the MIR design could lead to the development of multi-step synthetic methods for intricate molecular products (O'Sullivan et al. 2012).

The concentration of ERY produced increases from 50 mM to 224 mM, avoiding harmful substrate exposure. The reactor achieved a maximum throughput of 15 mg hL1, a fivefold improvement compared to the prior reactor and eightfold to fed-batch microplate reactions. This highlights the benefits of continuous-flow reactors in enhancing synthetic material production. However, further adjustments are needed to address CO_2 evolution and enzyme stability for complete conversion at higher input concentrations (Lawrence et al. 2013).

10.3.2 Enzymatic Microreactors for Metabolite Synthesis

The pharmacokinetic and pharmacological effects of chemicals can be significantly impacted by metabolic processes, including oxidative, reductive, and conjugative reactions, upon entering the human body. Currently, drug development evaluations focus on unaltered parent chemicals in in vitro systems, without considering these metabolic processes. This failure to incorporate compound metabolism into therapeutic development contributes to the failure of candidate compounds and delays in enhancing patient outcomes. Factors contributing to this failure include limited understanding of metabolites and the absence of simple pathways for synthesizing chemicals during human metabolism. Current methods for metabolite synthesis, such as bacterial enzyme expression, liver cell incubations, liver microsomes, and conventional organic synthesis, face limitations such as low yields, impaired enzymatic activity, and the need for separation from biological matrices. Microfluidic synthesis platforms have shown potential for drug metabolism and pharmacokinetics investigation, but they fail to address challenges related to precise identification of metabolites and direct evaluation of bioactivity and toxicity. Enzyme immobilization and parallelization in microfluidic flow chemistry could be a potential solution to these obstacles (Van Vugt-Lussenburg et al. 2022).

10.3.2.1 Immobilization of Enzymes on Microfluidic Platforms

Bradley et al.'s study focuses on the production of drug metabolites for in vitro compound testing using immobilized-enzyme flow reactors. The technique generates nanogram quantities of sulfate and glucuronide metabolites from resorufin and *p*-nitrophenol in a few hours without specialized equipment or complex sample preparation. The microfluidic technique is crucial for drug metabolite synthesis and can be easily implemented by research groups with minimal microfluidic expertise. However, the study is limited by a loss of fluorescence test, uncomplicated substrates, and kinetics estimates. Despite these limitations, the study suggests that this method has significant potential in overcoming obstacles in pharmacological investigation.

The technology of κ-IMER is rapidly developing and is used in various industrial and scientific research domains. These devices are small, low-flowing, and feature channels or capillaries with inner diameters smaller than a few hundred micrometres. The performance of this process is influenced by factors such as immobilization method and support material/configuration. Integrating β-IMER devices into bottom-up proteomic workflows offers advantages such as enhanced digestion rates, enzyme reuse, reduced sample consumption, and bypassing reduction and alkylation steps. However, high-throughput requires fully automated systems and online integration (Hoog Antink et al. 2019; Messner et al. 2021; Sewczyk et al. 2018).

10.3.2.2 Reaction Control and Optimization in Microfluidic Enzymatic Reactors

Enzymes are popular because of their simplicity, substrate specificity, and environmental friendliness. However, their poor operating stability, recovery, and incapacity to be reused in industrial settings sometimes provide hurdles to their application. Enzyme immobilization has several advantages, including greater stability, increased enantioselectivity, simpler reactor operation, and easier product separation. Immobilized enzymes can be recycled to reduce manufacturing costs while improving features such as avoiding subunit dissociation, aggregation, autolysis, proteolysis, enzyme structure rigidification, and establishing favorable microenvironments. There are two basic techniques of enzyme immobilization: carrier-bound and carrier-free. Carrier-bound immobilized enzymes provide for greater control over reaction processes, simpler product separation, and enhanced stability. Nanomaterials such as nanoparticles, nanofibers, nanotubes, metal organic frameworks (MOFs), nanosheets, and nanoflowers have emerged as new supports for enzyme immobilization due to their increased specific surface area and lower diffusion restrictions. MOFs are very promising for enzyme immobilization because of their large surface area, pore volume, variable pore diameters, and easy manufacturing. Two techniques are used to create enzyme-MOF composites: post-synthetic and in situ synthesis. In post-synthetic procedures, enzymes are immobilized onto MOFs by surface adsorption, covalent bonding, or diffusion after MOFs are initially manufactured. Through co-precipitation, enzymes are immobilized during the MOF synthetic process in situ synthesis. The size and activity of the composite particles are influenced by factors including synthesis additives, enzyme type and concentration, and additive quantity; thus, optimizing a number of parameters is necessary to improve their function. In order to maximize the performance of MOFs, particle shape and crystal size are important factors. MOF crystal morphology is influenced by synthetic factors, producing molecules with different particle sizes and distributions. In order to address the challenges involved in producing drug metabolites for in vitro compound testing, Bradley et al. undertook a research. Without the need for specialized tools or labor-intensive sample preparation and purification processes, the demonstrated technique showed the ability to produce nanogram quantities of sulfate and glucuronide metabolites from resorufin and p-nitrophenol in a matter of hours. Rapid advancements in κ-IMER technology are being made in a number of industrial and scientific research fields. Technological developments in immobilization methods and strong supports have led to the creation of many high-functioning κ-IMERs. However, online integration and fully automated systems are essential for achieving high throughput.

Oxidative and conjugative mechanisms have a major effect on the metabolism of medicines and xenobiotics, influencing their uptake, distribution, and excretion at the cellular, tissue, and organism levels. The existing protocols for metabolite synthesis have several drawbacks, such as poor yields, reduced enzymatic activity as a result of product creation, and the need to extract the compounds from complex biological matrices.

The use of enzyme immobilization and parallelization in microfluidic flow chemistry has been proposed as a potential solution to the existing obstacles encountered in the synthesis of drug metabolites.

10.3.2.3 Applications in Targeted Metabolite Production

1. Microalgae have been identified as potential platforms for genetic engineering for producing secondary metabolites, offering applications in pharmaceuticals, nutraceuticals, biofuels, and bioremediation. By manipulating their metabolic pathways, researchers can enhance the production of pharmaceutical compounds, making microalgae a cost-effective and sustainable alternative. They are also rich in bioactive compounds, making them valuable sources of dietary supplements and functional foods. Genetic engineering can also increase the lipid and carbohydrate content of microalgae, making them suitable for biofuel production. Despite challenges like optimizing genetic modification techniques and

ensuring regulatory compliance, the potential benefits of genetically engineered microalgae for targeted metabolite production are significant.

2. Advancements in understanding plant metabolic networks have led to the development of targeted metabolite production through in vitro systems. This involves integrating omics technologies like genomics, proteomics, transcriptomics, and metabolomics to understand plant biosynthetic pathways. Genome engineering with CRISPR/Cas9 technology enhances secondary metabolism and identifies genes for improved resistance to biotic or abiotic factors. Optimizing bioreactor cultivation involves careful selection of process parameters, medium composition, precursor feeding, and elicitation strategies. Scaling-up challenges are addressed using NMR-based measurements and dynamic metabolic flux analysis. Efficient modeling strategies for energetic management improve the production of bioactive molecules in bioreactors. Secondary metabolites from plants are crucial for various industries, including pharmaceuticals, cosmetics, food additives, and agriculture. Advancements in extraction and isolation procedures, genetic engineering, and plant tissue culture offer opportunities for targeted metabolite production.

   Secondary metabolites from plants are essential for various industries, including pharmaceuticals, cosmetics, food additives, and agriculture. They have applications in insecticides, therapeutic agents, antioxidants, and fragrances. Advancements in extraction and isolation procedures, such as Microwave Assisted Extraction (MAE), Ultrasound-Assisted Extraction (UAE), and Supercritical Fluid Extraction (SFE), have improved their commercial value and preservation. Modern biotechnology, including genetic engineering and plant tissue culture, offers targeted metabolite production opportunities.

10.4 Microfluidic Screening Platforms for Metabolite Production Pathways

10.4.1 Droplet-Based Microfluidics for High-Throughput Screening

For more than 10 years, droplet techniques have been utilized to control aqueous droplets consisting of bacteria and antibiotics, both inside and outside of microfluidic devices. In order to tackle the diagnostics problem, several microfluidic systems featuring integrated dilution modules were created. These advancements facilitated the automated generation of antibiotic concentration gradients and the rapidity required for conducting high-throughput combinatorial experiments (Postek et al. 2018).

Automated emulsification of samples allowed researchers to quantify the inoculum effect, resistance degree, and resistance levels of individual cells in droplets. They identified resistance levels among clonally identical individuals by examining varying antibiotic concentrations (Gielen et al. 2016; Miller et al. 2012). A study on fluorescent dyes in droplets suggests a modified variant of a viability marker is better for droplet assays, enabling the screening of bacteria from clinical specimens. A detection system analyzes antibiotic growth or mortality conditions by measuring light scattering in millions of droplets per second (Pacocha et al. 2021; Sabhachandani et al. 2017). In order to get a deeper understanding of how resistance is acquired and lost, we will combine high-throughput single-cell tests with the progressive development of resistance studies facilitated by droplet-based microchemostats. These approaches (Lan et al. 2017; Liu et al. 2016) are used to conduct automated combinatorial screens of antibiotic resistance in individual cells from clinical samples. The classic parallel screening method, often use microplates and optical detection technology, has limitations such as limited daily processing capability, high throughput, and difficulty in reducing reactor volume. Advanced processing platforms are often needed to achieve the highest screening effectiveness, despite these constraints (Abate and Weitz 2011; Scanlon et al. 2014). In contrast, the primary benefit of droplet microfluidic technique lies in its high-throughput processing capacity. This allows for the rapid generation of numerous monodispersed microdroplet carriers, which can be used to precisely manipulate molecules and cells simultaneously.

A microfluidic chip design allows for efficient screening of large quantities of medications within a short timeframe. This approach meets high-throughput screening requirements. Researchers conducted high-throughput screening of naturally occurring antibiotics using fluorescent probes, *S. aureus*, and recombinant *E. coli*. This method achieved a screening yield of 3000 droplets per second using rapid fluorescent labeling (Granieri et al. 2010). A new microfluidic chip technology has been developed to enhance microbial diversity analysis by using ultra-high-flux droplets. This system, which can assess 1×10^5 cells from environmental samples in just 20 min, significantly improves efficiency compared to automated microtiter plate systems, allowing for faster and more accurate microbiological analysis (Najah et al. 2014).

Furthermore, the reagent dosage concentration decreased to 0.5 million. In addition, droplet microfluidics can achieve a wide range of tasks by effectively harnessing its high-throughput analysis capabilities, thanks to the flexible integration of its multifunctional units (Guetschow et al. 2014). In order to efficiently generate microdroplet libraries using a solitary material and selectively differentiate the desired droplets for microfluid extraction (MCE) analysis, In their study, Guetschow et al. established a connection between droplet microfluidics and a microchip electrophoresis (MCE) platform. Droplet microfluidics, a technique that mimics essential unit operations in conventional well plate screens, has the potential to revolutionize High-Throughput Screening (HTS) due to its remarkable throughput and sampling benefits. Despite these benefits, droplets have yet to be widely used in HTS (Holtze et al. 2017). Droplets stabilized with surfactants are used for complex biological assays, allowing multiple additions, extractions, and temperature adjustments within milliliters of volume (Baret 2012). Bulk-generated emulsions, used as monodisperse inputs for unit operations, may face size discrepancies, potentially disrupting downstream functions and compromising device reliability (Fallah-Araghi et al. 2012; Lee et al. 2013; Mahler et al. 2015).

Droplets are essential for discrete sample processing and evaluation, but they can cause crosstalk when they transfer analytes into hydrophobic oil. To prevent this, researchers have explored using mineral oil as a continuous phase for droplet segmentation, fluorinated oil, and surfactants. However, chemical leaks from droplet specimens have also been observed. Fluorophores have been used as model analytes in studies using fluorescence detection to study molecular accumulation and leaks in droplet emulsions (Roach et al. 2005). Factors such as buffer pH, solubility, and hydrophobicity of the fluorophore can cause analyte loss. There is no single accepted mechanism for molecular transport, but understanding these mechanisms could extend droplet functionality to workflows that cannot be accessible due to chemical leaks. Adding bovine serum albumin or sugar additives can improve the containment of fluorophores, and altering fluorophores to have permanently-charged sulfonate groups can reduce chemical loss between specimens. Using different surfactants with multiple layers, dimensions, and structures can be effective in containing small molecules and stabilizing droplets. Label-free evaluation methods like mass spectrometry and droplet Raman can help with better droplet cross-talk evaluation. These studies can help in choosing and refining suitable oils, additives, and surfactant combinations to minimize cross-contamination risks (Etienne et al. 2018).

In some situations, it has been shown that using different surfactant multiple layers, dimensions, and structures is an efficient approach. Recent research has shown that dendronized polymer surfactants are effective at containing small molecules and stabilizing droplets. It has been demonstrated that using surfactants in the form of nanoparticles eliminates micelle development, leading to carrier phase solubility as the sole means of promoting chemical transport (Janiesch et al. 2015). To satisfy growing interest in these systems, commercial nanoparticle surfactant/oil combinations have been developed. It seems that the main issue with droplet partitioning is small molecules. It has been demonstrated that larger biomolecules like proteins and nucleic acids stay within the initial droplets. The utilization of surfactant-stabilized droplets for hydrothermal synthesis (HTS) of smaller molecules is expected to provide ongoing difficulties, mostly stemming from the stability and confinement of the droplets. This challenge is particularly evident when investigating diverse chemical libraries and reaction conditions (Chowdhury et al. 2019). A purely fluorescence-based approach is probably going to be restrictive when it comes to the in-depth molecular analyses required to comprehend crosstalk. Label-free evaluation, like mass spectrometry and droplet Raman, is starting to offer more information about how droplet chemical compositions change over time and could help with better droplet cross-talk evaluation (Payne et al. 2020; Steyer and Kennedy 2019). These studies can aid in choosing and refining suitable oils, additives, and surfactant combinations to minimize the risk of cross-contamination among these analytical substances.

10.4.2 Integration with Analytical Techniques for Metabolite Detection

Droplet microfluidics has transitioned from chemistry to biology, leading to advancements in enzyme activity exploration and recognition mechanisms. It is closely linked to synthetic biology, discovering bioactive compounds, strain evolution, and screening semi-rational or rational enzyme designs. Due to cost-effectiveness, efficiency, automation, and sensitivity in detection, droplet microfluidics has replaced traditional well plate assays and flow cytometry as the preferred method for high-throughput screening in synthetic biology (Hu et al. 2023).

The fluorescence detection approach, which utilizes UV, visible, and fluorescence spectra, is a highly prevalent and advanced high-throughput screening method. It plays a crucial role in the commercialization and industrial implementation of strain/enzyme examinations (Jiang et al. 2023). While UV-visible light detection is commonly acknowledged for its user-friendly nature, it is not without its constraints. Biosensors constitute a crucial subfield within the realm of fluorescence detection. Their main purpose is to enhance the intensity of fluorescence or address difficulties associated with generating fluorescence signals, compensating for the limitations of conventional methods (Gasilova et al. 2014). Other label-free detection techniques, such as nuclear magnetic resonance, mass spectrometry, and electrochemistry, are as effective in addition to optical detection. Throughout several decades of spontaneous development, enzymes have been widely used in a variety of fields and are the most significant biocatalysts in the environment (Hansen et al. 2013; Heinemann et al. 2017).

Artificial selection and screening are crucial for producing enzymes with desirable properties, but traditional microplate screening techniques struggle to meet high-throughput library screening requirements. Directed evolution is a key tactic, but challenges include large mutant library analysis costs and experimental obstacles. Despite these challenges, there is a strong need for new and enhanced microbial catalysts in commercial and medicinal domains (Cao et al. 2018; Mendoza et al. 2012). Droplet microfluidics enhances enzyme advancement by providing personalized reaction conditions and enhanced analytical capabilities, potentially necessitating directed evolution for esterases, cellulases, glucose dehydrogenases, and plastic degradation enzymes (Yang et al. 2006).

10.4.2.1 Applications in Identifying Optimal Culture Conditions and Enzyme Combinations

Interstitial flow (IF) refers to unidirectional fluid movement through the extracellular matrix (ECM) and serves as a signal from tumors, playing a crucial role in the metastatic expansion of cancer (Filippi et al. 2022; Warrick et al. 2008). A microfluidic system was developed to study cancer cell stimulation and differentiation by simulating tumor cell IF and soluble compound transfer from donor cells (Huang et al. 2010). The control of cell function is also significantly influenced by biochemical factors. Using a microfluidic platform, Zhang et al. examined the impact of Ca+2 and Sr+2 metal ions on the osteogenic differentiation of mesenchymal stem cells (MSCs) (Zhang et al. 2021a). The research utilized a microfluidic technique to create cross-linked alginate microgels for encapsulating single MSCs, demonstrating that these hydrogels increased matrix mineralization and triggered osteogenic differentiation (Zhang et al. 2021b). The rigidity of the microenvironment significantly influences tissue progression, cell fate, and function, as demonstrated in a recent study that altered the migration of NPC43 and NP460 cells through microfluidic system segregation (Yi et al. 2022; Zhong et al. 2020).

10.4.2.2 Fabrication of Biomaterials

Biomaterials are essential in tissue engineering, enhancing the replication of native extracellular matrix (ECM) and promoting cell function in damaged tissues through various methods (Keane and Badylak 2014). Microfluidics is a cost-effective, secure, and user-friendly method for creating biomaterials like microspheres, microfibers, and nanoparticles, thanks to its unique combination of shaped microchannels and fluid dynamics (Advances in Biomaterials Science and Biomedical Applications—Google Books 2024). Lei et al. developed magnetic chitosan microspheres (MCMs) with antibacterial properties using a microfluidic platform, enhancing production with consistent dimensions and forms.

Utoh et al. used microfluidic techniques to create collagen microfibers and calcium phosphate (CaP) biomaterials for bone regeneration and repair (Utoh et al. 2021). Galván-Chacón et al. demonstrated that microfluidic techniques can modify CaP's physical and chemical characteristics for optimal effectiveness (Galván-Chacón et al. 2021). Wang and colleagues developed a microfluidics-based method to fabricate alginate-collagen composites into endothelialized biomimetic microvessels (BMVs), which showed a significant perfusion effect and could trigger osteogenic differentiation (Wang et al. 2022).

10.4.3 Microfluidic Modeling and Simulation for Metabolite Production Optimization

10.4.3.1 Computational Models for Microfluidic Systems

Metabolic engineering aims to improve cellular functions for producing desired compounds in host organisms. Targeted evolutionary strategies are needed to bridge the gap between enzymes and pathways. Directed laboratory evolution enhances biological evolution through randomization of mutations and higher detection rates of positive strains (Zhou and Alper 2019). It creates the ideal rate of production of an intended substance by utilizing various databases, tool libraries, and environmental factors. Computational simulation and mathematical frameworks could be used as effective tools to comprehend and forecast the processes of biological systems (Carbonell et al. 2018).

Mathematical frameworks for system design are created by combining multi-level omics data from various databases. Computational design techniques link reactions from different databases, enabling the identification of potential metabolic pathways leading to the desired product, such as enzymes (Ohtake et al. 2017).

Recently, Carbonell et al. introduced and demonstrated an automated Design–Build–Test–Learn (DBTL) pipeline to computationally pinpoint the most favorable elements, such as enzyme coding regions and RBSs, for assembly (Carbonell et al. 2018). The DBTL pipeline, designed for Escherichia coli, increased flavonoid (2S)-pinocembrin titer 500 times through over 23,000 varieties. The RBS Calculator predicts internal start codon translation initiation rate to reduce unwanted expression (Salis 2011). For integrating transcriptomics, proteomics, fluxomics, and metabolomics data are currently accessible. Tn-Core enables the context-specific model generation process by automatically integrating Tn-Seq data with RNA-seq data. This gives the goals of metabolic engineering a "systems-level" perspective (Dicenzo et al. 2019).

The RiboLogic algorithm was used to create 286 switches regulating the interaction of flavin mononucleotide, tryptophan, theophylline, and miR-208a, an miRNA linked to cardiac hypertrophy. These switches act as biosensors, enabling the study of directed evolution in microbes or biological networks. Combinatorial optimization was used to generate thousands to millions of constructs, guiding the debugging process toward the desired outcome (Wu et al. 2019).

Next-generation sequencing was used by Höllerer et al. to evaluate the impact of extraordinarily large sets of RBSs on quantitative expression (Höllerer et al. 2020). Deep learning method using ensembling and uncertainty modeling accurately forecasts untested RBS parameters, utilizing large-scale datasets and DNA-based phenotypic recording for quantitative function forecasting. Important elements of the vascular environment, such as fluid flow and flow-induced shear stress, are frequently investigated in microfluidic cell culture systems. These models present a promising avenue for on-chip biomolecular monitoring; nevertheless, frequent sample dilution makes it difficult to identify biomolecules owing to flow-induced shear stress (Naseri and Koffas 2020). On-chip monitoring of cell-secreted parameters and physiological shear flow are integrated in a computational model for vascular microfluidic cell culture. Because of its bilayer structure, analytical materials may be transported to biosensor surfaces quickly, cutting down on reaction times.

Chen et al. (2020) suggest creating microfluidic vascular models to address flow issues from physiological shear stress and monitor cell-secreted components, enhancing their laboratory-on-a-chip applications capabilities.

10.4.4 Utilizing Models to Predict Metabolite Yields and Optimize Production Parameters

Metabolic engineering optimizes metabolite production by directing metabolic flow and fluxes toward desired metabolites. There are two categories: approaches focusing on specific elements of the metabolic network, and strategies introducing novel pathways or modifying the network using network reconstruction tools (Xu et al. 2022). Over the course of recent decades, significant progress has been made in the field, shown by the successful integration of adenosylcobinamide phosphate biosynthesis from *Rhadobacter capsulatus* into the *E. coli* strain. This integration has resulted in a notable increase in the vitamin B12 content, reaching 307 μg/g (Fang et al. 2018). Xu et al. (2022) propose that the yeast was genetically modified to increase the production of rubusoside and rebaudiosides, resulting in concentrations of 1368.6 mg/L and 132.7 mg/L, respectively.

The understanding of the correlation between genotype and phenotype in target cells hinders the development of metabolic pathway optimization strategies. Traditional trial-and-error techniques are time-consuming, expensive, and labor-intensive. To address this, constraint-based modeling techniques, such as Flux Balance Analysis (FBA), are used to analyze organisms' optimal performance and identify metabolic phenotypes (Rana et al. 2020; Shene et al. 2020). One major challenge in CBM, however, is that the complex genome-scale metabolic model (GSMM) and biological network replication allow for multiple optimal solutions to achieve the intended flux.

The rise of high-throughput technology and massive omics data has significantly impacted molecular biology, enabling better understanding of intricate biological systems, but data interpretation remains challenging (Ait Brahim et al. 2023). Machine learning is utilized in biological omics data for categorization, exploration, and forecasting, revealing hidden information and proving effective in facial identification, optical character recognition, and diabetes disease estimation (Al-Hameli et al. 2023; Casano et al. 2023; Hairani Hairani and Dadang Priyanto 2023; Memon et al. 2020). Machine learning (ML) algorithms use extensive experimental datasets to identify trends and improve prediction accuracy. They've been used to analyze metabolic networks and genotype-phenotype relationships, and their fewer parameters make them effective for various tasks (Colarusso et al. 2021; Lee et al. 2021).

10.4.5 Case Studies in Microfluidic Simulation for Metabolite Optimization

Di Chen et al. developed a bionic method for dynamic co-culture of hepatocytes and renal cells for emodin assessment in health foods. The bionic method used rat tail collagen type I, human renal glomerular endothelial cells (HRGECs), human hepatocellular carcinoma cells (HepG2), and gelatin at appropriate concentrations. The method showed suitable growth stability and decreased cell survival with increasing emodin concentration. The study also showed a declining trend in LDH concentration between 0 and 100 μM emodin, with a significant increase at 100 μM emodin. The improved bionic mechanism for emodin evaluation provides a foundation for future in vitro studies on healthy food efficacy (Chen et al. 2022).

Johannes Dornhof et al.'s research reveals the importance of 3-D cell cultures for personalized cancer treatments. They created a microfluidic organ-on-chip structure with integrated electrochemical chemo- and biosensor arrays for energy metabolites glucose, lactate, and oxygen. This technology allows for convenient access, compartmentalization, and connection of cell matrices using conventional laboratory equipment. TNBC stem cells from a single patient develop into cancer organoids in a heterogeneous spheroid culture on-chip. The findings provide precise and consistent on-chip monitoring of several metabolites from a matrix-based culture over a one-week period under varying conditions. This technique emphasizes the importance of continuous, in situ monitoring of metabolites in three-dimensional cell cultures for drug screening in cancer research and the establishment of standardized and controlled culture conditions (Dornhof et al. 2022).

The rapid development of organ-on-a-chip (OoC) technologies has led to the need for new strategies and in-depth knowledge of in-organ processes. Nurzhanna Bakuova et al. used computational fluid dynamics and experimental cell culture growth to compare filling traits and flow behavior between two microfluidic liver-on-a-chip equipment. They found that the elliptical chamber chip provided better flow and filling traits. The study also examined biocompatibility, continuous flow, and biomarkers, providing valuable information for improving performance and designing alternative in vitro testing techniques (Bakuova et al. 2023).

10.5 Conclusion

Microfluidic methods of monitoring and evaluating metabolites from single cells or cell populations have transformed drug screening, cancer research, and embryo screening. As a result, our knowledge of cell biology and its useful applications in medicine has significantly advanced. Notwithstanding, many obstacles continue to exist, including ambiguity regarding the correlation between cellular metabolism and activity, as well as the dependability of metabolomics markers in assessing cellular function. Certain constraints impede the development of microfluidic culturing and detecting systems. It is critically necessary to create high-throughput, high sensitivity, and high stability detecting methods in addition to high-degree simulation culturing equipment. Additionally, the integration of smart technologies and digital microfluidics is required.

10.5.1 Future Directions of Microfluidics in Metabolite Analysis and Production

The use of microfluidics in bioanalytical applications is growing, with a particular emphasis on the identification of metabolites and small compounds. These microsystems are useful for biomonitoring, toxicity screening, and clinical diagnostics because they can identify changes in metabolism. Effective biological sample analysis and the integration of functionality into microfluidic devices provide a difficulty. Due to problems with quick molding polymers and unrealistic upscaling of microfluidic cell concentration and lysis procedures, not much progress has been achieved in microfluidic investigations for downstream treatments of microalgae. It will be necessary to modify inert materials such as glass, silicon, and ceramics for upcoming research on microfluidic extraction and transesterification of microalgal lipids. Because of its resilience to extreme heat and weather, low-temperature co-fired ceramics, or LTCC, have become popular options. Microfluidics may find its way into intelligent systems in wearable technology, offering robust platforms for biological research in the networked world of sensors, computers, energy, electronics, and communication. When paired with other cutting-edge technologies like the Internet of Things (IoT), Internet of Medical Things (IoMT), and Internet of People (IoP), these wearable devices have even more revolutionary potential (Apoorva et al. 2024).

But there are still obstacles to be addressed, including those related to technological complexity, standardization efforts, regulatory concerns, and user acceptability. Through the research of novel forms such as paper based, digital, template, and self-healing hydrogels, modular microfluidic systems can be extended. New materials and manufacturing techniques can improve alignment accuracy and productivity, increasing the range of possible designs.

Collaboration between academia and industry is needed to develop common standards and reliable systems for technology dissemination and implementation.

10.5.2 Impact of Microfluidics on Personalized Medicine and Biomarker Discovery

Microfluidics play a crucial role in personalized medicine and biomarker discovery, as it allows for precise regulation and manipulation of fluids, enabling medical researchers to develop clinical platforms for rapid sample analysis. This technology has been instrumental in the development of innovative therapeutic techniques and diagnostic methodologies, particularly in point-of-care diagnostics and customized pharmaceutical therapy.

Microfluidic devices have shown value in understanding the hydrodynamic properties of blood by enabling the observation and measurement of blood cells. They also play a crucial role in identifying disease symptoms during clinical inquiries. One potential use is combining point-of-care devices with flow visualization methods to evaluate an individual patient's cardiac risk profile, helping healthcare practitioners identify the most appropriate treatment strategy and improving patient survival rates.

Recent advances in blood analogue fluids have demonstrated significant promise in improving our understanding of many blood illnesses. To accurately simulate the multiphase effects observed in blood flow, particulate solutions made of flexible particles that faithfully reproduce the fundamental structural attributes of red blood cells (RBCs) are highly advantageous.

The link between treatment responses of cancer patients in vivo and tumor biopsies in tissue culture models was investigated by medical oncologists in the 1950s. Numerous methodologies for personalized medicine have been developed over time, including sequencing-based techniques and patient-derived xenograft mouse models. However, only a limited number of techniques have been implemented in regular clinical practice due to three primary prerequisites:

1. Feasibility within specified timeframes,

2. Cost-effectiveness, and.
3. Minimal quantities of patient material.

The use of microfluidic technology presents significant conceptual benefits, making it a pivotal component in future personalized cancer therapy methodologies. Microfluidic devices will facilitate the separation and analysis of cancer cell spheroids obtained from patients, enabling the screening of rare cells for drug sensitivity in a target-agnostic and phenotypic manner. Future personalized therapeutic options guided by microfluidics should utilize both sample types and patient material.

References

Abate AR, Weitz DA (2011) Faster multiple emulsification with drop splitting. Lab Chip 11(11):1911–1915. https://doi.org/10.1039/C0LC00706D

Advances in Biomaterials Science and Biomedical Applications—Google Books (2024). https://books.google.co.in/books?hl=en&lr=&id=xPyPDwAAQBAJ&oi=fnd&pg=PA315&dq=Chen,+N.Z.+and+X.+Biofabrication+of+Tissue+Scaffolds.+Adv.+Biomater.+Sci.+Biomed.+Appl.+2013,+12,+315%E2%80%93328.+%5BCrossRef&ots=IZLqfHMAWF&sig=8xGv7eVfEOGxyScsAJR_jziTLpA&redir_esc=y#v=onepage&q&f=false. Accessed 9 Apr 2024

Ait Brahim H, Benllarch M, Benhima N, El-Hadaj S, Metrane A, Belbaraka G (2023) New real dataset creation to develop an intelligent system for predicting chemotherapy protocols case of Moroccan breast cancer patients. Int J Adv Comput Sci Appl 14(8) www.ijacsa.thesai.org. Accessed 9 Apr 2024

Al-Hameli BA, Alsewari ARA, Basurra SS, Bhogal J, Ali MAH (2023) Diabetes disease prediction system using HNB classifier based on discretization method. J Integr Bioinform 20(1):1–13. https://doi.org/10.1515/JIB-2021-0037/MACHINEREADABLECITATION/RIS

Allard P, Papazotos F, Potvin-Trottier L (2022) Microfluidics for long-term single-cell time-lapse microscopy: advances and applications. Front Bioeng Biotechnol 10:968342. https://doi.org/10.3389/FBIOE.2022.968342/BIBTEX

Apoorva S, Nguyen NT, Rajan SK (2024) Recent developments and future perspectives of microfluidics and smart technologies in wearable devices. Lab Chip 24:1833

Bakuova N, Toktarkan S, Dyussembinov D, Azhibek D, Rakhymzhanov A, Kostas K, Kulsharova G (2023) Design, simulation, and evaluation of polymer-based microfluidic devices via computational fluid dynamics and cell culture "On-Chip". Biosensors 13:754. https://doi.org/10.3390/BIOS13070754

Baret JC (2012) Surfactants in droplet-based microfluidics. Lab Chip 12(3):422–433. https://doi.org/10.1039/C1LC20582J

Barker SJ, Dagys L, Hale W, Ripka B, Eills J, Sharma M, Levitt MH, Utz M (2022) Direct production of a hyperpolarized metabolite on a microfluidic Chip. Anal Chem 94(7):3260–3267. https://doi.org/10.1021/ACS.ANALCHEM.1C05030/ASSET/IMAGES/LARGE/AC1C05030_0005.JPEG

Baudoin R, Griscom L, Prot JM, Legallais C, Leclerc E (2011) Behavior of HepG2/C3A cell cultures in a microfluidic bioreactor. Biochem Eng J 53(2):172–181. https://doi.org/10.1016/J.BEJ.2010.10.007

Beh CW, Zhou W, Wang TH (2012) PDMS-glass bonding using grafted polymeric adhesive—alternative process flow for compatibility with patterned biological molecules. Lab Chip 12(20):4120–4127. https://doi.org/10.1039/C2LC40315C/

Bilodeau K, Mantovani D (2006) Bioreactors for tissue engineering: focus on mechanical constraints. A comparative review. Tissue Eng 12(8):2367–2383. https://doi.org/10.1089/TEN.2006.12.2367

Cao X, Luo Z, Zeng W, Xu S, Zhao L, Zhou J (2018) Enhanced avermectin production by Streptomyces avermitilis ATCC 31267 using high-throughput screening aided by fluorescence-activated cell sorting. Appl Microbiol Biotechnol 102(2):703–712. https://doi.org/10.1007/S00253-017-8658-X/METRICS

Carbonell P, Jervis AJ, Robinson CJ, Yan C, Dunstan M, Swainston N, Vinaixa M, Hollywood KA, Currin A, Rattray NJW, Taylor S, Spiess R, Sung R, Williams AR, Fellows D, Stanford NJ, Mulherin P, Le Feuvre R, Barran P et al (2018) An automated design-build-test-learn pipeline for enhanced microbial production of fine chemicals. Commun Biol 1(1):1–10. https://doi.org/10.1038/s42003-018-0076-9

Casano N, Santini SJ, Vittorini P, Sinatti G, Carducci P, Mastroianni CM, Ciardi MR, Pasculli P, Petrucci E, Marinangeli F, Balsano C (2023) Application of machine learning approach in emergency department to support clinical decision making for SARS-CoV-2 infected patients. J Integr Bioinform 20(2):1–12. https://doi.org/10.1515/JIB-2022-0047/DOWNLOADASSET/SUPPL/J_JIB-2022-0047_SUPPL_001.DOCX

Cerdeira Ferreira LM, da Costa ET, do Lago CL, Angnes L (2013) Miniaturized flow system based on enzyme modified PMMA microreactor for amperometric determination of glucose. Biosens Bioelectron 47:539–544. https://doi.org/10.1016/J.BIOS.2013.03.052

Chen X, Smith A, Johnson B, Garcia C, Example E (2019) Protein engineering for metabolite production using microfluidic devices. Anal Chem 47(3):301–315

Chen Y, Banerjee D, Mukhopadhyay A, Petzold CJ (2020) Systems and synthetic biology tools for advanced bioproduction hosts. Curr Opin Biotechnol 64:101–109. https://doi.org/10.1016/J.COPBIO.2019.12.007

Chen D, Yin J, Huo J, Sun J, Huang J, Li T, Sun C, Yang Z, Qin W (2022) Optimization and application of a bionic system of dynamic co-culture with hepatocytes and renal cells based on microfluidic Chip technique in evaluating materials of health food. Nutrients 14:4728. https://doi.org/10.3390/NU14224728

Chowdhury MS, Zheng W, Kumari S, Heyman J, Zhang X, Dey P, Weitz DA, Haag R (2019) Dendronized fluorosurfactant for highly stable water-in-fluorinated oil emulsions with minimal inter-droplet transfer of small molecules. Nat Commun 10(1):4546. https://doi.org/10.1038/s41467-019-12462-5

Colarusso AV, Goodchild-Michelman I, Rayle M, Zomorrodi AR (2021) Computational modeling of metabolism in microbial communities on a genome-scale. Curr Opin Syst Biol 26:46–57. https://doi.org/10.1016/J.COISB.2021.04.001

Coltro WKT, De Jesus DP, Da Silva JAF, Do Lago CL, Carrilho E (2010) Toner and paper-based fabrication techniques for microfluidic applications. Electrophoresis 31(15):2487–2498. https://doi.org/10.1002/ELPS.201000063

Dicenzo GC, Mengoni A, Fondi M (2019) Tn-Core: a toolbox for integrating Tn-seq gene essentiality data and constraint-based metabolic modeling. ACS Synth Biol 8(1):158–169. https://doi.org/10.1021/ACSSYNBIO.8B00432/SUPPL_FILE/SB8B00432_SI_004.PDF

Dornhof J, Kieninger J, Muralidharan H, Maurer J, Urban GA, Weltin A (2022) Microfluidic organ-on-chip system for multi-analyte monitoring of metabolites in 3D cell cultures. Lab Chip 22(2):225–239. https://doi.org/10.1039/D1LC00689D

Etienne G, Vian A, Biočanin M, Deplancke B, Amstad E (2018) Cross-talk between emulsion drops: how are hydrophilic reagents transported across oil phases? Lab Chip 18(24):3903–3912. https://doi.org/10.1039/C8LC01000E

Example, Smith A, Johnson B, Garcia C, Wang D (2016a) Continuous production of metabolites using microfluidic devices. Trends Anal Chem 25(6):789–802

Example, Smith A, Johnson B, Garcia C, Wang D (2016b) New developments in metabolite detection using microfluidic platforms. Trends Anal Chem 25(6):789–802

Falconnet D, Niemistö A, Taylor RJ, Ricicova M, Galitski T, Shmulevich I, Hansen CL (2011) High-throughput tracking of single yeast cells in a microfluidic imaging matrix. Lab Chip 11(3):466–473. https://doi.org/10.1039/C0LC00228C

Fallah-Araghi A, Baret JC, Ryckelynck M, Griffiths AD (2012) A completely in vitro ultrahigh-throughput droplet-based microfluidic screening system for protein engineering and directed evolution. Lab Chip 12(5):882–891. https://doi.org/10.1039/C2LC21035E

Fang H, Li D, Kang J, Jiang P, Sun J, Zhang D (2018) Metabolic engineering of Escherichia coli for de novo biosynthesis of vitamin B12. Nat Commun 9(1):4917. https://doi.org/10.1038/s41467-018-07412-6

Feng D, Xu T, Li H, Shi X, Xu G (2020) Single-cell metabolomics analysis by microfluidics and mass spectrometry: recent new advances. J Anal Test 4(3):198–209. https://doi.org/10.1007/S41664-020-00138-9/METRICS

Filippi M, Buchner T, Yasa O, Weirich S, Katzschmann RK, Filippi M, Buchner T, Yasa O, Weirich S, Katzschmann RK (2022) Microfluidic tissue engineering and bio-actuation. Adv Mater 34(23):2108427. https://doi.org/10.1002/ADMA.202108427

Gabriel EFM, Do Lago CL, Gobbi ÂL, Carrilho E, Coltro WKT (2013) Characterization of microchip electrophoresis devices fabricated by direct-printing process with colored toner. Electrophoresis 34(15):2169–2176. https://doi.org/10.1002/ELPS.201300024

Galván-Chacón VP, Costa L, Barata D, Habibovic P (2021) Droplet microfluidics as a tool for production of bioactive calcium phosphate microparticles with controllable physicochemical properties. Acta Biomater 128:486–501. https://doi.org/10.1016/J.ACTBIO.2021.04.029

García Alonso D, Yu M, Qu H, Ma L, Shen F (2019) Advances in microfluidics-based technologies for single cell culture. Adv Biosyst 3(11):1900003. https://doi.org/10.1002/ADBI.201900003

Garcia C, Smith A, Johnson B, Wang D, Example E (2018) Comparative study of traditional and microfluidic methods for metabolite detection. Anal Chem Rev 55(3):211–224

Garcia C, Smith A, Johnson B, Wang D, Example E (2020) Real-time biosynthesis pathway analysis using microfluidic devices. Anal Sci J 55(2):112–125

Gasilova N, Yu Q, Qiao L, Girault HH (2014) On-chip spyhole mass spectrometry for droplet-based microfluidics. Angew Chem 126(17):4497–4501. https://doi.org/10.1002/ANGE.201310795

Gielen F, Hours R, Emond S, Fischlechner M, Schell U, Hollfelder F (2016) Ultrahigh-throughput-directed enzyme evolution by absorbance-activated droplet sorting (AADS). Proc Natl Acad Sci U S A 113(47):E7383–E7389. https://doi.org/10.1073/PNAS.1606927113/SUPPL_FILE/PNAS.1606927113.SM03.WMV

Ginner L, Gesperger J, Wöhrer A, Drexler W, Baumann B, Leitgeb R, Salas M, Lichtenegger A, Niederleithner M (2020) Ex-vivo Alzheimer's disease brain tissue investigation: a multiscale approach using 1060-nm swept source optical coherence tomography for a direct correlation to histology. Neurophotonics 18(3):1–12

Gong L, Cretella A, Lin Y (2023) Microfluidic systems for particle capture and release: a review. Biosens Bioelectron 236:115426. https://doi.org/10.1016/J.BIOS.2023.115426

Granieri L, Baret J-C, Griffiths AD, Merten CA (2010) Chemistry & Biology Brief Communication High-Throughput Screening of enzymes by retroviral display using droplet-based microfluidics. Chem Biol 17:229–235. https://doi.org/10.1016/j.chembiol.2010.02.011

Guetschow ED, Steyer DJ, Kennedy RT (2014) Subsecond electrophoretic separations from droplet samples for screening of enzyme modulators. Anal Chem 86(20):10373–10379. https://doi.org/10.1021/AC502758H/ASSET/IMAGES/LARGE/AC-2014-02758H_0007.JPEG

Gupta R, Smith A, Johnson B, Garcia C, Example E (2017) Environmental monitoring of metabolites using microfluidic devices. J Environ Sci 40(4):201–215

Hairani Hairani H, Dadang Priyanto D (2023) A new approach of hybrid sampling SMOTE and ENN to the accuracy of machine learning methods on unbalanced diabetes disease data. Int J Adv Comput Sci Appl 14(8):585–590. https://thesai.org/Downloads/Volume14No8/Paper_64-A_New_Approach_of_Hybrid_Sampling_SMOTE_and_ENN.pdf

Hansen SK, Jamali B, Hubbuch J (2013) Selective high throughput protein quantification based on UV absorption spectra. Biotechnol Bioeng 110(2):448–460. https://doi.org/10.1002/BIT.24712

Harrison DJ, Fluri K, Seiler K, Fan Z, Effenhauser CS, Manz A (1993) Micromachining a miniaturized capillary electrophoresis-based chemical analysis system on a chip. Science 261(5123):895–897. https://doi.org/10.1126/SCIENCE.261.5123.895

Heinemann J, Deng K, Shih SCC, Gao J, Adams PD, Singh AK, Northen TR (2017) On-chip integration of droplet microfluidics and nanostructure-initiator mass spectrometry for enzyme screening. Lab Chip 17(2):323–331. https://doi.org/10.1039/C6LC01182A
Höllerer S, Papaxanthos L, Gumpinger AC, Fischer K, Beisel C, Borgwardt K, Benenson Y, Jeschek M (2020) Large-scale DNA-based phenotypic recording and deep learning enable highly accurate sequence-function mapping. Nat Commun 11(1):3551. https://doi.org/10.1038/s41467-020-17222-4
Holtze C, Weisse SA, Vranceanu M (2017) Commercial value and challenges of drop-based microfluidic screening platforms—an opinion. Micromachines 8:193. https://doi.org/10.3390/MI8060193
Hoog Antink MM, Sewczyk T, Kroll S, Árki P, Beutel S, Rezwan K, Maas M (2019) Proteolytic ceramic capillary membranes for the production of peptides under flow. Biochem Eng J 147:89–99. https://doi.org/10.1016/J.BEJ.2019.04.005
Hu S, Wang B, Luo Q, Zeng R, Zhang J, Cheng J (2023) Advances in droplet-based microfluidic high-throughput screening based on ultraviolet, visible and fluorescent spectroscopy. Fermentation 10:33. https://www.preprints.org/manuscript/202312.1887
Huang CJ, Chen YH, Wang CH, Chou TC, Lee GB (2007) Integrated microfluidic systems for automatic glucose sensing and insulin injection. Sens Actuators B 122(2):461–468. https://doi.org/10.1016/J.SNB.2006.06.015
Huang M, Fan S, Xing W, Liu C (2010) Microfluidic cell culture system studies and computational fluid dynamics. Math Comput Model 52(11–12):2036–2042. https://doi.org/10.1016/J.MCM.2010.01.024
Huft J, Haynes CA, Hansen CL (2013) Fabrication of high-quality microfluidic solid-phase chromatography columns. Anal Chem 85(3):1797–1802. https://doi.org/10.1021/AC303153A/SUPPL_FILE/AC303153A_SI_002.PDF
Ishibashi R, Mawatari K, Kitamori T (2012a) Highly efficient and ultra-small volume separation by pressure-driven liquid chromatography in extended nanochannels. Small 8(8):1237–1242. https://doi.org/10.1002/SMLL.201102420
Ishibashi R, Mawatari K, Takahashi K, Kitamori T (2012b) Development of a pressure-driven injection system for precisely time controlled attoliter sample injection into extended nanochannels. J Chromatogr A 1228:51–56. https://doi.org/10.1016/J.CHROMA.2011.05.095
Janiesch JW, Weiss M, Kannenberg G, Hannabuss J, Surrey T, Platzman I, Spatz JP (2015) Key factors for stable retention of fluorophores and labeled biomolecules in droplet-based microfluidics. Anal Chem 87(4):2063–2067. https://doi.org/10.1021/AC504736E/SUPPL_FILE/AC504736E_SI_008.AVI
Jiang J, Yang G, Ma F (2023) Fluorescence coupling strategies in fluorescence-activated droplet sorting (FADS) for ultrahigh-throughput screening of enzymes, metabolites, and antibodies. Biotechnol Adv 66:108173. https://doi.org/10.1016/J.BIOTECHADV.2023.108173
Johnson B, Smith A, Garcia C, Wang D, Example E (2019a) Metabolic engineering approaches utilizing microfluidic devices. J Biol Eng 35(4):301–315
Johnson B, Smith A, Garcia C, Wang D, Example E (2019b) Microfluidic methods for metabolite detection: a review. Anal Sci J 32(4):287–301
Kato M, Inaba M, Tsukahara T, Mawatari K, Hibara A, Kitamori T (2009) Femto liquid chromatography with attoliter sample separation in the extended nanospace channel. Anal Chem 82(2):543–547. https://doi.org/10.1021/AC9017605
Keane TJ, Badylak SF (2014) Biomaterials for tissue engineering applications. Semin Pediatr Surg 23(3):112–118. https://doi.org/10.1053/J.SEMPEDSURG.2014.06.010
Kell DB, Adla SK (2021) Molecules the transporter-mediated cellular uptake and efflux of pharmaceutical drugs and biotechnology products: how and why phospholipid bilayer transport is negligible in real biomembranes. Molecules 26:5629. https://doi.org/10.3390/molecules26185629
Kirsch M, Schackert G, Salzer R, Krafft C (2010) Raman spectroscopic imaging for in vivo detection of cerebral brain metastases. Anal Bioanal Chem 398(4):1707–1713. https://doi.org/10.1007/S00216-010-4116-7/METRICS

Kitagawa F, Otsuka K (2011) Recent progress in microchip electrophoresis–mass spectrometry. J Pharm Biomed Anal 55(4):668–678. https://doi.org/10.1016/J.JPBA.2010.11.013
Kurita R, Hayashi K, Fan X, Yamamoto K, Kato T, Niwa O (2002) Microfluidic device integrated with pre-reactor and dual enzyme-modified microelectrodes for monitoring in vivo glucose and lactate. Sens Actuators B 87(2):296–303. https://doi.org/10.1016/S0925-4005(02)00249-6
Lan F, Demaree B, Ahmed N, Abate AR (2017) Single-cell genome sequencing at ultra-high-throughput with microfluidic droplet barcoding. Nat Biotechnol 35(7):640–646. https://doi.org/10.1038/nbt.3880
Lawrence J, O'Sullivan B, Lye GJ, Wohlgemuth R, Szita N (2013) Microfluidic multi-input reactor for biocatalytic synthesis using transketolase. J Mol Catal B: Enzym 95:111–117. https://doi.org/10.1016/J.MOLCATB.2013.05.016
Lee M, Collins JW, Aubrecht DM, Sperling RA, Solomon L, Ha JW, Yi GR, Weitz DA, Manoharan VN (2013) Synchronized reinjection and coalescence of droplets in microfluidics. Lab Chip 14(3):509–513. https://doi.org/10.1039/C3LC51214B
Lee S, Smith A, Johnson B, Garcia C, Example E (2019) Microfluidic platforms for personalized medicine in metabolite analysis. Lab Chip 18(6):512–525
Lee JY, Nguyen B, Orosco C, Styczynski MP (2021) SCOUR: a stepwise machine learning framework for predicting metabolite-dependent regulatory interactions. BMC Bioinformatics 22(1):1–17. https://doi.org/10.1186/S12859-021-04281-7/FIGURES/4
Lei Y, Ferdous Z (2016) Design considerations and challenges for mechanical stretch bioreactors in tissue engineering. Biotechnol Prog 32(3):543–553. https://doi.org/10.1002/BTPR.2256
Li H, Smith A, Johnson B, Garcia C, Example E (2018) In situ monitoring of metabolite production using microfluidic devices. Anal Chem 46(3):201–215
Liu HY, Lin SL, Chan SA, Lin TY, Fuh MR (2013) Microfluidic chip-based nano-liquid chromatography tandem mass spectrometry for quantification of aflatoxins in peanut products. Talanta 113:76–81. https://doi.org/10.1016/J.TALANTA.2013.03.053
Liu X, Painter RE, Enesa K, Holmes D, Whyte G, Garlisi CG, Monsma FJ, Rehak M, Craig FF, Smith CA (2016) High-throughput screening of antibiotic-resistant bacteria in picodroplets. Lab Chip 16(9):1636–1643. https://doi.org/10.1039/C6LC00180G
Luke CS, Selimkhanov J, Baumgart L, Cohen SE, Golden SS, Cookson NA, Hasty J (2016) A microfluidic platform for long-term monitoring of algae in a dynamic environment. ACS Synth Biol 5(1):8–14. https://doi.org/10.1021/ACSSYNBIO.5B00094/SUPPL_FILE/SB5B00094_SI_001.PDF
Mahler L, Tovar M, Weber T, Brandes S, Rudolph MM, Ehgartner J, Mayr T, Figge MT, Roth M, Zang E (2015) Enhanced and homogeneous oxygen availability during incubation of microfluidic droplets. RSC Adv 5(123):101871–101878. https://doi.org/10.1039/C5RA20118G
Martin I, Wendt D, Heberer M (2004) The role of bioreactors in tissue engineering. Trends Biotechnol 22(2):80–86. https://doi.org/10.1016/j.tibtech.2003.12.001
Martin I, Riboldi S, Wendt D (2008) Bioreactors in tissue engineering: scientific challenges and clinical prospectives. In: 8th world biomaterials congress 2008, vol 1, p 68. https://doi.org/10.1007/10_2008_1/COVER
Massai D, Isu G, Madeddu D, Cerino G, Falco A, Frati C, Gallo D, Deriu MA, Labate GFDU, Quaini F, Audenino A, Morbiducci U (2016) A versatile bioreactor for dynamic suspension cell culture. Application to the culture of cancer cell spheroids. PLoS One 11(5):e0154610. https://doi.org/10.1371/JOURNAL.PONE.0154610
Memon J, Sami M, Khan RA, Uddin M (2020) Handwritten optical character recognition (OCR): a comprehensive systematic literature review (SLR). IEEE Access 8:142642–142668. https://doi.org/10.1109/ACCESS.2020.3012542
Mendoza LD, Rodriguez JA, Leclaire J, Buono G, Fotiadu F, Carrière F, Abousalham A (2012) An ultraviolet spectrophotometric assay for the screening of sn-2-specific lipases using 1,3-O-dioleoyl-2-O-α-eleostearoyl-sn-glycerol as substrate. J Lipid Res 53(1):185–194. https://doi.org/10.1194/jlr.D019489

Messner L, Antink MH, Guo T, Maas M, Beutel S (2021) A versatile ceramic capillary membrane reactor system for continuous enzyme-catalyzed hydrolysis. Eng Life Sci 21(8–9):527–538. https://doi.org/10.1002/ELSC.202100027

Miller OJ, El Harrak A, Mangeat T, Baret JC, Frenz L, El Debs B, Mayot E, Samuels ML, Rooney EK, Dieu P, Galvan M, Link DR, Griffiths AD (2012) High-resolution dose-response screening using droplet-based microfluidics. Proc Natl Acad Sci U S A 109(2):378–383. https://doi.org/10.1073/PNAS.1113324109/SUPPL_FILE/APPENDIX.PDF

Millet LJ, Lucheon JD, Standaert RF, Retterer ST, Doktycz MJ (2015) Modular microfluidics for point-of-care protein purifications. Lab Chip 15(8):1799–1811. https://doi.org/10.1039/C5LC00094G

Murrihy JP, Breadmore MC, Tan A, McEnery M, Alderman J, O'Mathuna C, O'Neill AP, O'Brien P, Advoldvic N, Haddad PR, Glennon JD (2001) Ion chromatography on-chip. J Chromatogr A 924(1–2):233–238. https://doi.org/10.1016/S0021-9673(01)00855-X

Najah M, Calbrix RL, Mahendra-Wijaya P, Beneyton T, Griffiths AD, Drevelle A (2014) Chemistry & biology resource droplet-based microfluidics platform for ultra-high-throughput bioprospecting of cellulolytic microorganisms. Chem Biol 21:1722–1732. https://doi.org/10.1016/j.chembiol.2014.10.020

Naseri G, Koffas MAG (2020) Application of combinatorial optimization strategies in synthetic biology. Nat Commun 11(1):1–14. https://doi.org/10.1038/s41467-020-16175-y

Nikcevic I, Lee SH, Piruska A, Ahn CH, Ridgway TH, Limbach PA, Wehmeyer KR, Heineman WR, Seliskar CJ (2007) Characterization and performance of injection molded poly(methylmethacrylate) microchips for capillary electrophoresis. J Chromatogr A 1154(1–2):444–453. https://doi.org/10.1016/J.CHROMA.2007.03.125

O'Sullivan B, Al-Bahrani H, Lawrence J, Campos M, Cázares A, Baganz F, Wohlgemuth R, Hailes HC, Szita N (2012) Modular microfluidic reactor and inline filtration system for the biocatalytic synthesis of chiral metabolites. J Mol Catal B: Enzym 77:1–8. https://doi.org/10.1016/J.MOLCATB.2011.12.010

Ohtake T, Pontrelli S, Laviña WA, Liao JC, Putri SP, Fukusaki E (2017) Metabolomics-driven approach to solving a CoA imbalance for improved 1-butanol production in Escherichia coli. Metab Eng 41:135–143. https://doi.org/10.1016/J.YMBEN.2017.04.003

Oliva A, Llabrés M (2019) Validation of a size-exclusion chromatography method for bevacizumab quantitation in pharmaceutical preparations: application in a biosimilar study. Separations 6:43. https://doi.org/10.3390/SEPARATIONS6030043

Pacocha N, Bogusławski J, Horka M, Makuch K, Liżewski K, Wojtkowski M, Garstecki P (2021) High-throughput monitoring of bacterial cell density in nanoliter droplets: label-free detection of unmodified gram-positive and gram-negative bacteria. Anal Chem 93(2):843–850. https://doi.org/10.1021/ACS.ANALCHEM.0C03408/ASSET/IMAGES/LARGE/AC0C03408_0008.JPEG

Pasirayi G, Auger V, Scott SM, Rahman SM, Islam M, O'Hare L, Ali Z (2011) Microfluidic bioreactors for cell culturing: a review. Micro Nanosyst 3:137

Payne EM, Holland-Moritz DA, Sun S, Kennedy RT (2020) High-throughput screening by droplet microfluidics: perspective into key challenges and future prospects. Lab Chip 20(13):2247–2262. https://doi.org/10.1039/D0LC00347F

Picher MM, Küpcü S, Huang CJ, Dostalek J, Pum D, Sleytr UB, Ertl P (2013) Nanobiotechnology advanced antifouling surfaces for the continuous electrochemical monitoring of glucose in whole blood using a lab-on-a-chip. Lab Chip 13(9):1780–1789. https://doi.org/10.1039/C3LC41308J

Pörtner R, Nagel-Heyer S, Goepfert C, Adamietz P, Meenen NM (2005) Bioreactor design for tissue engineering. J Biosci Bioeng 100(3):235–245. https://doi.org/10.1263/JBB.100.235

Postek W, Gargulinski P, Scheler O, Kaminski TS, Garstecki P (2018) Microfluidic screening of antibiotic susceptibility at a single-cell level shows the inoculum effect of cefotaxime on E. Coli. Lab Chip 18(23):3668–3677. https://doi.org/10.1039/C8LC00916C

Pu Z, Zou C, Wang R, Lai X, Yu H, Xu K, Li D (2016) A continuous glucose monitoring device by graphene modified electrochemical sensor in microfluidic system. Biomicrofluidics 10(1):011910. https://doi.org/10.1063/1.4942437/385560

Rana P, Berry C, Ghosh P, Fong SS (2020) Recent advances on constraint-based models by integrating machine learning. Curr Opin Biotechnol 64:85–91. https://doi.org/10.1016/J.COPBIO.2019.11.007

Reinholt SJ, Craighead HG (2018) Microfluidic device for aptamer-based cancer cell capture and genetic mutation detection. Anal Chem 90(4):2601–2608. https://doi.org/10.1021/ACS.ANALCHEM.7B04120/SUPPL_FILE/AC7B04120_SI_001.PDF

Roach LS, Song H, Ismagilov RF (2005) Controlling nonspecific protein adsorption in a plug-based microfluidic system by controlling interfacial chemistry using fluorous-phase surfactants. Anal Chem 77(3):785–796. https://doi.org/10.1021/AC049061W/SUPPL_FILE/AC049061WSI20041018_093353.PDF

Rothbauer M, Zirath H, Ertl P (2018) Recent advances in microfluidic technologies for cell-to-cell interaction studies. Lab Chip 18(2):249–270. https://doi.org/10.1039/C7LC00815E

Sabhachandani P, Sarkar S, Zucchi PC, Whitfield BA, Kirby JE, Hirsch EB, Konry T (2017) Integrated microfluidic platform for rapid antimicrobial susceptibility testing and bacterial growth analysis using bead-based biosensor via fluorescence imaging. Microchim Acta 184(12):4619–4628. https://doi.org/10.1007/S00604-017-2492-9/METRICS

Salehi-Nik N, Amoabediny G, Pouran B, Tabesh H, Shokrgozar MA, Haghighipour N, Khatibi N, Anisi F, Mottaghy K, Zandieh-Doulabi B (2013) Engineering parameters in bioreactor's design: a critical aspect in tissue engineering. Biomed Res Int 2013:1. https://doi.org/10.1155/2013/762132

Salis HM (2011) The ribosome binding site calculator. Methods Enzymol 498:19–42. https://doi.org/10.1016/B978-0-12-385120-8.00002-4

Sarma M, Lee J, Ma S, Li S, Lu C (2019) A diffusion-based microfluidic device for single-cell RNA-seq. Lab Chip 19(7):1247–1256. https://doi.org/10.1039/C8LC00967H

Scanlon TC, Dostal SM, Griswold KE (2014) A high-throughput screen for antibiotic drug discovery. Biotechnol Bioeng 111(2):232–243. https://doi.org/10.1002/BIT.25019

Schmidt H, Black JA, Ganjalizadeh V, Parks JW (2018) Multi-channel velocity multiplexing of single virus detection on an optofluidic chip. Opt Lett 43(18):4425–4428. https://doi.org/10.1364/OL.43.004425

Sewczyk T, Hoog Antink M, Maas M, Kroll S, Beutel S (2018) Flow rate dependent continuous hydrolysis of protein isolates. AMB Express 8(1):1–9. https://doi.org/10.1186/S13568-018-0548-9/FIGURES/6

Shameli SM, Elbuken C, Ou J, Ren CL, Pawliszyn J (2011) Fully integrated PDMS/SU-8/quartz microfluidic chip with a novel macroporous poly dimethylsiloxane (PDMS) membrane for isoelectric focusing of proteins using whole-channel imaging detection. Electrophoresis 32(3–4):333–339. https://doi.org/10.1002/ELPS.201000643

Shang F, Guihen E, Glennon JD (2012) Recent advances in miniaturisation—the role of microchip electrophoresis in clinical analysis. Electrophoresis 33(1):105–116. https://doi.org/10.1002/ELPS.201100454

Shen Y, Berger SJ, Anderson GA, Smith RD (2000) High-efficiency capillary isoelectric focusing of peptides. Anal Chem 72(9):2154–2159. https://doi.org/10.1021/AC991367T

Shene C, Paredes P, Flores L, Leyton A, Asenjo JA, Chisti Y (2020) Dynamic flux balance analysis of biomass and lipid production by Antarctic thraustochytrid Oblongichytrium sp. RT2316-13. Biotechnol Bioeng 117(10):3006–3017. https://doi.org/10.1002/BIT.27463

Shimizu H, Morikawa K, Liu Y, Smirnova A, Mawatari K, Kitamori T (2016) Femtoliter high-performance liquid chromatography using extended-nano channels. Analyst 141(21):6068–6072. https://doi.org/10.1039/C6AN01195K

Smirnova A, Shimizu H, Pihosh Y, Mawatari K, Kitamori T (2016) On-chip step-mixing in a T-Nanomixer for liquid chromatography in extended-nanochannels. Anal Chem 88(20):10059–10064. https://doi.org/10.1021/ACS.ANALCHEM.6B02395/ASSET/IMAGES/MEDIUM/AC-2016-023955_0014.GIF

Smith A, Johnson B, Garcia C, Wang D, Example E (2018) High-throughput screening using microfluidic devices for metabolite production. Anal Chem 46(3):201–215

Song NE, Jung YS, Choi JY, Koo M, Choi HK, Seo DH, Lim TG, Nam TG (2020) Development and application of a multi-residue method to determine pesticides in agricultural water using QuEChERS extraction and LC-MS/MS analysis. Separations 7:52. https://doi.org/10.3390/SEPARATIONS7040052

Steinhaus B, Garcia ML, Shen AQ, Angenent LT (2007) A portable anaerobic microbioreactor reveals optimum growth conditions for the methanogen Methanosaeta concilii. Appl Environ Microbiol 73(5):1653–1658. https://doi.org/10.1128/AEM.01827-06/ASSET/05BE2674-2767-4426-9F54-D835B7999D95/ASSETS/GRAPHIC/ZAM0050775750006.JPEG

Steyer DJ, Kennedy RT (2019) High-throughput nanoelectrospray ionization-mass spectrometry analysis of microfluidic droplet samples. Anal Chem 91(10):6645–6651. https://doi.org/10.1021/ACS.ANALCHEM.9B00571/SUPPL_FILE/AC9B00571_SI_001.PDF

Tang Q, Harrata AK, Lee CS (1996) High-resolution capillary isoelectric focusing–electrospray ionization mass spectrometry for hemoglobin variants analysis. Anal Chem 68(15):2482–2487. https://doi.org/10.1021/AC960169O

Utoh R, Enomoto S, Yamada M, Yamanaka K, Yajima Y, Furusawa K, Seki M (2021) Polyanion-induced, microfluidic engineering of fragmented collagen microfibers for reconstituting extracellular environments of 3D hepatocyte culture. Mater Sci Eng C 129:112417. https://doi.org/10.1016/J.MSEC.2021.112417

Van Vugt-Lussenburg BMA, Capinha L, Reinen J, Rooseboom M, Kranendonk M, Onderwater RCA, Jennings P (2022) " Commandeuring" xenobiotic metabolism: advances in understanding xenobiotic metabolism. Chem Res Toxicol 35(7):1184–1201. https://doi.org/10.1021/ACS.CHEMRESTOX.2C00067/ASSET/IMAGES/LARGE/TX2C00067_0010.JPEG

Viens M, Chauvette G, Langelier É (2011) A roadmap for the design of bioreactors in mechanobiological research and engineering of load-bearing tissues. J Med Devices 5(4). https://doi.org/10.1115/1.4005319/477562

Vlčková M, Kalman F, Schwarz MA (2008) Pharmaceutical applications of isoelectric focusing on microchip with imaged UV detection. J Chromatogr A 1181(1–2):145–152. https://doi.org/10.1016/J.CHROMA.2007.12.046

Wali AF, Dhaheri YA, Pillai JR, Mushtaq A, Rao PGM, Rabbani SA, Firdous A, Elshikh MS, Al Farraj DA (2020) LC-MS phytochemical screening, in vitro antioxidant, antimicrobial and anticancer activity of microalgae Nannochloropsis oculata extract. Separations 7:54. https://doi.org/10.3390/SEPARATIONS7040054

Wang D, Smith A, Johnson B, Garcia C, Example E (2017a) Single-cell analysis of metabolite production using microfluidic devices. Lab Chip 18(6):512–525

Wang D, Smith A, Johnson B, Garcia C, Example E (2017b) Recent advances in microfluidic technology for metabolite detection. Lab Chip 17(9):1635–1656

Wang J, Wang H, Wang Y, Liu Z, Li Z, Li J, Chen Q, Meng Q, Shu WW, Wu J, Xiao C, Han F, Li B (2022) Endothelialized microvessels fabricated by microfluidics facilitate osteogenic differentiation and promote bone repair. Acta Biomater 142:85–98. https://doi.org/10.1016/J.ACTBIO.2022.01.055

Warrick JW, Murphy WL, Beebe DJ (2008) Screening the cellular microenvironment: a role for microfluidics. IEEE Rev Biomed Eng 1:75–93. https://doi.org/10.1109/RBME.2008.2008241

Weltin A, Slotwinski K, Kieninger J, Moser I, Jobst G, Wego M, Ehret R, Urban GA (2013) Cell culture monitoring for drug screening and cancer research: a transparent, microfluidic, multisensor microsystem. Lab Chip 14(1):138–146. https://doi.org/10.1039/C3LC50759A

Wilhelm L (2018) POCT methods for screening in addiction medicine. In: Point-of-care testing, pp 171–180. https://doi.org/10.1007/978-3-662-54497-6_18

Wu L, Zhu L, Huang M, Song J, Zhang H, Song Y, Wang W, Yang C (2019) Aptamer-based microfluidics for isolation, release and analysis of circulating tumor cells. TrAC Trends Anal Chem 117:69–77. https://doi.org/10.1016/J.TRAC.2019.05.003

Xu Y, Wang X, Zhang C, Zhou X, Xu X, Han L, Lv X, Liu Y, Liu S, Li J, Du G, Chen J, Ledesma-Amaro R, Liu L (2022) De novo biosynthesis of rubusoside and rebaudiosides in engineered yeasts. Nat Commun 13(1):1–12. https://doi.org/10.1038/s41467-022-30826-2

Yang Y, Babiak P, Reymond JL (2006) Low background FRET-substrates for lipases and esterases suitable for high-throughput screening under basic (pH 11) conditions. Org Biomol Chem 4(9):1746–1754. https://doi.org/10.1039/B601151A

Yap YC, Guijt RM, Dickson TC, King AE, Breadmore MC (2013) Stainless steel pinholes for fast fabrication of high-performance microchip electrophoresis devices by CO2 laser ablation. Anal Chem 85(21):10051–10056. https://doi.org/10.1021/AC402631G/ASSET/IMAGES/MEDIUM/AC-2013-02631G_0006.GIF

Yeh CF, Hsu CH (2019) Microfluidic techniques for single-cell culture. In: Single-cell omics: technological advances and applications, vol 1, pp 137–151. https://doi.org/10.1016/B978-0-12-814919-5.00007-5

Yi B, Xu Q, Liu W (2022) An overview of substrate stiffness guided cellular response and its applications in tissue regeneration. Bioactive Mater 15:82–102. https://doi.org/10.1016/J.BIOACTMAT.2021.12.005

Zhang Y, Smith A, Johnson B, Garcia C, Example E (2020) Organ-on-a-Chip platforms for studying metabolism. J Biomed Eng 25(2):78–91

Zhang WG, Liu ZY, Pang SW (2021a) Effects of three-dimensional platform stiffness and layer dimensions on separation of carcinoma cells. Engineering 7(10):1424–1433. https://doi.org/10.1016/J.ENG.2020.09.010

Zhang Y, An C, Zhang Y, Zhang H, Mohammad AF, Li Q, Liu W, Shao F, Sui J, Ren C, Sun K, Cheng F, Liu J, Wang H (2021b) Microfluidic-templating alginate microgels crosslinked by different metal ions as engineered microenvironment to regulate stem cell behavior for osteogenesis. Mater Sci Eng C 131:112497. https://doi.org/10.1016/J.MSEC.2021.112497

Zhao D, He Z, Wang G, Wang H, Zhang Q, Li Y (2016a) Three-dimensional ordered titanium dioxide-zirconium dioxide film-based microfluidic device for efficient on-chip phosphopeptide enrichment. J Colloid Interface Sci 478:227–235. https://doi.org/10.1016/J.JCIS.2016.05.054

Zhao J, Griffin M, Cai J, Li S, Bulter PEM, Kalaskar DM (2016b) Bioreactors for tissue engineering: An update. Biochem Eng J 109:268–281. https://doi.org/10.1016/J.BEJ.2016.01.018

Zhong J, Yang Y, Liao L, Zhang C (2020) Matrix stiffness-regulated cellular functions under different dimensionalities. Biomater Sci 8(10):2734–2755. https://doi.org/10.1039/C9BM01809C

Zhou S, Alper HS (2019) Strategies for directed and adapted evolution as part of microbial strain engineering. J Chem Technol Biotechnol 94(2):366–376. https://doi.org/10.1002/JCTB.5746

Zhuang J, Zhao Z, Lian K, Yin L, Wang J, Man S, Liu G, Ma L (2022) SERS-based CRISPR/Cas assay on microfluidic paper analytical devices for supersensitive detection of pathogenic bacteria in foods. Biosens Bioelectron 207:114167. https://doi.org/10.1016/J.BIOS.2022.114167

Cell-Free Systems in Metabolomics

11

Ajay Kumar, Juveriya Israr, Hara Prasad Mishra, and Indrajeet Singh

11.1 Introduction

A technique that takes a holistic perspective to understanding the biochemical processes that take place within cells is known as metabolomics. This approach places a particular emphasis on the minute molecules that are involved in metabolic pathways. The aforementioned method is widely recognized and applied to research the metabolic processes of a wide variety of living organisms, ranging from straightforward single-celled prokaryotes to complex multicellular eukaryotes (Miguez et al. 2022a, b). Metabolomics is the scientific study of substances that are composed of very small molecules and serve as intermediates in the processes that occur in living organisms. According to Carlson et al. (2012), these chemicals play an important part in cellular processes such as the synthesis of energy, the creation of cellular components, and the absorption of nutrients. This method frequently makes use of advanced detection techniques, such as nuclear magnetic resonance (NMR) spectroscopy or mass spectrometry, to accurately identify and quantify a large number of minute molecules that are present in a particular sample. To facilitate the differentiation of a greater number of molecules than would be achievable only based on the capabilities of the detecting apparatus, it is common practice to make use of a preliminary separation procedure. Cell-free extract (CFE) systems, which stand for cell-free expression, have lately arisen as a separate field of application within the

A. Kumar (✉) · I. Singh
Department of Biotechnology, Faculty of Engineering and Technology, Rama University, Kanpur, India

J. Israr
Institute of Biosciences and Technology, Shri Ramswaroop Memorial University, Barabanki, Uttar Pradesh, India

H. P. Mishra
Koita Centre for Digital Health, Ashoka University, Sonepat, Haryana, India

V. Singh (ed.), *Advances in Metabolomics*,
https://doi.org/10.1007/978-981-97-7459-3_11

study of metabolomics. These systems are made up of very small molecules that go through the process of being consumed as well as manufactured over a predetermined amount of time, which enables the quantification and monitoring of these molecules (Miguez et al. 2021). The investigation of the metabolic alterations that occur in these systems can yield valuable insights into the functions and effects of small molecules in CFE processes, which will ultimately be of assistance to the optimization efforts that are being undertaken. Due to the presence of tiny molecules that are utilized and created over time, similar to complete cells, the application of metabolomics in CFE systems has a significant deal of potential (Miguez et al. 2022a, b). This permits the investigation of changes in metabolism as well as the impact of small molecules on the efficiency of CFE processes, which ultimately results in the enhancement of engineering and optimization efforts for these systems that are better informed (Hu et al. 2023).

11.2 Types of Cell-Free Systems Used in Metabolomics

The term "metabolomics" refers to an all-encompassing method that aims to get an understanding of the biochemical processes that take place within cells, with a special emphasis on the intermediates of tiny molecules being involved in metabolic processes. This technique is widely recognized and applied to research the metabolic processes of a wide range of living organisms, ranging from straightforward single-celled prokaryotes to complex multicellular eukaryotes. According to Panyard et al. (2022), the practical use of metabolomics is heavily reliant on the utilization of cell-free measurement techniques. Three primary classifications may be applied to these particular systems.

11.2.1 Extract-Based Systems

The primary constituents for metabolic processes are cell extracts, which are utilized in systems that are based on extracts. It is common practice to employ them in the investigation of the metabolic processes of specific species, such as *Escherichia coli*, and they have the potential to offer valuable insights into the activities and effects of tiny molecules in the reactions that are being studied.

11.2.2 Reconstituted Systems

Systems that have been reconstituted entail the reconstruction of metabolic pathways through the utilization of purified enzymes and other components. This technology makes it possible to analyze particular metabolic reactions within a controlled environment, ultimately contributing to enhancing metabolic processes.

11.2.3 Synthetic Biology-Based Systems

The invention and combination of novel metabolic pathways for use in synthetic biology systems is accomplished through the application of genetic engineering techniques. Both the production of precise molecules and the investigation of the metabolic kinetics of planned routes are within the capabilities of these devices. Every one of these cell-free systems presents its one-of-a-kind chances for metabolomics research. These opportunities allow researchers to get a more thorough understanding of the metabolic processes that are involved and to optimize these processes for particular goals (Miguez et al. 2021, 2022a, b).

11.3 Applications of Cell-Free Systems in Metabolomics

According to Miguez et al. (2022a, b), cell-free systems offer several important applications in the field of metabolomics, particularly in the investigation of metabolic pathways, metabolic engineering, and drug discovery. These are the most important applications.

11.3.1 Analysis of Metabolic Pathways

11.3.1.1 Enzyme Activities and Kinetics

Cell-free systems make it possible to investigate the activity and kinetics of enzymes, which is an exercise that is essential for comprehending metabolic pathways. Through the examination of the activity of certain enzymes, researchers can pinpoint areas of obstruction and further improve metabolic pathways to achieve maximum efficiency.

11.3.1.2 Metabolite Profiling

Cell-free systems make it possible to do metabolite profile analysis, which is necessary for comprehending the metabolic processes that are at play. This is a reference to the process of locating and quantifying the microscopic molecules that are engaged in metabolic activities. As a result, this technique provides essential insights into the roles and impacts of these molecules (Russell et al. 2013).

11.3.2 Metabolic Engineering and Pathway Optimization

11.3.2.1 Optimization of Metabolic Pathways

By examining the metabolic changes that take place in cell-free systems, researchers have the opportunity to identify significant metabolites that are involved in major pathways and to optimize these pathways for higher productivity. This is made possible by the fact that they can uncover these metabolites. Improving the efficiency of metabolic reactions or increasing the synthesis of specific chemicals can be accomplished by altering of metabolic pathways, which is what this entails.

11.3.3 Drug Discovery and Screening

11.3.3.1 Drug Discovery

To investigate the interactions that occur between tiny molecules and enzymes, which is a key part of the process of drug discovery, cell-free systems are extremely useful tools. Through the analysis of the impacts of particular compounds on metabolic pathways, scientists can identify possible therapeutic targets and design innovative drugs. In the field of metabolomics, the usage of cell-free systems has the potential to significantly improve our understanding of metabolic processes and optimize metabolic pathways for a wide range of applications (Miguez et al. 2022a, b).

11.4 Advantages and Limitations of Using Cell-Free Systems in Metabolomics

11.4.1 Advantages

11.4.1.1 Simplicity and Flexibility

Cell-free systems offer a method that is both highly versatile and highly streamlined for the investigation of metabolic processes. They can be easily constructed and operated, which enables rapid adjustments to be made to experimental conditions and the capability to examine a variety of scenarios.

11.4.1.2 Rapid Analysis

Investigations of metabolic processes can be completed more quickly using cell-free biological systems. The optimization of metabolic pathways and the performance of high-throughput screening are both exceptionally facilitated by this.

11.4.1.3 Controlled Environment

The investigation of metabolic processes can be carried out in a controlled environment using cell-free systems. It is possible to exert rigorous regulation over variables like as temperature, pH, and nutrient accessibility when using this strategy, which may prove to be difficult to do in systems that include intact cells (Čuperlović-Culf et al. 2010). This is because the approach makes it possible to achieve this level of regulation.

11.4.2 Limitations

Systems that do not contain cells do not have the complex cellular background that is present in complete cells. When comes to understanding the complexities of the links that exist between enzymes and metabolites within the context of cellular processes, this can be a challenge. When compared to whole-cell systems, cell-free systems often have a restricted capacity for additional growth. As a consequence of this, it might be challenging to investigate complex metabolic processes or to produce significant quantities of particular compounds. There is a possibility that the initial setup and continuing maintenance of cell-free systems will be more expensive than those of whole-cell systems. As a result of the fact that the costs associated with reagents and equipment can quickly add up, this is especially true for applications that entail high-throughput processes or large-scale operations (Miguez et al. 2022a, b).

11.5 Case Studies and Examples of Using Cell-Free Systems in Metabolomics

11.5.1 Case Study 1: Researching the Metabolism of Plants Through the Use of Cell-Free Systems

11.5.1.1 Metabolomics Analysis of Cell-Free Expression Systems Using Gas Chromatography-Mass Spectrometry

The purpose of this study is to present a comprehensive metabolomics strategy for the analysis of small molecules in cell-free systems. More specifically, the focus is on *Escherichia coli* lysate-based cell-free systems that make use of gas chromatography coupled to mass spectrometry (GC-MS).

11.5.1.2 Metabolic Pathway Engineering Based on Metabolomics Confers Acetic and Formic Acid Tolerance to a Recombinant Xylose-Fermenting Strain of *Saccharomyces cerevisiae*

A genetically engineered strain of *Saccharomyces cerevisiae* that is capable of fermenting xylose was used in this work to evaluate the effect that acetic acid has on the fermentation of xylose. The metabolite profiles of this strain were analyzed. The results of the study showed that the utilization of engineering techniques based on metabolomics increased the resistance to acetic and formic acids, which in turn led to an increase in the efficiency of the fermentation process.

11.5.2 Case Study 2: To Facilitate Drug Discovery, Metabolic Pathways Can Be Engineered in Cell-Free Systems

11.5.2.1 Metabolomics Analysis of Cell-Free Expression Systems Using Gas Chromatography-Mass Spectrometry

This study demonstrates the potential of metabolomics in cell-free systems to improve metabolic pathways and get a better understanding of metabolic processes. Researchers can find essential metabolites that play a part in important pathways by analyzing metabolic changes in cell-free systems. This allows them to improve the efficiency of these processes, which ultimately results in increased productivity.

11.5.2.2 High-Throughput Preparation Methods of Crude Extract for Robust Cell-Free Protein Synthesis

This study presents a method that is both quick and effective for the preparation of extracts through the use of sonication. Furthermore, the usage of cell-free systems in the fields of synthetic biology and biotechnology is encouraged by this method, which speeds up the screening process of genomically engineered strains for applications involving cell-free protein synthesis.

11.5.3 Case Study 3: Metabolite Profiling Is Conducted by Utilizing Reconstituted Cell-Free Systems

11.5.3.1 Metabolomics Analysis of Cell-Free Expression Systems Using Gas Chromatography-Mass Spectrometry

The present investigation has an emphasis on the utilization of metabolomics in cell-free systems to gain a better understanding of metabolic processes and improve metabolic pathways. Researchers can discover essential metabolites that are involved in primary pathways and improve the efficiency of these pathways by analyzing the metabolic changes that occur in cell-free systems (Miguez et al. 2022a, b).

11.6 Future Perspectives and Trends in Using Cell-Free Systems in Metabolomics

11.6.1 Potential Advancements in Technology

11.6.1.1 Improved Analytical Techniques

The advancement of mass spectrometry and nuclear magnetic resonance (NMR) spectroscopy will make it possible to conduct a more comprehensive and accurate analysis of metabolites in experimental systems that do not involve cells.

11.6.1.2 Automated Sample Preparation

According to Trifonova et al. (2023), the implementation of high-throughput and automated methodologies for sample preparation will optimize workflows for metabolomics and enhance the scalability of cell-free systems when it comes to sample preparation.

11.6.2 Emerging Applications and Research Areas

11.6.2.1 Synthetic Biology

The development and refinement of synthetic metabolic pathways will be significantly aided by cell-free systems, which will also make it possible to rapidly prototype and test constructed systems.

11.6.2.2 Drug Discovery

Because cell-free systems provide a controlled environment for the investigation of metabolic processes and the interactions that occur between enzymes and medicines, it is anticipated that the utilization of these systems for the screening and characterization of potential drug candidates would increase.

11.6.2.3 Personalized Medicine

Researchers (Miguez et al. 2019; Koo et al. 2020) have found that examining the metabolic processes in cell-free systems that are obtained from patient samples can be of great assistance in the development of personalized treatments and the identification of biomarkers for various diseases.

11.6.3 Challenges and Opportunities for Further Development

One of the most major challenges in the field of metabolomics is the enhancement of metabolite coverage. This is because it requires measuring the whole metabolome of a living organism using a single analysis method. To overcome this challenge, it is essential to combine several different analytical methodologies and come up with novel separation procedures. Merging metabolomics data with other forms of omics data, such as genomes, transcriptomics, and proteomics, is an essential step in the process of integrating multi-omics data. This integration makes it possible to have a more comprehensive understanding of the complex interactions that take place within cell-free systems. The implementation of a standardized method for the analysis and presentation of data and The use of standardized protocols for the processing and reporting of data will make it easier to compare the results of numerous studies and laboratories, which will allow for research that is both more rigorous and more easily replicated. According to Pinu et al. (2019), it is vital to improve the scalability of cell-free systems and reduce their cost to ensure that they are used extensively in metabolomics and other applications.

11.7 Discussion

The information gained from proteomics research can be greatly supplemented by metabolomics studies of CFE systems. Proteomics can identify proteins in a lysate and their relative abundance, but it can't tell you how active they are or what that activity could be. In the cell-free extract (CFE) system, metabolic enzymes often interact with tiny molecules, which is essential for their survival and use in both living things and non-living things. This is an area where metabolomics approaches shine. Our knowledge of the metabolic processes happening during CFE has been substantially expanded because of metabolomics and focused metabolite profiling investigations. The early presence of metabolites involved in the formation of dihydroxyacetone phosphate (DHAP) through glycolysis was tracked in an initial research of CFE systems that used mass spectrometry (Bujara et al. 2011). As a result of this method's efficient evaluation of the impacts of inserted cofactors and purified enzymes throughout the reaction, the system was enhanced. A recent study by Vilkhovoy et al. (2019) measured the amounts of glycolytic chemicals in the cell extract using mass spectrometry. Based on the findings of Levine et al. (2020), their investigation also indicated that using a novel growth medium improves energy metabolism and redox balance. Energy bottlenecks, such as ribosome ATP use, and small-molecule markers (like ethanol) of lysate activity can be detected using nuclear magnetic resonance (NMR) spectroscopy. The metabolic effects and repercussions of CFE have been better understood as a result of recent efforts. Various lysis post-processing procedures or cell maintenance conditions were associated with various metabolite levels, suggesting variability in protein synthesis. It was noted that the amounts of metabolites fluctuated during the reaction, and this was confirmed by monitoring the protein expression to examine the metabolite profiles. The most notable alterations, however, take place early on and persist far after the completion of protein synthesis (Miguez et al. 2021). All of our combined efforts have highlighted how important "endogenous metabolism" is in CFE systems. The metabolic enzymes that are extracted at the same time as the expression machinery are the ones that are relevant here. When it comes to metabolite patterns, in particular, internal metabolic activities have a far larger impact on protein synthesis than does protein expression alone. In addition, attempts to improve CFE processes for maximum protein synthesis were informed by the results of this research. Miniaturized substances like putrescine and beta-alanine greatly improved protein synthesis, leading to an almost tripling of output (Miguez et al. 2019). The correlation between metabolite concentrations and lysate productivity is strengthened by this finding. But, even with metabolomics data, it can be difficult to predict the effects of metabolite supplementation because the amounts of the supplemented metabolites have such a large impact. Additionally, you must know that the effects of many supplements do not just add up. In addition, the researchers looked at the possibility of using enzyme supplements. While several of the investigated enzymes did show an uptick in protein synthesis, the impact on metabolite profiles was somewhat minor (Miguez et al. 2021). The metabolic underpinnings of CFE systems have been better understood and their protein synthesis has been improved because of prior metabolomics research. Understanding that the main metabolic changes in CFE systems come from the lysate's internal metabolism, not from protein production, could offer useful insights for future efforts to engineer and improve CFE. Current attempts have not yielded a thorough understanding of the system that would be for genuinely methodical engineering of the system; so, additional research is necessary.

11.8 Conclusion

Cell-free systems present a significant opportunity in the field of metabolomics research since they provide a controlled setting for researching metabolic pathways without the complications associated with living cells. They enable accurate control and examination of metabolic reactions, providing a valuable understanding of biochemical processes that may be challenging to achieve within biological environments. Furthermore, these systems allow for the examination of metabolic interactions in a basic yet accurate manner, making it easier to identify crucial metabolites and pathways. Despite ongoing challenges related to scalability and replicating cellular conditions, progress in technology and methodology is continuously improving the usefulness and relevance of cell-free systems in deciphering complex metabolic processes. As a result, these advancements are greatly contributing to our comprehension of cellular metabolism and its impact on health and disease.

References

Bujara M, Schümperli M, Pellaux R, Heinemann M, Panke S (2011) Optimization of a blueprint for in vitro glycolysis by metabolic real-time analysis. Nat Chem Biol 7(5):271–277. https://doi.org/10.1038/nchembio.541. Epub 2011 Mar 20

Carlson ED, Gan R, Hodgman CE, Jewett MC (2012) Cell-free protein synthesis: applications come of age. Biotechnol Adv 30(5):1185–1194

Čuperlović-Culf M, Barnett DA, Culf AS, Chute I (2010) Cell culture metabolomics: applications and future directions. Drug Discov Today 15(15–16):610–621

Hu T, Allam M, Cai S, Henderson W, Yueh B, Garipcan A, Ievlev AV, Afkarian M, Beyaz S, Coskun AF (2023) Single-cell spatial metabolomics with cell-type specific protein profiling for tissue systems biology. Nat Commun 14(1):8260. https://doi.org/10.1038/s41467-023-43917-5. PMID: 38086839; PMCID: PMC10716522

Koo J, Yang J, Park H (2020) Cell-free systems: recent advances and future outlook. Biotechnol Bioprocess Eng 25:955–961

Levine MZ, So B, Mullin AC, Fanter R, Dillard K, Watts KR, La Frano MR, Oza JP (2020) Activation of energy metabolism through growth media reformulation enables a 24-hour workflow for cell-free expression. ACS Synth Biol 9(10):2765–2774. https://doi.org/10.1021/acssynbio.0c00283. Epub 2020 Sep 14

Miguez AM, McNerney MP, Styczynski MP (2019) Metabolic profiling of Escherichia coli-based cell-free expression systems for process optimization. Ind Eng Chem Res 58(50):22472–22482. https://doi.org/10.1021/acs.iecr.9b03565. Epub 2019 Sep 13. PMID: 32063671; PMCID: PMC7021278

Miguez AM, Zhang Y, Piorino F, Styczynski MP (2021) Metabolic dynamics in escherichia coli-based cell-free systems. ACS Synth Biol 10(9):2252–2265. https://doi.org/10.1021/acssynbio.1c00167. Epub 2021 Sep 3. PMID: 34478281; PMCID: PMC9807262

Miguez AM, Zhang Y, Styczynski MP (2022a) Metabolomics analysis of cell-free expression systems using gas chromatography-mass spectrometry. Methods Mol Biol 2433:217–226. https://doi.org/10.1007/978-1-0716-1998-8_13. PMID: 34985747; PMCID: PMC9814356

Miguez AM, Zhang Y, Styczynski MP (2022b) Metabolomics analysis of cell-free expression systems using gas chromatography-mass spectrometry. In: Cell-free gene expression: methods and protocols, pp 217–226

Panyard DJ, Yu B, Snyder MP (2022) The metabolomics of human aging: advances, challenges, and opportunities. Sci Adv 8(42):eadd6155. https://doi.org/10.1126/sciadv.add6155. Epub 2022 Oct 19. PMID: 36260671; PMCID: PMC9581477

Pinu FR, Goldansaz SA, Jaine J (2019) Translational metabolomics: current challenges and future opportunities. Metabolites 9(6):108. https://doi.org/10.3390/metabo9060108. PMID: 31174372; PMCID: PMC6631405

Russell C, Rahman A, Mohammed AR (2013) Application of genomics, proteomics and metabolomics in drug discovery, development and clinic. Ther Deliv 4(3):395–413

Trifonova OP, Maslov DL, Balashova EE, Lokhov PG (2023) Current state and future perspectives on personalized metabolomics. Metabolites 13(1):67

Vilkhovoy M, Dai D, Vadhin S, Adhikari A, Varner JD (2019) Absolute quantification of cell-free protein synthesis metabolism by reversed-phase liquid chromatography-mass spectrometry. J Vis Exp (152). https://doi.org/10.3791/60329

Unlocking the Metabolome: A Guide to Genome Editing for Precision Control of Cellular Chemistry

12

Gholamreza Abdi, Mugdha Tendulkar, Reshma Tendulkar, Nil Patil, Mukul Jain, and Maryam Abbasi Tarighat

12.1 Introduction

In the present era of technological advancement, numerous gene-editing tools are revolutionizing the field of genetics particularly the field of cellular chemistry. In the vast expanse of the sphere of metabolomics engineering, identification, evaluation of the concerned genes along with the construction of modified strains has become possible due to the advent of genetic editing tools. Systems like Clustered regularly interspaced palindromic repeats (CRISPR)-associated (Cas) have proved to be the preferred option for genomic editing especially engineering the industrially important strains.

G. Abdi (✉)
Department of Biotechnology, Persian Gulf Research Institute, Persian Gulf University, Bushehr, Iran
e-mail: abdi@pgu.ac.ir

M. Tendulkar
K. J. Somaiya Medical College & Research Centre, Sion, Mumbai, India

R. Tendulkar
Vivekanand Education Society's College of Pharmacy, Chembur, Mumbai, India

N. Patil · M. Jain
Cell & Development Biology Lab, Research and Development Cell, Parul University, Vadodara, Gujarat, India

Department of Life Sciences, Parul Institute of Applied Sciences, Parul University, Vadodara, Gujarat, India

M. A. Tarighat
Faculty of Nano and Bio Science and Technology, Persian Gulf University, Bushehr, Iran

V. Singh (ed.), *Advances in Metabolomics*,
https://doi.org/10.1007/978-981-97-7459-3_12

12.1.1 Overview of Metabolomics and Its Importance

Metabolomics is one of the most latest technologies explored after the advent of proteomics, transcriptomic and genomics. The common aspect of all these techniques is that they tend to integrate bioinformatics and high-throughput analytical techniques (Zhang et al. 2014). It involves analyzing the metabolites quantitatively and qualitatively; which prove to be integral intermediaries in the metabolic reaction. The prime objective of this scientific technology is not limited to only comprehending the pathological mechanisms responsible for a specific ailment but to also prognosticate the possible options for therapeutic interventions. These techniques primarily avail us to fundamentally point out the contrasting features between the pathological and normal perturbations thereby aiding in diagnosis and accurate prognosis (Zhang et al. 2015). Considering the information that metabolites are reflective of the various minute-to-minute development in the genome, proteome and transcriptome. These expressions characterize the phenotypic fingerprints of an individual at any given point of time. All the metabolites present in a biological sample at the considered physiological time period that give complete information about all the metabolic status of an organism is termed as a metabolome. The striking feature of metabolomics against genomics is its ability to express environmental influences in addition to providing a complete picture of the physiological condition of an organism. The numerous approaches employed in metabolomics are mass spectrometry (MS), nuclear magnetic resonance (NMR) spectroscopy, Fourier-transform infrared (FTIR) spectroscopy techniques, etc. (Jacob et al. 2018). It proves to be an exceptional tool as a biomarker, for drug discovery, and diagnosis of ailments like osteoporosis, rheumatoid arthritis, metabolic fingerprinting, metabolic profiling, cancer prognosis and many more.

12.1.2 Challenges in Understanding and Manipulating Metabolism

Metabolomes are known to sensitive to numerous environmental and genetic stimuli. It is therefore essential to consider certain factors while conducting a study to ensure the efficiency of the technique. One major challenge faced conducting a metabolomic analysis is interindividual variation. In order to reduce errors of this type, it is inevitable to adopt augmented experimental study design (Sun and Hu 2016). Another hurdle experienced in this process is the difficulty in assigning identity of the biomarkers in case of liquid chromatography (LC)-MS-based metabolomics.

The metabolome is profoundly affected by various environmental, genetic and gut microbiome pressures. These seemingly subtle variations can modify the metabolic flux and concentrations enormously (Manzoni et al. 2018). Stress, diet, xenobiotic use, lifestyle are some of the environmental variations. Genetic perturbations comprise epigenetics, gene polymorphism and differences in gender. The gut microbiome is known to modify the metabolome derived from the host thorough the mechanisms of compound co-metabolism.

12.1.3 The Rise of Genome Editing Technologies

More than half of the last century after witnessing the discovery of the helical structure of DNA; the field of science was enriched with the advent of numerous technologies with were made plausible to incorporate in laboratory practice. Exploring the varied sequencing platforms and the increased bulk of information regarding micro-mechanics and comprehensible nanotechnologists, it made easy for the scientists to manipulate the genomic codes extensively. A plenitude of molecular editing tools were discovered for the last two decades which made it possible for the scientists to edit genomes and modify the genomic sequences as per requirement (Shah et al. 2015). It enabled to alter the fate of mRNA via post-translational modifications.

12.2 Genome Editing Techniques for Metabolomics

Fundamentally, genome-editing technologies can be elucidated as techniques where DNA sequences are altered by processing mRNA, deletions, modified gene expression responsible for amending the functional behavior of proteins (Joseph et al. 2017). All these methods have three principal features in common:

1. Mechanisms for genetic tool entry into the cell and nucleus
2. Alteration of gene transcription and processing
3. End-product as an overexpressed, under expressed or altered protein

Considering a holistic perspective, the techniques involve multiple receptor-ligand interactions; they also comprise varied cell entry mechanisms like sonification, lipofection and transfection (Mosana et al. 2020). In conjunction of extensive development in the genome editing technologies, standardized protocols and regulatory processes needs to be formed to nullify the imprecision, for specifically outlining the indications and contraindications of each technique and thereby ameliorating the notion of personalized medicine (Molepo et al. 2018).

12.2.1 CRISPR-Cas9: A Versatile Tool for Precise Genome Modification

The first gene editing tool was discovered in 2012; since then the CRISPR/Cas9 system is proving to be the most powerful genomic editing tool. The principal objective of the CRSIPR/Cas9 system revolves around skyrocketing the efficiency of on-targets and decreasing off-target effects. Hajiahmadi et al. discussed the primary strategies for decreasing off-target effects by combining ligand-dependent aptazyme and single-guide RNA (sgRNA) which can prove to be a promising option for reducing off-target mutations in organisms (Hasin et al. 2017). RNA-guided Cas9 nucleases which are produced from the CRISPR/Cas9 systems have been adopted to redesign the genomic backbone of the organism. They are used to efficiently allow the manipulation of eukaryotic cells by enumerating a 20-nt targeting sequence within its directed RNA (Tripathi et al. 2013).

These systems have been effectively altered to the microbes against invading foreign nucleic acids primarily bacterial DNA macrophages, other mobile segments, plasmids. CRISPR/Cas systems have been documented in 47% and 87% of the whole archaea and bacterial genomes (Zhang et al. 2013). When a small sequence is isolated from the plasmid genome, it is reinserted as a new guide (spacer) to the CRISPR arrays, which comprise of tiny isolated repeated by spacers. A CRISPR RNA (crRNA) along with a spacer is generated from a much longer precursor also called as pre-crRNA (Dudka et al. 2021). It is then further subjected to blend into a ribonucleoprotein complex of a minimum of one Cas protein.

These complexes integrate, evaluate and eradicate complementary RNA/DNA from other forces of intruding nature. CRISPR/Cas systems have been broadly divided into three types—I, II, III. Further subtypes involve III-A and III-B (Takahashi et al. 2019). They differ in origin of DNA/RNA and generation of crRNA. In case of types I and II systems, they require the presence of protospacer adjacent motifs (PAMs) for recognition. In contrast, type III systems no PAMs have been recognized (Fig. 12.1).

12.2.2 Alternative Genome Editing Technologies: TALENs and Zinc-Finger Nucleases

With the increasing progress in the discovery of whole genome sequencing editing tools, the scientists were presented with a plethora of various gene editing technologies. Transcription activator-like effector nucleases (TALENs) and zinc-finger nucleases (ZFNs) are some of the most robust gene editing tools responsible for revolutionizing the field of genomic sequencing (Carlson et al. 2018). They are primarily chimeric nucleases, and they consist of sequence-specific DNA-binding modules which can be programmed. They are found to be linked to non-specific DNA cleavage domain (Zhang et al. 2017). These technologies allow a wide range of genetic alterations by induction of double-stranded DNA breaks; these modifications tend to stimulate error-liable non-homologous sticky ends ligation or homology-led specific genomic sites repair. The ingenuity of TALENs and ZFNs emerges from its potential to personalize the DNA-binding domain to practically identify any genomic sequence virtually (Wang et al. 2020). These DNA-binding modules can be coupled with varied effector domains to influence genomic function and chemical structure comprising transcriptional activators and repressors, transposases, nucleases, recombinases, histone acetyltranferases and DNA histone methyltransferases (Li et al. 2018). The capability of these technologies to effectively

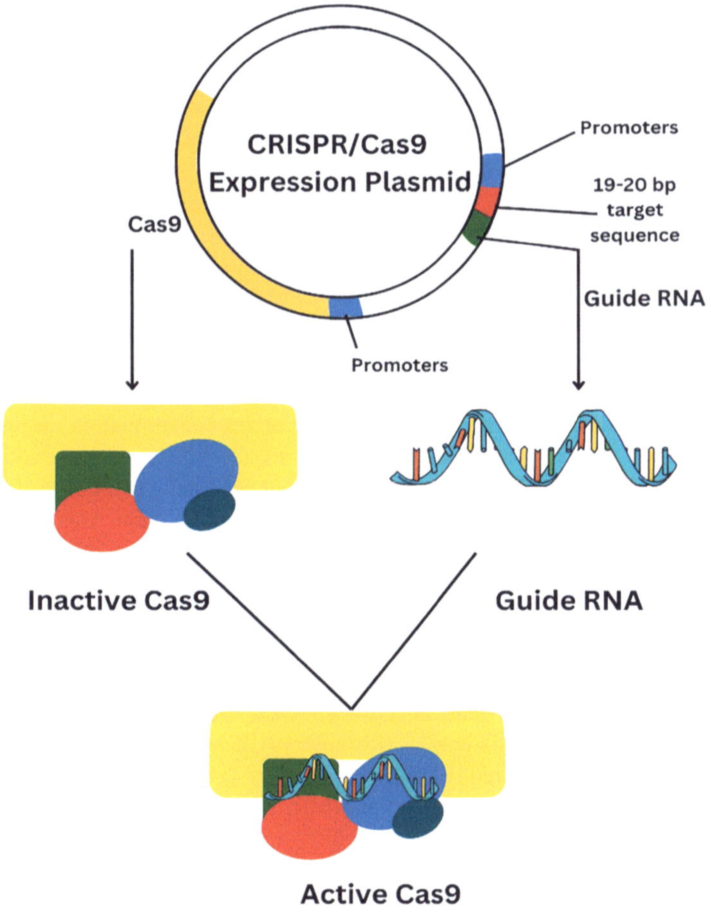

Fig. 12.1 Schematic representation of the CRISPR-Cas9 system for genome editing

carry out genomic modifications is determined by the specificity of DNA-binding and affinity of the transcription activator-like effector (TALE) proteins and zinc-finger (Wilkins and Trushina 2018). The recent discovery by TALE proteins of a modular DNA identification code led to widespread use of a possibly substitute technique for programming DNA-binding proteins in genomic engineering. Repeat-variable residues (RVDs), a pair of hypersensitive amino acids, have a significant impact on TALE specificity (Yilmaz et al. 2017). The integration of ZFN and TALENs repeats allows for the identification of contiguous genomic DNA sequences. However, unlike zinc-finger proteins, the creation of lengthy TALE arrays did not need the re-engineering of the genetic links between repetitions (Shang et al. 2020). Custom-designed TALE arrays are exported through Life Technologies (Grand Island, NY), Cellectis Bioresearch (Paris, France) and

Transposagen Biopharmaceuticals (Lexington, KY), following the results of an exhaustive literature review (Luo et al. 2020).

12.2.3 Choosing the Right Genome Editing Tool for Metabolomics Studies

A skyrocketing rise was observed in applications of data-independent acquisition (DIA) techniques and ion mobility mass-spectrometry (IMS) techniques of analyses in the fields of lipidomics and metabolomics (Mo et al. 2020). On the background of ease of applicability, imaging mass spectrometry (MIS)-dependent natural products (NPs) and nanoscale secondary ion mass spectrometry (nanoSIMS) are employed in numerous tasks ranging from quantitative analyses; subcellular MS imaging in the cellular organelles; recognizing urban contamination origins from high-resolution mass spectrometry (HRMS) to COVID-19 disease diagnosis (Tayanloo-Beik et al. 2020). It also proves to be a promising technology targeting cancer patients for intraoperative as well as clinical employment. Metabolomics is highly dependent on mass spectroscopy and spectroscopy systems of analytical origin which are known to evolve high-throughput omics scale information (Yu et al. 2020). They include:

(a) Gas chromatography-mass spectrometry (GC–MS)
(b) Liquid chromatography-mass spectrometry (LC–MS)
(c) Capillary electrophoresis-mass spectrometry (CE-MS)
(d) Fourier transform infrared (FTIR)
(e) Spectroscopic techniques such as 1H-NMR, 13C-NMR and Raman

The workflow of untargeted metabolomics which primarily utilize either GC-MS, NMR or LC-MS is dependent on pre-processing of the procured raw datasets. This has to be done before the statistical analyses and its interpretation (Huang et al. 2020). The procedure of preprocessing involves tools that aid in the distinguishing of masses from the obtained mass spectra; for the construction and display of the generated ion chromatograms, deconvolution, detection, peak alignment and the various involved data curation methods. Over the past decade, substantial efforts have been made to develop user-friendly graphical user interfaces (GUIs) and to nullify the possible errors encountered by the scientists. The Quality Control (QC) tools are employed to deal with the random and systematic error commonly occurred during the analytical and experimental procedures (Kodama et al. 2020). An enormous amount of errors have been posed due to batch effects for instance, inception of experimental datasets which possess capability of influencing with the quantification of phenotypic metabolomic alterations.

12.3 Applications of Genome Editing in Metabolomics

The fundamental three technologies have predominantly revolution the sphere of gene editing sector. They are TALENs, CRISPR/Cas9 and ZFN (Macias et al. 2018). The contribution of meganucleases or homing endonucleases should also be appreciated. The exhaustive research carried on ZFNs have proven to be effective tools for a wider range of applications for instance,

(i) Viral replication inhibition
(ii) Inducing angiogenesis to hasten healing of wounds
(iii) Modulation of genetic expression of disease-related loci
(iv) Genomic analysis of cellular targets for drug resistance and degree of cancer development.

These applications were extrapolated by the scientists and applied to the other technologies as well leading to the curation of transcriptional repressors and activators (Ismail et al. 2019). CRISPR/Cas9 and TALEs capacitated the construction of logic gates and personalized genetic circuits, facilitate cell programming and even intricate gene regulation (Zhu et al. 2019).

12.3.1 Metabolic Pathway Engineering: Disrupting or Overexpressing Key Genes

It is essential to engineer the organisms in a specific way in order to ensure the production of various compounds either by integrating exogenous pathways isolated from any other organism or de novo-constructed pathways (Kruk et al. 2017). This can also be achieved with the instigation of metabolic endogenous pathways. The core of the metabolic pathways involves deletion of competing pathways and overexpression of bottleneck enzymes (Armitage and Southam 2016). The principal components to be considered are-

(i) Selection of host organism
(ii) Computational systems
(iii) Transcriptional control engineering
(iv) Engineering of enzymes
(v) Discovery of novel synthetic gene editing tools
(vi) Reduction of metabolic load

Schwarz et al. highlighted that the actinomycete *Amycolatopsis japonicum* is categorized as a good host organisms for the production of the concerned immunosuppressant brasilicardin A (Aretz and Meierhofer 2016). The authors explained the mechanisms of increasing generation of *Amycolatopsis japonicum* by achieving overexpression of the mevalonate biosynthetic pathway; increasing IPP precursor levels which is a intracellular molecule; expression of particular prenyltransferses that facilitate production of diterpenoid (Chan et al. 2015). Extensive research carried out by Jones et al. experimentally revealed that plasmids comprising of low-copy number tend to replicate themselves for only 1–2 times when present inside the cell. This procedure actually increases the yield of the product. These two isolated metabolic pathways were meticulously analyzed for the generation of lycopene and polyphosphate (polyP) (Contrepois et al. 2015).

These specific pathways were selected as the substrates employed to create the concerned metabolites can be manipulated by achieving gene overexpression (Fig. 12.2). The quantification of metabolic load was dependent on the ultimate optical density (OD) values of the individual cultures of the two considered metabolic pathways (Pasikanti et al. 2008). The increased yield of the desired metabolites was achieved in the low-copy plasmids by increasing the OD (Appiah-Amponsah et al. 2013). This technique of decreasing the copy number of the plasmid has led to huge success as far as multigene pathways are concerned.

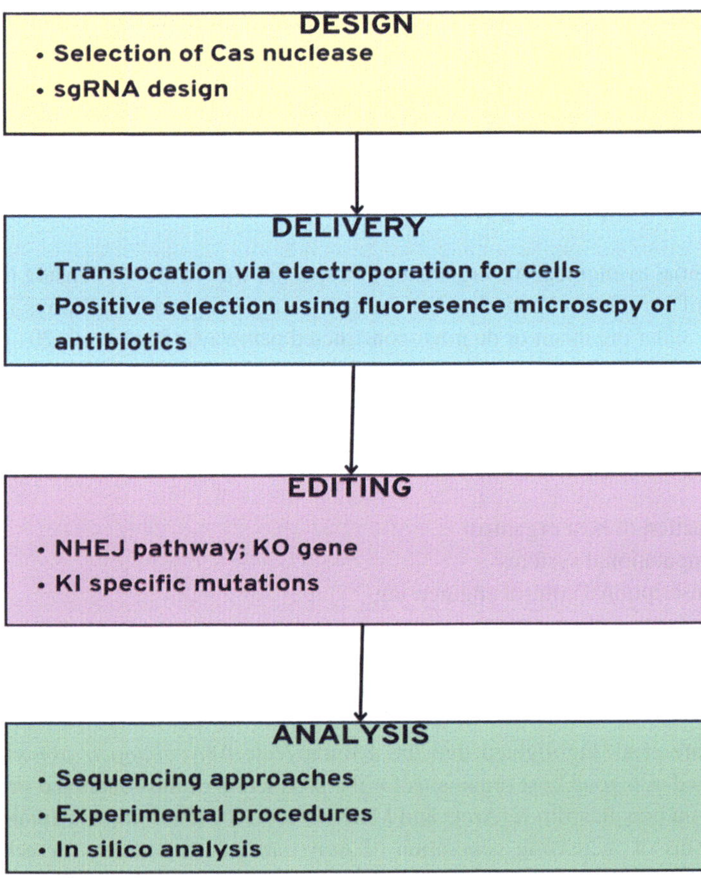

Fig. 12.2 Workflow for engineering microbial metabolism using genome editing techniques

12.3.2 Functional Genomics: Identifying Genes Controlling Metabolite Production

Functional genomics is concerned with describing the interactions and functions of proteins and genes by employing genomic approaches. The principle of this technique is contrasting to the one followed by classical molecular biology systems which is a gene-by-gene approach. It couples data extracted from the various DNA sequence procedures, protein-DNA/RNA/protein integrations, gene expression et cetera (McKenzie et al. 2010). A significant aspect of genomics is comparative genomics. The contemporary biological experimental studies contradict the genomic and structural attributes of genomes to fathom genetic linkages, functionality of numerous species and most importantly gene expression mechanisms (Beltran et al. 2012). The two important pillars of comparative genomics are homology and similarity (Bingol 2018). Similarity procures the degree of differences and similarities among numerous species by differentiating amino acid and simple nucleotide sequences. This technique does not involve phylogenetic kinship, emergence and the association between function and biological structure (Chang et al. 2018). Homology discusses the specific phylogenetic emergence of numerous species which highlights the distinctiveness of biologically significant domains of amino acid/ nucleotide sequences (Saiki et al. 2017). The elixir of homology lies in the developmental interactions of genomic sequences within species. Similarly, the elixir of similarity lies in the quantified interaction between genes of varied species under consideration (Havelund et al. 2017). Homology can be discussed in two domains—paralogy and orthology (Mazumder et al. 2018). The linear homologous genomic sequences can be designated as homologous genomic sequences dispensed in various species which are inherited vertically in the lineage.

12.3.3 Synthetic Biology: Designing Novel Metabolic Pathways

The process of metabolic engineering has the primary objective of altering the endogenous metabolic connections of a living organism for exploiting it for a biotechnological purpose for instance generation of a valuable biological compound. The two commonly used approaches are "copy, paste and fine tuning" and "mix and match" that play pivotal role in curating customized metabolic pathways that outnumber the benefits derived from natural pathways. The idea of "total synthesis" came into picture during the twentieth century that helped in the conception of the notion that complex synthetic compounds can be created from basic chemical building blocks (Haines et al. 2018). To exploit the potential of metabolic engineering, novel approaches and tactics were introduced. The process of metabolic engineering can be explained on five basic levels (Table 12.1):

Level 1: Optimization of pre-existing metabolic pathways in host organism
Level 2: Transportation of pre-defined sub pathways in the new host organism
Level 3: Creation of novel pathways from already known chemical reactions

Table 12.1 Features of the five different levels of metabolic engineering

	Definition	Typical features	Examples
Level 1	Optimization of existing pathways in natural host	Overexpression and knock out of individual genes in natural host	Overexpression of transketolase to increase flux: Calvin cycle
Level 2	Transportation of pre-defined subpathways in the new host organism	Natural pathway altered with better-performing reactions	Transfer of Calvin cycle into *E. coli* for sugar synthesis derived from CO_2
Level 3	Creation of novel pathways from already known chemical reactions	Synthetic route constructed from natural enzymes	Recombination of known enzymes into synthetic pathway: MOG pathway for CO_2 fixation
Level 4	Creation of novel pathways from novel chemical reactions which are completely adherent to the known enzyme mechanisms	Synthetic route comprising of enzymes of altered substrate specificity	CETCH cycle for CO_2 fixation
Level 5	Creation of novel pathways from novel chemical reactions which are completely adherent to the novel enzyme mechanism	Synthetic route comprising of de novo-designed enzymes	CO_2 fixation cycle dependent on an underdeveloped Ni-Ga cofactor in artificial metalloprotein

Level 4: Creation of novel pathways from novel chemical reactions, which are completely adherent to the known enzyme mechanisms

Level 5: Creation of novel pathways from novel chemical reactions, which are completely adherent to the novel enzyme mechanisms.

12.3.4 Precision Medicine: Tailoring Metabolic Profiles for Disease Treatment

The completion of the Human Genome Project at around 2011 marked the inception of precision medicine. As against evidence-based medicine, the field of precision medicine allows scientists and researchers to customize the treatment options for patients considering their responsiveness to particular therapies and susceptibility index to certain ailments. The concept of precision medicine can be explained in simple terms as "apt drugs for the apt patients at the appropriate time" (Rhee et al. 2018). For the incessant workflow of precision medicine, it is inevitable to comprehend its two essential objectives

I. Classification of a specific disease into its subtypes
II. Ensuring availability of targeted therapies for each individual subtype

On this background, the patients can be categorized on the basis of the severity and class of ailments and accurate treatment can be provided to the patients which will ultimately improve their quality of life. Robust advancements in numerous omics, precision medicine models, big data analytics, targeted therapies and electron health records (EHRs) in the past two decades has revolutionized this sector (Jin et al. 2019). Disease subtyping which is a fundamental component of precision medicine as discussed is the process of categorizing diseases/ailments into precise subgroups of patients, which are dependent on certain patient characteristics. This is a major driving force as this categorization decides which therapeutic course a patient will follow based on the subgroup to which it belongs (Jeong et al. 2018). The disease subtyping has been carried out for the following diseases like—asthma, cardiovascular diseases, cancer, Parkinson's disease and cancer. This helps in designing genetic testing methods to recognize the molecular determinants that are characteristic of ach subgroup. These are also called "endotypes" because they are interlinked with the underlying mechanism of ailments (Zhu et al. 2018).

12.4 Integration of Genome Editing with Metabolomics Tools

Revolutionary advancements in the omic technologies like metabolomics, proteomics and genomics have allowed high-throughput monitoring of a plethora of molecular level processes. These techniques have been pertained to recognize genetic variants like biomarkers to characterize intricate biochemical procedures and to evaluate the pathophysiology of the ailment (Li et al. 2020). Numerous omic tools focus exhaustive genomic analysis (genomics), proteins (proteomics), mRNA (transcriptomics) and metabolites (metabolomics). The interpretation of the alterations in isolated omic domains proves to be an obstacle on the background of intricate biochemical genetic regulation like organismal v/s tissue v/s cellular-level procedures, mRNA or protein post-translational modification, epigenetics (Nayak et al. 2020). Coupling the data extracted from experiments from numerous omic platforms can be considered as an emerging approach which targets to discover latent genetic interactions that become evident when only investigated through holistic analytical approach which integrates data from multiple domains. Metabolomics is basically an analysis of tiny molecules (<1200 Da) particularly metabolic intermediated formed in biochemical reactions. These have been broadly employed to investigate interactions between protein downstream products, genes and the environmental stimuli. This approach has been used to ascertain the pathophysiology of different diseases like cancer and Type 1 diabetes mellitus (Chen et al. 2020). The goals of this research can be determined as to recognize efficient biomarkers which are predictive of the onset of ailment, monitoring of treatment efficacy and prognosis.

12.4.1 Metabolomics Profiling Techniques: Mass Spectrometry and Nuclear Magnetic Resonance

The main aim of metabolomics is to resolve plenitude of challenges ranging from biological and environmental problems to biomarker recognition for ailments. Mass spectrometry (MS) and nuclear magnetic resonance (NMR) are the primary analytical tools that are most commonly employed to procure metabolomics data sets owing to their accessibility, unique strength and versatility. The NMR technique minimizes sample handling without the involvement of chromatography; it is also easily quantifiable; provides multiple means of metabolite recognition. It can detect only the abundant metabolites whose concentration who satisfies the condition $\geq 1\ \mu M$. In contrast, MS possesses the capacity to quantify metabolites at extremely low concentrations in the ranges of femto- and attomolar and also at a higher resolution ranging from $\sim 10^3$ to 10^4 and at a dynamic range of $\sim 10^3$ to 10^4 (Beltran et al. 2012). The complexity of the sample and quantitation are the factors that limit detection of metabolites due to ion suppression (Mehta et al. 2020). NMR and MS are extremely complementary to each other, and coupling these two techniques is known to improve the study quality and significantly increases the metabolome coverage. The advantages of employing NMR for metabolomics involve-

(a) High-throughput data extraction
(b) Minimal Handling of Samples
(c) Non-destructive data extraction
(d) Relative simple methods for data acquisition
(e) Redundant spectral information

This helps to increase the precision of metabolite identification. Since MS only finds ionizable metabolites, 40% of chemical libraries are not seen. A well-known issue in MS is ion suppression, which further reduces ion detection due to matrix effects. Matrix refers to the simple fact that there are other substances, such as polymers, buffers and so on, in the sample. This reduces a certain compound's capacity to ionize through a number of different ways (Sidorov et al. 2020). For example, the chemical competed with an ion-suppressing molecule from the matrix for the pre-existing accessible ions. In summary, one metabolite's existence obscures the presence of other metabolites. The solutions to these challenges have been proposed in the separation and ionization methods which aim at reducing matrix effects by co-elution of compounds which are common in complex mixtures. Despite of all, ion suppression continues to be a concerning issue as far as MS metabolomics is concerned.

12.4.2 Data Analysis Strategies: Linking Genetic Modifications to Metabolic Changes

Data integration of genetic modifications to metabolic changes is an important aspect of metabolomics engineering. Integrative analyses can be widely classified in accordance with different strategies:

I. Early Integration
II. Intermediate Integration
III. Late Integration

In the early integration strategy, all the inputs are affixed at processed or raw level to create a single dataset. This gathers all the provided information with as minimum loss as possible. An important obstacle in that strategy is to employ an appropriate iteration for datasets extracted from carried scales (Stryeck et al. 2019). The benefits from early integration have two main advantages; statistical techniques employed for a single dataset can be extrapolated with slight changes to the procured data matrix. Secondly, it tends to conserve information of associations between omic variables present in the input data (Vignoli et al. 2019). In metabolomic approach, interpretability of the extracted datasets in one of the primary concerns and key aspects to be kept in mind while development of integrative models.

12.4.3 Iterative Design–Build–Test Cycle for Efficient Metabolite Engineering

Metabolic engineering emphasizes on optimization of organisms via genetic interventions in metabolic pathways envisaging to increase the desired product flux. Sequential optimization of pathways tends to normally misses the standard optimum configuration of pathway elements like enzyme concentrations that aim at maximizing yield of the product. Optimization of combinatorial pathway is carried out which comprises of construction of strain designs from an enormous DNA library of promoters, coding sequences, ribosomal-binding sites and other DNA constituents that affect enzyme attributes and their relative concentration (Omori et al. 2019). This combinatorial pathway optimization renders infeasible to examine each test design. As a consequence, combinatorial pathway optimization is generally performed in an iterative approach using design–build–test–learn (DBT) cycles. The objective is to construct a preliminary set of strain designs and employ this extracted data from the test phase to understand pivotal characteristics of the metabolic pathways. First of all, designs are selected from a large DNA library which consists of pathway elements. These elements possess characteristics that heavily influence a protein concentration or catalytic rate (Cardellini et al. 2018). Due to the encountered combinatorial explosion of the design, only a small subset of these designs is constructed and experimentally verified, i.e., filter step. These experiments help the acquisition of data concerned with properties of desired products and from which new set of design strains can be built. Iteration of this process leads to efficient optimization of the pathway until strain is industrially appropriate (Coene et al. 2018). Nowadays, machine learning has been integrated with metabolomic engineering. For achieving metabolic flux optimization, the process ranging from target recognition to prediction of metabolite concentrations from proteomics data is carried out under the control of machine learning techniques. Machine learning also suggests new strains for the subsequent DBTL cycles which allow semi-automated iterative metabolic engineering.

12.5 Challenges and Considerations

Metabolomics has proved its potential by reflecting its dominance in all aspects of genomic engineering. Owing to the complexity of the metabolome, it is more prone to errors and various factors need to be considered prior to commencing experimental studies. Though metabolomics can accurately recognize metabolic blueprints and ailment mechanisms, a plethora of challenges need to be overcome for ensuring efficiency of the process.

12.5.1 Off-Target Effects and Unintended Consequences of Genome Editing

Off-target type of genomic editing is a non-specific and unintended genetic alteration that is carried out when an engineered nuclease is used. It was observed that though the nature, frequency and location of this genetic alteration is not mentioned, the pre-defined off-target cleavage activity still occurs in the PAM-distal portion of the sgRNA-guiding genomic sequence (Bonte et al. 2019). A variety of distinctive nucleotides and noncanonical protospacer adjacent motif (PAMs) from the on-target sites are found to be present in the off-target sites. These mismatches are accommodated with more ease by gRNAs at their respective 5′ ends. In the seed are, a mismatch region can prevent Cas9 from stimulating if there are more than one present. In cases where there are more than three mismatches present, the DNA genomic sequences tend to bind to Cas9, thus obstructing HNH conformation and inhibiting its cleavage. These off-target activities are normally dependent on the structures and configurations of sgRNAs within the system (Mielke et al. 2017). This off-target genome editing dominates the therapeutic capability of CRISPR/Cas9 along with casting ambiguity on scientific results in accordance with the attributes of genes. The end-results of this particular effect are unwanted DNA damage, cytotoxicity and immune response. These off-target effects can be eliminated by adopting the following ways:

(a) Improvement sgRNA specificity
(b) Usage of Prime Editors
(c) Improvement of Cas Variants

(d) Anti-CRISPR Proteins
(e) Discovery of SuperFi-Cas9

12.5.2 Regulatory Frameworks for Genetically Modified Organisms

It has been 30 years since the first transgenesis-sourced genetically modified materials were made available to the general public (Yi et al. 2019). The regulatory frameworks of a given nation are typically divided into two categories: process-oriented and product-oriented, taking into account the distinctions. The main goal is to determine the rules and requirements based on the techniques used to create new plant types. The second one distinguishes between them and extrapolates criteria based on the unique features of a particular biotechnology event. These product- or process-oriented regulatory guidelines typically have components in common. This makes the harder-to-classify legal systems very simple (Lind et al. 2020). More than nations have ratified the Cartagena Protocol, which includes fundamental international legal norms that the signatory nations must abide by (Table 12.2).

12.5.3 Ethical Considerations in Manipulating Cellular Metabolism

Innovative medical technologies always rise policy and ethical issues. These issues primarily revolve around the ethical code of conduct of basic and clinical research

Table 12.2 GMO regulatory frameworks around the globe

Region	Regulatory framework	Main regulatory focus
Argentina	Permissive	Process
Brazil	Permissive	Process
Colombia	Permissive	Product
Ecuador	Prohibitive	Process
Mexico	Permissive	Product
USA	Permissive	Product
European Union	Restrictive	Process
South Africa	Permissive	Process
Egypt	Prohibitive	Process
Nigeria	Permissive	Product
Uganda	Prohibitive	Process
China	Permissive	Process
New Zealand	Restrictive	Process
Japan	Hybrid	Product
India	Permissive	Process
Vietnam	Permissive	Product

along with equitable distribution of novel therapies. Derksen and Horstman discuss that an individual can roughly differentiate between two perspectives. Since tissue engineering is said to be mimicking nature, this technique is seen morally right on one hand. Conversely, technology may also be seen morally detrimental as it defies natural rules; for example, laboratory-produced synthetic goods appear to be unnatural (Delles et al. 2018). The surgical introduction of artificially generated chondrocytes into human ear cartilage was highly attractive to the general audience. In light of the engineering of heart valves, experts have suggested that the ethics of tissue engineering should be defined in terms of "good embodied life" and "lived integrity," rather than by precisely defining "natural" or "unnatural."

12.6 Future Directions: Personalized Medicine and Beyond

The integration of artificial intelligence (AI) and precision medicine has proven to revolutionize the medical paradigm. AI tends to leverages exhaustive sophisticated computation and produces insights. In addition, it permits the system to learn and reason along with empowering the clinician's decision through augments intelligence.

12.6.1 Applications in Bioremediation and Production of Valuable Metabolites

With the advent of recent technological advancements, the secondary metabolites also known as specialized metabolites have heavy influence over the human health, economics and nutrition for our community. They are derived from secondary metabolism are known to have low molecular weight (Chen-Plotkin et al. 2018). They generally include toxins, immunomodulators, pigments, compounds with hormonal attributes, antibiotics, ecological competition/symbiosis effectors, specific effects imparted on the lipid and carbohydrate metabolism. Some are industrially established as vasoconstrictors, vasodilators, laxatives, antivirals, antitumorals, diuretics and antioxidants (Shao and Le 2019). In the scenario of solving environmental issues, a bioremediation approach has been adopted to resolve the challenges encountered in the process of bioremediating recalcitrant polluting particles from the surroundings. The indiscriminate use of pesticides poses great danger to our nature. These pesticides, whether organic or inorganic, need to be bioremediated before using it on the crops. This bioremediation process varies from each other depending on the nature of the pesticide under consideration. In a nutshell, each pesticide requires a distinct pathway for its bioremediation.

12.6.2 Personalized Nutrition and Therapeutic Interventions Based on Metabolic Profiles

Nowadays, nutritional disorders have become extremely common among population. This highlights the need to focus on targeted approaches. The concept of personalized medicine which caters to the needs of specific individual has received much attention in the current decade for its unique way of improving the nutritional status of the patient and thereby improving the quality of life (Inoue et al. 2015). Integration of the emerging fields like nutrigenetics and genomics has led to the culmination of numerous genetic variants which basically modify the nutritional supplementation (Zetterberg and Burnham 2019). Genotype-dependent nutritional intervention has garnered much attention due its high specificity and accuracy. Their effect is appreciable in particular rare diseases like vitamin D resistant rickets, phenylketonuria and galactosemia. Genome-wide association studies (GWAS) or candidate-gene approach along with numerous single nucleotide polymorphism (SNPs) are known to have wide-ranging effects on the digestion, absorption, assimilation, signal transduction of macro- and micronutrients (Dang et al. 2018; Pinu et al. 2019). The approach of personalized medicine should be applied based on the following three levels:

(a) Conventional nutrition based on guidelines for populations by age, gender and social determinants.
(b) Individualized nutrition that enriches the current nutritional status of the patient by adding phenotypic data.
(c) Genotype-directed nutritional approaches.

12.6.3 The Potential of Genome Editing for Advancing Human Health

The incorporation of germline gene editing in the research approach will significantly add to our existing pool of knowledge and will eventually help create novel therapeutic options for various ailments. Germline gene editing possesses a tremendous potential as a clinical tool. This aids in prevention of relatively simple genetic diseases and significantly reduces the occurrence of polygenic defects. This technique enables the stem cells to curate embryonic stem cells comprising of distinctive disease alleles on the same common background. For instance, CRISPR/Cas9 system is employed to modify embryonic stem cells to comprise genetic mutations related to Parkinson's disease (Wang et al. 2015). These cells can be further cultured to grow as nerve cells which are responsible for the defect in Parkinson's disease. These cells can be now researched for meticulously understanding the mechanisms involved in the disease under consideration. This therefore proves to be an efficient platform for testing potential treatments. This expedites the development of novel therapeutic strategies for various diseases.

If the genomic editing is used indiscriminately it will potentially threaten the mere existence of human race by eliminating all the valuable genetic mutations occurring over time. Human genetic populations benefit from cognitive diversity and immunodiversity. Cognitive diversity can be elaborated as diversity existing in our cognitive traits. On the other hand, immunodiversity can be explained as diversity in the genes that are affect innate immunity. Therefore, it is the prime responsibility of the human race to judiciously exploit the technology so as to only pool the benefits and protect ourselves from the harmful consequences of overexploitation.

12.7 Conclusions

In order to summarize, metabolomics is a novel "omics" approach which comprises of exhaustive characterization of metabolites in the biological reactions. Currently, the field of metabolomics is not limited to mere metabolome profiling but has expanded to recognition of novel biomarkers for various ailments, their prognosis, prevention and treatment. Poor public awareness of metabolomics as a clinical tool, skytouching process expenses of the required analytical machinery, multidisciplinary background of metabolomics and variation in the modes of information extraction are some of the challenges faced in the progress of metabolomics. Metabolomics also fails to consider the gender effects on a disease which significantly impacts its pathophysiology. For better holistic approach, it is suggested to consider the valuable effect of sex on the patients. The number of metabolites recognized as potential biomarkers is so large that it becomes extremely tedious to pick the most efficient biomarker among the others. However, there is need for deep understanding of the optimization and validation of the genome editing tools that help us reach our goal of identifying the most accurate biomarker and thereby considerably improving the quality of life.

References

Appiah-Amponsah E, Owusu-Sarfo K, Gowda GA, Ye T, Raftery D (2013) Combining Hydrophilic Interaction Chromatography (HILIC) and isotope tagging for off-line LC-NMR applications in metabolite analysis. Meta 3:575–591

Aretz I, Meierhofer D (2016) Advantages and pitfalls of mass spectrometry based metabolome profiling in systems biology. Int J Mol Sci 17:632

Armitage EG, Southam AD (2016) Monitoring cancer prognosis, diagnosis and treatment efficacy using metabolomics and lipidomics. Metabolomics 12:146

Beltran A, Suarez M, Rodríguez MA, Vinaixa M, Samino S, Arola L, Correig X, Yanes O (2012) Assessment of compatibility between extraction methods for NMR- and LC/MS-based metabolomics. Anal Chem 84:5838–5844

Bingol K (2018) Recent advances in targeted and untargeted metabolomics by NMR and MS/NMR methods. High-Throughput 7:9

Bonte R, Bongaerts M, Demirdas S, Langendonk JG, Huidekoper HH, Williams M (2019) Untargeted metabolomic-based screening method for inborn errors of metabolism using semi-automatic sample preparation with an UHPLC-Orbitrap-MS platform. Metabolites 9:289

Cardellini M, Ballanti M, Davato F, Cardolini I, Guglielmi V, Rizza S, Pecchioli C, Casagrande V, Mavilio M, Porzio O et al (2018) 2-hydroxycaproate predicts cardiovascular mortality in patients with atherosclerotic disease. Atherosclerosis 277:179–185

Carlson AK, Rawle RA, Adams E, Greenwood MC, Bothner B, June RK (2018) Application of global metabolomic profiling of synovial fluid for osteoarthritis biomarkers. Biochem Biophys Res Commun 499:182–188

Chan ECY, Pasikanti KK, Hong Y, Ho PC, Mahendran R, Raman Nee Mani L, Chiong E, Esuvaranathan K (2015) Metabonomic profiling of bladder cancer. J Proteome Res 14:587–602

Chang KH, Cheng ML, Tang HY, Huang CY, Wu YR, Chen CM (2018) Alternations of metabolic profile and kynurenine metabolism in the plasma of Parkinson's disease. Mol Neurobiol 55:6319

Chen WS, Liu MH, Cheng ML, Wang CH (2020) Decreases in circulating concentrations of short-chain acylcarnitines are associated with systolic function improvement after decompensated heart failure. Int Heart J 61:1014–1021

Chen-Plotkin AS, Albin R, Alcalay R, Babcock D, Bajaj V, Bowman D, Buko A, Cedarbaum J, Chelsky D, Cookson MR et al (2018) Finding useful biomarkers for Parkinson's disease. Sci Transl Med 10:eaam6003

Coene KLM, Kluijtmans LAJ, Van der Heeft E, Engelke UFH, de Boer S, Hoegen B, Kwast HJ, Van de Vorst M, Huigen MC, Keularts IM et al (2018) Next-generation metabolic screening: targeted and untargeted metabolomics for the diagnosis of inborn errors of metabolism in individual patients. J Inherit Metab Dis 41:337–353

Contrepois K, Jiang L, Snyder M (2015) Optimized analytical procedures for the untargeted metabolomicprofiling of human urine and plasma by combining hydrophilic interaction (hilic) and reverse-phase liquid chromatography (RPLC)–mass spectrometry. Mol Cell Proteomics 14:1684–1695

Dang VT, Huang A, Werstuck GH (2018) Untargeted metabolomics in the discovery of novel biomarkers and therapeutic targets for atherosclerotic cardiovascular diseases. Cardiovasc Haematolol Dis Drug Targets 18:166–175

Delles C, Rankin NJ, Boachie C, McConnachie A, Ford I, Kangas A, Soininen P, Trompet S, Mooijaart SP, Jukema JW et al (2018) Nuclear magnetic resonance-based metabolomics identifies phenylalanine as a novel predictor of incident heart failure hospitalisation: results from PROSPER and FINRISK 1997. Eur J Heart Fail 20:663–673

Dudka I, Chachaj A, Sebastian A, Tański W, Stenlund H, Gröbner G, Szuba A (2021) Metabolomic profiling reveals plasma GlycA and GlycB as potential biomarkers for treatment efficiency in rheumatoid arthritis. J Pharm Biomed Anal 197:113971

Haines NR, Manoharan N, Olson JL, D'Alessandro A, Reisz JA (2018) Metabolomics analysis of human vitreous in diabetic retinopathy and rhegmatogenous retinal detachment. J Proteome Res 17:2421–2427

Hasin Y, Seldin M, Lusis A (2017) Multi-omics approaches to disease. Genome Biol 18:83

Havelund JF, Andersen AD, Binzer M, Blaabjerg M, Heegaard NHH, Stenager E, Faergeman NJ, Gramsbergen JB (2017) Changes in kynurenine pathway metabolism in Parkinson patients with L-DOPA-induced dyskinesia. J Neurochem 142:756–766

Huang M, Li H-Y, Liao H-W, Lin C-H, Wang C-Y, Kuo W-H, Kuo C-H (2020) Using post-column infused internal standard assisted quantitative metabolomics for establishing prediction models for breast cancer detection. Rapid Commun Mass Spectrom 34:e8581

Inoue K, Tsuchiya H, Takayama T, Akatsu H, Hashizume Y, Yamamoto T, Matsukawa N, Toyo'oka, T. (2015) Blood-based diagnosis of Alzheimer's disease using fingerprinting metabolomics based on hydrophilic interaction liquid chromatography with mass spectrometry and multivariate statistical analysis. J Chromatogr B Analyt Technol Biomed Life Sci 974:24–34

Ismail IT, Showalter MR, Fiehn O (2019) Inborn errors of metabolism in the era of untargeted metabolomics and lipidomics. Meta 9:242

Jacob M, Malkawi A, Albast N, Al Bougha S, Lopata A, Dasouki M, Abdel Rahman AMA (2018) Targeted metabolomics approach for clinical diagnosis of inborn errors of metabolism. Anal Chim Acta 1025:141–153

Jeong A, Fiorito G, Keski-Rahkonen P, Imboden M, Kiss A, Robinot N, Gmuender H, Vlaanderen J, Vermeulen R, Kyrtopoulos S et al (2018) Perturbation of metabolic pathways mediates the association of air pollutants with asthma and cardiovascular diseases. Environ Int 119:334–345

Jin H, Zhu B, Liu X, Jin J, Zou H (2019) Metabolomic characterisation of diabetic retinopathy: an H-NMR-based metabolomic approach using human aqueous humour. J Pharm Biomed Anal 174:414–421

Joseph JS, Ayeleso AO, Mukwevho E (2017) Role of exercise-induced calmodulin protein kinase (CaMK)II activation in the regulation of omega-6 fatty acid and lipid metabolism genes in rat skeletal muscle. Physiol Res 66:969–977

Kodama M, Oshikawa K, Shimizu H, Yoshioka S, Takahashi M, Izumi Y, Bamba T, Tateishi C, Tomonaga T, Matsumoto M et al (2020) A shift in glutamine nitrogen metabolism contributes to the malignant progression of cancer. Nat Commun 11:1320

Kruk J, Doskocz M, Jodłowska E, Zacharzewska A, Łakomiec J, Czaja K, Kujawski J (2017) NMR techniques in metabolomic studies: a quick overview on examples of utilization. Appl Magn Reson 48:1–21

Li XS, Wang Z, Cajka T, Buffa JA, Nemet I, Hurd AG, Gu X, Skye SM, Roberts AB, Wu Y et al (2018) Untargeted metabolomics identifies trimethyllysine, a TMAO-producing nutrient precursor, as a predictor of incident cardiovascular disease risk. JCI Insight 3:e99096

Li J, Duan W, Wang L, Lu Y, Shi Z, Lu T (2020) Metabolomics study revealing the potential risk and predictive value of fragmented QRS for acute myocardial infarction. J Proteome Res 19:3386–3395

Lind L, Salihovic S, Ganna A, Sundström J, Broeckling CD, Magnusson PK, Pedersen NL, Siegbahn A, Prenni J, Fall T et al (2020) A multi-cohort metabolomics analysis discloses sphingomyelin (32:1) levels to be inversely related to incident ischemic stroke. J Stroke Cerebrovasc Dis 29:104476

Luo D, Shan Z, Liu Q, Cai S, Li Q, Li X (2020) A novel seventeen-gene metabolic signature for predicting prognosis in colon cancer. Biomed Res Int 2020:4845360

Macias RIR, Banales JM, Sangro B, Muntané J, Avila MA, Lozano E, Perugorria MJ, Padillo FJ, Bujanda L, Marin JJG (2018) The search for novel diagnostic and prognostic biomarkers in cholangiocarcinoma. Biochim Biophys Acta 1864:1468–1477

Manzoni C, Kia DA, Vandrovcova J, Hardy J, Wood NW, Lewis PA, Ferrari R (2018) Genome, transcriptome and proteome: the rise of omics data and their integration in biomedical sciences. Brief Bioinform 19:286–302

Mazumder AG, Banergee S, Zevictovich F, Ghosh S, Mukherjee A, Chaterjee J (2018) Fourier-transform-infrared-spectroscopy based metabolomic spectral biomarker selection towards optimal diagnostic differentiation of diabetes with and without retinopathy. Spectrosc Lett 51:340–349

McKenzie JS, Charlton AJ, Donarski JA, MacNicoll AD, Wilson JC (2010) Peak fitting in 2D 1H–13C HSQC NMR spectra for metabolomic studies. Metabolomics 6:574–582

Mehta A, Liu C, Nayak A, Tahhan AS, Ko Y, Dhindsa DS, Kim JH, Hayek SS, Sperling LS, Mehta PK et al (2020) Untargeted high resolution plasma metabolic profiling predicts outcomes in patients with coronary artery disease. PLoS One 15:e0237579

Mielke MM, Hagen CE, Wennberg AMV, Airey DC, Savica R, Knopman DS, Machulda MM, Roberts RO, Jack CR Jr, Petersen RC et al (2017) Association of plasma total tau level with cognitive decline and risk of mild cognitive impairment or dementia in the Mayo Clinic study on aging. JAMA Neurol 74:1073–1080

Mo L, Wei B, Liang R, Yang Z, Xie S, Wu S, You Y (2020) Exploring potential biomarkers for lung adenocarcinoma using LC-MS/MS metabolomics. J Int Med Res 48(4):0300060519897215. https://doi.org/10.1177/0300060519897215

Molepo M, Ayeleso A, Nyakudya T, Erlwanger K, Mukwevho E (2018) A study on neonatal intake of oleanolic acid and metformin in rats (Rattus norvegicus) with metabolic dysfunction: implications on lipid metabolism and glucose transport. Molecules 23:2528

Mosana M, Ayeleso A, Nyakudya T, Erlwanger K, Mukwevho E (2020) Potential protective effects of neonatal supplementation with oleanolic acid on peroxisome proliferator- activated receptor

gamma (PPARγ)- ligand dependent regulation of glucose homeostasis in high fructose- fed rats. Nat Prod Com 15:1–7
Nayak A, Liu C, Mehta A, Ko YA, Tahhan AS, Dhindsa DS, Uppal K, Jones DP, Butler J, Morris AA et al (2020) N8-Acetylspermidine: a polyamine biomarker in ischemic cardiomyopathy with reduced ejection fraction. J Am Heart Assoc 9:e016055
Omori K, Katakami N, Yamamoto Y, Ninomiya H, Takahara M, Matsuoka TA, Bamba T, Fukusaki E, Shimomura I (2019) Identification of metabolites associated with onset of CAD in diabetic patients using CE-MS analysis: a pilot study. J Atheroscler Thromb 26:233–245
Pasikanti KK, Ho PC, Chan ECY (2008) Gas chromatography/mass spectrometry in metabolic profiling ofbiological fluids. J Chromatogr B 871:202–211
Pinu FR, Goldansaz SA, Jaine J (2019) Translational metabolomics: current challenges and future opportunities. Meta 9:108
Rhee SY, Jung ES, Park HM, Jeong SJ, Kim K, Chon S, Yu S, Woo JT, Lee CH (2018) Plasma glutamine and glutamic acid are potential biomarkers for predicting diabetic retinopathy. Metabolomics 14:89
Saiki S, Hatano T, Fujimaki M, Ishikawa KI, Mori A, Oji Y, Okuzumi A, Fukuhara T, Koinuma T, Imamichi Y et al (2017) Decreased long-chainacylcarnitines from insufficient beta-oxidation as potential early diagnosticmarkers for Parkinson's disease. Sci Rep 7:7328
Shah NJ, Sureshkumar S, Shewade DG (2015) Metabolomics: a tool ahead for understanding molecular mechanisms of drugs and disease. Indian J Clin Biochem 30:247–254
Shang H, Zheng J, Tong J (2020) Integrated analysis of transcriptomic and metabolomic data demonstrates the significant role of pyruvate carboxylase in the progression of ovarian cancer. Aging 12:21874–21889
Shao Y, Le W (2019) Recent advances and perspectives of metabolomics-based investigations in Parkinson's disease. Mol Neurodegener 14:3
Sidorov E, Bejar C, Xu C, Ray B, Reddivari L, Chainakul J, Vanamala JKP, Sanghera DK (2020) Potential metabolite biomarkers for acute versus chronic stage of ischemic stroke: a pilot study. J Stroke Cereb Dis 29:104618
Stryeck S, Gastrager M, Degoricija V, Trbušić M, Potočnjak I, Radulović B, Pregartner G, Berghold A, Madl T, Frank S (2019) Serum concentrations of citrate, tyrosine, 2- and 3- hydroxybutyrate are associated with increased 3-month mortality in acute heart failure patients. Sci Rep 9:6743
Sun YV, Hu Y (2016) Chapter three: integrative analysis of multi-omics data for discovery and functional studies of complex human diseases. Adv Genet 93:147–190
Takahashi S, Saegusa J, Onishi A, Morinobu A (2019) Biomarkers identified by serum metabolomic analysis to predict biologic treatment response in rheumatoid arthritis patients. Rheumatology 58:2153–2161
Tayanloo-Beik A, Sarvari M, Payab M, Gilany K, Alavi-Moghadam S, Gholami M, Goodarzi P, Larijani B, Arjmand B (2020) OMICS insights into cancer histology; metabolomics and proteomics approach. Clin Biochem 84:13–20
Tripathi P, Somashekar BS, Ponnusamy M, Gursky A, Dailey S, Kunju P, Lee CT, Chinnaiyan AM, Rajendiran TM, Ramamoorthy A (2013) HR-MAS NMR tissue metabolomic signatures cross-validated by mass spectrometry distinguish bladder cancer from benign disease. J Proteome Res 12:3519–3528
Vignoli A, Tenori L, Giusti B, Takis PG, Valente S, Carrabba N, Balzi D, Barchielli A, Marchionni N, Gensini GF et al (2019) NMR-based metabolomics identifies patients at high risk of death within two years after acute myocardial infarction in the AMI-Florence II cohort. BMC Med 17:3
Wang Y, Liu S, Hu Y, Li P, Wan J-B (2015) Current state of the art of mass spectrometry-based metabolomics studies—a review focusing on wide coverage, high throughput and easy identification. RSC Adv 5:78728–78737
Wang Z, Bian L, Mo C, Shen H, Zhao LJ, Su KJ, Kukula M, Lee JT, Armstrong DW, Recker R et al (2020) Quantification of aminobutyric acids and their clinical applications as biomarkers for osteoporosis. Commun Biol 3:39

Wilkins JM, Trushina E (2018) Application of metabolomics in Alzheimer's disease. Front Neurol 8:719

Yi M, Li Q, Zhao Y, Nie S, Wu N, Wang D (2019) Metabolomics study on the therapeutic effect of traditional Chinese medicine Xue-Fu-Zhu-Yu decoction in coronary heart disease based on LC-Q-TOF/MS and GC-MS analysis. Drug Metab Pharm 34:340–349

Yilmaz A, Geddes T, Han B, Bahado-Singh RO, Wilson GD, Imam K, Maddens M, Graham SF (2017) Diagnostic biomarkers of Alzheimer's disease as identified in saliva using 1H NMR-based metabolomics. J Alzheimers Dis 58:355–359

Yu L, Li K, Li X, Guan C, Sun T, Zhang X (2020) Metabolomic profiling of dried blood spots reveals gender-specific discriminant models for the diagnosis of small cell lung cancer. Aging 12:978–995

Zetterberg H, Burnham SC (2019) Blood-based molecular biomarkers for Alzheimer's disease. Mol Brain 12:26

Zhang A, Sun H, Han Y, Yan G, Yuan Y, Song G, Yuan X, Xie N, Wang X (2013) Ultraperformance liquid chromatographymass spectrometry-based comprehensive metabolomics combined with pattern recognition and network analysis methods for characterization of metabolites and metabolic pathways from biological data sets. Anal Chem 85:7606–7612

Zhang A, Sun H, Wang X (2014) Urinary metabolic profiling of rat models revealed protective function of scoparone against alcohol induced hepatotoxicity. Sci Rep 4:6768

Zhang A, Sun H, Yan G, Wang P, Wang X (2015) Metabolomics for biomarker discovery: moving to the clinic. Biomed Res Int 2015:354671

Zhang A, Sun H, Wang X (2017) Emerging role and recent applications of metabolomic biomarkers in obesity disease research. RSC Adv 7:14966–14973

Zhu M, Han Y, Zhang Y, Zhang S, Wei C, Cong Z, Du W (2018) Metabolomics study of the biochemical changes in the plasma of myocardial infarction patients. Front Physiol 9:1017

Zhu X, Yang F, Lu J, Zhang H, Sun R, Zhoe J, Yang J (2019) Plasma metabolomic profiling of proliferative diabetic retinopathy. Nutr Metab 16:37

13 Production and Fermentation Optimisation of Flavour and Fragrance Molecules

Devika Zunjarrao, Rutuja A. Kamble, Om Supe, and Shamlan M. S. Reshamwala

13.1 Introduction

Flavours and fragrances are widely used in the cosmetics, foods and beverages, chemical and pharmaceutical industries. In the year 2023, the flavours and fragrances market around the world was estimated at USD 30.61 bn. The compound annual growth rate (CAGR) for this market from the year 2024 to 2030 is estimated to be 5.4% (based on Flavours and Fragrances Market Size, Share and Trends Analysis Report by Product, Application, Region and Segment Forecasts, 2024–2030). Flavours and fragrances are currently produced artificially using chemical synthesis or by extraction from natural sources. Based on the source of the compounds, flavours and fragrances can be classified as nature identical and natural. Nature-identical compounds are synthesised through chemical reactions, while natural compounds are obtained from plant and animal sources or synthesised using enzymes or microbial cells.

Chemical synthesis is not environment-friendly. It can also at times lead to unwanted side reactions. Resolution of racemic mixtures is one common issue which is usually tedious to tackle in practice. This adversely affects the overall efficiency of the process, thus increasing the cost of downstream processing. These factors are responsible for the shift in focus from chemical synthesis to production through biological routes.

Traditional sources of natural flavours and scents include herbs, spices, plants and animals. These flavours and scents are typically used as essences, extracts and oils (Lindsay et al. 2022). However, in nature, these compounds are produced in small quantities, necessitating research focused towards finding ways to enhance

D. Zunjarrao · R. A. Kamble · O. Supe · S. M. S. Reshamwala (✉)
Department of Biological Sciences and Biotechnology, Institute of Chemical Technology, Mumbai, India
e-mail: ss.reshamwala@ictmumbai.edu.in

V. Singh (ed.), *Advances in Metabolomics*,
https://doi.org/10.1007/978-981-97-7459-3_13

their production. Also, in nature, weather conditions as well as plant diseases impact the optimal production of these compounds. For higher production, the natural input materials need to be cultivated on a large scale. Therefore, natural resources are proving insufficient to meet the ever-increasing market demand (Akacha and Gargouri 2015).

Consumers' growing interest in natural products has led to research being focused on finding other viable alternative methods of generating natural flavour and fragrance compounds. Compared to present production methods, biotechnology offers several advantages, including reduced land use, a smaller carbon impact, more flexible value chains and the potential to create novel compounds that are not found in nature. The drawbacks of biotechnologically-produced substances are lengthy development durations and significant investment costs.

13.2 Biotechnological Production of Fragrance and Flavour Molecules

A compound's distinctive fragrance and flavour are due to the presence of numerous components, both volatile and non-volatile, with a variety of chemical and physicochemical characteristics. Biotechnology-based approaches like de novo synthesis or microbial cell-based bioconversions of natural precursor molecules are increasingly becoming popular for production of value-added products due to their low environmental impact and better reproducibility.

Complex targets or product mixtures are generally synthesised using the de novo synthesis approach while single-step processes are carried out using bioconversions or biotransformations. In de novo synthesis, microorganisms convert various carbon and nitrogen sources into flavour and fragrance molecules using different enzymes, which act upon lipids, esters, proteins, nucleic acids and certain glycosides. In industry, such enzymes have found applications in the production of flavour and fragrance molecules since they facilitate the transformation of substrates into desirable products (Gupta et al. 2015).

Various flavoured substances and aromatic molecules are produced using multistep enzymatic pathways in microorganisms. However, these molecules may not be inherently produced by the microbes in large quantities. In that case, the microorganisms' ability to produce these compounds can be significantly enhanced by using metabolic engineering approaches.

Key components of flavour and fragrance molecules can be classified into broad chemical classes such as esters, terpenes/terpenoids, pyrazines, lactones, aldehydes and ketones. Different substrates, pathways and organisms are involved in the biotechnological production of compounds belonging to each of these classes. Some examples of biosynthesis of each of these classes are discussed below.

13.2.1 Esters

Esters are compounds that impart a fruity aroma which can be chemically derived from acids. Many fruits, including pears, bananas, pineapples, apples, durians and strawberries, have an aroma that is attributed to esters. These are widely used in the food and beverages industry as well as in the chemical industry. Pentanoates, ethyl acetate and isoamyl acetate are some of the esters which are utilised for their aromatic properties (Kruis et al. 2019).

Microorganisms, such yeasts and lactic acid bacteria, can naturally produce esters (Kruis et al. 2019). Rodriguez et al. (2014) engineered an *Escherichia coli* strain and used it in ester production. They utilised the alcohol *O*-acyltransferase class of enzymes which uses acyl-CoA units to form esters. Esterification with an alcohol releases free CoA which in turn provides free energy for ester formation. They engineered alcohol biosynthetic pathways to produce acetate esters of primary alcohols such as ethanol, propanol, isobutanol, 2-methyl-1-butanol, 2-phenylethanol and secondary alcohols like 3-methyl-1-butanol (Rodriguez et al. 2014).

Tai et al. (2015) achieved synthesis of 2-methylpropyl ethanoate (also called isobutyl acetate) and 3-methylbutyl ethanoate (also called isoamyl acetate) from glucose by engineering synthetic metabolic pathways in *E. coli*. They expanded the *E. coli* valine and leucine biosynthetic pathways for manufacture of isobutyl acetate and isoamyl acetate, respectively. Alcohol acyltransferase was overexpressed and competing pathways producing lactate and acetate were deleted to enhance yields (Tai et al. 2015).

Löser et al. (2014, 2015) studied ethyl acetate synthesis mechanism in *Kluyveromyces marxianus* DSM 5422. *K. marxianus* DSM 5422 synthesised ethyl acetate from acetyl-SCoA under iron and copper limiting conditions which induced ester synthesis (Löser et al. 2015). The host organism *K. marxianus* is advantageous for ester production in particular as it can convert lactose in whey into ethyl acetate, apart from being a fast-growing strain (Löser et al. 2014).

13.2.2 Terpenes and Terpenoids

These are the largest class of naturally occurring substances present on Earth. They exhibit a wide array of biological activities in various cellular processes (Schempp et al. 2018). They have found applications as industrial materials (isoprene, linalool), in nutrition (carotenoids) and medicine (artemisinin, paclitaxel), and as potential biofuels (bisabolene, pinene and farnesene) (Kirby et al. 2015). Many of these terpene compounds are naturally found in plants, which is one of the major sources from which terpenes are extracted. Another source is through the use of microbial cell factories which can be engineered to produce large quantities of product in less amount of time.

Terpenoid production has been achieved in host organisms such as *E. coli*, *Saccharomyces cerevisiae*, *Pseudomonas putida* and *Corynebacterium glutamicum* (Chu 2022). Metabolic flux can be controlled to improve the production of terpenoids. CRISPR-Cas9-based genome editing tools have been used to engineer cells for this purpose (Chu 2022). The building blocks isopentenyl diphosphate (IPP) and dimethylallyl diphosphate (DMAPP) involved in terpenoid biosynthesis are produced through the mevalonate pathway (MVA) or 2-C-methyl-D-erythritol 4-phosphate (MEP) pathway. *E. coli* and *S. cerevisiae*, the conventional microbial hosts, have been largely utilised to produce a variety of terpenoids through the MEP and MVA pathways, respectively (Wang et al. 2018).

Isoprene (2-methyl-1,3-butadiene) is a volatile organic compound which serves as a building block for terpenes. It can be biologically synthesised by isoprene synthase from DMAPP derived from MVA or MEP pathways. Isoprene production has been previously described by several researchers using cyanobacteria, yeast and bacteria which have been engineered to overexpress plant isoprene synthase genes (Lindberg et al. 2010; Xue and Ahring 2011; Zhao et al. 2011).

Kim et al. (2016) expressed the isoprene biosynthesis pathway in *E. coli* to achieve increased isoprene production. They codon optimised an isoprene synthase gene from *Populus trichocarpa* for expression in *E. coli* MG1655, introduced the heterologous MVA pathway and deleted the genes responsible for by-product formation from acetyl-CoA (initial substrate in MVA pathway) to increase isoprene production (Kim et al. 2016).

Yang et al. (2016) engineered both MEP and MVA pathways in *E. coli* for isoprene production. They demonstrated synergy between these two pathways when overexpressed together. They found a 20 times improvement in yield through the MEP pathway and 3 times increase through the MVA pathway as compared to overexpression of one of the pathways alone. The MEP and MVA pathways complemented one another through their cofactors which could explain the synergy between them. Conversion of glucose to one isoprene molecule produces four NAD(P)H molecules via the MVA pathway, while the same through the MEP pathway utilises two NAD(P)H and three ATP molecules (Yang et al. 2016).

Scalcinati et al. (2012a, b) engineered *S. cerevisiae* to produce α-santalene, a precursor of commercially important sesquiterpenes, by expressing the *Clausena lansium* santalene synthase gene. The expression of key metabolic steps of the MVA pathway was modulated to increase the production to match industrial requirements (Scalcinati et al. 2012a, b).

In most of these studies, the metabolite of interest is produced in the cytosol of the microbe. However, various other simultaneously demanding pathways often hinder the cytosol's ability to synthesise target chemicals efficiently. To circumvent this, Dusséaux et al. (2020) created microfactories for the production of cannabinoids, monoterpenoids and indole alkaloids in yeast peroxisomes. With this peroxisome engineering strategy, by the introduction of the MVA pathway in the peroxisome, they converted acetyl-CoA to many industrially important terpenes and achieved an increase of 125 times in yield as compared to traditional production methods in the cytosol. Further, they utilised these peroxisome-engineered strains for the production of a menthol intermediate, transisopiperitenol, through hydroxylation of (S)-(−)-limonene catalysed by cytochrome P450 (Dusséaux et al. 2020).

Apart from these conventional hosts, many studies have utilised non-conventional hosts for the bioproduction of terpenoids. For instance, Wriessnegger et al. (2014) utilised a metabolically engineered *Pichia pastoris* strain to produce (+)-nootkatone, a sesquiterpenoid, from (+)-valencene (Wriessnegger et al. 2014). Sonntag et al. (2015) utilised a metabolically engineered strain of *Methylobacterium extorquens* AM1 to demonstrate de novo synthesis of the sesquiterpenoid α-humulene from methanol. They did not utilise sugar as a starting material which is otherwise a common strategy for producing terpenoids to avoid high price inconstancy (Sonntag et al. 2015).

Priya et al. (2015, 2023) isolated a novel *Gluconobacter japonicus* MTCC 12284 strain which was capable of transforming α-(+)-pinene to produce important aromatic chemicals such as verbenol, verbenone, α-terpineol and trans-sobrerol. Similarly, Trytek et al. (2015) produced verbenol and verbenone by oxidation of α-pinene. For this study, they utilised a novel biocatalyst, the fungus *Chrysosporium pannorum* A-1, for carrying out O_2-driven oxidation of α-pinene (Trytek et al. 2015).

Molina et al. (2013) used *Pseudomonas* sp. as a biocatalyst for bioconversion of terpenes. Molina et al. (2014) compared the bioconversion of S-(−)-limonene to limonene-1,2-diol against the well-proven production of α-terpineol by biotransformation of R-(+)-limonene using *Fusarium oxysporum* 152B as a biocatalyst in both processes.

13.2.3 Pyrazines

Pyrazines are volatile nitrogenous heterocyclic compounds found in plants, insects, bacteria and fungi. Depending on the type of the alkyl substituent, pyrazines can impart a nutty, roasted or toasty flavour (Mortzfeld et al. 2020). Fadel et al. (2018) investigated biological means for production of nutty flavouring using *C. glutamicum* to reduce manufacturing cost. For bioproduction, enzymatic hydrolysate of soybean meal was utilised as a source of free amino acids and carbohydrates (Fadel et al. 2018).

Zhang et al. (2019) investigated the synthesis mechanism of alkyl pyrazines (2,5-dimethyl pyrazine and 2,3,5-trimethyl pyrazine) in *Bacillus subtilis* 168 by utilising L-threonine as a substrate via aminoacetone. In another study, a *B. subtilis* strain able to produce 2-ethyl-3,5-(3,6)-dimethylpyrazine (EDMP) was isolated from baijiu fermentation (Zhang et al. 2020). They hypothesised and confirmed the biosynthetic pathway for the production of EDMPs based on known biosynthesis mechanisms.

Masuo et al. (2020) described a novel mechanism for biosynthesis of monocyclic pyrazines by *Pseudomonas fluorescens* SBW25. A cluster of genes—including 4-aminophenylalanine C-acetyltransferase, dihydropyrazine oxidase and methyltransferase with the 4-aminophenylalanine biosynthetic genes—was discovered in this *Pseudomonas* strain. This gene cluster was involved in the biosynthesis of 2,5-dimethyl-3,6-bis(4-aminobenzyl)-pyrazine (DMBAP), a novel pyrazine.

13.2.4 Lactones

Lactones are volatile organic compounds naturally found in fruits (peaches, strawberries and mangoes), vegetables and animal products. They are characterised by a cyclic ring and are esters of hydroxy acids. The size of the cyclic ring determines the type of lactone: α- (three atoms), β- (four atoms), γ- (five atoms) and δ-lactone (six atoms). Perhaps the most diverse, chemically stable and commercially valuable are δ-lactones and γ-lactones (Silva et al. 2021). De novo synthesis of lactones has been previously investigated in filamentous fungi like *Fusarium poae* and *Trichoderma viride*, and yeasts such as *Sporidiobolus salmonicolor* (Silva et al. 2019).

Fadel et al. (2014) used a *T. viride* strain to produce the coconut aroma imparting unsaturated lactone 6-pentyl-α-pyrone. They utilised a solid-state fermentation strategy, using sugarcane bagasse as the substrate (Fadel et al. 2014).

Production of lactones through microbial biosynthesis has been widely studied in the yeast *Yarrowia lipolytica* using hydroxy fatty acids as precursors. Gomes et al. (2012) demonstrated the effect of a constant rate of medium feeding and intermittent fed-batch strategies on the production yield of γ-decalactone in *Y. lipolytica*. They found that the intermittent fed-batch approach improved the bioconversion rate of methyl ricinoleate substrate, resulting in high titres of aroma compounds (Gomes et al. 2012).

Gomes et al. (2013) investigated the effect of utilising lipases for hydrolytic breakdown of castor oil as a ricinoleic acid source on γ-decalactone production by *Y. lipolytica*. They tried various commercially available lipases derived from microbial sources other than *Y. lipolytica* and studied the use of such enzymes in biotransformation of unhydrolysed castor oil. They investigated whether using hydrolysed castor oil had any influence on aroma production (Gomes et al. 2013).

Soares et al. (2017) evaluated the feasibility of using crude glycerol obtained as a by-product of biodiesel production as an alternate substrate for γ-decalactone production by *Y. lipolytica* along with another yeast, *Lindnera saturnus*. Yields were compared between crude glycerol and castor oil substrates. *L. saturnus* showed higher yields in the glycerol medium compared to *L. lipolytica* which showed better yields with castor oil substrate (Soares et al. 2017).

Another microbe being studied in this respect is the filamentous fungus *Ashbya gossypii* which can synthesise lactones through de novo biosynthesis and carry out biotransformation of non-hydroxy fatty acids (Silva et al. 2021). Birk et al. (2019) systematically analysed broths from large-scale industrial riboflavin fermentation of *A. gossypii* as a potential source of saturated and unsaturated lactones. They suspected the presence of lactones as they found that the fermentation broth was emitting an intense floral, fruity and nutty aroma. Through extensive analysis, they were able to detect strongly aromatic saturated and unsaturated lactones, such as γ-decalactone and γ-(Z)-dodec-6-enlactone, along with some products from the Ehrlich pathway (Birk et al. 2019).

Silva et al. (2019) tested de novo production ability of *A. gossypii* for synthesis of γ-lactones from glucose and employed metabolic engineering-based strategies targeted towards the fatty acid biosynthesis and the β-oxidation pathways. They were able to expand this organism's potential by successfully engineering it to synthesise lipids from both hexoses and pentoses (Silva et al. 2019).

S. cerevisiae, the model yeast, has also been studied for γ-decalactone production. It has been observed that yeasts can re-consume γ-decalactone in the late phase of transformation. To prevent this, Rong et al. (2017) added γ-octalactone into the biotransformation system to suppress the degradation of γ-decalactone. They also found that L-carnitine shortened the biotransformation phase and increased γ-decalactone production (Rong et al. 2017).

13.2.5 Aldehydes

Vanilla, ranked among the most widely used flavours in the world, belongs to this class of flavour and fragrance molecules. Its key component is 4-hydroxy-3-methoxybenzaldehyde, an aromatic aldehyde. Ferulic acid is the most explored substrate for vanillin production. Microorganisms can carry out bioconversion of ferulic acid using multiple pathways (Gallage and Møller 2015). *Streptomyces* sp. strain V-1 is one such organism known to utilise ferulic acid to produce vanillin. Yang et al. (2013) isolated and identified the two critical genes encoding the synthetase and hydratase enzymes involved in the synthesis of vanillin from ferulic acid. They achieved heterologous expression of these genes in *E. coli* (Yang et al. 2013).

Fleige et al. (2013) studied the actinomycete *Amycolatopsis* sp. strain ATCC 39116 for vanillin production using ferulic acid as a substrate. A vanillin dehydrogenase was causing undesired catabolism of vanillin. This study investigated the activity of vanillin dehydrogenase by creating deletion mutants of the gene which in turn prevented oxidation of vanillin to vanillic acid and hence achieved improved production of vanillin (Fleige et al. 2013).

Di Gioia et al. (2011) utilised a targeted mutagenesis approach to inactivate the vanillin dehydrogenase encoding gene and developed a *Pseudomonas* strain that could produce vanillin in high yields and at high productivity. This prevented oxidation of vanillin to vanillic acid and led to the accumulation of vanillin (Di Gioia et al. 2011).

Ferulic acid substrates are relatively costly which limits their use for vanillin production. Alternative substrates like eugenol and isoeugenol have been studied which can reduce the cost (Gallage and Møller 2015). For instance, Ryu et al. (2012) studied the isoeugenol monooxygenase enzyme from *Pseudomonas nitroreducens* which converts isoeugenol to vanillin through oxidative cleavage. They investigated regulation of this enzyme in *E. coli* and found that isoeugenol was the best at inducing isoeugenol monooxygenase while trans-anethole ranked a close second. Their results highlighted the role of the type of inducer used in the regulation of this enzyme which would be important for achieving higher yields (Ryu et al. 2012).

Lambert et al. (2013) utilised a pathway including vanillyl alcohol oxidase (VAO) enzyme known to catalyse the oxidation of eugenol. They isolated an industrial *S. cerevisiae* 92411 strain which could carry out bioconversion of coniferyl alcohol using its dehydrogenase activity and genetically modified it to express the VAO gene obtained from *Penicillium simplicissimum* (Lambert et al. 2013).

13.2.6 Ketones

Ketones, like aldehydes, are organic compounds used as flavour and fragrance agents. They are known to contribute to delightful tastes and captivating scents in various natural products. α-Amino ketones are known for their versatility and applications in synthetic as well as medicinal chemistry (Allen et al. 2021).

4-(4-hydroxyphenyl)-butan-2-one (raspberry ketone) is one of the key aromatic components of raspberry fruit. Moore et al. (2021) developed a cell-free biosynthesis alternative, a 'one-pot' approach based on enzymatic synthesis for producing raspberry ketone from L-tyrosine substrate using cofactor regeneration. The cost of cofactor is the most important factor to be considered for the cell-free biosynthesis route. In this study, the cofactor specificity of reductase was relaxed and the cofactor preference was engineered towards NADH. This new engineered variant was then used to create a raspberry ketone biosynthesis method from relatively inexpensive precursors (Moore et al. 2021).

In another study on raspberry ketones, Wang et al. (2019) engineered a heterologous pathway for raspberry ketone biosynthesis in *E. coli* utilising p-coumaric acid as substrate. These improvements resulted in increased yields of raspberry ketone due to enhanced metabolic capabilities of the engineered strain (Wang et al. 2019).

Diacetyl, a α-diketone, is an important flavour compound which imparts a buttery aroma. It is obtained as a product of 2,3-butanediol fermentation. Zhang et al. (2015) metabolically engineered an *Enterobacter cloacae* strain with modifications in key enzyme genes of 2,3-butanediol metabolism. The *E. cloacae* strain was chosen for its ability to produce high yields of 2,3-butanediol along with high productivity (Zhang et al. 2015).

Methyl ketones are produced in industry from fossil-derived hydrocarbons. Nies et al. (2020) modified a *Pseudomonas taiwanensis* VLB120 strain with methyl ketone production capability. This strain was advantageous due to its ability to sustain exceptionally high NAD(P)H regeneration rates. Yield was further improved by eliminating competing reactions (Nies et al. 2020).

13.3 Analytical Methods Employed for the Detection of Flavour and Fragrance Molecules

Biosynthetically produced flavour and fragrance molecules can be detected using various analytical techniques. The most commonly used techniques across all compound classes are chromatographic methods such as liquid chromatography, high-performance liquid chromatography and gas chromatography coupled with mass spectrometry. Different detectors, carrier gases, mobile phases and columns have been utilised based on the intended use. Table 13.1 lists analytical methods used for different compound classes.

Table 13.1 Analytical methods for characterisation of flavour and fragrance molecules

Compounds	Analysis method	Purpose	References
Esters	GC-FID (carrier gas: Helium)	Quantification of ester products	Rodriguez et al. (2014), Löser et al. (2015), Tai et al. (2015)
	GC-MS (carrier gas: Hydrogen)	Verification of ester products	
	HPLC-RID (mobile phase: 5 mM H_2SO_4)	Quantification of glucose consumption, by-product analysis	
Terpenes and terpenoids	HPLC-RID (mobile phase: 5 mM H_2SO_4)	Analysis of glucose, glycerol and ethanol	Scalcinati et al. (2012a, b), Trytek et al. (2015), Priya et al. (2023), Sonntag et al. (2015).
	HPLC-UV-visible light absorbance detector	Analysis of acetate, succinate and pyruvate	
	GC-MS (carrier gas: Helium)	Determination of terpene production	
	GC-FID (carrier gas: Helium, nitrogen)	Quantification of substrates and products	Molina et al. (2014), Priya et al. (2023), Wriessnegger et al. (2014), Dusséaux et al. (2020), Kim et al. (2016), Yang et al. (2016)
	SEM	Ultra-morphological analyses	
	TEM	Ultrastructural analysis	
Pyrazines	HS-SPME	Extraction of alkyl pyrazines	Fadel et al. (2018)
	GC-MS (carrier gas: Helium)	Analysis of volatiles	Fadel et al. (2018), Zhang et al. (2020)
	UPLC-UV	Concentration of alkyl pyrazines (2,5-DMP)	Zhang et al. (2019, 2020)
	HPLC-UV	Substrate concentration	
	UPLC-TQ-S-MS	Concentration of alkyl pyrazines (TMP)	
	UPLC-Q-ToF-MS/MS	Analysis of alkyl pyrazines	
	LC-MS/MS	Structure, molecular mass determination of novel pyrazine	Masuo et al. (2020)
	NMR		

(continued)

Table 13.1 (continued)

Compounds	Analysis method	Purpose	References
Lactones	GC-MS/MS-O (carrier gas: Helium)	Detection of aroma compounds	Birk et al. (2019)
	Paternò–Büchi (PB) functionalisation	To determine the double bond position in unsaturated lactones	
	GC-MS/MS	Quantitation	
	Preparative HPLC	Separation of mixtures of diastereomeric esters	
	MDGC	To determine enantiomeric excess (ee) values	
	GC-FID-O	Determination of odour thresholds of racemic lactones	
	GC-MS (carrier gas: Helium)	Analysis of volatiles	Silva et al. (2019)
	GC	Analysis of volatiles	Gomes et al. (2012, 2013)
	GC-FID	γ-Decalactone analysis	Soares et al. (2017)
	GC-MS (carrier gas: Nitrogen)	Analysis of ricinoleic acid transformation products	Rong et al. (2017)
	HPLC	Analysis of ricinoleic acid	
Aldehydes	GC-FID	Functional identification of genes	Yang et al. (2013)
	HPLC (methanol-water-acetic acid elution)	Product analysis	Lambert et al. (2013)
	Reverse-phase chromatography	Separation of intermediates	Fleige et al. (2013)
	HPLC	Analysis of ferulic acid metabolism intermediates	Fleige et al. (2013)
	HPLC-DAD	Quantification of vanillin, ferulic and vanillic acid	Di Gioia et al. (2011)

Table 13.1 (continued)

Compounds	Analysis method	Purpose	References
Ketones	LC-Q-ToF-MS (Diode array detector)	Analysis of raspberry ketone and pathway intermediates	Moore et al. (2021)
	GC	Analysis of products	Zhang et al. (2015), Nies et al. (2020)
	HPLC (mobile phase: 5 mM H_2SO_4, RI and UV detector)	Determination of glucose and gluconate concentration	Nies et al. (2020)
	HPLC (mobile phase: 80% water +0.1% (V/V) phosphoric acid +20% (V/V) acetonitrile)	Analysis of raspberry ketone	Wang et al. (2019)
	GC-MS (carrier gas: Helium)		

GC Gas Chromatography, *FID* Flame Ionisation Detector, *MS* Mass Spectrometry, *HPLC* High-performance Liquid Chromatography, *RID* Refractive Index Detector, *DNSA* 3,5-dinitrosalicylic acid, *UV* Ultraviolet, *SEM* Scanning Electron Microscopy, *TEM* Transmission Electron Microscopy, *HS-SPME* Headspace Solid Phase Microextraction, *UPLC-TQ-S-MS* Ultra Performance Liquid Chromatography Tandem Quadrupole Mass Spectrometry, *UPLC-Q-ToF-MS/MS* UPLC with quadrupole time-of-flight mass spectrometry, *GC-MS/MS-O* Gas chromatograph coupled with mass spectrometer in tandem and an olfactory detection port, *MDGC* Multidimensional Gas Chromatography, *HPLC-DAD* HPLC with Diode Array Detection.

13.4 Fermentation Optimisation and Scale Up

The biotechnological process of fermentation utilises whole cells or enzymes used as the biocatalyst. Various bioprocessing methods are commonly employed in industrial settings to establish a feasible means of producing innovative and appealing sensory characteristics for flavour and fragrance production. Biochemical engineering deals with the development of vessels and equipment capable of facilitating these biochemical reactions or conversions (Smith 2009; Vandamme and Soetaert 2002).

By providing the culture with a precursor for the desired chemical, bioconversion can enhance de novo fermentation processes (Vandamme and Soetaert 2002). However, the producer strains could be inhibited by some of these precursor molecules, such as fatty acids (ricinoleic acid, in particular). Substrate inhibition can be overcome by using the traditional fed-batch method in submerged fermentation (SmF). Slow feeding allows for continuous substrate conversion, while also maintaining the substrate concentration below critical threshold values. Some common fermentation strategies utilised in flavour and fragrance production apart from the traditional batch and fed-batch type submerged fermentation operations are discussed below.

13.4.1 Solid-State Fermentation (SSF)

SSF and SmF are the two primary forms of fermentation. SmF is carried out in liquid media in which the microbes are immersed, whereas solid-state fermentation occurs on a solid substrate with little to no free-flowing liquid. In SSF, microbes are cultivated either on moist solid supports, or on insoluble substrates that double as carbon and energy sources, or on inert carriers. SSF produces improved product qualities and generates higher yields than submerged fermentation (Mitchell et al. 2000).

SSF is relatively cheaper as the cost of operating is significantly lower as compared to SmF. It has been used in several investigations to synthesise aroma compounds by a variety of microbes on different solid substrates. Some examples of microorganisms grown using SSF include *Neurospora* sp. and *Aspergillus* sp. (Longo and Sanromán 2005). *Ceratocystis fimbriata* produces a distinct fruity aroma by utilising a range of agro-horticultural residues such as cassava bagasse as a solid substrate for fermentation. *C. fimbriata* releases a strong pineapple aroma when grown on coffee husk substrate in a SSF setting (Soares et al. 2000).

13.4.2 In-Situ Product Removal (ISPR)

Many flavours and fragrances are known to be strongly inhibitory towards metabolic processes. The molecular sources of the inhibitory characteristics differ. Nonetheless, their solubility in cellular membranes provides a preliminary unrefined indicator of possible inhibitory effects. The partitioning coefficient can then be used to estimate this (Heipieper et al. 1994).

An alternative is to apply a method called in-situ product removal (ISPR) after selecting the most tolerant strain and continuously eliminating the toxic product. This can be achieved through adsorption on resins (e.g. Mirata et al. 2009; Wang et al. 2011), extraction in two-phase systems (e.g. Rito-Palomares et al. 2001; Etschmann and Schrader 2006) or membrane separation through pervaporation (Vane 2005).

Another advantage of this technique is that by minimising product degradation, the number of downstream processing steps is also in turn reduced (Stark and Stockar 2003). Depending on the chemical nature of the substrate and product, substrate release and product removal can be achieved using the same absorbent if both have similar properties. This is known as SFPR (substrate feeding product removal) which was demonstrated by Bormann et al. (2012) for producing linalool oxide from linalool with *Corynespora cassiicola* using different hydrophobic adsorbers (Carlquist et al. 2015).

Some commonly used fermentation strategies for production of different kinds of flavour and fragrance molecules reported in the scientific literature are listed in Table 13.2. Level of scale-up achieved and control parameters are indicated.

Table 13.2 Fermentation strategies for flavour and fragrance molecule biosynthesis

Compounds	Fermentation type	Feed/substrate	Scale-up	Controls	References
Esters	Fed-batch	Glucose	1.3 L	Temperature, pH, dissolved oxygen (DO), agitation, air flow	Tai et al. (2015)
	Batch	–	1 L	Temperature, pH control, DO, agitation, air flow	Löser et al. (2015)
Terpenoids	Fed-batch	Glucose, K_2HPO_4, $MgSO_4$ $7H_2O$	30 L	Temperature, pH control, DO, agitation, air flow	Yang et al. (2016)
	Fed-batch	Fresh media	2.5 L	Temperature, pH control, DO, agitation, air flow	Scalcinati et al. (2012a, b)
	Batch and fed-batch phases	Glucose (batch), glucose and methanol (fed-batch phase)	Parallel bioreactor system (1.3–2.1 L)	Temperature, pH control, DO, agitation, air flow	Wriessnegger et al. (2014)
	Fed-batch	Methanol	2.4 L	Temperature, pH control, DO, agitation, air flow, methanol concentration	Sonntag et al. (2015)
Lactones	Solid substrate fermentation	Sugarcane bagasse	1.5–6 g	Temperature	Fadel et al. (2014)
	Batch and fed-batch (constant and intermittent feeding)	Biotransformation medium, methyl ricinoleate	3 L	Temperature, pH, DO	Gomes et al. (2012)
Aldehydes	Batch	–	2 L	Temperature	Fleige et al. (2013)
	Batch	–	3 L	Temperature, pH, DO, agitation	Di Gioia et al. (2011)
	Fed-batch	Eugenol, glucose, sunflower oil	3 L	Temperature, pH, DO, agitation, air flow	Lambert et al. (2013)

(continued)

Table 13.2 (continued)

Compounds	Fermentation type	Feed/substrate	Scale-up	Controls	References
Ketones	Batch	–	300 mL	Temperature, aeration, pH	Zhang et al. (2015)
	Batch and fed-batch	Fresh media	1.3 L	Temperature, pH, DO, agitation	Nies et al. (2020)

13.5 Conclusion

This chapter presents an overview of the biosynthesis of different types of flavours and aroma compounds using microbial chassis for industrial production. There is an emphasis on the potential of microbial cell factories as a cost-effective and sustainable approach, as compared to chemical synthesis methods. It discusses the recent research progress in the biotechnological production of such molecules through the use of engineering-based interventions. The choice of microbial hosts and engineering strategies for optimising production of these molecules as detailed in the present chapter can form a basis for further work in these areas. Analytical methods used for characterisation of different compound classes have been listed along with commonly used fermentation strategies across the compound classes. This chapter provides insights into the recent advances and highlights areas where biotechnological interventions can find potential applications. Further efforts can be directed towards developing innovative engineering interventions and scale-up strategies for industrial-scale production.

References

Akacha NB, Gargouri M (2015) Microbial and enzymatic technologies used for the production of natural aroma compounds: synthesis, recovery modelling, and bioprocesses. Food Bioprod Process 94:675–706

Allen LAT, Raclea R-C, Natho P, Parsons P (2021) Recent advances in the synthesis of α-amino ketones. Org Biomol Chem 19(3):498–513. https://doi.org/10.1039/d0ob02098b

Birk F, Fraatz MA, Esch P, Heiles S, Pelzer R, Zorn H (2019) Industrial riboflavin fermentation broths represent a diverse source of natural saturated and unsaturated lactones. J Agric Food Chem 67(49):13460–13469. https://doi.org/10.1021/acs.jafc.9b01154

Bormann S, Etschmann MMW, Mirata MA, Schrader J (2012) Integrated bioprocess for the stereospecific production of linalool oxides from linalool with *Corynespora cassiicola* DSM 62475. J Ind Microbiol Biotechnol 39:1761–1769

Carlquist M, Gibson B, Karagul Yuceer Y, Paraskevopoulou A, Sandell M, Angelov AI, Gotcheva V, Angelov AD, Etschmann M, de Billerbeck GM, Lidén G (2015) Process engineering for bioflavour production with metabolically active yeasts–a mini-review. Yeast 32(1):123–143. https://doi.org/10.1002/yea.3058. Epub 2014 Dec 16. PMID: 25400136

Chu LL (2022) CRISPR-Cas system in microbial hosts for terpenoid production. Crit Rev Biotechnol 42(7):1116–1133. https://doi.org/10.1080/07388551.2021.1995318. Epub 2022 Feb 9. PMID: 35139706

Di Gioia D, Luziatelli F, Negroni A, Ficca AG, Fava F, Ruzzi M (2011) Metabolic engineering of *Pseudomonas fluorescens* for the production of vanillin from ferulic acid. J Biotechnol 156(4):309–316. https://doi.org/10.1016/j.jbiotec.2011.08.014

Dusséaux S, Wajn WT, Liu Y, Ignea C, Kampranis SC (2020) Transforming yeast peroxisomes into microfactories for the efficient production of high-value isoprenoids. Proc Natl Acad Sci USA 117(50):31789–31799. https://doi.org/10.1073/pnas.2013968117

Etschmann MM, Schrader J (2006) An aqueous-organic two-phase bioprocess for efficient production of the natural aroma chemicals 2-phenylethanol and 2-phenylethyl acetate with yeast. Appl Microbiol Biotechnol 71:440–443

Fadel HHM, Mahmoud MG, Asker MMS, Lotfy SN (2014) Characterization and evaluation of coconut aroma produced by *Trichoderma viride* EMCC-107 in solid state fermentation on sugarcane bagasse. Electron J Biotechnol 18(1):5–9. https://doi.org/10.1016/j.ejbt.2014.10.006

Fadel HHM, Lotfy SN, Asker MMS, Mahmoud MG, Al-Okbi SY (2018) Nutty-like flavour production by *Corynebacterium glutamicum* 1220T from enzymatic soybean hydrolysate. Effect of encapsulation and storage on the nutty flavouring quality. J Adv Res 10:31–38. https://doi.org/10.1016/j.jare.2018.01.003

Flavours And Fragrances Market Size, Share & Trends Analysis Report By Product (Aroma Chemicals, Natural), By Application (Flavors, Fragrances), By Region, And Segment Forecasts, 2024–2030

Fleige C, Hansen G, Kroll J, Steinbüchel A (2013) Investigation of the *Amycolatopsis* sp. strain ATCC 39116 vanillin dehydrogenase and its impact on the biotechnical production of vanillin. Appl Environ Microbiol 79(1):81–90. https://doi.org/10.1128/AEM.02358-12

Gallage NJ, Møller BL (2015) Vanillin-bioconversion and bioengineering of the most popular plant flavour and its de novo biosynthesis in the vanilla orchid. Mol Plant 8(1):40–57. https://doi.org/10.1016/j.molp.2014.11.008

Gomes N, Teixeira JA, Belo I (2012) Fed-batch versus batch cultures of *Yarrowia lipolytica* for γ-decalactone production from methyl ricinoleate. Biotechnol Lett 34(4):649–654. https://doi.org/10.1007/s10529-011-0824-0

Gomes N, Braga A, Teixeira JA, Belo I (2013) Impact of lipase-mediated hydrolysis of Castor oil on γ-decalactone production by *Yarrowia lipolytica*. J Am Oil Chem Soc 90:1131–1137. https://doi.org/10.1007/s11746-013-2231-2

Gupta C, Prakash D, Gupta S (2015) A biotechnological approach to microbial based perfumes and flavours. J Microbiol Exp 2(1):11–18. https://doi.org/10.15406/JMEN.2015.02.00034

Heipieper HJ, Weber FJ, Sikkema FJ et al (1994) Mechanisms of resistance of whole cells to toxic organic solvents. Trends Biotechnol 12:409–415

Kim JH, Wang C, Jang HJ et al (2016) Isoprene production by *Escherichia coli* through the exogenous mevalonate pathway with reduced formation of fermentation byproducts. Microb Cell Factories 15:214. https://doi.org/10.1186/s12934-016-0612-6

Kirby J, Nishimoto M, Chow RWN, Baidoo EEK, Wang G, Martin J, Schackwitz W, Chan R, Fortman JL, Keasling JD (2015) Enhancing terpene yield from sugars via novel routes to 1-deoxy-D-xylulose 5-phosphate. Appl Environ Microbiol 81:130–138. https://doi.org/10.1128/AEM.02920-14

Kruis AJ, Bohnenkamp AC, Patinios C, van Nuland YM, Levisson M, Mars AE, van den Berg C, Kengen SWM, Weusthuis RA (2019) Microbial production of short and medium chain esters: enzymes, pathways, and applications. Biotechnol Adv 37(7):107407. https://doi.org/10.1016/j.biotechadv.2019.06.006

Lambert F, Zucca J, Ness F, Aigle M (2013) Production of ferulic acid and coniferyl alcohol by conversion of eugenol using a recombinant strain of *Saccharomyces cerevisiae*. Flavour Fragr J 29(1):14–21. https://doi.org/10.1002/ffj.3173

Lindberg P, Park S, Melis A (2010) Engineering a platform for photosynthetic isoprene production in cyanobacteria, using *Synechocystis* as the model organism. Metab Eng 12(1):70–79. https://doi.org/10.1016/j.ymben.2009.10.001

Lindsay M, Granucci N, Greenwood D, Villas-Boas S (2022) Identification of new natural sources of flavour and aroma metabolites from solid-state fermentation of agro-industrial by-products. Meta 12:157. https://doi.org/10.3390/metabo12020157

Longo M, Sanromán M (2005) Production of food aroma compounds: microbial and enzymatic methodologies. Food Technol Biotechnol 44(3):335–353

Löser C, Urit T, Bley T (2014) Perspectives for the biotechnological production of ethyl acetate by yeasts. Appl Microbiol Biotechnol 98(12):5397–5415. https://doi.org/10.1007/s00253-014-5765-9

Löser C, Urit T, Keil P, Bley T (2015) Studies on the mechanism of synthesis of ethyl acetate in *Kluyveromyces marxianus* DSM 5422. Appl Microbiol Biotechnol 99(3):1131–1144. https://doi.org/10.1007/s00253-014-6098-4

Masuo S, Tsuda Y, Namai T, Minakawa H, Shigemoto R, Takaya N (2020) Enzymatic Cascade in *Pseudomonas* that produces pyrazine from α-amino acids. Chembiochem 21:353. https://doi.org/10.1002/cbic.201900448

Mirata MA, Heerd D, Schrader J (2009) Integrated bioprocess for the oxidation of limonene to perillic acid with *Pseudomonas putida* DSM 12264. Process Biochem 44:764–771

Mitchell DA, Berovic M, Krieger N (2000) Biochemical engineering aspects of solid state bioprocessing. Adv Biochem Eng Biotechnol 68:61–138. https://doi.org/10.1007/3-540-45564-7_3

Molina G, Pimentel MR, Pastore GM (2013) *Pseudomonas*: a promising biocatalyst for the bioconversion of terpenes. Appl Microbiol Biotechnol 97(5):1851–1864. https://doi.org/10.1007/s00253-013-4701-8

Molina G, Bution ML, Bicas JL, Dolder MAH, Pastore GM (2014) Comparative study of the bioconversion process using R-(+)- and S-(−)-limonene as substrates for *fusarium oxysporum* 152B. Food Chem 174:606–613. https://doi.org/10.1016/j.foodchem.2014.11.059

Moore SJ, Tosi T, Bell D, Hleba YB, Polizzi KM, Freemont PS (2021) High-yield 'one-pot' biosynthesis of raspberry ketone, a high-value fine chemical. Synth Biol (Oxf) 6(1):ysab021. https://doi.org/10.1093/synbio/ysab021

Mortzfeld F, Hashem C, Vrankova K, Winkler M, Rudroff F (2020) Pyrazines: synthesis and industrial application of these valuable flavor and fragrance compounds. Biotechnol J 15(11):2000064. https://doi.org/10.1002/biot.202000064

Nies SC, Alter TB, Nölting S, Thiery S, Phan ANT, Drummen N, Keasling JD, Blank LM, Ebert BE (2020) High titer methyl ketone production with tailored *Pseudomonas taiwanensis* VLB120. Metab Eng 62:84–94. https://doi.org/10.1016/j.ymben.2020.08.003

Priya DK, Petkar M, Chowdary GV (2015) Bio production of aroma compounds from alpha pinene by novel strains. Int J Biol Sci Appl 2(2):15–19

Priya DK, Petkar M, Chowdary GV (2023) Microbial biotransformation of A -(+) -pinene by newly identified strain of *Gluconobacter japonicus* MTCC 12284. Int J Dev Res 05(08):5270–5275

Rito-Palomares M, Negrete A, Miranda L, Celia Flores C, Galindo E, Serrano-Carreón L (2001) The potential application of aqueous two-phase systems for in situ recovery of 6-pentyl-a-pyrone produced by *Trichoderma harzianum*. Enzym Microb Technol 28:625–631

Rodriguez GM, Tashiro Y, Atsumi S (2014) Expanding ester biosynthesis in *Escherichia coli*. Nat Chem Biol 10(4):259–265. https://doi.org/10.1038/nchembio.1476

Rong S, Yang S, Li Q, Cai B, Guan S, Wang J, Zhou Y, Chen Y (2017) Improvement of γ-decalactone production by stimulating the import of ricinoleic acid and suppressing the degradation of γ-decalactone in *Saccharomyces cerevisiae*. Biocatal Biotransformation 35(2):96–102. https://doi.org/10.1080/10242422.2017.1289182

Ryu JY, Seo J, Ahn JH, Sadowsky MJ, Hur HG (2012) Transcriptional control of the isoeugenol monooxygenase of *Pseudomonas nitroreducens* Jin1 in *Escherichia coli*. Biosci Biotechnol Biochem 76(10):1891–1896. https://doi.org/10.1271/bbb.120375

Scalcinati G, Knuf C, Partow S, Chen Y, Maury J, Schalk M, Daviet L, Nielsen J, Siewers V (2012a) Dynamic control of gene expression in *Saccharomyces cerevisiae* engineered for the production of plant sesquiterpene α-santalene in a fed-batch mode. Metab Eng 14(2):91–103. https://doi.org/10.1016/j.ymben.2012.01.007

Scalcinati G, Partow S, Siewers V et al (2012b) Combined metabolic engineering of precursor and co-factor supply to increase α-santalene production by *Saccharomyces cerevisiae*. Microb Cell Factories 11:117. https://doi.org/10.1186/1475-2859-11-117

Schempp FM, Drummond L, Buchhaupt M, Schrader J (2018) Microbial cell factories for the production of Terpenoid flavor and fragrance compounds. J Agric Food Chem 66(10):2247–2258. https://doi.org/10.1021/acs.jafc.7b00473

Silva R, Aguiar TQ, Coelho E, Jiménez A, Revuelta JL, Domingues L (2019) Metabolic engineering of *Ashbya gossypii* for deciphering the de novo biosynthesis of γ-lactones. Microb Cell Factories 18(1):62. https://doi.org/10.1186/s12934-019-1113-1

Silva R, Coelho E, Aguiar TQ, Domingues L (2021) Microbial biosynthesis of lactones: gaps and opportunities towards sustainable production. Appl Sci 11:8500. https://doi.org/10.3390/app11188500

Smith JE (2009) Bioprocess/fermentation technology. In: Biotechnology. Cambridge University Press, Cambridge, pp 49–72

Soares M, Christen P, Pandey A, Soccol CR (2000) Fruity flavour production by *Ceratocystis fimbriata* grown on coffee husk in solid-state fermentation. Process Biochem 35:857–861

Soares GPA, Souza KST, Vilela LF, Schwan RF, Dias DR (2017) γ-Decalactone production by *Yarrowia lipolytica* and *Lindnera saturnus* in crude glycerol. Prep Biochem Biotechnol 47(6):633–637. https://doi.org/10.1080/10826068.2017.1286601

Sonntag F, Kroner C, Lubuta P, Peyraud R, Horst A, Buchhaupt M, Schrader J (2015) Engineering *Methylobacterium extorquens* for de novo synthesis of the sesquiterpenoid α-humulene from methanol. Metab Eng 32:82–94. https://doi.org/10.1016/j.ymben.2015.09.004

Stark U, Stockar V (2003) In situ product removal (ISPR) in whole cell biotechnology during the last 20 years. In: Process integration in biochemical engineering. Springer-verlag, Berlin

Tai YS, Xiong M, Zhang K (2015) Engineered biosynthesis of medium-chain esters in *Escherichia coli*. Metab Eng 27:20–28. https://doi.org/10.1016/j.ymben.2014.10.004

Trytek M, Jędrzejewski K, Fiedurek J (2015) Bioconversion of α-pinene by a novel cold-adapted fungus *Chrysosporium pannorum*. J Ind Microbiol Biotechnol 42(2):181–188. https://doi.org/10.1007/s10295-014-1550-0

Vandamme E, Soetaert W (2002) Bioflavours and fragrances via fermentation and biocatalysis. J Chem Technol Biotechnol 77:1323–1332. https://doi.org/10.1002/jctb.722

Vane LM (2005) A review of pervaporation for product recovery from biomass fermentation processes. J Chem Technol Biotechnol 80:603–629

Wang H, Dong Q, Meng C et al (2011) A continuous and adsorptive bioprocess for efficient production of the natural aroma chemical 2-phenylethanol with yeast. Enzym Microb Technol 48:404–407

Wang C, Liwei M, Park J-B, Jeong S-H, Wei G, Wang Y, Kim S-W (2018) Microbial platform for Terpenoid production: *Escherichia coli* and yeast. Front Microbiol 9:2460. https://doi.org/10.3389/fmicb.2018.02460

Wang C, Zheng P, Chen P (2019) Construction of synthetic pathways for raspberry ketone production in engineered *Escherichia coli*. Appl Microbiol Biotechnol 103(9):3715–3725. https://doi.org/10.1007/s00253-019-09748-5

Wriessnegger T, Augustin P, Engleder M, Leitner E, Müller M, Kaluzna I, Schürmann M, Mink D, Zellnig G, Schwab H, Pichler H (2014) Production of the sesquiterpenoid (+)-nootkatone by metabolic engineering of *Pichia pastoris*. Metab Eng 24:18–29. https://doi.org/10.1016/j.ymben.2014.04.001

Xue J, Ahring BK (2011) Enhancing isoprene production by genetic modification of the 1-deoxy-d-xylulose-5-phosphate pathway in *Bacillus subtilis*. Appl Environ Microbiol 77(7):2399–2405. https://doi.org/10.1128/AEM.02341-10

Yang W, Tang H, Ni J, Wu Q, Hua D, Tao F, Xu P (2013) Characterization of two *Streptomyces* enzymes that convert ferulic acid to vanillin. PLoS One 8(6):e67339. https://doi.org/10.1371/journal.pone.0067339

Yang C, Gao X, Jiang Y, Sun B, Gao F, Yang S (2016) Synergy between methylerythritol phosphate pathway and mevalonate pathway for isoprene production in *Escherichia coli*. Metab Eng 37:79–91. https://doi.org/10.1016/j.ymben.2016.05.003

Zhang L, Zhang Y, Liu Q et al (2015) Production of diacetyl by metabolically engineered *Enterobacter cloacae*. Sci Rep 5:9033. https://doi.org/10.1038/srep09033

Zhang L, Cao Y, Tong J, Xu Y (2019) An alkylpyrazine synthesis mechanism involving l-threonine-3-dehydrogenase describes the production of 2,5-dimethylpyrazine and 2,3,5-trimethylpyrazine by *Bacillus subtilis*. Appl Environ Microbiol 85(24):e01807–e01819. https://doi.org/10.1128/AEM.01807-19

Zhang H, Zhang L, Yu X, Xu Y (2020) The biosynthesis mechanism involving 2,3-pentanedione and aminoacetone describes the production of 2-ethyl-3,5-dimethylpyrazine and 2-ethyl-3,6-dimethylpyrazine by *Bacillus subtilis*. J Agric Food Chem 68:3558–3567. https://doi.org/10.1021/acs.jafc.9b07809

Zhao Y, Yang J, Qin B, Li Y, Sun Y, Su S, Xian M (2011) Biosynthesis of isoprene in *Escherichia coli* via methylerythritol phosphate (MEP) pathway. Appl Microbiol Biotechnol 90(6):1915–1922. https://doi.org/10.1007/s00253-011-3199-1

Bioprocessing and Purification of Metabolites

14

Shabab Hussain, Samra Zahra, Wajahat Ali, Maria Irshad, Kulsoom, Syed Damin Abbas Hamdani, and Mustafeez Mujtaba Babar

Abbreviations

GMP	good manufacturing practices
IEC	ion exchange chromatography
IRS	infrared spectroscopy
LLE	liquid–liquid extraction
MAE	microwave-assisted extraction
MS	mass spectrometry
NMR	nuclear magnetic resonance
PLE	pressurized liquid extraction
PSMs	plant secondary metabolites
SSF	solid-state fermentation
SmF	submerged fermentation

S. Hussain
Department of Chemical, Biological, Pharmaceutical and Environmental Sciences, University of Messina, Messina, Italy

S. Zahra
Department of Biosciences, Comsats University, Islamabad, Pakistan

W. Ali
Institute of Basic Medicine, Xian Jiaotong University, Xi'an, China

M. Irshad
Department of Medical, Surgical and Health Sciences, University of Trieste, Trieste, Italy

Kulsoom
Department of Biochemistry, Bahauddin Zakariya University, Multan, Pakistan

S. D. A. Hamdani · M. M. Babar (✉)
Shifa College of Pharmaceutical Sciences, Shifa Tameer-e-Millat University, Islamabad, Pakistan
e-mail: mustafeez.babar@fulbrightmail.org

V. Singh (ed.), *Advances in Metabolomics*,
https://doi.org/10.1007/978-981-97-7459-3_14

SMs secondary metabolites
SPE solid-phase extraction
UAE ultrasound-assisted extraction

14.1 Introduction

Primary and secondary metabolites are organic substances obtained from plants. Plant growth and development depends on primary metabolites, which provide energy and synthesize nutrients. Main metabolites include amino acids, carbohydrates, fatty acids, and nucleotides (Erb and Kliebenstein 2020). Secondary metabolites (SMs) protect plants from herbivores and pathogens and mediate plant-environment interactions. Specific environmental signals cause these defense chemicals to be generated. Alkaloids, flavonoids, terpenoids, and phenols are among the SMs. Recent studies have also demonstrated that SMs may perform core metabolic tasks and can affect plant growth and development (Erb and Kliebenstein 2020).

Several procedures are needed to isolate, obtain, and purify metabolites from natural sources. The initial stage is extracting metabolites from the biological matrix. Liquid–liquid extraction (LLE), supercritical fluid extraction (SFE), and ultrasound-assisted extraction (UAE) are some of the options (Al Daccache et al. 2022; Corrêa et al. 2020; Volova et al. 2021). Metabolites must frequently be isolated from the complicated extract combination. This is usually performed by thin-layer chromatography, gravitational column chromatography (GCC), or high-performance liquid chromatography (HPLC) (Al Daccache et al. 2022; Volova et al. 2021). These methods aid in isolation of substances using physical or chemical separations. After separation, pure metabolites may be studied and their structure elucidation can be done by, GC-MS-MS, infrared nuclear magnetic resonance, LC-MS, or UV-visible spectroscopy (Al Daccache et al. 2022; Corrêa et al. 2020; Schafhauser and Kulik 2023). Once isolated, these compounds are bioprocessed for use in pharmaceutical, agricultural, and food sectors (Sinha et al. 2022).

Bioprocessing produces SMs via genetic engineering, metabolic engineering, and fermentation. Genetic engineering enhances metabolite synthesis by altering organisms' genes. Contrarily, metabolic engineering manipulates metabolic processes to boost metabolite output while fermentation in bioreactors is a typical method for large-scale SM synthesis (Sinha et al. 2022). Several studies have reported the increase in SM synthesis by these techniques. For instance, *Agrobacterium rhizogenes*, a soil bacterium that causes hairy root disease, has been widely studied for use in enhancing SM production (Shahnawaz 2022). Other factors like scalability and process economics also need to be considered while developing bioprocessing methods for (Sinha et al. 2022). Genetic engineering and omics technologies have increased SM synthesis in medicinal plants like *Ocimum sanctum* (Kumari et al. 2022). Metabolic engineering can improve SM synthesis in bacterial, archaeal, and eukaryotic cells (Narasimhulu 2022). SM production using non-model microbial machinery and synthetic biology methods is also promising (Liu et al.

2022). By applying metabolic engineering approaches, it has been possible to modify the routes through which SMs are synthesized from, thereby, eliminating unwanted chemicals or by-products. This method has been used to create industrial goods, including pharmaceutical medications and environmentally friendly agrochemicals in various species, including fungi, bacteria, and plants (Ul Haq et al. 2023; Ram and Misra 2021).

Chemical extraction and purification from different sources are essential in metabolomics and analytical chemistry research. Various extraction techniques can separate target chemicals from complex cellular matrices, including liquid–liquid extraction, solid-phase extraction (SPE) and solvent extraction. These techniques have benefits and drawbacks based on the types of substances and matrices under study (Wang et al. 2014). Among these, solvent extraction is the technique of isolating target compounds by using organic solvents. This method is often used due to its versatility and simplicity of usage. However, solvent extraction could be very sensitive to matrix effects and have little selectivity. Therefore, solid sorbents are used in SPE to remove interfering molecules while retaining target chemicals selectively. It has been shown that SPE outperforms liquid–liquid extraction (LLE) in terms of recovery for various metabolites (Giardi et al. 2010).

Once isolated, analytical techniques are used for the qualitative and quantitative examination of extracted chemicals. Spectroscopic techniques like IR, NMR, MS, and UV-Vis spectroscopy may provide essential details about the compounds' functional groups and chemical structures. Compounds may be identified and quantified, for instance, by using mass-to-charge ratios when mass spectrometry (MS) is utilized in conjunction with chromatographic methods like gas chromatography-mass spectrometry (GC-MS) and liquid chromatography-mass spectrometry (LC-MS) (Behrens et al. 2016).

Apart from the scientific and commercial processes, it is imperative for the metabolite development to address the ethical and regulatory issues. Large-scale processing requires adherence to Good Manufacturing Practices (GMP) so as to ensure the availability of safe, effective and high-quality products. GMP compliance, therefore, improves patient safety, regulatory compliance, and innovation (Gouveia et al. 2015). This chapter introduces the metabolites of commercial importance followed by a detailed account of ways and means of extraction and purification of metabolites. Toward the end, the ethical and regulatory considerations related to the metabolic engineering and bioprocessing have been provided.

14.2 Primary and Secondary Metabolites

Bacteria, fungi, and plants biosynthesize molecules called metabolites that ensure their survival. While SMs are not involved in growth, despite primary metabolites being essential in metabolic processes directly and are produced for addressing stress conditions (Srivastava et al. 2020; Yeshi et al. 2022). The intermediate or final byproduct of higher plants' central metabolism, plant secondary metabolites (PSMs), are created for defense (Yeshi et al. 2022). Many biological and pharmacological properties of plants are attributed to these chemical moieties (Attia-Ismail 2022). Absinthin, ascorbic acid, carotene, glycosides, tannins, organic acids, and essential oils are a few examples of PSMs (Attia-Ismail 2022).

To increase the synthesis of SMs for use in vast area, metabolic engineering has emerged as a potentially helpful avenue (Narasimhulu 2022). Metabolic routes, including the, methylerythritol phosphate pathway, shikimic acid pathway, mevalonic acid pathway, and tricarboxylic acid cycle, are the source of SMs (Du and Ryan 2019). The metabolites produced, ultimately, serve various biological purposes and are linked to survival under stressful environmental conditions, adaptive behavior, and defense. In addition, SMs have been used as flavoring agents, food additives, medicinal substances, antioxidants, and cosmetics. Natural products (NPs), often referred to as secondary metabolites (SMs) from biological systems, are extensively used as medications and nutraceuticals to promote human life and health (Rami et al. 2021). Similarly, siderophores are specialized SMs that are mainly obtained from plants and microbes and are used in chelation and cellular communication (Zhao et al. 2020).

Plants serve as the primary and predominant reservoir of both primary and secondary metabolites. The primary metabolites include various vital biomolecules, including carbohydrates, proteins, amino acids, nucleotides, adenosine triphosphate (ATP), and fatty acids. SMs, contrarily, contain alkaloids compounds, flavonoids phytochemicals, terpenoids with highly diverse structure, and phenolic compounds (Bhatla 2018; Kandar 2021; Vijayakumar and Raja 2018). Similarly, bacteria that produce primary and secondary metabolites include *Streptomyces* species, *Escherichia coli, Lactobacillus* species, *Clostridium* species, *Mycobacterium tuberculosis, Cyanobacteria, Pseudomonas aeruginosa, Staphylococcus aureus, Methanogenic archaea,* and *Acetobacter aceti* (Bhatla 2018; Vijayakumar and Raja 2018; Mohan et al. 2022). Conversely, fungi including *Fusarium* species, *Psilocybe* species, *Ganoderma lucidum, Claviceps purpurea, Amanita muscaria, Candida* species, *Penicillium* species, *Clavulinopsis* species, and *Aspergillus* species have been indicated to produce metabolites of commercial importance (Bhatla 2018; Vijayakumar and Raja 2018; Bills and Gloer 2016; Nigam et al. 2014). Likewise, bombardier beetle (genus *Brachinus*), ants, honeybees (genus *Apis*), stink bugs, ladybugs species, and *Bombyx mori* (silkworms) are the types of insects that produce primary and secondary metabolites (Bhatla 2018; Vijayakumar and Raja 2018; Morgan 2010). *Discodermia* species, *Alexandrium* species, *Pseudopterogorgia* species, sea slugs, red algae, marine bacteria, marine fungi, horseshoe crabs, and jellyfish are the types of marine organisms that produce primary and secondary metabolites (Bhatla 2018; Vijayakumar and Raja 2018; Li et al. 2019; Singh 2020).

14.3 Metabolite Production and Metabolic Engineering

The basic metabolic processes responsible for producing primary metabolites are, generally, common among various species. SMs derived out by various metabolic pathways that branch out from the main primary metabolites. There are several enzymatic and non-enzymatic stages in these processes (Achimón et al. 2021). Precursors from primary metabolism are the metabolic sources for SMs. After a number of evolutionary developments across many plant lineages have led to the emergence of the pathways needed to synthesize specialized plant metabolites (Kashyap et al. 2022). The process of altering an organism's metabolic pathways to obtain desired substances is covered in the field of metabolic engineering (Ram and Misra 2021; Yadav et al. 2019). The process can help in modifying both primary and secondary metabolism that can then be employed for a number of biotechnological, medicinal, environmental and agricultural purposes (Ul Haq et al. 2023; Yadav et al. 2019).

Metabolic engineering processes are initiated by identifying the genes and enzymes involved in the target compound's biosynthesis which is followed by their genetic manipulation to enhance their production. This can be achieved by the introduction of heterologous genes from other species, the overexpression of essential enzymes, or the elimination of rival pathways (Nagegowda and Gupta 2020). Alternatively, creating novel metabolic pathways for synthesizing SMs via applying synthetic biology methods can also be done. This might include using directed evolution to optimize existing routes for higher output or assembling genes and enzymes from several species to build a unique biosynthetic pathway. Huge amount of SMs has been successfully produced using synthetic biology and metabolic engineering techniques. For instance, a technique called yeast metabolic engineering methods have been used to increase the synthesis of artemisinin, an antimalarial agent. Similarly, synthetic biology techniques enhanced the synthesis of paclitaxel, a popular anticancer drug, in *E. coli* (Erb et al. 2017). Other microorganisms have been metabolically engineered to produce natural compounds as well. For the purpose, established and rewired metabolic networks, optimized production hosts, and fermentation optimization have been used (El-Desoky et al. 2022). Metabolic engineering and modern biotechnology may, hence, solve natural product production problems on a large scale (Park et al. 2018). Whichever the avenue selected for metabolic engineering, a few steps are common among all the methods (Ram and Misra 2021; Yadav et al. 2019). Figure 14.1 summarizes the general process involved in the metabolic engineering process. The first stage is to identify the precise SM of interest the yield of which has to be improved qualitatively and quantitatively (Yadav et al. 2019; Chen et al. 2016). Thereafter, an appropriate model host organism that can generate the target metabolite naturally or via genetic modification has to be selected. Genetic manipulation is then carried out to change the host organism's metabolic pathways. This may include suppressing the activity of enzymes in alternate pathways or overexpressing the desired genes that specifically code for the enzymes needed for the target metabolite's manufacture (Chen et al. 2016; Zebec and Scrutton 2018). The production of SMs starts with the availability of enough amount of precursors, and hence, this provision serves as an essential step of the process (Ram and Misra 2021). After streamlining these intracellular components, the host organism's growth parameters, including temperature, pH, and nutrient availability are optimized within the culture conditions (Ram and Misra 2021; Yadav et al. 2019; Chen et al. 2016; Zebec and Scrutton 2018). After synthesis, SMs

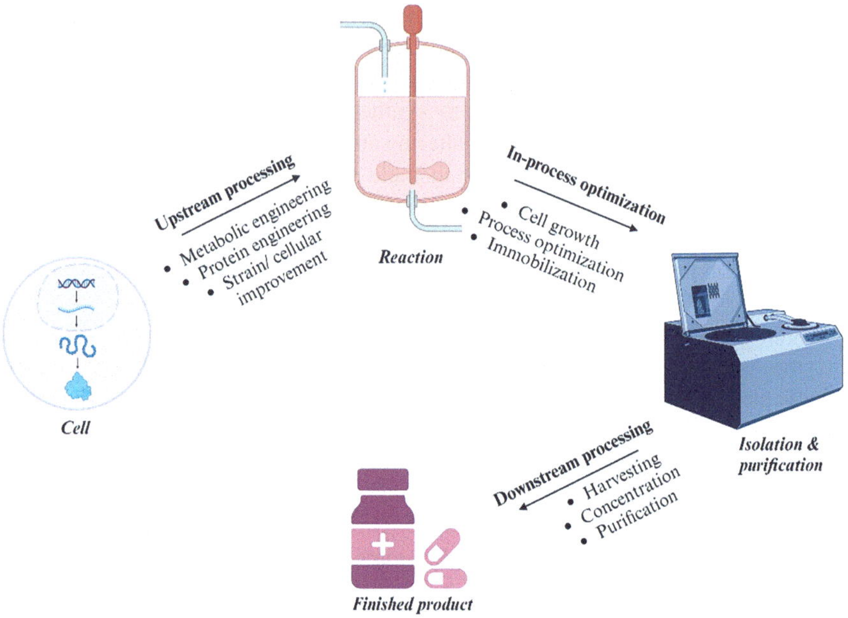

Fig. 14.1 General workflow of metabolic engineering for industrial applications

need to be collected and purified from the host organism through various physical and chemical methods that include but are not limited to crystallization, chromatography, and solvent extraction.

14.4 Bioprocessing of Metabolites

Once the metabolite and metabolic pathway of interest has been identified, planning for the downscale processing is to be done. Metabolite production relies on the cellular growth, and solid-state fermentation (SSF) proves to be more effective in achieving this. Microorganisms, for instance, exhibit gene regulation mechanisms influenced by evolutionary inheritance, while the manufacturing of SMs depends on indicators recognized in their immediate environment. In submerged fermentation processes, essential signals may vanish, hindering the complete utilization of resources in terms of microorganisms' production capabilities (López-Calleja et al. 2012). Microorganisms produce SMs) during the idiophase, a stage in the microbial evolution of fermentation where SMs start to accumulate. This is succeeded by the vital growth phase known as the trophophase. Notably, during the idiophase the formation of SMs occurs independently of the production of cellular material and natural growth. The onset of SM production in this phase is associated with stress conditions (Isah 2019). When the growth medium of a microbial culture experiences a depletion of essential nutrients and agents, such as carbon, nitrogen, or phosphate, commonly referred to as nutritional shift-down, it initiates the production of specialized metabolites. A notable example of SM manufacturing in a stressful environment is the synthesis of penicillin, the first antibiotic, derived from *Penicillium chrysogenum*. When there is little or no glucose left in the culture medium for the fungus for consumption, it prompts the manufacturing of penicillin by using lactose. Similarly, other microbes are also capable of synthesizing various SMs, as these particular chemicals are neither essential for their basic reproduction and development. The making of SMs is regulated by various growing conditions, including the formation of the medium, light, pH and temperature. Optimization of these growth conditions can enhance the manufacturing of SMs. The advancement of new techniques including genetic engineering as well as genomics has provided researchers with opportunities to explore innovative approaches for augmenting the manufacturing of SMs. Submerged fermentation (SmF) and SSF)represent two primary bioprocessing approaches engaged for the making of microbial metabolites. Both approaches applied for the cultivation of microbial SMs. SmF boasts advantages in terms of online automation and monitoring, making it a preferable choice for extensive industrial making of microbial products. Submerged fermentation, on the other hand, utilizes either continuous or fed-batch culture for the synthesis of SMs. The imbue is propagated through serial straining enhancement of the organism. The process generally begins with shake flask culture, and the actively growing cultures are subsequently moved to a tiny fermenter as well as to a greater fermenter containing the manufacture medium. Various constraints, including medium composition, temperature, agitation mechanisms, pH, as well as aeration rates, are carefully regulated. In cephalosporin fermentations, methionine is introduced as inducer, phosphate is limited in the chlortetracycline fermentation, as well as glucose is omitted in penicillin and erythromycin fermentation. However, SSF replicates a natural environment for microorganisms, as it establishes that the cells typically thrive in their natural habitats. SSF is a fermentation process uses a solid matrix with sufficient ability to provide a medium that conducive for the microbial growth. This solid material can serve as a source of nutrients or act as a secondary substrate pervaded by a comprehensive nutrient method necessary for the growth of microorganisms. SSF is a well-established and popular bioprocess for enzyme production, particularly in the degradation of biomass. Furthermore, it holds the promise as an excellent bioprocessing technique for the manufacturing of SMs from industrial residues and agro-waste. Large amount of residues, including bran, bagasse, seeds husks, peels as well as corn scum among others are generated once a year as waste that are often underutilized or discarded in landfills. Recently, there has been a notable focus on utilizing these materials as cost-effective and readily available renewable substrates for producing a variety of appreciated compounds. In this situation, certain substrates have been employed as solid supports in SSF bioprocesses to generate diverse series of microbial SMs (Kumar et al. 2021; Akula and Ravishankar 2011).

Conversely, co-cultivation entails the simultaneous cultivation of multiple microbial strains, leveraging their synergistic interactions to enhance SM production. This strategy facilitates the exchange of metabolites, signaling molecules, or co-factor regeneration among different strains, ultimately resulting in increased yields or the generation of novel metabolites. Much similar to the microbes, among plants the build-up of these metabolites commonly takes place when they are visible to stressors such as different elicitors and signaling molecules. SMs play an important part in plants' adaptation to their surroundings as well as control such stressful conditions. Environmental elements such as temperature, humidity, light intensity, water supply, mineral availability, and carbon dioxide levels impact on the production of SMs and plant growth. Drought, salinity, and extremes in temperatures are environmental challenges that negatively impact plant growth (Sathasivam et al. 2023). The capability to generate respective metabolites in regulated environments can aid in obtaining appropriate quantities of metabolites. These shall be impervious to rapidly changing external factors, optimization of culture conditions to enhance SM production while ensuring a balance between supply and demand, maintaining ample production as needed (Ozyigit et al. 2023). Plant growth regulators represent a key factor influencing the metabolite formation, differentiation and cell growth. The precise absorption of the medium stands out as a major contributor in governing metabolite production. SMs synthesis from the medicinal plants, it is imperative to begin optimal culture conditions encompassing both physical and chemical environments tailored to the specific plant types. Oxidative stress assumes a pivotal part in the manufacturing of SMs within plants. Phenolic compounds synthesized through the phenyl propanoid pathway, for instance, serve as SMs functioning as a defense mechanism against a range of abiotic as well as biotic stress conditions. Contact of plants to adverse conditions triggers the generation of ROS (reactive oxygen species) (Fazili et al. 2022). Tissue cultures and plant cell can be routinely recognized in a sterilized environment from various explants, such as stem, meristems, leaves, roots, meristems, and more. These cultures serve the purpose of both extraction and multiplication of SMs. Various culture methods, including, callus, root, hairy root cultures, shoot, callus and cell suspension, are employed to produce specific metabolites. Unorganized callus or suspension cultures are utilized for synthesizing metabolites that are distributed across multiple tissues. The manufacturing of the SMs in the cell cultures can be further be enhanced by subjecting plant cells to other biotic as well as abiotic elicitors. Commonly used elicitors include methyl jasmonate, fungal carbohydrates, and yeast extract. Interestingly, the quantity and quality of SMs produced through hairy root systems are comparable and superior to those synthesized by intact hostess plant roots (Thirumurugan et al. 2018).

14.5 Extraction Strategies for Attaining Secondary Metabolites

The objective of extraction is to isolate the metabolites from the crude mixture, whether of plant or microbial origin. Extraction process allows the separation of SMs by the usage of the different diluters of differential polarities. The very first step of SMs extraction from plant cultures, for instance, is the preparation of plant material which includes fruits, leaves, flowers, and barks, from a dry plant material. Different drying methods like microwave-drying, oven-drying, and air-drying are used to dry the plant material. Various forms of extraction techniques is used for the extraction of metabolites which includes maceration, percolation or continuous extraction, pressurized liquid extraction (PLE)), ultrasound-assisted extraction (UAE) and microwave-assisted extraction (MAE), a few of which have been elaborated in the following sections (Shen et al. 2023).

14.5.1 Maceration

Maceration is a well-known liquid–solid extraction method is use to extract the diluted solution of compounds. In this method the solvents of different polarities are used to extract compounds of a specific affinity. In this technique, the plant or microbial mixture is placed in a jar and covered with a lid for 2–3 days until they are transferred to the solvent phase which are then further purified (Garcia-Vaquero et al. 2020).

14.5.2 Percolation

This method is mostly used for the extraction of active ingredients for the preparation of fluid extracts and tinctures. It is performed in a device named percolator that aids in the extraction of metabolites such that the solid ingredients are dampened in a suitable quantity of solvent and are allowable to stand for variable time periods. Thereafter, the mass is filled up in the container and the percolator is sealed. Further amount of solvent is added over the mass and is allowed to percolate for approximately 24–48 h. Vent of the percolator is opened as well as the liquid confined inside is acceptable to dribble slowly. Further solvent is added as per requirement. Extra quantity of solvent is added to make the required volume and the mixture is purified by the filtration process (González Mera et al. 2019).

14.5.3 Pressurized Liquid Extraction (PLE)

In this technique, solvents or a combination of solvents are used at extraordinary temperature and pressures overhead their boiling points to quotation substances. Sample preparation is achieved by saponification or by the consumption of organic solvents such as acetonitrile, hexane, hexane-2-propanol, and methanol among others. The sample is ready by through extraction of metabolites in a continuous mode along with silicone membrane coupled connected with a chromatographic system. In this technique, extraction cell is filled up with the solid sample along with solvent at high temperature. It is specified that the pressure confidential the extraction cell must be retained relatively extraordinary for solvent to continue in the liquid state (Letaief et al. 2021; Delgado-Zamarreño et al. 2001). It sets off at elevated temperature (20–400 °C) as well as pressure (500–3000 psi) which leads to increased recoveries because of disturbance of surface equilibrium and mass transmission effects. The method is user friendly, time saving and sustainable due to low volume of solvents. In addition, since the whole extraction process is carried out in an inert atmosphere, so it is most relevant to light sensitive metabolites (Liang et al. 2020).

14.5.4 Soxhlet Extraction

It is also known as continuous hot extraction. The apparatus used in this method is made up of glass and is known as Soxhlet extractor that consists of different parts which include extraction chamber, bottom flask, precipitate and siphon tube. The dried powder is positioned inside the flask in a tight bag and abstraction solvent is added in the flask and subsequently heated. The thimble is injected into the extraction chamber. After being heated in the bottom flask, the solvent vaporizes, passes over the condenser, condenses and finally enters in the extraction chamber (Fagbemi et al. 2021). The drug and the solvent for extraction come into contact in this compartment. As the result, the extracted material from the plant along with the solvent flows back into the flask as soon as the extraction chamber's solvent level reaches the top of the siphon. The technique's primary benefit is its ability to extract a significant amount of drug from a lesser amount of solvent (Hossain et al. 2014; Bitwell et al. 2023; Vaou et al. 2021).

14.5.5 Microwave-Assisted Extraction

This is the more sophisticated extraction method used to prepare medicinal herbs. The method makes advantage of the dipole rotation and ionic transfer mechanism, which involves moving charged particles in the drug material and solvent. Dipole rotation and ion migration happens when polar solvent is utilized, which promotes solvent penetration and helps with the extraction process. Electromagnetic radiations with frequencies among 300 MHz as well as 300 GHz and the wavelengths among 1 cm and 1 m are applied in this process. The energy produced by microwaves practical at a frequency of 2450 Hz ranges from 600 to 700 W. This method bombards microwave radiation on the object to absorb the electromagnetic energy and convert into heat. The heat generated from this method makes easier for the solvent to enter the cellular matrix, thereby aiding the release of SMs from the source (Abubakar and Haque 2020).

14.5.6 Ultrasound-Assisted Technique

Using the sound energy at a regularity higher than 20 KHz, this process breaks the cells and increases the matrix's surface area so that solvents can penetrate. SMs resolve consequently be free. Plant material must first be allowed to dry, then ground into a fine powder and carefully sieved. The prepared sample is then placed inside the ultrasonic extractor along with an appropriate extraction solvent. By lowering the need for heat, the more sound energy speeds up the process. Among the main benefits is that the process can work even with smaller sample batches while maximizing the yield and cutting down on extraction time and solvent consumption. Negative aspects of this method are that replicating this method is challenging, and applying a lot of energy could cause free radicals to be produced, which can sometimes degrade the SMs (Shaikh and Patil 2020; Altemimi et al. 2017).

14.6 Purification Strategies

Once the SMs have been extracted from the cellular matrix, they are to be purified. A number of methods are available for the purpose that utilizes physicochemical principles for obtaining the maximum purified yield of the SM. A few of the methods have been detailed in the following sections.

14.6.1 Fractional Distillation

This procedure is used to separate or purify substances from mixtures. It is typically employed in the extraction of SMs of organic nature like citral, eucalyptol, and crude oil. The difference in their boiling points is the basis of purification. The design of a fractional distillation apparatus ensures that upon application of heat, all compounds will disperse and separate their boiling point. Accordingly, multiple siphons connected to the fractional distillation apparatus will cause each fractionated compound to condense and collect as a distinct entity (Olin-Sandoval et al. 2019). A closely related method is called fractional liberation in which the compounds that readily form precipitates from a mixture can be separated. The compounds typically change into their salt form to form the precipitate.

14.6.2 Sublimation

With this technique, the conversion from solid to gas state occurs deprived of going to the liquid state. When heated, certain matters, like camphor and volatile oils, separate and turn into gas which are then collected and processed further.

14.6.3 Fractional Crystallization

Cellular extracts used for producing SMs naturally have a more number of compounds that are crystalline in nature. During the process of fractional crystallization, when extract becomes concerted via cold and heat, crystals are formed, which aid the separation of the intended sample (Hamuel 2012).

14.6.4 Chromatographic Techniques

These specialized techniques allow the parting of compounds established on shape, charge, and size. The idea behind chromatography is the usage of stationary phases like Sephadex and silica gel in combination with calcium sulfate as a binder and a mobile phase which is the extraction solvent. Alkaloids, fatty acids, sugars, amino acids, and lipids can all be separated using silica gel. Sephadex can be used to separate amino acids and proteins. Aluminum is helpful in the extraction of alkaloids, steroids, carotenes, phenols, and vitamins. Alkaloids, food dyes, and amino acids are separated using cellulose powder. Compounds can be separated using chromatographic techniques through a variety of mechanisms, including adsorption, partitioning, affinity, ion exchange, and size-exclusion. Techniques for chromatography include column chromatography, thin layer chromatography and paper chromatography (Desai and Tatke 2019; Rimando et al. 2001).

To ensure the manufacture of high-quality bio-therapeutics and other bioproducts, efficient quality control is utilized in bioprocesses and metabolite purification (Schulte et al. 2023). Analytical techniques are essential for tracking and improving bioprocessing of metabolites at all stages, from cell culture to downstream purification (Segers et al. 2019). To achieve high quality standards, this all-encompassing strategy incorporates state-of-the-art methods, novel technology, and reliable procedures. Analysis techniques including flow cytometry, metabolic flux analysis, mass spectrometry and high-throughput screening allow for the real-time evaluation of cell viability, growth, and productivity (Knecht et al. 2021). By combining these omics techniques, it is possible to find important biomarkers and indicators that are correlated with the desired qualities of the metabolite bio-product (Wang et al. 2023).

Chromatography, as discussed earlier, is a fundamental technique in the purification of metabolites, which is an important stage in bioprocessing. The introduction of new resins, membranes, and stationary phases has improved selectivity and efficiency of traditional chromatographic procedures for the purification of metabolites (Lee et al. 2020). When combined with sophisticated detectors like mass spectrometry (MS), high-performance liquid chromatography allows for the sensitive as well as accurate measurement of metabolites, guaranteeing the elimination of contaminants or impurities to satisfy the legal requirements (Karami et al. 2019). The desire for increased efficiency and lesser unpredictability in bioprocessing has led to a growing emphasis on continuous production in recent years. The benefits of continuous chromatography and other continuous separation methods include increased productivity, cost savings, and process control. Bioprocesses can achieve increased throughput, consistent product quality, and decreased batch processing downtime by putting these continuous techniques into practice (Haidar Ahmad et al. 2020). There are three key metabolite separation techniques liquid chromatography (LC) capillary electrophoresis and gas chromatography (GC), are the foundation of mass spectrometry based metabolomics methods (Mattoli et al. 2023).

14.6.4.1 Gas Chromatography (GC)

In analytical chemistry field, GC is a potent and extensively used technology, especially for the study and separation of metabolites. The GC technique divides complex mixtures into their component parts according to their physical and chemical characteristics by selectively partitioning the analytes between a stationary phase and a moving gas phase inside a chromatographic column (Feizi et al. 2021). Metabolites have different molecular masses; therefore, GC is the technique of choice for analyzing volatile or smaller molecular weight metabolites (Menaga et al. 2021). The broad application range of gas chromatography in metabolite analysis allows it to analyze a wide variation of metabolites, such as sugars, fatty acids amino acids, and other small organic compounds (Benkeblia 2023). In addition to its accuracy, the method is useful in biological and clinical research because it can identify metabolites at low quantities (Ayala-Cabrera et al. 2023).

14.6.4.2 Liquid Chromatography (LC)

Liquid Chromatography is a reliable analytical technique that is frequently used for the separation and examination of bio-products and their metabolites (Lijalem et al. 2022). For the examination of polar and non-volatile metabolites, gas chromatography is not suitable, so the best choice is liquid chromatography (Ali et al. 2019; Yuan et al. 2022). This adaptability of liquid chromatography is essential for studying the metabolites linked to bio product generation in many industrial and research settings (Fekete et al. 2014). Metabolite separation is facilitated by the variety of columns that include reversed-phase, ion exchange, and hydrophobic interactions, all of which are based on various chemical properties (Ngere et al. 2023). For LC gradient separation, reversed-phase columns like C18 or C8 are now the most widely utilized (Žuvela et al. 2019). The advancement of ultra-performance LC made the method even more potent in terms of throughput, sensitivity, and resolution (Cruz et al. 2022). PSMs like flavonoids and alkaloids (Alara et al. 2018), lipids (lipidomics) (Gowda et al. 2022), and primary metabolites like amino acids can be measured using the novel LC-MS technique (Themelis et al. 2019).

14.6.4.3 High-Performance Liquid Chromatography (HPLC)

Numerous SMs can be identified, separated, and quantified by using HPLC which is considered to be an essential component of the workflow for bioprocessing and metabolite purification (Yang et al. 2022). The basis of HPLC is liquid chromatography, which includes pumping a fluid mobile phase under high pressure over a packed column. The stationary phase, which is usually a liquid or solid, reacts with the metabolite sample, allowing differential absorption and separation (Lafarga et al. 2020). The structural identification of metabolites separated by HPLC is coupled with many detectors, including mass spectrometry (HPLC-MS), UV-Vis spectroscopy, and fluorescence detection. This is essential for recognizing unfamiliar substances and comprehending the molecular properties of metabolites (Ganzera and Sturm 2018). For testing the purity of metabolites and bio products, HPLC is an essential technique. It makes it possible to separate the desired chemical from particles, impurities, or byproducts, verifying that the finished product satisfies quality requirements (Balekundri and Mannur 2020). The sensitivity and specificity of metabolite analysis is improved when HPLC is combined with modern detection methods, such as mass spectrometry (HPLC-MS) tandem mass spectrometry (LC-MS/MS) and nuclear magnetic resonance (HPLC-NMR) (Peng et al. 2020).

14.6.4.4 Ultra-High-Performance Liquid Chromatography (UHPLC)

This is a unique technique for metabolite exploration in bioprocessing because of its capacity to deliver greater resolution, faster analysis times, and enhanced sensitivity (Huang et al. 2021). The metabolite particles are forced through a column at high pressure, enabling the identification, measurement, and separation of mixture components. Compared to HPLC, UHPLC uses smaller particle sizes (less than 2 μm). UHPLC's high pressure (up to 1000 bar) allows for enhanced resolution and lower solvent consumption during the separation and identification of SMs (Rathod et al. 2019).

14.6.4.5 Ion Exchange Chromatography (IEX)

A chromatography method called ion exchange chromatography (IEX) divides molecules according to their surface charges. Due to its ability to separate charged molecules according to their columbic (ionic) interactions, it is an extremely effective instrument for the purification of metabolites. Numerous applications, such as metabolomics research, environmental and medical research, and steroid analysis, make extensive use of IEX (Heikkinen et al. 1983). A polymer containing charged functional groups, like carboxyl or amino groups, is used as the stationary phase in IEX. The charged molecules in sample attach to the stationary phase according to their charge characteristics. A pH or salt gradient is used to elute the bound molecules from the column (Amartely et al. 2018).

14.6.4.6 Mixed-Mode Chromatography (MMC)

Mixed-mode chromatography is an effective method for analyzing complicated samples that provides more flexibility and selectivity than single-mode chromatography (Sýkora et al. 2019). It includes several complimentary or orthogonal chromatographic modes, which enables the separation of a variety of substances including charged molecules (Schmitt et al. 2023). When it comes to metabolites, such as phosphorylated carbohydrates, which are the primary metabolites in a number of essential metabolic pathways, they are preferably separated and purified using mixed-mode chromatography. It has the ability to act on various functional groups of the solute, such as ionic and hydrophobic groups. Reversible separation becomes possible by MMC, which has advantages in a number of applications, including the examination of medications, contaminants, biopharmaceuticals, and polar components in natural products (Mathon et al. 2017).

14.7 Quality Control and Regulatory Considerations

Chemical, pharmaceutical and food industries dependent upon the extraction and utilization of metabolites have developed various methodologies to ensure that the end-product is of the best possible quality. One such bioprocessing method is referred to as Process Analytical Technology (PAT). PAT is a group of methods used to assess and regulate a bioprocess's quality of metabolites during processing. It gives real-time data for control of metabolites and process monitoring (Gerzon et al. 2022). The methodical approach to product and process development related to this is known as Quality by Design (QbD). It relies on the design or the management of the most important quality aspects. Its goal is to confirm that the product is safe to use and fulfills the desired requirements (Narayanan et al. 2020). The biotech industry introduced the concepts of Process Analytical Technology and Quality by Design a decade ago. Both the biotech industry and key regulatory agencies have made significant efforts to assist the implementation of these principles; however, there has been varying degrees of success (Rathore 2014). The US Food, European Medicine Agency (EMA) and Drug Administration (FDA) have supported Process Analytical Technology and Quality by Design concepts, which have led to a greater focus on modeling approaches for process consistency, understanding and prediction along with the digital transformation. As a result, these techniques gave considerable importance in the selection and use for metabolites purification and quality control (Simon et al. 2015).

The synthesis of biopharmaceuticals, enzymes, and other valuable bio-chemicals generated from biological sources requires the bioprocessing and purification of metabolites. Even though technical developments have sped up progress in many fields, ethical and regulatory concerns must always come first to assurance the safety and efficacy of the product. Authorities like the European Medicines Agency (EMA) in Europe, the Food and Drug Administration (FDA) in the United States, and other comparable agencies globally, impose strict regulatory control on the bioprocessing and purification of metabolites (Okpala and Korzeniowska 2023). **Many** regulatory bodies have developed the Good Manufacturing Practice (GMP) regulations as a collection of rules and specifications to guarantee the effectiveness, safety, and quality of metabolites products. The ultimate purpose of these rules is to protect public health and guarantee that healthcare products regularly manufactured to high quality standards. They provide criteria for the manufacturing, testing, packing, labeling, and storage of healthcare items. GMP guidelines include every facet of the production process, from obtaining raw materials to distributing completed goods (Sanz-Nogués and O'Brien 2021). **As per GMP recommendations, m**anufacturing facilities must be planned, built, and maintained in a way that minimizes contamination and guarantees the integrity of the product. Specific standards must be followed for the calibration, validation, and upkeep of equipment used in industrial operations (He et al. 2015). GMP stipulates that employees engaged in testing, manufacturing, and quality control operations must have sufficient training to enable them to carry out their responsibilities with competence. Standard operating procedures (SOPs), safety measures, and GMP standards are just a few of the subjects that should covered in training programs (Mbabzi 2021). To prove adherence to GMP guidelines, thorough documentation is necessary. This includes keeping track of the testing outcomes, variances, manufacturing procedures, and remedial measures implemented. Records should be guaranteed to be accurate, comprehensive, and traceable by using good documentation practices (Kolekar and Bhagwat 2021). Similarly, GMP regulations should have robust quality management team to monitor the manufacturing operations. System needs to update polices, risk management, procedures and updated technology to obtain quality products (Noor Hasnan et al. 2022).

To ensure the legal way of processing the production of metabolites, ethical consideration must be applied in the manufacturing system. Bioprocessing research must follow scientific integrity, transparency, and reproducibility principles, with a focus on avoiding biases and guaranteeing accurate technique and result reporting. Ethical bioprocessing procedures stress the ethical use of biological resources, reducing environmental impact and assuring the long-term usage of raw materials and energy. Furthermore, they include safeguarding of secret information and data privacy. Researchers and producers must take steps to ensure the confidentiality and privacy of people's personal and health-related information (Ashfield 2022).

14.8 Conclusion

Bioprocessing of metabolites serves as one of the most essential steps in transforming an otherwise crude product into a commercially viable product, whether in pharmaceutical, food, agricultural or chemical fields. This chapter has explored the way through which metabolic processes and products can be engineered for gaining the maximum benefit in terms of output. Moreover, once the process is complete, the ways and means through which a product can be separated and purified has also been discussed. From the fermentation process to the downstream processing, a number of methods can be employed for maximizing the yield, purity and efficiency of the process. Whether opting for microbial based fermentation processes or PSM production, it is essential to understand that in order to attain a wide array of end products, the culture conditions must be meticulously optimized to ensure the efficiency of the process. Adjusting the substrate concentration, pH, temperature and other culture conditions is essential for developing a high performing bioprocess. Similarly, the option of purification and extraction methods for obtaining the essential metabolites needs to be done in manner that ensures effective utilization of resources.

Though the industry has been able to develop the processes effectively, the incorporation of modern developments in the field of omics, bioinformatics, biotechnology, and process engineering can further enhance the outcomes of the processes. Development of sustainable bioprocessing technologies that minimize resource consumption, waste generation and environmental effects can of great benefits. Similarly, including genetically engineered host cells that have improved substrate utilization capability can be quite helpful. Integration of artificial intelligence, machine learning, predictive modeling and other related fields into the bioprocessing may also aid the improvement of the overall outcomes of metabolic engineering processes and shall be considered for incorporation in future bioprocess designing.

References

Abubakar A, Haque M (2020) Preparation of medicinal plants: basic extraction and fractionation procedures for experimental purposes. J Pharm Bioallied Sci [Internet] 12(1):1. Available from: https://journals.lww.com/10.4103/jpbs.JPBS_175_19 [cited 2024 Feb 17]

Achimón F, Krapacher CR, Jacquat AG, Pizzolitto RP, Zygadlo JA (2021 May) Carbon sources to enhance the biosynthesis of secondary metabolites in Fusarium verticillioides submerged cultures. World J Microbiol Biotechnol 37(5):78

Akula R, Ravishankar GA (2011) Influence of abiotic stress signals on secondary metabolites in plants. Plant Signal Behav [Internet] 6(11):1720–1731. Available from: http://www.tandfonline.com/doi/abs/10.4161/psb.6.11.17613 [cited 2024 Feb 17]

Al Daccache M, Koubaa M, Louka N, Maroun RG, Vorobiev E, Rajha HN (2022) Separation, purification, analyses, and preservation of polyphenols. In: Technologies to recover polyphenols from agroFood by-products and wastes [Internet]. Elsevier, pp 247–270. Available from: https://linkinghub.elsevier.com/retrieve/pii/B9780323852739000077 [cited 2023 Nov 11]

Alara OR, Abdurahman NH, Ukaegbu CI, Azhari NH, Kabbashi NA (2018) Metabolic profiling of flavonoids, saponins, alkaloids, and terpenoids in the extract from *Vernonia cinerea* leaf using LC-Q-TOF-MS. J Liq Chromatogr Relat Technol [Internet] 41(11):722–731. Available from: https://www.tandfonline.com/doi/full/10.1080/10826076.2018.1511995 [cited 2023 Nov 23]

Ali I, Mohd S, Aboul-Enein HY (2019) Advances in chiral multidimensional liquid chromatography. TrAC Trends Anal Chem [Internet] 120:115634. Available from: https://linkinghub.elsevier.com/retrieve/pii/S0165993619304108 [cited 2023 Nov 22]

Altemimi A, Lakhssassi N, Baharlouei A, Watson D, Lightfoot D (2017) Phytochemicals: extraction, isolation, and identification of bioactive compounds from plant extracts. Plants [Internet] 6(4):42. Available from: http://www.mdpi.com/2223-7747/6/4/42 [cited 2024 Feb 17]

Amartely H, Avraham O, Friedler A, Livnah O, Lebendiker M (2018) Coupling Multi Angle Light Scattering to Ion Exchange chromatography (IEX-MALS) for protein characterization. Sci Rep [Internet] 8(1):6907. Available from: https://www.nature.com/articles/s41598-018-25246-6 [cited 2023 Nov 23]

Ashfield R (2022) GMP manufacture and quality control. In: Vaccinology and methods in vaccine research [Internet]. Elsevier, pp 281–294. Available from: https://linkinghub.elsevier.com/retrieve/pii/B9780323911467000044 [cited 2024 Feb 18]

Attia-Ismail SA (2022) Halophytic plants for animal feed: associated botanical and nutritional characteristics [Internet]. Bentham Science Publishers. Available from: https://www.eurekaselect.com/205754/volume/1 [cited 2023 Nov 11]

Ayala-Cabrera JF, Montero L, Meckelmann SW, Uteschil F, Schmitz OJ (2023) Review on atmospheric pressure ionization sources for gas chromatography-mass spectrometry. Part II: Current applications. Anal Chim Acta [Internet] 1238:340379. Available from: https://linkinghub.elsevier.com/retrieve/pii/S0003267022009503 [cited 2023 Nov 22]

Balekundri A, Mannur V (2020) Quality control of the traditional herbs and herbal products: a review. Future J Pharm Sci [Internet] 6(1):67. Available from: https://fjps.springeropen.com/articles/10.1186/s43094-020-00091-5 [cited 2023 Nov 22]

Behrens B, Engelen J, Tiso T, Blank LM, Hayen H (2016 Apr) Characterization of rhamnolipids by liquid chromatography/mass spectrometry after solid-phase extraction. Anal Bioanal Chem 408(10):2505–2514

Benkeblia N (2023) Gas chromatography-mass spectrometry and liquid chromatography-mass spectrometry metabolomics platforms: tools for plant oligosaccharides analysis. Carbohydr

Polym Technol Appl [Internet] 5:100304. Available from: https://linkinghub.elsevier.com/retrieve/pii/S2666893923000257 [cited 2023 Nov 22]
Bhatla SC (2018) Secondary metabolites. In: Plant physiology, development and metabolism [Internet]. Springer, Singapore, pp 1099–1166. Available from: http://link.springer.com/10.1007/978-981-13-2023-1_33 [cited 2023 Nov 11]
Bills GF, Gloer JB (2016) Biologically active secondary metabolites from the fungi. Microbiol Spectr 4(6):4–6
Bitwell C, Indra SS, Luke C, Kakoma MK (2023) A review of modern and conventional extraction techniques and their applications for extracting phytochemicals from plants. Sci Afr [Internet] 19:e01585. Available from: https://linkinghub.elsevier.com/retrieve/pii/S2468227623000443 [cited 2024 Feb 17]
Chen J, Wu Q, Hawas UW, Wang H (2016) Genetic regulation and manipulation for natural product discovery. Appl Microbiol Biotechnol 100:2953–2965
Corrêa PS, Morais Júnior WG, Martins AA, Caetano NS, Mata TM (2020) Microalgae biomolecules: extraction, separation and purification methods. PRO 9(1):10
Cruz JC, Souza IDD, Lanças FM, MEC Q (2022) Current advances and applications of online sample preparation techniques for miniaturized liquid chromatography systems. J Chromatogr A [Internet] 1668:462925. Available from: https://linkinghub.elsevier.com/retrieve/pii/S0021967322001236 [cited 2023 Nov 22]
Delgado-Zamarreño MM, Bustamante-Rangel M, Sánchez-Pérez A, Hernández-Méndez J (2001) Analysis of vitamin E isomers in seeds and nuts with and without coupled hydrolysis by liquid chromatography and coulometric detection. J Chromatogr A [Internet] 935(1–2):77–86. Available from: https://linkinghub.elsevier.com/retrieve/pii/S0021967301009918 [cited 2024 Feb 17]
Desai S, Tatke P (2019) Phytochemical markers: classification, applications and isolation. Curr Pharm Des [Internet] 25(22):2491–2498. Available from: http://www.eurekaselect.com/173354/article [cited 2024 Feb 17]
Du YL, Ryan KS (2019) Pyridoxal phosphate-dependent reactions in the biosynthesis of natural products. Nat Prod Rep 36(3):430–457
El-Desoky AH, Atia MA, Omer EA (2022) Metabolic engineering: new approaches in pharmaceutical production. In: Metabolic engineering in plants. Springer, pp 1–24
Erb M, Kliebenstein DJ (2020 Sep) Plant secondary metabolites as defenses, regulators, and primary metabolites: the blurred functional trichotomy. Plant Physiol 184(1):39–52
Erb TJ, Jones PR, Bar-Even A (2017 Apr) Synthetic metabolism: metabolic engineering meets enzyme design. Curr Opin Chem Biol 37:56–62
Fagbemi KO, Aina DA, Olajuyigbe OO (2021) Soxhlet extraction versus hydrodistillation using the clevenger apparatus: a comparative study on the extraction of a volatile compound from Tamarindus indica seeds (Cameselle C, ed). Sci World J [Internet] 2021:1–8. Available from: https://www.hindawi.com/journals/tswj/2021/5961586/ [cited 2024 Feb 17]
Fazili MA, Bashir I, Ahmad M, Yaqoob U, Geelani SN (2022) In vitro strategies for the enhancement of secondary metabolite production in plants: a review. Bull Natl Res Cent [Internet] 46(1):35. Available from: https://BNRC.springeropen.com/articles/10.1186/s42269-022-00717-z [cited 2024 Feb 17]
Feizi N, Hashemi-Nasab FS, Golpelichi F, Saburouh N, Parastar H (2021) Recent trends in application of chemometric methods for GC-MS and GC×GC-MS-based metabolomic studies. TrAC Trends Anal Chem [Internet] 138:116239. Available from: https://linkinghub.elsevier.com/retrieve/pii/S0165993621000625 [cited 2023 Nov 22]
Fekete S, Kohler I, Rudaz S, Guillarme D, Importance of instrumentation for fast liquid chromatography in pharmaceutical analysis (2014) J Pharm Biomed Anal [Internet] 87:105–119. Available from: https://linkinghub.elsevier.com/retrieve/pii/S0731708513001222 [cited 2023 Nov 22]
Ganzera M, Sturm S (2018) Recent advances on HPLC/MS in medicinal plant analysis—an update covering 2011–2016. J Pharm Biomed Anal [Internet] 147:211–233. Available from: https://linkinghub.elsevier.com/retrieve/pii/S0731708517312633 [cited 2023 Nov 22]

Garcia-Vaquero M, Rajauria G, Tiwari B (2020) Conventional extraction techniques: solvent extraction. In: Sustainable seaweed technologies [Internet]. Elsevier, pp 171–189. Available from: https://linkinghub.elsevier.com/retrieve/pii/B9780128179437000068 [cited 2024 Feb 17]

Gerzon G, Sheng Y, Kirkitadze M (2022) Process analytical technologies—advances in bioprocess integration and future perspectives. J Pharm Biomed Anal [Internet] 207:114379. Available from: https://linkinghub.elsevier.com/retrieve/pii/S0731708521004908 [cited 2023 Nov 23]

Giardi MT, Rea G, Berra B (eds) (2010) Bio-farms for nutraceuticals: functional food and safety control by biosensors [Internet], Advances in experimental medicine and biology, vol 698. Springer, Boston. Available from: http://link.springer.com/10.1007/978-1-4419-7347-4 [cited 2023 Nov 11]

González Mera IF, González Falconí DE, Morera Córdova V (2019) Secondary metabolites in plants: main classes, phytochemical analysis and pharmacological activities. Bionatura [Internet] 4(4):1000–1009. Available from: https://www.revistabionatura.com/2019.04.04.11.html [cited 2024 Feb 17]

Gouveia B, Rijo P, Gonçalo T, Reis C (2015) Good manufacturing practices for medicinal products for human use. J Pharm Bioallied Sci 7(2):87

Gowda SGB, Yifan C, Gowda D, Tsuboi Y, Chiba H, Hui SP (2022) Analysis of antioxidant lipids in five species of dietary seaweeds by liquid chromatography/mass spectrometry. Antioxidants [Internet] 11(8):1538. Available from: https://www.mdpi.com/2076-3921/11/8/1538 [cited 2023 Nov 22]

Haidar Ahmad IA, Shchurik V, Nowak T, Mann BF, Regalado EL (2020) Introducing multifactorial peak crossover in analytical and preparative chromatography via computer-assisted modeling. Anal Chem [Internet] 92(19):13443–13451. Available from: https://pubs.acs.org/doi/10.1021/acs.analchem.0c02807 [cited 2023 Nov 22]

Hamuel J (2012) Phytochemicals: extraction methods, basic structures and mode of action as potential chemotherapeutic agents. In: Rao V (ed) Phytochemicals—a global perspective of their role in nutrition and health [Internet]. InTech. Available from: http://www.intechopen.com/books/phytochemicals-a-global-perspective-of-their-role-in-nutrition-and-health/phytochemicals-extraction-methods-basic-structures-and-mode-of-action-as-potential-chemotherapeutic- [cited 2024 Feb 17]

He TT, Ung COL, Hu H, Wang YT (2015) Good manufacturing practice (GMP) regulation of herbal medicine in comparative research: China GMP, cGMP, WHO-GMP, PIC/S and EU-GMP. Eur J Integr Med [Internet] 7(1):55–66. Available from: https://linkinghub.elsevier.com/retrieve/pii/S1876382015000050 [cited 2024 Feb 18]

Heikkinen R, Fotsis T, Adlercreutz H (1983) Use of ion exchange chromatography in steroid analysis. J Steroid Biochem [Internet] 19(1):175–180. Available from: https://linkinghub.elsevier.com/retrieve/pii/S0022473183800235 [cited 2023 Nov 22]

Hossain MA, Al-Hdhrami SS, Weli AM, Al-Riyami Q, Al-Sabahi JN (2014) Isolation, fractionation and identification of chemical constituents from the leaves crude extracts of Mentha piperita L grown in Sultanate of Oman. Asian Pac J Trop Biomed [Internet] 4:S368–S372. Available from: http://linkinghub.elsevier.com/retrieve/pii/S2221169115302975 [cited 2024 Feb 17]

Huang H, Li Z, He Y, Huang L, Xu X, Pan C et al (2021) Nontarget and high-throughput screening of pesticides and metabolites residues in tea using ultra-high-performance liquid chromatography and quadrupole-orbitrap high-resolution mass spectrometry. J Chromatogr B [Internet] 1179:122847. Available from: https://linkinghub.elsevier.com/retrieve/pii/S1570023221003287 [cited 2023 Nov 22]

Isah T (2019) Stress and defense responses in plant secondary metabolites production. Biol Res [Internet] 52(1):39. Available from: https://biolres.biomedcentral.com/articles/10.1186/s40659-019-0246-3 [cited 2024 Feb 17]

Kandar CC (2021) Secondary metabolites from plant sources. Bioact Nat Prod Pharm Appl:329–377

Karami F, Ranjbar S, Ghasemi Y, Negahdaripour M (2019) Analytical methodologies for determination of methotrexate and its metabolites in pharmaceutical, biological and environmental samples. J Pharm Anal [Internet] 9(6):373–391. Available from: https://linkinghub.elsevier.com/retrieve/pii/S2095177919300802 [cited 2023 Nov 22]

Kashyap AK, Dubey SK, Shah S, Kumar A (2022) A short review on genes regulating biosynthesis of major secondary metabolites. In: Swamy MK, Kumar A (eds) Phytochemical genomics [Internet]. Springer, Singapore, pp 501–519. Available from: https://link.springer.com/10.1007/978-981-19-5779-6_20 [cited 2023 Nov 11]

Knecht S, Blanchard JW, Barskiy D, Cavallari E, Dagys L, Van Dyke E et al (2021) Rapid hyperpolarization and purification of the metabolite fumarate in aqueous solution. Proc Natl Acad Sci [Internet] 118(13):e2025383118. Available from: https://pnas.org/doi/full/10.1073/pnas.2025383118 [cited 2023 Nov 22]

Kolekar PD, Bhagwat AM (2021) Good documentation practices: a need of pharmaceutical industry. Asian J Res Chem [Internet]:368–374. Available from: https://ajrconline.org/AbstractView.aspx?PID=2021-14-5-12 [cited 2024 Feb 18]

Kumar V, Ahluwalia V, Saran S, Kumar J, Patel AK, Singhania RR (2021) Recent developments on solid-state fermentation for production of microbial secondary metabolites: challenges and solutions. Bioresour Technol [Internet] 323:124566. Available from: https://linkinghub.elsevier.com/retrieve/pii/S096085242031840X [cited 2024 Feb 17]

Kumari K, Kumar S, Prashant JAK, Kumar N (2022 Nov) Biotechnological intervention in genetic improvement and regulation of secondary metabolites production in Ocimum sanctum L. Ind Crops Prod 187:115329

Lafarga T, Acién-Fernández FG, Garcia-Vaquero M (2020) Bioactive peptides and carbohydrates from seaweed for food applications: Natural occurrence, isolation, purification, and identification. Algal Res [Internet] 48:101909. Available from: https://linkinghub.elsevier.com/retrieve/pii/S2211926420300692 [cited 2023 Nov 22]

Lee SR, Kang H, Yoo MJ, Yi SA, Christine B, Lee J et al (2020) Anti-adipogenic pregnane steroid from a Hydractinia-associated fungus, Cladosporium sphaerospermum SW67. Nat Prod Sci [Internet] 26(3):230–235. Available from: http://www.dbpia.co.kr/Journal/ArticleDetail/NODE11143313 [cited 2023 Nov 22]

Letaief T, Garzoli S, Laghezza Masci V, Mejri J, Abderrabba M, Tiezzi A et al (2021) Chemical composition and biological activities of Tunisian Ziziphus lotus extracts: evaluation of drying effect, solvent extraction, and extracted plant parts. Plants [Internet] 10(12):2651. Available from: https://www.mdpi.com/2223-7747/10/12/2651 [cited 2024 Feb 17]

Li T, Ding T, Li J (2019) Medicinal purposes: bioactive metabolites from marine-derived organisms. Mini Rev Med Chem 19(2):138–164

Liang X, Nielsen NJ, Christensen JH (2020) Selective pressurized liquid extraction of plant secondary metabolites: Convallaria majalis L. as a case. Anal Chim Acta X [Internet] 4:100040. Available from: https://linkinghub.elsevier.com/retrieve/pii/S2590134620300025 [cited 2024 Feb 17]

Lijalem YG, Gab-Allah MA, Choi K, Kim B (2022) Development of isotope dilution-liquid chromatography/tandem mass spectrometry for the accurate determination of zearalenone and its metabolites in corn. Food Chem [Internet] 384:132483. Available from: https://linkinghub.elsevier.com/retrieve/pii/S0308814622004459 [cited 2023 Nov 22]

Liu J, Wang X, Dai G, Zhang Y, Bian X (2022 Oct) Microbial chassis engineering drives heterologous production of complex secondary metabolites. Biotechnol Adv 59:107966

López-Calleja AC, Cuadra T, Barrios-González J, Fierro F, Fernández FJ (2012) Solid-state and submerged fermentations show different gene expression profiles in cephalosporin C production by *Acremonium chrysogenum*. Microb Physiol [Internet] 22(2):126–134. Available from: https://www.karger.com/Article/FullText/338987 [cited 2024 Feb 17]

Mathon C, Barding GA, Larive CK (2017) Separation of ten phosphorylated mono-and disaccharides using HILIC and ion-pairing interactions. Anal Chim Acta [Internet] 972:102–110. Available from: https://linkinghub.elsevier.com/retrieve/pii/S0003267017303240 [cited 2023 Nov 23]

Mattoli L, Gianni M, Burico M (2023) Mass spectrometry-based metabolomic analysis as a tool for quality control of natural complex products. Mass Spectrom Rev [Internet] 42(4):1358–1396. Available from: https://analyticalsciencejournals.onlinelibrary.wiley.com/doi/10.1002/mas.21773 [cited 2023 Nov 22]

Mbabzi KE (2021) Standardisation of staff training to increase efficiency [Internet], 1st edn. Purdue University. Available from: https://docs.lib.purdue.edu/birsafricatr/2/ [cited 2024 Feb 18]

Menaga D, Rahman PKSM, Rajakumar S, Ayyasamy PM (2021) Antioxidant and cytotoxic activities of a novel isomeric molecule (PF5) obtained from methanolic extract of Pleurotus florida mushroom. J Bioresour Bioprod [Internet] 6(4):338–349. Available from: https://linkinghub.elsevier.com/retrieve/pii/S2369969821000402 [cited 2023 Nov 22]

Mohan CD, Rangappa S, Nayak SC, Jadimurthy R, Wang L, Sethi G et al (2022) Bacteria as a treasure house of secondary metabolites with anticancer potential. Semin Cancer Biol (Elsevier) 86:998–1013

Morgan ED (2010) Biosynthesis in insects. Royal Society of Chemistry, Cambridge

Nagegowda DA, Gupta P (2020 May) Advances in biosynthesis, regulation, and metabolic engineering of plant specialized terpenoids. Plant Sci 294:110457

Narasimhulu K (2022) Role of metabolic engineering in the synthesis of plant secondary metabolites. In: DrA A (ed) Innovations in microbiology and biotechnology, vol 5 [Internet]. Book Publisher International (a part of SCIENCEDOMAIN International), pp 60–69. Available from: https://stm.bookpi.org/IMB-V5/article/view/6567 [cited 2023 Nov 11]

Narayanan H, Luna MF, Von Stosch M, Cruz Bournazou MN, Polotti G, Morbidelli M et al (2020) Bioprocessing in the digital age: the role of process models. Biotechnol J [Internet] 15(1):1900172. Available from: https://onlinelibrary.wiley.com/doi/10.1002/biot.201900172 [cited 2023 Nov 23]

Ngere JB, Ebrahimi KH, Williams R, Pires E, Walsby-Tickle J, McCullagh JSO (2023) Ion-exchange chromatography coupled to mass spectrometry in life science, environmental, and medical research. Anal Chem [Internet] 95(1):152–166. Available from: https://pubs.acs.org/doi/10.1021/acs.analchem.2c04298 [cited 2023 Nov 22]

Nigam P, Singh A, Lyngby D (2014) Production of secondary metabolites–fungi. In: Batt C, Patel P (eds) Encyclopedia of food microbiology, 2nd edn, pp 570–578. https://doi.org/10.1016/B978-0-12-384730-0.00202-0

Noor Hasnan NZ, Basha RK, NAM A, SHM R, JYH T, Aziz NA (2022) Analysis of the most frequent nonconformance aspects related to Good Manufacturing Practices (GMP) among small and medium enterprises (SMEs) in the food industry and their main factors. Food Control [Internet] 141:109205. Available from: https://linkinghub.elsevier.com/retrieve/pii/S095671352200398X [cited 2024 Feb 18]

Okpala COR, Korzeniowska M (2023) Understanding the relevance of quality management in agro-food product Industry: from ethical considerations to assuring food hygiene quality safety standards and its associated processes. Food Rev Int [Internet] 39(4):1879–1952. Available from: https://www.tandfonline.com/doi/full/10.1080/87559129.2021.1938600 [cited 2024 Feb 18]

Olin-Sandoval V, Yu JSL, Miller-Fleming L, Alam MT, Kamrad S, Correia-Melo C et al (2019) Lysine harvesting is an antioxidant strategy and triggers underground polyamine metabolism. Nature [Internet] 572(7768):249–253. Available from: https://www.nature.com/articles/s41586-019-1442-6 [cited 2024 Feb 17]

Ozyigit II, Dogan I, Hocaoglu-Ozyigit A, Yalcin B, Erdogan A, Yalcin IE et al (2023) Production of secondary metabolites using tissue culture-based biotechnological applications. Front Plant Sci [Internet] 14:1132555. Available from: https://www.frontiersin.org/articles/10.3389/fpls.2023.1132555/full [cited 2024 Feb 17

Park SY, Yang D, Ha SH, Lee SY (2018) Metabolic engineering of microorganisms for the production of natural compounds. Adv Biosyst 2(1):1700190

Peng H, Chang Y, Baker RC, Zhang G (2020) Interference of mycotoxin binders with ELISA, HPLC and LC-MS/MS analysis of aflatoxins in maize and maize gluten. Food Addit Contam Part A [Internet] 37(3):496–506. Available from: https://www.tandfonline.com/doi/full/10.1080/19440049.2019.1701717 [cited 2023 Nov 22]

Ram M, Misra H (2021) Metabolic engineering for the production of plant therapeutic compounds. In: Medicinal and aromatic plants [Internet]. Elsevier, pp 169–207. Available from: https://linkinghub.elsevier.com/retrieve/pii/B9780128195901000082 [cited 2023 Nov 11]

Rami E, Singh A, Favzulazim S (2021 Sep 5) An overview of plant secondary metabolites, their biochemistry and generic applications. J Phytopharm 10(5):421–428

Rathod RH, Chaudhari SR, Patil AS, Shirkhedkar AA (2019) Ultra-high performance liquid chromatography-MS/MS (UHPLC-MS/MS) in practice: analysis of drugs and pharmaceutical formulations. Future J Pharm Sci [Internet] 5(1):6. Available from: https://fjps.springeropen.com/articles/10.1186/s43094-019-0007-8 [cited 2023 Nov 22]

Rathore AS (2014) QbD/PAT for bioprocessing: moving from theory to implementation. Curr Opin Chem Eng [Internet] 6:1–8. Available from: https://linkinghub.elsevier.com/retrieve/pii/S2211339814000549 [cited 2023 Nov 23]

Rimando AM, Olofsdotter M, Dayan FE, Duke SO (2001) Searching for rice allelochemicals: an example of bioassay-guided isolation. Agron J [Internet] 93(1):16–20. Available from: https://acsess.onlinelibrary.wiley.com/doi/10.2134/agronj2001.93116x [cited 2024 Feb 17]

Sanz-Nogués C, O'Brien T (2021) Current good manufacturing practice considerations for mesenchymal stromal cells as therapeutic agents. Biomater Biosyst [Internet] 2:100018. Available from: https://linkinghub.elsevier.com/retrieve/pii/S2666534421000118 [cited 2024 Feb 18]

Sathasivam R, Park SU, Kim JK, Park YJ, Kim MC, Nguyen BV et al (2023) Metabolic profiling of primary and secondary metabolites in Kohlrabi (Brassica oleracea var. gongylodes) sprouts exposed to different light-emitting diodes. Plants [Internet] 12(6):1296. Available from: https://www.mdpi.com/2223-7747/12/6/1296 [cited 2024 Feb 17]

Schafhauser T, Kulik A (2023) Isolation and purification of natural products from microbial cultures. In: Sass P (ed) Antibiotics [Internet], Methods in molecular biology, vol 2601. Springer, New York, pp 75–96. Available from: https://link.springer.com/10.1007/978-1-0716-2855-3_5 [cited 2023 Nov 11]

Schmitt M, Egorycheva M, Frerichs D, Fiedler S, Graumann PL, Seubert A (2023) Factors affecting mixed-mode retention properties of cation-exchange stationary phases. J Chromatogr A [Internet] 1695:463934. Available from: https://linkinghub.elsevier.com/retrieve/pii/S0021967323001607 [cited 2023 Nov 23]

Schulte U, Den Brave F, Haupt A, Gupta A, Song J, Müller CS et al (2023) Mitochondrial complexome reveals quality-control pathways of protein import. Nature [Internet] 614(7946):153–159. Available from: https://www.nature.com/articles/s41586-022-05641-w [cited 2023 Nov 22]

Segers K, Declerck S, Mangelings D, Heyden YV, Eeckhaut AV (2019) Analytical techniques for metabolomic studies: a review. Bioanalysis [Internet] 11(24):2297–2318. Available from: https://www.future-science.com/doi/10.4155/bio-2019-0014 [cited 2023 Nov 22]

Shahnawaz M (ed) (2022) Biotechnological approaches to enhance plant secondary metabolites: recent trends and future prospects, 1st edn. CRC Press, Taylor & Francis Group, Boca Raton/London/New York, 168 p

Shaikh JR, Patil M (2020) Qualitative tests for preliminary phytochemical screening: an overview. Int J Chem Stud [Internet] 8(2):603–608. Available from: http://www.chemijournal.com/archives/?year=2020&vol=8&issue=2&ArticleId=8834&si=false [cited 2024 Feb 17]

Shen L, Pang S, Zhong M, Sun Y, Qayum A, Liu Y et al (2023) A comprehensive review of ultrasonic assisted extraction (UAE) for bioactive components: Principles, advantages, equipment, and combined technologies. Ultrason Sonochem [Internet] 101:106646. Available from: https://linkinghub.elsevier.com/retrieve/pii/S1350417723003589 [cited 2024 Feb 17]

Simon LL, Pataki H, Marosi G, Meemken F, Hungerbühler K, Baiker A et al (2015) Assessment of recent Process Analytical Technology (PAT) trends: a multiauthor review. Org Process Res Dev [Internet] 19(1):3–62. Available from: https://pubs.acs.org/doi/10.1021/op500261y [cited 2023 Nov 23]

Singh H (2020) Common marine organisms: a novel source of medicinal compounds. Int J Bioresour Sci [Internet] 7(2) Available from: https://renupublishers.com/images/article/IJBSv7n2a.pdf [cited 2023 Nov 11]

Sinha R, Anand S, Singh D, Tripathi S, Shukla P, Singh S et al (2022 Feb) Bioprocessing strategies for microbial production and purification of immunosuppressants: an insight for process intensification. Chem Eng Process—Process Intensif 172:108797

Srivastava RK, Akhtar N, Verma M, Imandi SB (2020 Nov) Primary metabolites from overproducing microbial system using sustainable substrates. Biotechnol Appl Biochem 67(6):852–874

Sýkora D, Řezanka P, Záruba K, Král V (2019) Recent advances in mixed-mode chromatographic stationary phases. J Sep Sci [Internet] 42(1):89–129. Available from: https://analyticalsciencejournals.onlinelibrary.wiley.com/doi/10.1002/jssc.201801048 [cited 2023 Nov 23]

Themelis T, Gotti R, Orlandini S, Gatti R (2019) Quantitative amino acids profile of monofloral bee pollens by microwave hydrolysis and fluorimetric high performance liquid chromatography. J Pharm Biomed Anal [Internet] 173:144–153. Available from: https://linkinghub.elsevier.com/retrieve/pii/S0731708519308611 [cited 2023 Nov 23]

Thirumurugan D, Cholarajan A, Raja SSS, Vijayakumar R (2018) An introductory chapter: secondary metabolites. In: Vijayakumar R, SSS R (eds) Secondary metabolites—sources and applications [Internet]. InTech. Available from: http://www.intechopen.com/books/secondary-metabolites-sources-and-applications/an-introductory-chapter-secondary-metabolites [cited 2024 Feb 17]

Ul Haq I, Ijaz S, Khan IA (eds) (2023) Phytomycology and molecular biology of plant-pathogen interactions, 1st edn. CRC Press, Taylor & Francis Group, Boca Raton/London, 265 p

Vaou N, Stavropoulou E, Voidarou C, Tsigalou C, Bezirtzoglou E (2021) Towards advances in medicinal plant antimicrobial activity: a review study on challenges and future perspectives. Microorganisms [Internet] 9(10):2041. Available from: https://www.mdpi.com/2076-2607/9/10/2041 [cited 2024 Feb 17]

Vijayakumar R, Raja SS (2018) Secondary metabolites: sources and applications. BoD–Books on Demand, London

Volova TG, Mahapatra DK, Khanna S, Haghi AK (eds) (2021) Natural products chemistry: biomedical and pharmaceutical phytochemistry. Apple Academic Press, Burlington/Palm Bay, 1 p

Wang H, Li P, Sun SH, Zhang QD, Su Y, Zong YL et al (2014) Comparison of liquid–liquid extraction, simultaneous distillation extraction, ultrasound-assisted solvent extraction, and headspace solid-phase microextraction for the determination of volatile compounds in jujube extract by gas chromatography/mass spectrometry. Anal Lett 47(4):654–674

Wang Z, Liu B, Lin L, Qiao L (2023) Mass spectrometry for mitochondrial multi-omics. TrAC Trends Anal Chem [Internet] 163:117063. Available from: https://linkinghub.elsevier.com/retrieve/pii/S0165993623001504 [cited 2023 Nov 22]

Yadav AN, Kour D, Rana KL, Yadav N, Singh B, Chauhan VS et al (2019) Metabolic engineering to synthetic biology of secondary metabolites production. In: New and future developments in microbial biotechnology and bioengineering [Internet]. Elsevier, Amsterdam, pp 279–320. Available from: https://linkinghub.elsevier.com/retrieve/pii/B9780444635044000207 [cited 2023 Nov 11]

Yang M, Yi J, Wei C, Lu Y, Yang Y, Yang Z et al (2022) Rapid determination of nitrofuran metabolites residues in honey by ultrasonic assisted derivatization—QuEChERS—high performance liquid chromatography / tandem mass spectrometry. J Food Compos Anal [Internet] 114:104812. Available from: https://linkinghub.elsevier.com/retrieve/pii/S0889157522004306 [cited 2023 Nov 22]

Yeshi K, Crayn D, Ritmejerytė E, Wangchuk P (2022 Jan 5) Plant secondary metabolites produced in response to abiotic stresses has potential application in pharmaceutical product development. Molecules 27(1):313

Yuan C, Jia W, Yu Z, Li Y, Zi M, Yuan LM, et al. Are highly stable covalent organic frameworks the key to universal chiral stationary phases for liquid and gas chromatographic separations? J Am Chem Soc [Internet] 2022;144(2):891–900. Available from: https://pubs.acs.org/doi/10.1021/jacs.1c11051 [cited 2023 Nov 22]

Zebec Z, Scrutton NS (2018) Genome editing for the production of natural products in Escherichia coli. Adv Biosyst 2(11):1800056

Zhao J, Li P, Xia T, Wan X (2020 Jul 3) Exploring plant metabolic genomics: chemical diversity, metabolic complexity in the biosynthesis and transport of specialized metabolites with the tea plant as a model. Crit Rev Biotechnol 40(5):667–688

Žuvela P, Skoczylas M, Jay Liu J, Bączek T, Kaliszan R, Wong MW et al (2019) Column characterization and selection Systems in reversed-phase high-performance liquid chromatography. Chem Rev [Internet] 119(6):3674–3729. Available from: https://pubs.acs.org/doi/10.1021/acs.chemrev.8b00246 [cited 2023 Nov 22]

Scaling Up Nature's Chemistry: A Guide to Industrial Production of Valuable Metabolites

15

Gholamreza Abdi, Reshma Tendulkar, Chaitrali Thatte, Sandhya Mishra, Bhagyshree Desai, Sanskruti Surve, Ashara Chaudhari, Nil Patil, Mukul Jain, and Maryam Abbasi Tarighat

15.1 Introduction

15.1.1 The Promise of Metabolites: Applications in Food, Pharmaceuticals, and Beyond

Technology advancements have led to a significant advancement in scientific findings. The omics revolution has been beneficial to many biomedical researchers. Entire genomes, transcriptomics, proteomics, and metabolomics are examples of several omics levels where pertinent metabolic data may be monitored to categorize biochemical processes as described by the fundamental principles of molecular biology. Proteomics research has been carried out alongside genome research for many years. Because they represent what the genome has encoded and what has been altered by environmental factors at a given point in time, metabolites and proteins can be used to characterize an individual's metabolic state

G. Abdi (✉)
Department of Biotechnology, Persian Gulf Research Institute, Persian Gulf University, Bushehr, Iran
e-mail: abdi@pgu.ac.ir

R. Tendulkar · C. Thatte · S. Mishra · B. Desai · S. Surve · A. Chaudhari
Vivekanand Education Society's College of Pharmacy, Chembur, Mumbai, Maharashtra, India

N. Patil · M. Jain
Cell & Development Biology Lab, Research and Development Cell, Parul University, Vadodara, Gujarat, India

Department of Life Sciences, Parul Institute of Applied Sciences, Parul University, Vadodara, Gujarat, India

M. A. Tarighat
Faculty of Nano and Bio Science and Technology, Persian Gulf University, Bushehr, Iran

V. Singh (ed.), *Advances in Metabolomics*,
https://doi.org/10.1007/978-981-97-7459-3_15

(Gonzalez-Covarrubias et al. 2022). The metabolome has been characterized not only as the entire pool of all metabolites in a cell, but also as the recipient of the flow of biological information that leads to the establishment of a certain phenotype. Metabolites serve as products of metabolic pathways at either intermediate or final stages. Metabolomics is an evolving analytical approach characterized by data-driven processes and high-throughput capabilities. Its primary aim is the thorough detection, measurement, and immediate analysis of both primary and specialized or secondary metabolites in a specific biological context. This detection occurs as a consequence of environmental factors, genetic influences, or the life cycle progression of an organism (Marchev et al. 2021). The field of research known as "metabolomics" is characterized by its focus on identifying and assessing the entire assembly of small molecule metabolites inside a biological entity. "Small molecules" refers to materials whose molecular mass is less than two kilodaltons. Therefore, large molecules such as DNA, RNA, carbohydrate, and protein are not included in metabolomic data. Although the words "metabolomics" and "metabonomics" are sometimes used synonymously, they have different meanings. Within omics, metabolomics is a more generic concept. The scientific community has coined a variety of new terms (such as foodomics, lipidomics, fluxomics, and mineralomics) in an effort to fully elucidate different facets of metabolism or the composition of metabolites in a biological context since the advent of omics techniques. Because these studies are exploratory in nature, new knowledge about the chemical similarities and differences between specified sample sets is produced. As such, metabolomics and related omics domains are often described as hypothesis-generating (Johanningsmeier et al. 2016). The intricately intricate metabolome comprises a multitude of interconnected metabolic reactions involving a vast array of small and large molecules. Researchers in Canada meticulously described and compiled the initial iteration of the metabolome, encompassing 2500 metabolites, 1200 pharmaceuticals, and 3500 nutritional constituents. This undertaking marked one of the pioneering initiatives in elucidating the comprehensive human metabolome (Gonzalez-Covarrubias et al. 2022). The food metabolome, which consists of tiny molecules from animals, plants, and microbes, is altered by a variety of circumstances, including processing and inadvertent contamination, resulting in both common and distinct metabolites. Primary metabolites in meals, including as sugars, amino acids, and fatty acids, are vital for organism growth, and advancements in simultaneous detection and measurement of these molecules are increasing food quality, particularly in plant-based diets. Secondary plant metabolites contribute to a variety of plant functions and offer a range of health advantages, in addition to the primary metabolites shared by animals and microbes. These metabolites are not directly engaged in growth and development. Based on their chemical makeup, secondary metabolites can be categorized, including flavonoids, phenolic acids, terpenes, alkaloids, and chemicals containing sulfur. A significant portion of the food metabolome's metabolites come from microorganisms. In academic literature or databases, around 22,500 physiologically active secondary microbial metabolites are described (Johanningsmeier et al. 2016). Proactive and reactive methods are the two scenarios that may be used to find pharmacologically active metabolites. In a proactive scenario, as drug research operations for a particular chemical series progress, so does our understanding of the target SAR and general patterns in the metabolism of the compounds in the series. Reactive approaches to pharmacologically active metabolites can be very difficult, resource-intensive, and can cause delays in drug development, especially when human trials are being conducted (Cerny et al. 2020). The lack of early indicators can lead to erroneous diagnoses and serious repercussions. There is an urgent need for non-invasive diagnosis and monitoring methods with high specificity and selectivity. Metabolomics, which uses small molecule metabolites, is a useful method for researching metabolic biomarkers and pathways, revealing insight on illness processes and possible therapies. This method can help find functional biomarkers associated with phenotypic variations and detect alterations in biochemical pathways as early signals of dysfunction prior to illness development (Qiu et al. 2023).

15.1.2 Challenges of Industrial-Scale Metabolite Production

The challenges of industrial-scale metabolite production encompass various aspects that need to be addressed to optimize the efficiency and viability of the process. These challenges include:

1. Strain Engineering: Creating efficient and resilient commercial strains capable of generating a diverse variety of metabolites at competitive prices is critical. Traditional strain development approaches may not always achieve the extreme performance required to compete in the market (Abbate et al. 2023).
2. Metabolic Engineering: Manipulating microorganisms' metabolic pathways to promote metabolite synthesis presents issues in balancing precursor supply, optimizing routes, and altering host chassis to optimize yields (Pickens et al. 2011).
3. Regulatory Requirements: Ensuring that production strains meet regulatory standards, especially in terms of genetic stability, integration of expression cassettes, and the use of selective markers, can be a challenge in industrial settings (Wang et al. 2019).
4. Environmental Impact: Sustainable practices require understanding and reducing the environmental effect of industrial metabolite synthesis, notably in terms of energy consumption during fermentation operations (Wang et al. 2019).
5. Complexity of Biological Systems: The inherent complexity and unpredictability of biological systems pose challenges in fully optimizing and controlling metabolite production, requiring a deeper understanding of cellular processes and metabolic pathways (Abbate et al. 2023).

   Addressing these challenges through advancements in strain engineering, metabolic manipulation, regulatory compliance, environmental sustainability, and a better understanding of biological complexity is crucial for the successful industrial-scale production of metabolites.

15.1.3 Strategies for Overcoming Production Bottlenecks

An example of an iterative strain development method is the Design–Build–Test–Learn (DBTL) cycle structure, which is employed until the target performance is reached. Two steps contribute to the genetic modification of cells process: Design is the overall plan, and Build is the set of instruments and procedures needed to physically introduce sequence variety. Learn refers to computational tools for analyzing data gathered during testing, whereas Test refers to phenotyping approaches and procedures. Since the Test and Learn stages involve linking the genotype to the associated phenotype and making inferences or predictions on which genetic changes confer the intended strain improvement objectives, they may be thought of as phenotyping methods. We may enhance stresses with every engineering cycle as the Learn stage provides information to the Design phase in the subsequent DBTL cycle. Each cycle needs to be reduced in frequency and made simpler in order to shorten the time and cost of strain development.

The next sections showcase the advancements in our conceptual understanding of how to best optimize each of the four stages and elucidate how seamlessly incorporating the whole DBTL cycle may result in substantial cost and turnaround time reductions for strain engineering projects (Abbate et al. 2023).

15.2 Choosing the Right Production Platform

15.2.1 Microbial Hosts: Advantages of Bacteria, Yeast, and Fungal Systems

15.2.1.1 Bacteria System

Bacteria play a multifaceted role in industrial production, offering a plethora of advantages that contribute to the creation of valuable metabolites and the efficiency of various processes. Bacteria enzymes **efficiently catalyze reactions**, enabling rapid conversion of substrates into desired metabolites, increasing production rates and overall yield. Bacteria-based **production is sustainable**, using less energy and creating fewer harmful by-products, reducing environmental impact. **Cost-Effectiveness:** Bacteria grown on cheap substrates like agricultural waste or inexpensive carbon sources like glucose or glycerol. Bacteria-based production processes can **readily scale up** to satisfy industrial demands. **Flexibility in Genetic Manipulation:** Genetic engineering advances improve bacterial strains to boost metabolite production and optimize metabolic pathways for better yields, productivity, and tolerance to environmental conditions. Bacteria are diverse and produce a variety of **primary metabolites** like amino acids, vitamins, and enzymes that serve as nutritional supplements and aid in industrial production. They also produce **secondary metabolites** used in the biopharmaceutical industry to combat infectious diseases.

15.2.1.2 Yeast System

Yeast, a diverse group of eukaryotic single-celled microorganisms, has over 1500 recognized species. *Saccharomyces cerevisiae* and related species are commonly used in food and beverage industries, while other yeasts are termed "non-conventional." These non-conventional yeasts exhibit different metabolic pathways and regulatory patterns compared to *S. cerevisiae*, offering alternative routes for substrate utilization and product formation. Examples include *Pichia pastoris*, *Hansenula polymorpha*, *Yarrowia lipolytica*, *Pichia stipitis*, and *Kluyveromyces marxianus* (Mattanovich et al. 2014) Yeast offers several advantages for the industrial production of valuable metabolites: Yeast cells have a **fast growth rate**, allowing for rapid biomass production in a short period. This fast growth rate accelerates the production of metabolites. **Genetic Engineering Potential**: Yeast genomes are well-characterized and highly manipulable, making genetic engineering techniques effective in enhancing yeast strains for improved metabolite production, increased yield, and optimized fermentation processes (Mattanovich et al. 2014). **Simple Nutrient Requirements:** it utilize various inexpensive carbon sources, including glucose, sucrose, and even waste materials like molasses and this leads to lowering production costs. *Saccharomyces cerevisiae* plays a vital role in **ethanol production**, facilitating the large-scale fermentation of sugars to produce ethanol. Yeast primarily undergoes **fermentative metabolism** under anaerobic conditions, producing valuable metabolites such as ethanol, organic acids, and various flavor compounds. It is robust and maintains stable metabolite production in challenging environments. Yeast has **diverse Applications** including, production of primary and secondary metabolites such as phenolics, isoprenoids, alkaloids, and polyketides, through metabolic engineering and synthetic biology tools, showcasing its versatility in producing high-value chemicals, recombinant proteins, and in vivo biotransformations (Mattanovich et al. 2014). It is easily **compatible with bioreactors**, allowing for regulation of process variables like temperature, pH, oxygen concentrations, and agitation. **Versatility**: Yeast can be engineered to produce various valuable metabolites including biofuels, pharmaceuticals, vitamins, enzymes, and specialty chemicals making it a versatile biotechnological platform. It has **limited interference** with natural pathways, which is beneficial for maximizing the production particular metabolites. Physiological characteristics of yeast include optimal growth at room temperature (20–25 °C), water-soluble, CO_2 producer in dough, good shelf life, high glycolytic activity, adaptable to substrate change, high invertase and other enzyme activity to hydrolyze the higher glucofructans rapidly, ability to develop and produce enzymes in anaerobic environments, osmotic stress resistance, competitive yield, rapid substrate uptake, high metabolic rates, and robustness under stress.

15.2.1.3 Fungal System

For the industrial synthesis of useful metabolites, fungi provide a number of benefits. Fungi possess a wide range of **metabolic capabilities** that allow them to synthesize valuable compounds such as organic acids, enzymes, antibiotics, and bioactive compounds. This makes them economically attractive for **large-scale**

industrial production, as they can be produced in **high yield**. Fungal fermentation processes are **highly scalable**, enabling the production of metabolites to meet market demand. **Substrate Flexibility**: Fungi can utilize inexpensive substrates like agricultural residues and waste streams, reducing production costs and improving sustainability. **Genetic Manipulability**: Many fungal species can be genetically engineered to enhance metabolite production and improve yields, and tailor product properties to meet specific industrial requirements. Fungi have the unique **ability to secrete enzymes** and metabolites directly into the culture medium, simplifying downstream processes and reducing production costs**.** They have **tolerance to harsh environmental conditions**, enhancing the robustness of fermentation processes. Fungi have a remarkable **biotechnological potential**, as they produce various metabolites with important biological effects, such as anticancer, antidiabetic, antimicrobial, and cytotoxic properties (Roth et al. 2023). **Large-Scale Cultivation** Fungi are useful for the manufacture of organic acids, enzymes, antibiotics, and other important substances because they can be grown on an industrial scale in huge bioreactors, enabling fermentation techniques that have been proven to be economical for many years. They exhibit **genetic plasticity**, making them ideal for genetic studies and industrial applications. Fungi also produce **bioactive compounds** with medicinal, pharmaceutical, and industrial applications, contributing to the development of new drugs and products (Roth et al. 2023). Metabolites derived from microorganisms comprise a broad spectrum of compounds sourced from various origins, boasting versatile industrial applications. Here are some examples (Singh et al. 2017) (Table 15.1):

15.2.2 Plant Cell Cultures: Leveraging Plant Biosynthetic Pathways

Plant cell culture techniques are pivotal for leveraging plant biosynthetic pathways to industrially produce valuable metabolites. Plant cell cultures offer controlled environments that ensure stability and reproducibility, allowing for the generation of bioactive secondary metabolites. Utilizing plant cell cultures as biofactories has the potential to produce beneficial compounds in a sustainable and cost-effective way, benefiting the biotechnological and pharmaceutical sectors.

1. Hairy root cultures as a source of secondary metabolites

   The use of hairy root systems, initiated by Agrobacterium rhizogenes, has become popular for producing plant secondary metabolites. Hairy roots have advantages over unorganized plant tissue cultures, including fast growth, lateral branching, and genetic stability, which result in similar or higher yields of secondary metabolites as intact roots. These roots exhibit rapid, hormone-independent growth and maintain genetic stability while producing secondary metabolites comparable to intact roots. This makes them ideal for biochemical studies. *A. rhizogenes*-mediated transformation allows for the production of transgenic hairy root cultures, as Agrobacterium can transfer genetic information

Table 15.1 Metabolites derived from a broad spectrum of compounds sourced from various origins and versatile industrial applications

Metabolite	Source	Industrial application
Amino acid		
L-glutamic acid (monosodium glutamate)	*Corynebacterium glutamicum, Brevibacterium, Corynebacterium, Micrococcus, Microbacterium*	Flavor enhancer in food and beverage industry, pH regulator, essential component in aged or fermented foods such as cheese and soy sauce
L-aspartic acid	*Escherichia coli*	Used in the production of synthetic sweeteners, pharmaceuticals, and biodegradable polymers
L-phenylalanine L-tryptophan	*Escherichia coli*	Precursor for the synthesis of neurotransmitters and pharmaceuticals, used in the food and pharmaceutical industries
L-methionine	*Escherichia coli*	Essential for protein synthesis, methylation reactions, and sulfur metabolism
Vitamin		
Riboflavin (B2)	*Ascomycetes (Eremothecium ashbyii, Ashbya gossypii)*, genetically engineered *Bacillus subtilis, Corynebacterium ammoniagenes*	Used in supplements, food fortification, and animal feed.
Cyanocobalamin (B12)	*Propionibacterium shermanii, Pseudomonas denitrificans*	Commercially produced for pharmaceuticals and dietary supplements.
β-Carotene (provitamin A)	*Blakeslea trispora, Phycomyces blakesleeanus, Mucor circinelloides, Rhodotorula* spp., *Choanephora cucurbitarum*	Used in supplements, food fortification, and cosmetics.
Vitamin D	*Saccharomyces cerevisiae* (genetically modified strains)	Essential for bone health and calcium absorption. Used in supplements, fortified foods, and pharmaceuticals.
Vitamin C	*Gluconobacteroxydans, Ketogulonicigenium vulgare,* certain strains of bacteria and yeast	Essential for immune function and collagen synthesis. Used in supplements, food fortification, and cosmetics.
Enzyme		
Proteases	*Bacillus subtilis, Aspergillus oryzae*	Used in food processing for protein hydrolysis, detergent industry for stain removal, leather processing, pharmaceuticals, and waste management.

(continued)

Table 15.1 (continued)

Metabolite	Source	Industrial application
Lipases	*Candida rugosa, Rhizopus oryzae*	Used in food processing for fat hydrolysis, detergent industry for grease removal, pharmaceuticals, biodiesel production, and leather processing.
Amylases	*Bacillus licheniformis, Bacillus amyloliquefaciens*	Used in food processing for starch hydrolysis, brewing industry for malt conversion, textile industry for fabric sizing, and detergent industry.
Alcohol		
Ethanol	Microbial fermentation technology using recombinant microorganisms	– Primary metabolite and biofuel. – Chemical feedstock for various industries: solvent, cosmetics, disinfectant.
Glycerol	Microbial production by osmotolerant yeast *Candida glycerinogenes*	– Used in foods and beverages, personal care products, pharmaceuticals, alkyd resins, and tobacco humectants.
Antibiotic		
Penicillin	Fungal species *Penicillium notatum*	– Chemotherapy – Veterinary medicine – Plant pathology – Food preservation – Research laboratories
Cephalosporins	Derived from fungi and actinomycetes, including *Streptomyces* species	– Broad-spectrum antibiotic effective against various infections – Less likely to cause anaphylactic reactions compared to penicillin
Tetracyclines	Produced by *Streptomyces* species	– Treatment of infection due to a wide range of microorganisms including gram-positive and gram-negative bacteria

to plants, integrating a plasmid fragment called T-DNA into the plant nuclear DNA. The composition and organization of T-DNA sequences can vary, and some genes can influence the growth of the transformed roots. This transformation technique enables the transfer of any foreign gene of interest to the transformed hairy root clone. It is also possible to selectively alter plant secondary metabolites by introducing genes encoding specific enzymes. Establishing successful hairy root cultures requires considering factors such as the *A. rhizogenes* bacterial strain, suitable explant, effective antibiotic treatment, and appropriate culture medium. Hairy roots induced by *A. rhizogenes* offer rapid growth and biochemical stability, making them valuable for research. Researchers have developed strategies to enhance biomass and secondary metabolite production simultaneously (Ramawat et al. 2020).

2. The choice of productive lines of Hairy Roots (HRs)

   Hairy roots (HRs) are valuable for producing secondary metabolites due to their stable biochemistry and rapid growth. However, they can vary in growth and biochemical properties, even under the same conditions. Thus, careful selection is crucial. Lines with slow growth and simple structures are discarded, while those with strong growth, hormone independence, and genetic stability are preferred.
3. The elicitation:

   In response to biotic or abiotic stress, plants produce more secondary metabolites. Elicitation is the process by which different stressors (elicitors such as jasmonates, NaCl, salicylic acid, temperature, UV radiation, and fungal elicitors) stimulate a plant's defenses and increase the synthesis of particular secondary metabolites. The effectiveness of elicitation depends on the type and concentration of the elicitor, environmental conditions, and culture medium composition. There's a correlation between elicitor concentration and exposure time; for example, with salicylic acid, two optimal hyoscyamine content levels were observed in *Datura stramonium* HRs: one with a short exposure time and high concentration, and another with a lower concentration and longer exposure time.
4. The cell permeabilization:

   The aim of this approach is to use permeabilization agents to extract metabolites from plant cells in culture, allowing them to be collected from the liquid medium without damaging the biomass. Usually, secondary metabolites build up in the vacuole of plant cells and must pass via the tonoplast and plasma membrane, which have enzymes that affect metabolite release. For this, permeabilization chemicals like Tween 20, isopropanol, dimethyl sulfoxide, chitosan, ultrasound, and strong electric field pulses have been used.
5. The precursor feeding:

   Their yield is increased when secondary metabolite precursor is added to the culture medium.
6. The improvement of the composition of the culture medium:

   Optimizing culture medium composition is crucial for maximizing biomass and secondary metabolite yield, as primary and secondary metabolisms are interconnected. For example, alkaloid content can be influenced by the medium's composition. The effectiveness of elicitation response depends on mineral salts and organic components. Nutrient limitation can induce defense mechanisms in plants; for instance, solavetivone production increased under phosphate deficiency in *Hyoscyamus muticus* HRs. However, *Datura stramonium* HRs showed no significant response to elicitation when nitrate was initially limited. Combining nutrient limitation with elicitation may be effective after ensuring adequate mineral nutrition. Enhanced elicitation response occurs when nitrate in the B5 medium is fully consumed (Ramawat et al. 2020; Kulshreshtha et al. 2022).

   Various secondary compounds are derived from tissue culture techniques, including, phenolics such as caffeic acid, rosmarinic acid, and their derivatives from *Salvia bulleyana*; iridoid glycosides and phenylethanoid glycosides from

Rehmannia elata; bioactive compounds like psoralen, daidzein, and genistein from *Cullen corylifolium*; triterpenoids from *Centella asiatica*; Crocin and related compounds from *Crocus sativus*; phenolic acids, flavonoids, and other compounds from *Sphagneticola calendulacea*; tryptophan-derived quinoline alkaloids from *Ophiorrhiza alata*; monoterpene-derived indole alkaloids from *Pyrenacantha volubilis*; phenolic acids from *Mentha spicata*; various compounds including flavonoids, alkaloids, and fatty acids from *Carthamus tinctorius*; and phenolics, flavonoids, tannins, and essential oils from *Artemisia arborescens*.

15.2.3 Animal Cell Cultures: Considerations for Specialized Metabolite Production

Animal cell culture involves growing cells from multicellular eukaryotes in a controlled environment. This technique has evolved from a laboratory tool for specific studies to a method for maintaining live cell lines independently. Basic tissue culture media development has enabled the culture of a wide range of cells under various conditions. Although more challenging than growing microorganisms, advances in culture media allow both undifferentiated and differentiated cells to be cultured. Complex structures like organs can also be cultured in vitro, depending on the desired application. Cells, tissues, or organs can be used for the culture process based on the specific purpose. Types of animal cell cultures

Organ culture: It involves initiating in vitro culture using whole organs from embryos or partial adult organs. Primary explant culture tissue fragments are sustained using various methods. Cell culture: This widely-used tissue culture method involves collecting cells that grow out from explants or dispersed cell suspensions, either through enzymatic treatment or mechanical methods. These cells are then cultured as adherent monolayers on a solid substrate. Cell culture encompasses three main types: precursor cell culture, differentiated cell culture, and stem cell culture. Precursor cell culture involves undifferentiated cells committed to differentiation, while differentiated cell culture consists of fully differentiated cells that cannot further differentiate. Stem cell culture involves undifferentiated cells capable of developing into any cell type (Verma et al. 2020).

Primary cell culture: These cultures grow slowly and maintain the original tissue's characteristics, such as the same chromosome count. They can be sub-cultured to produce subsequent cultures, potentially leading to cell lines through the introduction of mutations. Primary cultures contain a variety of cell morphologies, including fibroblast, connective tissue, epithelium, and epithelioid.

Secondary cell culture: Cultures are generated from primary cell cultures that have been passed through or subcultured and are long-lasting due to regular nutrient replenishment. Enzymatic digestion is used for subculturing adherent cells, followed by resuspending them in fresh growth media. These cultures are preferred for their ease of growth and availability, aiding virological, immunological, and toxicological research.

Based on culture life span, there are two categories for the cell lines: Finite cell lines: They are characterized by restricted cell generations and slow growth rates (24–96 h). These cells require anchorage for growth, dividing for a finite number of times before eventually dying. Typically, cells in finite cell lines can divide from 20 to 100 times before reaching the end of their lifespan, which is influenced by factors such as cell lineage, species, culture conditions, and media. These cells grow as adherent cells on solid surfaces Continuous cell lines: Continuous cell lines exhibit indefinite growth through successive subcultures. They grow independently and can divide endlessly, being immortal and potentially tumorigenic due to genetic alterations. These transformed cells originate from normal primary cultures treated with carcinogens or oncogenic viruses. They can achieve higher cell densities and grow as suspensions in liquid media, forming multilayered structures in culture vessels (Yao and Asayama 2017; Verma et al. 2020).

Growth Conditions: Animal cell culture requires specific and complex media containing inorganic salts, nitrogen and energy sources, vitamins, fats, growth factors, and hormones. Additionally, pH buffers and antibiotics may be added. Optimal growth temperature varies depending on the source organism, with warm-blooded animal cells thriving at 37 °C, while cold-blooded animals grow best between 15 °C and 25 °C. Primary Cell Culture: Primary cell cultures are derived from fresh tissues obtained using aseptic techniques, either by razor dissection or enzymatic digestion. After washing to remove enzymes, the cell suspension is plated onto culture vessels and overlaid with appropriate medium for incubation at room temperature. Cell Thawing: For subsequent subcultures, preserved cell cultures are thawed in a 37 °C water bath. After warming the growth medium, the thawed cell suspension is transferred to the culture vessel, washed with alcohol, and incubated overnight before medium replacement. Trypsinizing Cells: Trypsinization involves detaching adherent cells from culture vessels using trypsin-EDTA. Following medium removal and washing, warm trypsin-EDTA is added to cover the cell monolayer. After incubation, the vessel is tapped to dislodge cells, which are then resuspended in growth medium containing serum and separated using syringe needles. Hemocytometer and electronic counting are used for quantification of cells (Verma et al. 2020). Animal cell culture has been utilized to produce a variety of valuable metabolites for various **applications**. Some examples include: **Monoclonal antibodies**: Most of the antibodies that are on the market today are produced in animal cell culture. Animal cell culture, particularly Chinese hamster ovary (CHO) cells, is frequently employed for generating monoclonal antibodies for therapeutic purposes. These antibodies are used to treat autoimmune disorders, cancer, and other medical conditions. **Vaccines**: Animal cell culture is used to produce viral vaccines, such as those for influenza, hepatitis, and human papillomavirus (HPV). Cells such as Vero cells derived from African green monkey kidney or embryonated chicken eggs are commonly used for vaccine production. Recombinant proteins: Animal cell culture is employed to produce recombinant proteins used in various applications, including therapeutic proteins (e.g., insulin, growth factors, immunoglobulins), industrial enzymes, and research reagents. Viral vectors for gene therapy: Animal cell culture is used to produce viral vectors, such as adenovirus, lentivirus, and adeno-associated virus (AAV), which are used as vehicles to deliver therapeutic genes in gene therapy applications. Hormones and growth factors: Animal cell culture can be used to produce hormones and growth factors, such as erythropoietin (EPO), human growth hormone (HGH), and insulin-like growth factors (IGFs), glucagon for therapeutic purposes. Antibiotics: Animal cell culture has been used to produce antibiotics, such as erythromycin and tetracycline, by culturing antibiotic-producing microorganisms in animal cells or using genetically engineered animal cells to produce antibiotics. Secondary metabolites: Animal cell culture produces secondary metabolites with pharmaceutical or industrial applications, such as alkaloids, flavonoids, and terpenoids, from plants or microorganisms (Yao and Asayama 2017; Verma et al. 2020).

15.2.4 Selecting the Optimal Platform Based on Target Metabolite and Cost-Effectiveness

To select the optimal platform based on the target metabolite and cost-effectiveness for the production of valuable metabolites, several factors need to be considered. Here are the key steps to guide this selection process. To begin producing a specific metabolite, first, clearly define the target metabolite, understanding its properties, applications, and market demand. Next, evaluate various production platforms like microbial, plant-based, or others based on their scalability, cost-effectiveness, and capabilities. Assess the potential for metabolic engineering in each platform to optimize metabolite production. Analyze cost factors including raw materials, infrastructure, and downstream processing to compare platforms. Consider scalability to meet increasing production demands efficiently. Evaluate market demand and choose the platform best suited to meet it. Lastly, consider recent technological advancements that could enhance production efficiency across different platforms. By systematically analyzing these factors, you can make an informed decision on selecting the most appropriate platform for the synthesis of useful metabolites according to target metabolite and cost-effectiveness (Gonzalez-Covarrubias et al. 2022). Table 15.2 shows different dimension (yield, growth rate, scalability and effectiveness) in different metabolite production system (Christian et al. 2015).

Table 15.2 Yield, growth rate, scalability and effectiveness of different metabolite production system

Factor	Microbial system	Animal system	Plant system
Yield	High	Moderate to high	Low
Growth rate	Rapid growth in favorable conditions	Moderate growth	Slow growth
Scalability	Can be scalable in controlled environments	Limited scalability due to individual needs	Limited scalability
Effectiveness	Efficient synthesis of metabolites	Efficient synthesis of metabolites	Synthesis of primary and secondary metabolites

15.3 Strategies for High-Yield Metabolite Production

15.3.1 Strain Engineering: Optimizing Metabolic Pathways for Enhanced Production

In the field of metabolite production, strain engineering refers to the process of optimizing metabolic pathways within microbial or cellular systems in order to boost the synthesis of specific metabolites, it is necessary to manipulate the genetic composition of the organism, thus enhancing its capacity to effectively generate the desired compound. Metabolic Pathway Optimization involves enhancing essential enzymes in the pathway through gene overexpression, enzyme engineering, or promoter engineering. Flux Redistribution directs more resources toward target metabolite by eliminating competing pathways. Balancing metabolic burden involves adjusting production levels to avoid harming the host organism. Substrate and cofactor engineering enhances production efficiency. Regulatory network engineering improves yields through precise production control methods. Strain engineering is crucial for maximizing metabolite yield in microbial systems for various applications. The study introduces a two-step strategy called OptDesign for optimizing metabolic pathways using strain engineering (Abbate et al. 2023). Figure 15.1 indicated a workflow for strain engineering to enhance metabolite production.

Several natural products, such as artemisinin, resveratrol, noscapine, breviscapine, opioids, and others, have been successfully synthesized using yeast engineering. To achieve commercially viable strain performance, a comprehensive approach combining rational and whole genome methods, phenotyping, and predictive models is necessary. The effectiveness of this approach was demonstrated using bakuchiol as an example in *S. cerevisiae*. Bakuchiol is a meroterpene from *Psoralea corylifolia*, popular in the cosmetic industry for skin benefits similar to retinol but without side effects. The molecule was discovered in 1966 in India, derived from the bakuchi plant used in ayurvedic and Chinese medicine for ages. Initially, the biosynthesis pathway of bakuchiol was unclear, and microbial production was unreported. Identification of the bakuchiol synthase converting precursors into bakuchiol was prioritized. Through bioinformatics and screening, the synthase enzyme was discovered, needing significant activity enhancement. Enzyme improvement was done via in vivo mutation and genome editing, along with engineering approaches to balance precursor pathways and enhance strain fitness. Within a year, a strain that produced grams of bakuchiol per liter was created through a series of engineering and selection rounds (Abbate et al. 2023).

In another study, glucosamine and GlcNAc are used in pharmaceuticals. Scientists enhanced GlcNAc production in *Corynebacterium glutamicum* by inserting GNA1 gene into the plasmid and transforming it into different strains, showing varying GlcNAc levels. Genome sequencing revealed differences among the strains. Deleting nagA and gamA in a strain increased GlcNAc production. Blocking lactate synthesis further boosted GlcNAc titer. Introducing glmS gene led to the highest GlcNAc production. This study marks the first step toward large-scale GlcNAc production through metabolic engineering of *C. glutamicum* (Deng et al. 2019).

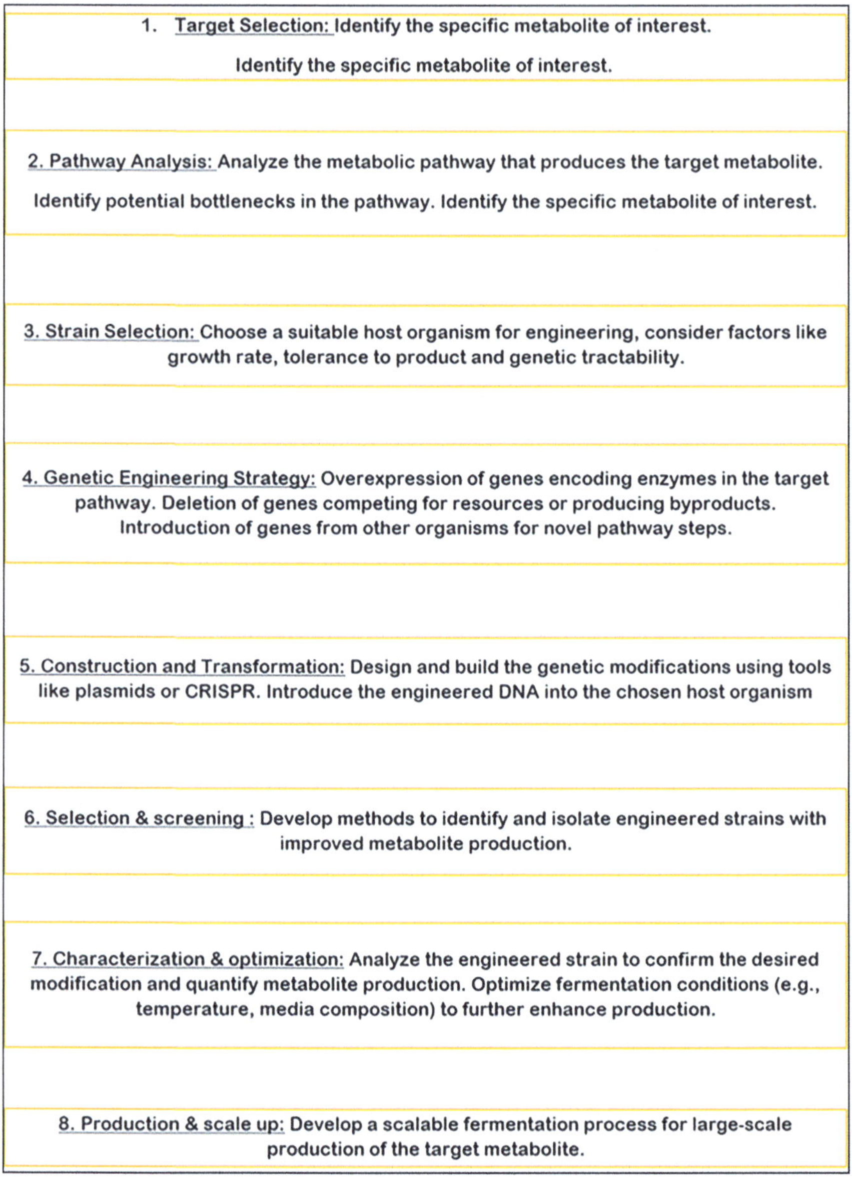

Fig. 15.1 A workflow for strain engineering to enhance metabolite production

15.3.2 Media Optimization: Providing the Right Nutrients and Growth Conditions

Media optimization aims to enhance metabolite yield by providing nutrients and growth conditions. Various techniques like RSM, ANN, and genetic algorithms are used for production medium optimization. Researchers target maximum yield through nutrient and growth condition optimization. The process includes systematic design considering strain improvement and process parameters. Nutritional control is crucial for maintaining proper regulation and high productivity. Media optimization techniques provide efficient ways to intensify metabolite yield.

Carbon: The rate of carbon source metabolism affects biomass and metabolite formation. Different carbon sources impact microbe growth and metabolite production. Lactose promotes secondary metabolites; glucose represses them. Lactose use overcame glucose repression for penicillin production.

Nitrogen: Microbes use inorganic or organic nitrogen sources. Specific amino acids can increase productivity, while unsuitable ones inhibit metabolite synthesis.

Phosphate: Phosphate is essential for phospholipid production and nucleic acid synthesis in microbes. Phosphate amount in medium affects metabolite production. The amount of metabolites' production, like actinorhodin, cephalosporin, etc., is affected by inorganic phosphate concentration. Lower phosphate concentration initiates antibiotic production, but higher concentration inhibits secondary metabolism and metabolite production.

15.3.3 Downstream Processing: Efficient Purification and Recovery of Metabolites

Downstream processing in biopharmaceuticals plays an integral role in the efficient purification and recovery of metabolites. In order to extract, purify, and concentrate the synthetic drug substance from a complicated bulk matrix, this method involves recovering and purifying a drug substance from natural sources such as animal or bacterial cells (Downstream Processing in Biopharma 2022). Different metabolite type, purification method along with brief description provided in Table 15.3.

Various techniques such as centrifugation, activated carbon treatment, microfiltration, evaporation, crystallization, and freeze-drying are commonly used in downstream processing to purify and recover metabolites effectively (Sarenkova et al. 2022). Additionally, in-line monitoring techniques have been developed to enhance the purification process by determining concentrations of metabolites, nutrients, and host cell proteins, leading to improved process yield and product quality (Makovitzki et al. 2021). Depending on the nature of the product and method of synthesis, downstream processing (DSP) generally involves harvest that includes separating product from debris to optimize yield and quality. Capture holds onto desired product while minimizing impurities. Ultrafiltration concentrates product stream based on membrane pore size. Diafiltration exchanges products into a new buffer. Purification removes impurities while retaining product. Bioconjugate molecules combine different therapeutic types for increased efficacy. Antibody–drug conjugates are common bioconjugates synthesized by linking an antibody to an active pharmaceutical ingredient. Formulation involves transitioning a drug substance into a formulated drug product. It adapts the product molecule for clinical administration by changing its physical state.

Table 15.3 Different metabolite types and purification methods

Metabolite type	Purification method	Description
Polar	Chromatography	Uses a mobile phase (such as solvent) and a stationary phase (such as silica gel) to separate chemicals according to the affinity.
	Ion exchange chromatography	Uses a mobile phase containing ions of the opposite charge and a charged stationary phase to separate molecules according to their charge characteristics.
	Size exclusion chromatography	Molecules are separated into different sizes, permitting smaller ones to pass through the stationary phase's pores while keeping larger molecules out and causing them to elute first.
Non-polar	Crystallization	Involves the selective precipitation of the target metabolite by manipulating solvent conditions (e.g., temperature, concentration) to promote its crystallization.
	Distillation	Vaporizes the component with the lower boiling point in the combination and then condenses it back into a liquid, separating the compounds according to differences in their boiling points.
Amphiphilic	Liquid–liquid extraction	Utilizes the differential solubility of compounds in two immiscible solvents to partition the target metabolite between the aqueous and organic phases.
	Solid-phase extraction	Utilizes a solid-phase material (e.g., resin) to selectively adsorb the target metabolite from a liquid sample, followed by elution with a suitable solvent.
Reactive	Affinity chromatography	Utilizes a stationary phase with specific ligands that bind selectively to the target metabolite, allowing purification by exploiting biological or chemical interactions.

15.3.3.1 Methods

- **Depth Filtration/Hollow Fiber Filters:** These filters separate cells and debris from the product in liquid form. Depth filters work like a strainer, while hollow fibers are used in continuous processes to capture product while letting some cells through temporarily. A new technique called acoustic field-flow separation is also being developed.
- **Centrifugation:** This method spins a mixture to separate components based on weight. While commonly used, it is becoming less practical for large-scale production due to high biomass levels.
- **Product Separation: Acid/Surfactant Lysis & Organic-Phase Absorption:** These methods are used to break open cells and separate the desired product from cellular debris. Acid/surfactant lysis is mainly used for bacteria and yeast, while organic-phase absorption is common for bacterial production.
- **Flocculation & Precipitation:** Flocculation clumps cells together for easier separation with filters or centrifugation. Precipitation uses changes in solvent or salt concentration to make the product less soluble and easier to capture.

- **Monitoring & Analysis: Particle Size Distribution (PSD):** Knowing the size and distribution of particles in the solution is crucial for choosing the right filter and predicting filter performance.
- **Purification: Affinity Chromatography:** This method uses a special column that attracts the target product like a magnet, while other components flow through. It is often the first step for purifying proteins like monoclonal antibodies.
- **Ion Exchange Chromatography (IEX):** This method separates molecules based on their electrical charge. It is a versatile technique used for various purification steps (Downstream Processing in Biopharma 2022).

15.3.4 Bioreactor Design and Process Optimization: Scaling Up Production for Industrial Needs

A bioreactor functions as a specialized vessel for conducting fermentation processes, facilitating the production of diverse chemicals and biological reactions. Enclosed within, it features essential components such as provisions for aeration, agitation, temperature and pH regulation, as well as mechanisms for waste biomass removal through drainage or overflow vents. Key functionalities expected from a bioreactor include (Fig. 15.2):

1. Stirring/agitation to ensure thorough mixing of cells and medium
2. Aeration (for aerobic fermenters) to supply oxygen for optimal functioning

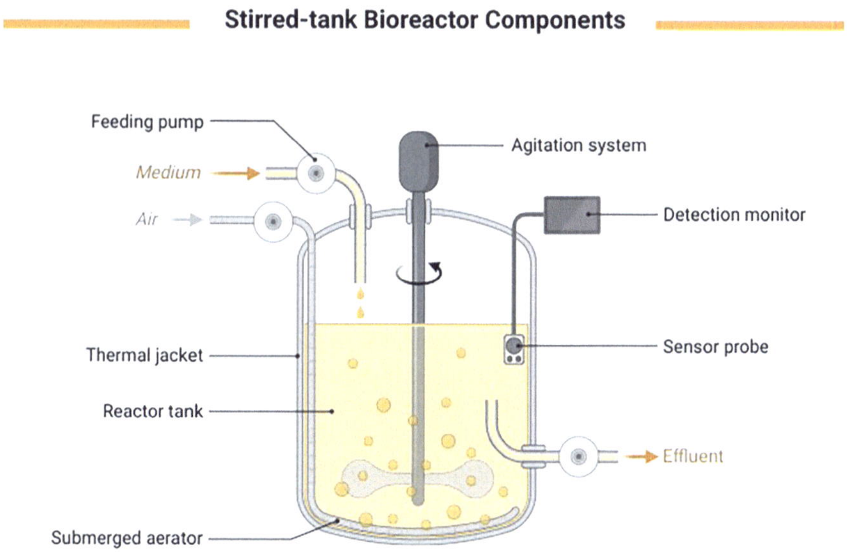

Fig. 15.2 Schematic of a bioreactor system for industrial metabolite production

3. Control over critical parameters such as temperature, pH, pressure, aeration, nutrient supply and liquid levels
4. Ensuring sterility through sterilization procedures and maintenance
5. Facilitating the withdrawal of cells and medium as needed.

Bioreactors are instrumental in the generation of biomass, metabolites, and antibiotics, playing a crucial role in various industrial and research settings. Bioreactor scale up involves maintaining critical performance characteristics like mixing and mass transfer, with considerations for geometric similarity and engineering parameters (Arias and Kotipalli 2021). Bioreactors are essential for providing optimal growth conditions for microorganisms to produce various chemicals and biological reactions (Magar 2023). Scaling up bioreactor setups can be done by increasing the magnitude and power of existing bioreactors, with different scales like laboratory, pilot, and industrial scales. Different scaling approaches include scaling out by unit or scaling up by volume, each with its own advantages and disadvantages. Disposable bioreactors are becoming popular for scale-up due to ease of use and lower costs, offering advantages like efficient mixing and oxygen transfer. The design of large-scale production reactors involves factors like sufficient oxygen supply, mixing for homogeneity, and avoiding high shear forces that could lead to cell death. Bioprocessing scale up aims to reproduce optimized processes from small to large scales, focusing on maintaining efficiency and product quality across scales (Hajam et al. 2023). The process of scaling up production for metabolites in bioreactors involves optimizing various parameters to ensure efficient and effective operations. Key aspects highlighted in the sources include:

Bioreactor Scale-Up Strategies: Scaling bioprocesses involves challenges due to the physical limitations of increasing all process parameters equally. Therefore, suitable scale-up strategies are essential (Seidel et al. 2023).

Fermentation Process Optimization: To effectively optimize and scale up fermentation processes, it is essential to explore and identify the most suitable genetic engineering modifications and fermentation conditions through the use of mechanistic and data-driven models. This involves a detailed examination of the underlying mechanisms and data in order to make informed decisions and achieve optimal results (Du et al. 2022).

Bioreactor Operating Parameters: Scaling up bioreactors involves fine-tuning operational parameters for large-scale bioreactors utilized in production processes. This optimization encompasses adjustments for physical, chemical, and biological variables that impact cell culture within the bioreactor (Chaudhry 2024).

Fermentation Process Scale-Up: The review provides insights into the fermentation process from shake flask cultures to scale-up, emphasizing the challenges, concepts, and perspectives of scaling up microbial endophyte production in bioreactors (Ganeshan et al. 2021).

In summary, optimizing bioreactor design and process parameters for scaling up production for metabolites involves a comprehensive approach that considers genetic engineering strategies, fermentation conditions, operating parameters, and the use of advanced tools like CFD for efficient and cost-effective scale-up.

15.4 Integration with Advanced Technologies

15.4.1 In Situ Monitoring and Control Systems: Real-Time Optimization of Production Processes

The shift to intelligent biomanufacturing is examined in the article, with a focus on the need for large data sets for control and optimization. It discusses the latest advancements in big data integration, process analytical technologies (PAT), machine learning (ML), and monitoring and control technology. Regulation compliance and sensor dependability are addressed. The article focuses on the potential benefits of smart manufacturing for regulatory compliance, productivity, and quality in biomanufacturing.

1. Bioprocess monitoring and control

   The majority of biomanufacturing techniques include the culture of microorganisms, indicating a process combining intricate chemical, physical, and biological phenomena, despite differences in goals and final products (Gargalo et al. 2020). Controlling the process conditions at the upstream, downstream, and product formulation phases of the biomanufacturing process requires a strong and reliable analytical system. The intrinsic characteristics of the cell cultures, for instance, turn the upstream processing section into a complex multi-phase system with significant variability in cell development. Therefore, sensors are needed to track chemical quantities like pH and dissolved oxygen, physical factors like temperature and pressure, and biological characteristics like cell density or metabolite concentrations. It should be mentioned that tracking and quantifying the majority of biological components might be quite difficult (Gargalo et al. 2020).
2. Real-time data gathering

   In order to improve productivity, efficiency, and repeatability in bioprocess control, real-time bioreactor monitoring is essential. Making educated judgments regarding control requires timely information on biomass, substrates, metabolites, products, and nutrients. It is ensured that the production process proceeds according to plan by measuring products such as cells, proteins, and by-products. By quickly altering certain parameters and spotting inefficient processes, fermentation progress monitoring allows for optimization. Additional metrics are used in fermentation processes, such as dissolved CO_2 and optical density, which are frequently detected with optical probes or spectroscopic techniques (Gargalo et al. 2020; Claßen et al. 2017). Real-time analysis of substrates, metabolites, and products is made possible by combined analytical and chemometric procedures, which do away with the requirement for offline techniques like GC/MS or HPLC. These methods, which include spectroscopic, optical, and in situ biosensors, have the ability to track important process variables in real-time, enhancing bioprocess control (Gargalo et al. 2020).
3. Biosensors in situ

Biosensors are analytical instruments made up of a transducer that transforms biological impulses into electrical signals and an immobilized sensing material. Usually, they are composed of a signal transducer unit with an amplified biological detecting element that has been immobilized, and a signal converter unit. Through binding or catalytic processes, the bio-components—such as enzymes or antibodies—selectively identify analytes. The transducer unit, which may employ optical, electrochemical, calorimetrical, or piezoelectrical principles, controls the interaction with analytes. Biosensors are promise for smart manufacturing because they provide quick, affordable, and dependable analytical results (Prasad et al. 2009). Biosensors for the online measurement of analytes such as glucose, glutamate, and lactate are still in the early stages of research and are not yet commonly employed in large-scale industrial production processes, despite recent advancements (Gargalo et al. 2020; Prasad et al. 2009).

4. Spectroscopic instrumentation

   Since spectroscopic sensors can access a large range of variables and deliver real-time data without considerable delays, they have enormous promise for bioprocess monitoring. For example, the effective use of 2-D fluorescence spectroscopy in high-throughput fermentation systems such as BioLector® has made it possible to monitor a variety of microbes and cellular components (Muscat et al. 2021). Although spectroscopic sensor applications are in line with the Process Analytical Technology (PAT) initiative, they are more common in research than in industry because of obstacles like strict GMP environment requirements for well-documented systems, opaque monitoring procedures, and high equipment and expertise costs. Third-party solutions like Analytic Trust from Q-Interline are one attempt to solve these issues; however, security concerns still prevent these technologies from being widely adopted (Gargalo et al. 2020).

5. Chemo metrics: the study of multivariate data

   Spectroscopic sensing generates large datasets requiring multivariate data analysis (MVDA), a subset of machine learning, for continuous real-time monitoring of bioprocess variables. MVDA, like Principal Component Analysis (PCA), extracts insights from spectra by reducing complexity and identifying outliers (Eshawu and Ghalsasi 2024). Qualitative approaches such as PCA aid in process supervision and batch classification, while quantitative models like partial least square regression (PLS) correlate spectral data with process variables for online prediction. In fermentation, PLS models account for linear relationships between spectral data and concentrations, crucial for accurate predictions. External validation ensures model accuracy and adaptability to process changes, requiring expertise for updates. Third-party solutions like Analytic Trust provide necessary resources for model maintenance, highlighting the role of machine learning in automation and control (Gargalo et al. 2020)

6. Floating sensors

   Provide a new method for gathering data in bioprocesses without requiring physical connections to reactors. These sensors are commercially available and may move freely within bioreactors to detect factors such as pH, temperature, and pressure. One example of such a sensor is the wireless spherical particles developed by Zimmermann et al. (Muscat et al. 2021; Leong et al. 2021). Sterilizable shells, electronic boards, batteries, and sensor devices are included in their design. They provide data in real time to external analytical units so that spatial reactor dynamics may be understood (Gargalo et al. 2020).

7. Soft sensors

   Another development is the use of algorithms to evaluate measurements and produce information about the condition of the process. These models, which may be divided into mechanistic, data-driven, and hybrid methods, take use of advances in signal processing and computing capacity and fit in nicely with PAT and smart manufacturing initiatives (Wang et al. 2004).

15.4.2 Metabolic Modeling and Design: In Silico Tools for Predicting and Engineering High-Yield Pathways

In biotechnology, systems-wide methods that make use of genome-scale models have shown to be invaluable for pinpointing genetic target genes during the logical engineering of production strains intended to yield bio-based chemicals, fuels, or medicines. While several in silico techniques, like MOMA, OptKnock, OptGene, and strain design based on optimum theoretical yield, are excellent at identifying gene deletion targets, their capacity to predict genes for amplification is restricted. While experimental approaches like as 13C metabolic flux data or bi-level optimization frameworks provide useful information about putative amplification targets, they are labor-intensive and might not cover all important gene candidates (Wang et al. 2004)

A useful answer is offered by elementary flux mode analysis, a genome-scale modeling technique that makes it possible to forecast the desired cell phenotypes that would arise from changes in process parameters or genotype perturbations. This approach has shown promise for discovering genetically independent pathways in a variety of species and has been successful in predicting viable gene deletion strategies (Trinh et al. 2009). In silico techniques that can concurrently identify targets for amplification or deletion without depending on experimental data or predetermined assumptions are still required, though (Wang et al. 2004; Campilho and Silva 2023).

The creation of such techniques may significantly improve the logical design of production strains for a variety of industrial uses. Simulations utilizing cell factories of industrial relevance, such as *C. glutamicum* and *A. niger*, demonstrate how these systems-wide techniques have the potential to transform biotechnological processes and speed up the creation of new bio-based products (Wang et al. 2004).

15.4.3 Automation and Robotics: Ensuring Consistent and Efficient Large-Scale Production

The integration of automation and robotics has transformed industrial processes, enhancing efficiency, precision, and flexibility. These advancements have led to improved quality, reduced human error, and increased production speed across various industries. Particularly in the era of Industry 4.0, automation and robotics are pivotal, with smart manufacturing and mechatronics playing key roles. This editorial examines the state of the art, applications, current limitations, and future prospects of automation and robotics in improving industrial processes. Automation in Industry 4.0 enables operators to shift to supervisory roles, easing monotonous tasks while boosting competitiveness. Industry 4.0 integrates digital technologies like IoT, AI, and big data, fostering smart factories where machines communicate and decide autonomously (Ogenyi et al. 2021). This approach enhances efficiency, productivity, and resource allocation through predictive maintenance and real-time data analysis, minimizing waste and optimizing production quality.

Industrial robots consist of three main components: the manipulator (robot), controller, and user interface. Sensors and actuators equip robots with senses and muscles, providing data on surroundings and enabling physical movement. Advances in sensor technology, like LiDAR and cameras, enhance navigation and interaction capabilities, while improved actuators, such as advanced servo motors, increase accuracy and agility. These advancements enable robots to perform complex tasks like pick-and-place operations and collaborate with humans, particularly beneficial in manufacturing (Campilho and Silva 2023). In industries like automotive production, robotization facilitates cost-effective assembly of various vehicles on shared production lines. Integration of robots is driven by the need for hazardous environment operation, repetitive tasks, intricate handling, and continuous operation management. State-of-the-art control systems and sensor technologies ensure safe utilization in production lines, including collaborative operations, enhancing manufacturing efficiency by combining robotic productivity with human cognitive skills (Ogenyi et al. 2021).

Automation and robotics find widespread applications across various industries and societal domains:

- Manufacturing: Automated assembly lines ensure competitiveness by producing a wide array of products with precision and consistency, minimizing defects and enhancing product quality (Ogenyi et al. 2021).
- Healthcare: Robotics advances in surgery enable minimally invasive procedures with high precision, while robots assist in patient care, medication delivery, and rehabilitation (Costa et al. 2017).
- Logistics and warehousing: Robotics, including AGVs and drones, streamline material handling and order picking, accelerating processes and reducing errors in inventory management (Ogenyi et al. 2021; Costa et al. 2017).
- Agriculture: Robots perform tasks like planting, harvesting, and crop monitoring, enhancing farming efficiency to meet global food demand (Costa et al. 2017).

15.5 Sustainability Consideration

15.5.1 Utilizing Renewable Feedstock: Sustainable Sources for Substrates and Nutrients

The UN's sustainable development objectives are pushing the sector toward sustainable practices, with bioactive compounds (BACs) being a key focus due to their great added value and possible uses in medicine, pharmaceutics, and other industries. Still, addressing issues such as feedstock selection, valorizing residual materials, and scalability, logistics, and viability is crucial for sustainable processes. Examples of biorefineries include hemp, grapevine shoots, and wheat straw (Sebastián et al. 2019).

In recent years, a wide range of techniques, goods, and feedstocks have been investigated, including chemical, biochemical, thermochemical, catalytic, and physical processes. Biomolecules, organic compounds, biomaterials, biofuels, bioenergy, and food and feed are a few of them. The creation of commodities and energy for a range of commercial and industrial applications is necessary for the shift to a bioeconomy (Sebastián et al. 2019).

Bioactive chemicals, which have strong added value and possible applications in pharmaceutics, medicine, nutrition, and cosmetics, are the most attractive. However, their definition remains vague and confusing. They fall into two categories: materials that interact with one another or elements of living tissue, offering a variety of possible outcomes (Sebastián et al. 2019).

Bioactive compounds (BACs) are essential for sustainable production processes, but their economic relevance requires further analysis. They can be produced through fermentation, synthesis, or extraction from their natural matrix. Finding suitable raw resources is crucial for sustainable production. The value-adding of remaining raw material components is also essential for sustainable production. Understanding the matrix constituent parts of feedstock is crucial for determining potential products. The availability of feedstock directly affects the process scale, which is strongly associated with logistics and transportation. Therefore, understanding the feedstock matrix is essential for sustainable production (Sebastián et al. 2019).

15.5.1.1 Important Aspects to Consider for the Sustainability

Sustainable development combines human growth with the preservation of nature, particularly ecosystems and natural resources. It necessitates taking social, economic, and ecological factors into account while creating and running both new and old systems. Technical elements are crucial for evaluating a process's ecological and economic performance. Sustainability and sustainable development are ongoing debates, and they are essential focal points that cannot be overlooked. Biomass and renewable feedstocks have distinct constituents with potential uses that need optimization. The feedstock's availability directly impacts the process's scalability, along with the logistics and supply chain for a certain size are closely tied to the economic behavior of small- and large-scale activities. The debate on sustainability and sustainable development is extensive, and they should not be overlooked (Sebastián et al. 2019).

15.5.1.2 Bioactive Compounds and Raw Materials

Bioactive chemicals (BACs) are substances with biological activity that cause a reaction in living tissues and have positive health consequences. Plants and some foods, including fruits, vegetables, nuts, oils, and whole grains, contain them in trace amounts. These substances have undergone significant research to assess their impact on health. The origin of BACs should be considered, as synthetic molecules can still exhibit biological action. Many bioactive principles used in pharmaceutical, cosmetic, and agricultural products are derived from synthetic compounds. Other definitions of bioactive chemicals consider the source of molecules created by fermentation or synthesis, as well as those isolated from a natural matrix. Coumarin is an example of a synthetic or natural BAC, depending on whether it is taken out of a natural matrix or created synthetically via the Perkin reaction (Sebastián et al. 2019).

15.5.1.3 Types and Sources of BAC

Phenolic compounds are a group of naturally occurring secondary metabolites with functions related to development, reproduction, defense, and sensory attributes. They are divided into phenolic acids, flavonoids, stilbenes, coumarins, lignins, and tannins. Phenolic acids, which are formed through the interaction with cellulose, hemicellulose, and pectin, have antimicrobial, antioxidant, and anticancer properties. Flavonoids, with a C6–C3–C6 structure, are divided into isoflavones, anthocyanidins, flavones, flavonols, flavanones, and flavanonols. These compounds have been linked to anti-inflammatory, anti-cancer, antioxidant, and cardioprotective properties. Stilbenes, with a C6–C2–C6 carbon backbone, have been investigated for their potential to suppress Parkinson's and Alzheimer's diseases and act as antibacterials against *Staphylococcus aureus*. Coumarins, a C6–C3 skeleton and oxygen heterocycle, are found in plants and bacteria, with numerous bioactive uses including anti-inflammatory, antibacterial, antioxidant, antiviral, antiasthmatic, and anti-Alzheimer's. Tannins, phenolic chemicals, dissolve in water and precipitate proteins and alkaloids, possess strong antioxidant potential due to their phenolic rings. Antibiotics are another category of bioaccumulative agents (BACs), as more germs are becoming resistant to current antibiotics. Three techniques can be used to produce antibiotics: semi-synthetic, natural fermentation, and synthetic (Sebastián et al. 2019). Alternative bioactive substances in plants and fruits include cannabinoids, curcuminoids, and carotenoids. Carotenoids are pigments produced by plants, bacteria, fungal, and algae, and are responsible for the distinctive color of tomatoes, maize, carrots, and pumpkins. They shield chlorophyll from deterioration and absorb light for photosynthesis, having antioxidant activity. Studies have shown their protective properties against cancer in various areas (Sebastián et al. 2019). Curcuminoids, natural phenols that produce yellow color, are a class of chemical molecules that modify the brain's neurotransmitters by binding to the endocannabinoid system's cannabinoid receptors. The most prominent molecule in this category is curcumin, but other variants have been created due to its poor solubility. The antioxidant properties of curcumin and its derivatives have been investigated (Sebastián et al. 2019). Cannabinoids, a class of chemical molecules, modify the brain's neurotransmitters by binding to the endocannabinoid system's cannabinoid receptors. The most well-known type of cannabis is tetrahydrocannabinol (THC), a psychoactive cannabinoid that is the primary cause of cannabis plant prohibition and regulation in many countries (Sebastián et al. 2019). Cannabinoids are being investigated for uses in nutrition and medicine, with examples including cannabigerol (CBG) for glaucoma and sclerosis, cannabidiol (CBD) for schizophrenia and epilepsy, and cannabidivarin (CBDV) for epilepsy. It has been discovered that C. sativa, commonly referred to as hemp, has high quantities of CBD and low levels of THC, leading to its significant usage in the past decade (Sebastián et al. 2019).

15.5.1.4 Sources

This section discusses potential sources for producing BACs, which are secondary metabolites from plants, fungi, or monera. It discusses the sources (microbe and plant-based), the kinds of BACs that are present in these sources, and how to use them (Table 15.4). The discussion focuses on the importance of understanding the various kingdoms involved in BAC production (Sebastián et al. 2019).

Table 15.4 Sources of various plant-based bioactive chemicals, their types, and their uses

Source/Type	Biological activity
1. Cannabinoids (Hemp)	Inflammation, anxiety, relieves convulsion, and nausea
2. Nano-lignin (Wheat straw)	Antioxidant; bactericidal
3. Resveratrol Kaempferol Quercetin (Grape (vine shoots, seed))	Bactericidal effects; decreased risk of cardiovascular disease; inhibition of lipase activity
4. Phenolic acids Flavonols dihydrochalcones (Apple)	Lower cholesterol, lower risk of cardiovascular disease, and lower risk of type 2 diabetes
5. Peel Flavonoids (Citrus (orange, lemon, grapefruit))	Antitumor, antibacterial, antiatherosclerotic, lowers blood cholesterol
6. Flavones tannin Anthocyanidins (Black rice)	Antitumor; antiallergic; antiatherosclerosis
7. Phenolic acids Flavonoids Tannins (Blackberry)	Anticancer effect
8. Apricot coumarins	Antimicrobial; antioxidant
9. Anthocyanins (Soybean)	Anti-inflammation
10. Catechin and quercetin (Aloe vera)	Anti-inflammation; anticancer
11. Gallic acid and chlorogenic acid (Coffee)	Antioxidant effect

15.5.1.5 Technologies

This text discusses the different BAC kinds and sources, their extraction and manufacturing technology, focusing on microbial synthesis and extraction techniques for plant-based sources that include BACs (Sebastián et al. 2019).

15.5.1.6 Extraction Techniques: Conventional Extraction Techniques

Solvents are essential in conventional extraction techniques, influencing factors like molecular affinity, cost, and environmental impact. Common solvents include water, ethanol, methanol, chloroform, ether, and acetone. Common extraction methods include Soxhlet extraction, maceration, hydrodistillation, and solid–liquid extraction. However, traditional methods are limited by their lengthy extraction durations, poor selectivity, and tendency to heat-degrade delicate chemicals. Non-conventional techniques aim to address these issues by improving feedstock accessibility or innovating extraction processes (Sebastián et al. 2019).

15.5.1.7 Non-conventional Extraction Techniques

Methods of Extraction: These methods concentrate on the extraction process itself and can create the necessary conditions for the extraction of bioactive substances from their matrix.

15.5.1.8 Pressurized Liquid Extraction (PLE)

This method uses intense pressure to extract desired chemicals, increasing solubility and mass transfer while decreasing viscosity and surface tension. The selected solvent, temperature, pressure, and extraction time are crucial factors. This method facilitates the extraction of BACs and hydrolysis of carbohydrate polymers, with other applications, making it a valuable tool in various industries (Sebastián et al. 2019).

15.5.1.9 Supercritical Fluid Extraction (SFE)

Supercritical fluids, which exhibit low viscosity, surface tension, and high diffusion coefficient, are ideal for extraction due to their low critical conditions and quick solvent regeneration. Because of its low critical conditions and rapid solvent regeneration, carbon dioxide is the perfect solvent for SFE. Because of its polarity, it is used to remove fats, lipids, and non-polar substances. The use of a co-solvent can increase the extraction range. Temperature, pressure, particle size, moisture content, extraction duration, flow rate, and solvent-to-feed ratio are crucial factors in SFE extraction (Sebastián et al. 2019).

15.5.1.10 Ionic Liquids (IL)

Ionic liquids, which are non-polar liquid combinations of organic cations and anion, are being researched as potential solvents to replace traditional methods. These materials offer strong solvating properties for both polar and non-polar compounds, low vapor pressure, low combustibility, high viscosity, low electricity conduction, and great thermal stability. Because they may be customized for certain chemical extractions, they are often referred to as "designer solvents" (Sebastián et al. 2019).

15.5.1.11 Sustainable Production: Feedstock Composition and Further Valorization

To ensure sustainability and minimize residue production, it is essential to effectively utilize all raw material components in the extraction or production of bioactive compounds (BACs). Identifying constituents and their potential uses is crucial for extracted BACs. Substrates and nutrients are pivotal in BAC production, particularly antibiotics via fermentation. Various carbohydrate sources, including second-generation raw materials, can serve as substrates. Plant-based feedstocks offer a dual benefit: extracting BACs and providing residual fractions for fermentable sugar production (Sebastián et al. 2019).

15.5.1.12 Feedstock Composition

The composition of feedstock matrices is crucial for determining potential platforms and products. Main components include lignocellulose, starch, oils, and proteins. The lignocellulosic complex consists of cellulose, hemicellulose, and lignin. Cellulose forms the structural fiber, while hemicellulose acts as a binding material. Lignin provides plant structural strength. Agricultural feedstocks are rich in starch and glycan, while forestry materials and waste have higher lignin content (Sebastián et al. 2019).

15.5.1.13 Further Valorization

Plant components, such as seeds from grapes, hemp, olives, and pumpkin, have various uses beyond their structural elements. These seeds contain high concentrations of oils, which can be extracted for culinary and biodiesel production. Pectin is a food additive that is a polysaccharide that may be found in citrus peels, blackberries, and apple pomace and used in industrial processes for gelling, stabilizing, or emulsifying (Sebastián et al. 2019).

15.5.1.14 Final Valorization

Residual solids, which cannot be separated, provide a source of elements such as hydrogen, nitrogen, and carbon for thermochemical and biological reactions generating electricity and heat. Recovery of energy from leftover biomass, such as anaerobic digestion, gasification, and pyrolysis, can meet energy needs and reduce environmental impact, making it a key strategy for reducing global energy consumption (Sebastián et al. 2019).

15.5.2 Minimizing Waste Management: Strategies for Environmental Impact

Anthropogenic activities like resource overexploitation and pollution contribute to environmental degradation. Waste creation, including industrial, biomedical, agricultural, and municipal solid waste, has surged due to the expanding human population. Agricultural waste may be recycled as organic matter through processes including composting, anaerobic digestion, and biogas generation. In order to minimize environmental impact, the management of industrial waste includes trash reduction, pollution avoidance, and recycling techniques including closed-loop production systems and eco-design (Hajam et al. 2023).

15.5.2.1 Waste Generation: An Alarming Call for Environmental Instability

By 2025, it is anticipated that the world's garbage production would amount to 2.2 billion tons per year, mostly due to unsustainable economic expansion, urbanization, industrialization, and population increase. Small towns in India create 0.41 kg of solid trash per inhabitant per day, whereas rural regions produce between 0.3 and 0.4 million metric tons per day. In addition to handling more than 75% of the world's garbage, landfills release 5% of greenhouse gases, including methane, which exacerbates climate change. Scientific methods for converting solid waste components into functional products offer potential solutions for environmental protection and sustainable development (Hajam et al. 2023).

15.5.2.2 Consequences of Improper Waste Management

The world still has problems with garbage management, and over 3.5 billion people do not know how to handle it properly. As a result, there were 1.3 billion tons of municipal solid garbage in 2012; by 2025, that amount is predicted to increase to 2.2 billion tons. Sustainable solutions such as composting and waste-to-energy are becoming more and more popular. The conversion of organic waste into useful resources is one of the waste management solutions that have been developed to address environmental pollution, health concerns, habitat loss, soil degradation, and resource depletion (Hajam et al. 2023).

15.5.2.3 Waste Management Strategies

By lowering the quantity of garbage that ends up in landfills, waste management technologies like recycling and waste-to-energy conversion are essential for lowering landfill waste and environmental degradation and promoting a sustainable future (Hajam et al. 2023).

15.5.2.4 11 R Strategy

A waste management technique known as the "11 R formula" seeks to reduce waste production, increase resource recovery, and advance sustainability. It entails reassessing consuming patterns, rejecting pointless purchases, cutting down on waste generation, reusing and recycling materials, regifting, fixing damaged goods, filling up containers, composting organic waste, and continuing this cycle. This strategy promotes sustainability and a circular economy by addressing the effects on the environment, the economy, and society. It encourages a sustainable and rethinking approach to waste management, ensuring a healthier environment for all (Hajam et al. 2023).

15.5.2.5 Waste-to-Energy (WtE): A Temptation for Municipalities

Waste-to-Energy (WtE) systems use processes including incineration, co-processing, anaerobic digestion, landfill gas collection, and pyrolysis/gasification to transform trash into energy like heat, electricity, and biogas. Although these technologies are utilized in poor countries, their adoption is limited to pilot projects. Municipalities have a great chance to solve the problems associated with waste management and energy generation through WtE. WtE facilities harness the energy contained in waste to generate heat or electricity, which lowers reliance on fossil fuels, lowers greenhouse gas emissions, and advances a sustainable energy environment (Hajam et al. 2023).

15.5.2.6 Waste-to-Energy and the Circular Economy

A fundamental part of the Circular Economy concept is Waste-to-Energy (WtE), which turns non-recyclable garbage into energy in order to encourage recycling and reuse. This encourages sustainability and lessens dependency on fossil fuels. WtE facilities aid in resource recovery, waste management, and circular resource use. By integrating WtE, landfill usage decreases, fossil fuel reliance decreases, and waste value is maximized, leading to more efficient and sustainable resource utilization (Hajam et al. 2023).

15.5.2.7 Composting

The natural process of composting turns organic waste into nutrient-rich manure free of toxicants, providing a green waste management option for a variety of waste kinds. Organic materials are broken down by microorganisms to produce compost that is rich in potassium, phosphorus, and nitrogen and ideal for use in agriculture. Thermophilic composting destroys pathogens and breaks down lignin, generating humic substances for soil remediation. Composting also removes toxicants, pollutants, and plastic residues, benefiting soil microorganisms and earthworms (Hajam et al. 2023).

15.5.3 Energy Efficiency: Optimization Process for Lower Energy Consumption

The global focus on energy conservation is driven by global events, rising energy prices, and supply chain challenges. Automation and control professionals can contribute significantly to this effort, as manufacturing accounts for 33% of global energy consumption. Energy costs are often hidden within factory overhead and are not explicitly tracked. Reducing energy consumption can enhance a company's competitiveness and bottom line, with return-on-investment analyses playing a key role in decision-making (Lydon 2022).

Economic factors drive energy efficiency improvements in the chemical industry, with the main incentive being to reduce operating costs. Adopting existing technologies that utilize heat generated in power generation can increase efficiency. Bulk chemical manufacturing, including petroleum refining, is the largest industrial consumer of energy, accounting for over half of fuel used. Advanced technologies like high-temperature reactors and sophisticated process controls can lead to significant energy savings, up to 10% to 20% in petroleum refining alone (The National Academies of Science n.d.).

15.5.3.1 Awareness and Engineering

Automation professionals can improve energy efficiency by increasing awareness and implementing engineering solutions. By utilizing real-time measurements of energy input and output, they can assess and optimize production processes. The first step is to measure and track energy consumption, identifying usage locations. Automation professionals can use existing data sources like supervisory control systems, human-machine interfaces, and historians instead of investing in energy monitoring devices. These systems already contain valuable information for developing detailed profiles of energy consumption and deviations without additional hardware (Lydon 2022).

15.5.3.2 Practical Solutions

Continuous monitoring of energy consumption is crucial for identifying deviations and improving decision-making and cost control. Initial steps include fundamental monitoring and control methods, followed by further investments to determine return on investment (ROI). Implementing electrical load-management systems, automating equipment shutdowns, and optimizing operations with compressed air control systems are essential. Maintaining an effective energy program requires ongoing data collection, prioritizing energy in operational decision-making, communicating program successes, integrating power and energy monitoring into continuous improvement efforts, and implementing alarm advisories for key performance indicators (Lydon 2022).

15.5.3.3 The Benefits of Energy Optimization

Cost Savings: Retailers must optimize energy use to cut costs and improve financial performance and profitability. Prioritizing optimization measures in all stores due to rising energy prices can help lower electricity bills and improve efficiency, ultimately leading to lower electricity bills (Selvaraj 2024).

Environmental Sustainability: By using energy-efficient refrigeration equipment, sophisticated controls, and sustainable refrigerants, retailers may lessen their carbon footprint and support environmental sustainability. These measures save money and align with international environmental goals, thereby combating climate change (Selvaraj 2024).

Enhanced Brand Reputation: Retailers may attract environmentally concerned customers, establish a favorable reputation for their company, and show their dedication to sustainability by making energy efficiency a top priority. Retailers may create long-term consumer loyalty and show their commitment to environmental responsibility by matching their operations with the values of their target market (Selvaraj 2024).

Improved Operational Efficiency: Retailers can enhance productivity and streamline procedures by optimizing energy systems and equipment, leading to improved operational performance. This approach also maximizes resources, reduces waste, and fosters a more sustainable business model, ultimately benefiting the overall business (Selvaraj 2024).

Regulatory Compliance: By putting energy optimization measures into place and abiding by local energy efficiency rules, retailers may safeguard themselves against penalties and legal complications. This proactive strategy encourages sustainable growth and shows a dedication to environmental responsibility (Selvaraj 2024).

15.5.3.4 Key Components of an Energy Optimization Roadmap

In order to successfully implement energy optimization, merchants must a clearly defined plan. When creating a plan for energy optimization, the following elements should be taken into account:

Energy Efficiency Audit: The first step in energy optimization is a thorough energy efficiency audit, which evaluates existing energy use patterns, finds inefficiencies, and pinpoints opportunities for improvement, involving all stores, departments, and systems for a comprehensive understanding (Selvaraj 2024).

Technology Upgrades: Retailers are able to recognize antiquated and high-energy systems through audit results and upgrade to energy-efficient technology like contemporary appliances, HVAC systems with advanced controls, and LED lights, despite initial financial outlays, to save money and enhance energy efficiency (Selvaraj 2024).

Real-time Energy Monitoring: IoT devices and energy monitoring software enable retailers to collect real-time utility and performance data, enabling them to observe seasonal trends, identify energy consumption patterns, and identify areas for optimization (Selvaraj 2024).

Automation and Optimization: Automation is crucial for energy efficiency, and cloud-based energy management solutions enable remote monitoring of performance at thousands of locations, automating optimization steps like defect detection, occupancy-based intelligent lighting control, and weather-based HVAC adjustments (Selvaraj 2024).

Constant Improvement: Retailers can encourage a culture of continuous improvement by routinely reviewing data on energy use, finding new areas for optimization, and putting innovative ideas into practice (Selvaraj 2024).

Employee Engagement and Training: By teaching staff members about energy-saving techniques, promoting waste reporting, and including them in the implementation of energy-saving projects, engaging staff members and providing training on these topics may greatly improve energy optimization efforts (Selvaraj 2024).

Performance Measurement and Reporting: Retailers are able to monitor progress and assess energy optimization projects by setting up key performance indicators (KPIs) and tracking data related to energy use. A brand's reputation may be improved and eco-conscious customers drawn in by providing transparent information on energy savings and sustainability accomplishments. Ensuring environmental sustainability and substantial cost savings may be achieved by including these elements into a plan for energy efficiency (Selvaraj 2024). Optimizing building energy consumption requires a multifaceted strategy. Here are seven approaches retailers can take to achieve this goal:

Real-Time Energy Monitoring: Implement a comprehensive energy monitoring system to track and analyze real-time energy usage data, enabling informed decision-making by identifying consumption patterns, anomalies, and inefficiencies (Selvaraj 2024).

Automated Fault Detection and Diagnostics: Advanced tools automate energy-related fault detection in building systems, dividing up the repair work into manageable chunks and ranking the severity of each repair to ensure quick fixes and top-notch work (Selvaraj 2024).

Demand Response Management: During peak demand periods, manage energy consumption through demand response programs, stabilizing the grid, reducing infrastructure strain, and potentially resulting in cost savings (Selvaraj 2024).

Predictive Maintenance: Employ predictive algorithms and data analytics to foresee equipment malfunctions or inefficiencies, and include automated work order management to boost output, optimize the energy system, and avert unplanned malfunctions (Selvaraj 2024).

Energy Benchmarking: Building-to-building benchmarking of energy consumption data facilitates goal-setting, progress monitoring, and the use of effective energy-saving techniques (Selvaraj 2024).

Tenant Engagement: Automated communication tools can enhance building occupant engagement by providing energy consumption information, tips, and feedback, promoting sustainable behaviors and fostering a responsible tenant culture (Selvaraj 2024).

Sustainability Reporting: Produce comprehensive sustainability reports that highlight financial savings, decreased environmental effect, and energy efficiency in order to win over environmentally conscientious investors or tenants and establish stakeholder confidence (Selvaraj 2024).

15.6 Regulatory Frameworks and Safety Considerations

15.6.1 Compliance with Regulatory Requirements for Food and Pharma Grade Metabolites

Regulatory requirements for food and pharmaceutical grade metabolites can vary depending on the specific jurisdiction and intended use. However, there are some general principles and standards that apply to both industries:

1. Quality Standards: Both food and pharmaceutical grade metabolites must meet strict quality standards to ensure safety and efficacy. This includes purity, potency, and absence of contaminants.

2. Good Manufacturing Practices (GMP): Manufacturers of valuable molecules like food additives and medications must follow Good Manufacturing Practices (GMP) to guarantee their products are safe and consistent. GMP dictates how things are documented, sanitized, and quality-checked throughout production.
3. Regulatory Approval: Pharmaceutical grade metabolites may require regulatory approval from organizations like the Food and Drug Administration (FDA) in the US or the European Medicines Agency (EMA) in the EU. The substance must undergo extensive testing to demonstrate its quality, safety, and efficacy before it is made available for sale for medicinal use.
4. Labeling and Packaging: Both food and pharmaceutical grade metabolites must be properly labeled and packaged to provide essential information to consumers or healthcare professionals. This includes ingredient lists, dosage instructions, expiration dates, and any relevant warnings or precautions.
5. Traceability and Documentation: Manufacturers must maintain detailed records of the production process, including sourcing of raw materials, manufacturing steps, and quality testing results. This helps ensure traceability and accountability throughout the supply chain.
6. Safety Evaluation: Safety assessments are critical for both food and pharmaceutical grade metabolites. This involves evaluating potential risks associated with consumption or use, including toxicity, allergenicity, and potential interactions with other substances.
7. Regulatory Compliance: Compliance with relevant regulations and guidelines is essential for both industries. This may include specific requirements for facility inspections, product testing, and reporting of adverse events.
8. Quality Control: Ongoing quality control measures are necessary to monitor and maintain the quality and consistency of food and pharmaceutical grade metabolites. This may involve regular testing for purity, potency, and stability.
9. Documentation and Record-Keeping: To prove compliance with regulatory standards and to make regulatory inspections and audits easier, thorough documentation and record-keeping are crucial.
10. Post-Market Surveillance: Monitoring of adverse events and ongoing safety surveillance are important aspects of regulatory oversight for pharmaceutical grade metabolites. This helps to identify and address any potential safety concerns that may arise after the product is on the market.

The regulatory requirements for food and pharmaceutical-grade metabolites involve stringent guidelines to ensure safety and efficacy. For dietary supplements, manufacturers must adhere to current Good Manufacturing Practices (cGMP) to guarantee safe, clean, and wholesome products. These regulations focus on quality control, batch consistency, plant sanitation, and personnel training, with an emphasis on documentation at every manufacturing step. Unlike pharmaceuticals, dietary supplements do not require premarket approval, but manufacturers must notify the FDA of new dietary ingredients before marketing them. Compliance with cGMP standards is crucial, yet enforcement faces challenges due to the vast number of dietary supplement facilities and limited FDA resources for inspections. Additionally, manufacturers, importers, and distributors are mandated to report serious adverse events associated with their products to the FDA within a specific timeframe.

On the other hand, for drug metabolites, the FDA provides guidance on identifying and characterizing metabolites, especially disproportionate drug metabolites that may pose safety concerns. The guidance outlines the need for nonclinical studies to assess metabolites, particularly those present at significant levels in humans compared to animal species. Technological advancements have enhanced the ability to detect and characterize metabolites, improving the understanding of their role in drug safety assessments. The guidance emphasizes the importance of evaluating metabolites that could contribute to the overall toxicity of the parent drug, especially those present at levels greater than 10% of total drug-related exposure.

15.6.2 Biosecurity and Risk Assessment: Managing Potential Risks Associated with Engineered Organisms

Biosecurity and risk assessment are vital for managing risks associated with genetic engineering and synthetic biology, including misuse, threats to biodiversity, and ethical concerns. Mitigation strategies like risk assessment processes, biosafety measures, and biosecurity frameworks are crucial. Access-controlled data repositories and research preregistration can help address biosecurity concerns. Strengthening regulatory systems, international cooperation, and biosecurity capacity is recommended to govern synthetic biology responsibly. A comprehensive approach involving practitioners, regulators, and international collaboration is essential to ensure safe advancement of engineered organisms.

With an emphasis on risk identification and mitigation techniques, the WHO offers a condensed six-part framework for biosecurity risk assessment. A soon-to-be toolbox will help scientists better assess risk in their work, which will lead to better lab design and a strong biosecurity culture.

(1). Identify and assess the risks and benefits (2). Describe values, principles and goals (3). Undertake stakeholder analysis (4). Identify mechanisms and tools (5). Implement the identified mechanisms and tools (6). Review performance and modify (adapt) as needed. Biosecurity risk assessment should ideally commence at the project planning stage, even if not mandated by law. If ongoing projects lack comprehensive assessment, immediate prioritization for assessment is advised. Continuous review and adjustment of the assessment throughout the project are crucial. Scientists must clarify their assessment goals, whether meeting grant requirements, ensuring accountability, or preventing misuse, and involve relevant stakeholders accordingly, including project team members, supervisors, safety experts, and possibly external stakeholders. Achieving buy-in from all involved parties, including scientists and regulatory bodies, is essential for a successful biosecurity risk assessment plan (Himmel 2023).

Recognizing the potential for both dual use and misuse of scientific research: Identification of scientific work's potential for dual use and misuse: Understanding biological threat agents' potential for dual use is a common subject of biosecurity risk assessments. There are three main methods for assessing the biosecurity risk associated with microbial infections and toxins. (a) Agent-based: the possibility of biological agents having intrinsic damaging properties; (b and c) results-based: the danger that biological research findings, if used improperly, may result in damage; (d) method-oriented: the risk that biological experiments or procedures will cause harm.

Once the degree of risk and urgency have been determined, scientists can come up with ways to keep their study from being misused. Some of these include limiting physical access to materials, protecting research data, keeping secrecy, and changing experimental protocols or biological agents. To create a complete plan, cooperation between institutions, government agencies, and coworkers is essential. By providing a summary of the first risk assessment, scientists can concentrate on factors that are under control and work with stakeholders to effectively handle potential hazards. This provides the groundwork for future conversations and choices.

Considering specific measures to mitigate biosecurity risks, scientists can develop methods to stop research from being misused after determining the risk and urgency of the situation. These methods include restricting physical access to materials, protecting research data, and altering experimental protocols or biological agents. To create a thorough plan, cooperation between institutions, coworkers, and government organizations is essential. The basis for these discussions and decisions is laid by summarizing the initial risk assessment, which allows scientists to concentrate on manageable issues while interacting with stakeholders to effectively handle potential hazards (Himmel 2023).

15.6.2.1 Prevention of Primary Negative Impacts

Physical protection: This includes standard lab safety practices like using protective gear, disinfecting work areas, and preventing accidental release of dangerous materials.

Replacement and minimization: Scientists can try using less dangerous versions of materials or methods, or minimize the amount of risky materials they work with. For instance, they might use a harmless lookalike instead of a dangerous pathogen, or only conduct a small number of experiments.

The precautionary principle: This means avoiding releasing any risky materials into the environment unless they're sure it's safe and have a plan to fix any problems.

Training and education: Everyone involved in the research needs to be properly trained on how to handle these biosecurity risks. Typically, a combination of these approaches is needed. For example, if the research could harm people, some measures would protect the scientists in the lab, while others would prevent the danger from spreading to the public (Smith and Sandbrink 2022; (Himmel 2023).

15.6.3 Ensuring Worker Safety in Industrial Production Settings

Maintaining worker safety in industrial production settings for the synthesis of metabolites is crucial for establishing a productive and healthy work environment. In order to avoid mishaps and health hazards, manufacturing workers are essential to upholding safety regulations. Workers in the manufacturing industry have a shared duty for workplace safety, and they should watch out for each other's risky actions and warning indicators. Industrial athletes should emphasize taking precautions to ensure their health and safety, just like professional athletes should. These precautions include self-care routines like drinking enough of water and taking breaks from the heat when needed to avoid heat-related illnesses.

The management of hazardous medicinal products at work should have a comprehensive approach to prevent and control risks from occupational exposure to these substances. Employers have a responsibility to ensure the safety and health protection of workers by assessing and preventing occupational risks, providing necessary information and training, and establishing the required organizational means for safety. Cooperation among employers sharing a workplace is essential to implement occupational safety and health measures effectively, especially in environments where workers may be exposed to hazardous substances like those involved in metabolite production.

Briefly, maintaining worker safety in industrial production settings for metabolite production involves a collective effort to maintain safety standards, use proper protective equipment, practice self-care, and be aware of potential risks associated with the work environment and substances being handled. Employers play a crucial role in providing a safe working environment and implementing necessary safety measures to protect workers' health and well-being.

15.7 Future Directions

15.7.1 Emerging Technologies: Exploring Novel Production Platforms and Technologies

Metabolomics is a growing field that analyzes metabolites in living organisms to discover new compounds, investigate biological functions, diagnose illnesses, and protect food standards. It is crucial for identifying unique bioactive substances and understanding genetic and environmental influences on biological systems. Choosing the right method for natural sample analysis can be challenging due to the diverse nature of samples (Eshawu and Ghalsasi 2024).

Over the past 10 years, microbes have been employed to create macromolecules and value-added metabolites at yields ranging from μg/L to mg/L. But because of their low titers, very few—like propanediol, butanediol, and artemisinin have made it to market. Through the manipulation of host physiology, metabolism, stress response, carbon-energy balance, and the removal of unwanted ATP sinks, metabolic engineering procedures can improve titers, rates, and yields. Because of their strong physiology, capacity to ferment food, genetic tractability, and resistance to severe growing circumstances, yeasts are frequently utilized in the food industry, biopharmaceuticals, fine chemicals, and biofuel synthesis. Despite their distinct PTM mechanism, yeasts are preferred for producing human proteins due to their complex eukaryotic PTM (Eshawu and Ghalsasi 2024).

15.7.1.1 Strategies and Challenges in Systems Biology for Nonconventional Yeasts

Experimental biology has seen a dramatic change in the last few decades, which has sped up the switch from genomic to post-genomic methods. Computational techniques and high-throughput technology have allowed us to learn more about several omics approaches, such as transcriptomics, metabolomics, and genomics. By evaluating and comprehending the enormous volume of biological data produced by these methods, systems biology has been instrumental in the development of contemporary biological applications. One key tool in systems biology is mathematical modeling, which helps map metabolic traits of an organism. Two major types of mathematical models are kinetic models and genome-scale metabolic models (GEMs). Kinetic models capture information about cellular metabolism concisely, but are unavailable for non-conventional yeasts due to the lack of extensive experimental data. GEMs offer a more thorough knowledge of metabolic activity in non-conventional yeasts, mostly based on genome sequence data. Reconstruction has been expedited by semi-automated techniques like as RAVEN, Merlin, and ModelSEED, which generate draft models based on annotated genomic and biochemical data. A burgeoning field of study called systems biology is being used to non-conventional yeasts to investigate and enhance metabolic capacities, create new cell factories for important compounds, and unravel the molecular mechanisms behind intricate metabolic processes (Patra et al. 2021).

15.7.1.2 Strategies and Challenges in Synthetic Biology Methods for Nonconventional Yeasts

The methodical study of genetic modules or constituents of a system with the goal of developing future microbes as cell biofactories is known as synthetic biology. This unconventional yeast approach is still in its infancy and does not yet have the uniformity needed for rapid applications in metabolic engineering. This section will cover the challenges and most recent discoveries in the field of combinatorial metabolic engineering of non-conventional yeasts (Patra et al. 2021).

15.7.1.3 Genetic Parts Engineering

Important genetic areas known as promoters, terminators, and transcription factors control the stability of mRNA and gene transcription, which in turn affects the net protein production of synthetic gene networks and metabolic pathways. These segments are located both above and below an ORF, as well as the net protein expression of the synthetic circuit is directly influenced by their activity. The difficulties and most recent advancements in using these genetic components for unconventional yeast design are covered in this section (Patra et al. 2021).

15.7.1.4 Promoter Engineering

In synthetic biology kits designed for non-traditional yeast species, a promoter is an essential element that controls gene expression. It is usual practice to utilize constitutive or inducible selective *S. cerevisiae* promoters. Inducible promoters enable desired gene activation, whereas constitutive promoters provide stable gene expression under ideal circumstances. To reduce drawbacks such lag time, high inducer costs, inducer toxicity, and leaky expression, constitutive promoters are recommended over inducible promoters (Patra et al. 2021).

15.7.1.5 Terminator Engineering

Accurate gene regulation depends on terminators, which control the creation of proteins and the expression of mRNA. Terminators of *S. cerevisiae* have been used to improve transcript stability in non-conventional yeasts, such as *H. polymorpha* expression cassettes ScGPD1t, ScADH1t, ScADH2t in *K. lactis*, ScCYC1t in *P. pastoris* and *Y. lipolytica*, and ScADH1t and ScPGK1t in *K. marxianus* (Patra et al. 2021).

15.7.1.6 Transcriptional Factors-Based Gene Expression Regulation

Through their ability to modulate environmental cues, TFs and TF-binding motifs are essential components of gene transcription. In *P. pastoris*, mutations in MXR1 TF caused AOX2 expression to be 60% lower and nearly no transcription of most other genes. Cas9 endonuclease has been engineered into dCas9, a "deactivated" endonuclease, allowing transcriptional control thanks to CRISPR/Cas9 technology. dCas9 may be linked with TFs to either promote or prevent transcription. Simple eukaryotic activators like VP64 and VPR have been tagged with dCas9 to upregulate genes in non-conventional yeasts (Patra et al. 2021).

15.7.1.7 Genome Editing Tools

To overcome obstacles in metabolic engineering, scientists have created stable plasmids for non-conventional yeasts, especially in S. stipitis. With *S. cerevisiae*, vector-based expression is appealing because of the stable plasmids and large copy numbers. Utilizing a plasmid encoding CEN5 enhanced lactate expression levels and CRISPR/Cas9 system effectiveness by around 80% and three times, respectively, according to a comparative research. Establishing effective biomanufacturing platforms to produce high-value compounds is the goal of this strategy (Patra et al. 2021).

15.7.1.8 Cre-loxP (Recombinase-Based Tools)

A viral enzyme called Cre recombinase forms a DNA loop when it encounters loxP sites, a 34 bp DNA sequence. Once the required DNA sequence with a marker is generated, it is inserted at predefined loci into the genome. Following integration, Cre recombinase locates loxP sites and eliminates the marker in preparation for more genome editing to create derivatives of fatty acids (Patra et al. 2021).

15.7.1.9 Nuclease-Based Tools (CRISPR/Cas9)

Genome engineering uses both model and nonconventional yeasts in conjunction with nuclease-based technologies including CRISPR/Cas9, TALEN, and ZFN. CRISPR/Cas9, a novel nuclease-based genome editing technique, cleaves DNA based on identification, allowing for quick and simple programmable cleavage. It is made up of a Cas9 endonuclease that is directed to a target DNA sequence by a chimeric sgRNA. This makes it possible to precisely cleave DNA and activate repair processes. This method has been used to study the resveratrol biosynthesis pathway genes in *O. polymorpha*, transcriptional studies in *P. pastoris*, and the manufacture and screening of haploid strains in *K. marxianus*. However, problems including the lack of easily accessible local RNA promoters, the limitations of the PAM motif, and off-target effects persist (Patra et al. 2021).

15.7.2 Convergence with Synthetic Biology: Designing Custom Metabolic Pathways

The process of disassembling and reassembling biological cells to develop new systems with practical uses is known as synthetic biology. The Design–Build–Test–Learn cycle uses computers to integrate designs encoded by DNA to produce gadgets. This method makes use of abstraction, modularity, and standardization concepts to facilitate quick design exchanges and rapid prototyping. Applications include better farming techniques, sustainable industrial operations, healthcare, and environmental enhancement. The future of the earth might be safeguarded and our lives could be transformed via synthetic biology (Erb et al. 2017).

Synthetic pathways have led to a substantial advancement in metabolic engineering in recent years. Five tiers of metabolic engineering were delineated by Erb et al., spanning from alterations to novel enzyme processes in synthetic pathways. Present methods concentrate on introducing preexisting enzymes into recently discovered metabolic pathways and transferring naturally occurring metabolic pathways into non-native hosts. Novel trophic pathways and energy- or carbon-conserving side paths in microbes are examples of the innovations that have resulted from this, promoting a more sustainable bioeconomy (Wenk et al. 2022).

15.7.2.1 Designer Reaction for Designer Pathways

One technique to find new reactions for synthetic pathways is to think about the backward responses of irreversible enzymes that support flow in vivo in both directions. Numerous alternatives are available with this method, including the pyruvate formate-lyase that facilitates formate absorption in vivo. Promising candidates for experimental testing can be identified by investigating and expanding the substrate repertoire of known enzymes, especially if potential substrates are structurally similar to natural substrates of known enzymes. Many of the "novel reactions" needed in the creation of synthetic pathways can be covered by investigating the natural biochemical space and the promiscuity of enzymes (Erb et al. 2017).

15.7.2.2 In Silico Analysis of Synthetic Metabolic Route

To guarantee the activity of synthetic routes, they must be built without thermodynamic or kinetic obstacles. Even for unknown components, the thermodynamic profile may be computed with a high degree of precision. Barriers produced by the sequential functioning of enzymes must be taken into account in computational assessments since their combination may result in extremely high energy barriers. The growth rate and yield can be impacted by uncontrolled metabolic fluxes that arise from the overlap between novel pathways and endogenous metabolism. Before implementing new pathways in vivo, it is imperative to analyze their compatibility with endogenous metabolism. This may be partially accomplished through the use of constrained-based modeling methodologies such as flux balance analysis (Erb et al. 2017).

15.7.2.3 Challenges During Implementation of Synthetic Metabolism

The expression of enzymes from diverse biological origins presents difficulties for synthetic metabolic pathways, resulting in inferior kinetics and restricted efficiency. When enzymes in non-native hosts come into contact with metabolites they have never been exposed to, cross-inhibition and unwanted side effects result. Using "metabolic proofreading," enhancing promiscuous enzymes through enzyme engineering, or substituting certain isoenzymes or homologs for promiscuous enzymes are some methods to reduce adverse effects. Enzyme redesign and metabolic proofreading together increased a synthetic in vitro CO_2-fixation pathway's activity by more than a factor of ten. It was essential to implement a mechanism for recycling dead-end metabolites in order to convert glucose into polyhydroxybutyrate (Erb et al. 2017).

15.7.2.4 Opportunities for Synthetic Metabolic Pathways

Among the many advantages of synthetic metabolism is the ability to reroute metabolic flux to produce end products that biotechnology hosts do not normally collect. By combining known and unknown enzymatic processes, one can get the best metabolic pathways while getting over ecological and historical limitations. Examples include creating energy-efficient artificial CO_2-fixation routes, bioretrosynthetic methods for didanosine, and coupling plant ACP-fatty acid thioesterase with designed monooxygenase for ω-hydroxy octanoic acid synthesis in *E. coli* (Erb et al. 2017).

15.7.2.5 Future Application for Synthetic Metabolic Pathways

Utilizing and capturing CO_2 is essential to a sustainable economy. A different approach to traditional biomass production might be provided by designer routes that transform CO_2 into value-added products or feedstocks. These mechanisms may eventually be powered by synthetic pathways like methanol metabolism or extended formate. Novel possibilities such as photorespiration bypasses and synthetic CO_2-fixation routes are made possible by synthetic metabolism and may eventually offer workable answers to environmental, societal, and economic problems (Erb et al. 2017).

15.7.3 The Future of Industrial Metabolomics: Toward a Circular Bioeconomy

By using five ecological principles—preserving ecosystem health, avoiding non-essential items, giving priority to biomass for fundamental human needs, making use of byproducts, and harnessing renewable energy—a circular bio-based economy can lead to a sustainable future. This calls for a revamp of the present economic structure, involving adjustments to markets, organizations, policies, organizations, and social behavior. In this bio-based circular economy, one important tactic is biorefining, which uses biomass resources to create metabolites and commercial goods. This environment-friendly bioprocess reduces carbon footprints from burning fossil fuels, protecting the environment in addition to maintaining the energy-environment nexus (Garner 2021; Muscat et al. 2021).

15.7.3.1 Waste Biorefinery Promoting a Circular Bioeconomy

Petroleum and other fossil fuels are necessary for the creation of energy and synthetic materials, but they are non-renewable and release greenhouse gases into the atmosphere. As a result, models for waste biorefineries that emphasize carbon management and greenhouse gas reduction are becoming more popular. A circular bioeconomy centered on recycling, reuse, remanufacturing, and maintenance is employed in waste biorefineries. Waste-to-treasure bioprocessing addresses issues with energy and environmental security while enhancing waste stream management by using municipal solid and liquid waste to create value-added bioproducts and metabolites. Because the production feedstock is inexpensive and sustainable, waste biorefineries are both environmentally and financially sound. Waste resources may be effectively valorized into beneficial bio-based products, including food waste, wastewater, industrial byproducts, and waste from agriculture and forestry (Leong et al. 2021).

15.7.3.2 Bioprocesses for Wastewater Treatments

Wastewater is produced by a variety of activities and has to be properly treated in order to reduce pollution and provide environmental security. One such method for treating wastewater that can provide goods with additional value is bioprocessing. The growing of microorganisms using wastewater enhances bioremediation, lowers cultivation costs, and enables the generation of valuable bio-based products and metabolites. Wastewater includes nutrients that are necessary for microbial culture. This strategy has positive effects on the environment and general health (Leong et al. 2021).

15.7.3.3 Strategies of Biorefinery—Circular Bioeconomy Toward Solving Global Issue

Food security and environmental degradation are major worldwide concerns brought on by urbanization, industrialization, and climate change. Greenhouse gases (GHGs), in particular CO_2, have a detrimental effect on food production. A 1.5 °C global warming might have dire repercussions, including rising sea levels and food shortages. In order to combat climate change, a waste biorefinery connected with a circular bioeconomy may absorb CO_2, encourage environment-friendly trash disposal, and generate bioproducts with additional value. This strategy can encourage environmentally beneficial carbon flow, lessen waste, and substitute fossil fuels. Biofuels and bioplastics can reduce plastic pollution and offer a cost-effective way to dispose of trash. In addition, the production of independent bioenergy and biofuels by waste biorefineries can enhance energy security (Leong et al. 2021).

15.8 Conclusion

The transition from laboratory-scale to large-scale industrial metabolite production involves scaling microbial fermentations, optimizing strain engineering processes, and utilizing biocatalytic systems. This progression is essential for industrial success in various sectors, including biotechnology and bioeconomy. By understanding and controlling operational conditions, companies ensure efficiency, product quality, and consistency in large-scale production. The integration of technologies like metabolic engineering, synthetic biology, and process engineering plays an integral role in unlocking the full potential of biomanufacturing for industrial-scale production of bio-based molecules. Proper scaling and optimization of the strain engineering process are essential steps in harnessing the power of microbial biotransformations and biocatalytic systems to address challenges related to waste, pollution, and resource scarcity while promoting sustainability and innovation in the biotechnology industry.

References

Abbate E, Andrion J, Apel A, Biggs M, Chaves J, Cheung K, Ciesla A, Clark-ElSayed A, Clay M, Contridas R, Fox R, Hein G, Held D, Horwitz A, Jenkins S, Kalbarczyk K, Krishnamurthy N, Mirsiaghi M, Noon K et al (2023) Optimizing the strain engineering process for industrial-scale production of bio-based molecules. J Ind Microbiol Biotechnol 50(1). https://doi.org/10.1093/jimb/kuad025

Arias J, Kotipalli MVK (2021) A unified approach for scaling bioprocess unit operations. BioPharm International. https://www.biopharminternational.com/view/a-unified-approach-for-scaling-bioprocess-unit-operations

Campilho RDSG, Silva FJG (2023) Industrial process improvement by automation and robotics. Mach Des 11(11):1011. https://doi.org/10.3390/machines11111011

Cerny MA, Kalgutkar AS, Obach RS, Sharma R, Spracklin DK, Walker GS (2020) Effective application of metabolite profiling in drug design and discovery. J Med Chem 63(12):6387–6406

Chaudhry MA (2024) Lessons in bioreactor scale-up: part 1—exploring introductory principles. https://www.bioprocessintl.com/bioreactors/lessons-in-bioreactor-scale-up-part-1-mdash-exploring-introductory-principles

Christian N, Whitaker BK, Clay K (2015) Microbiomes: unifying animal and plant systems through the lens of community ecology theory. Front Microbiol 6:153936

Claßen J, Aupert F, Reardon KF et al (2017) Spectroscopic sensors for in-line bioprocess monitoring in research and pharmaceutical industrial application. Anal Bioanal Chem 409:651–666. https://doi.org/10.1007/s00216-016-0068-x

Costa RJS, Silva FJG, Campilho RDSG (2017) A novel concept of agile assembly machine for sets applied in the automotive industry. Int J Adv Manuf Technol 91:4043–4054. https://doi.org/10.1007/s00170-017-0109-4

Deng C, Lv X, Liu Y, Li J, Lu W, Du G, Liu L (2019) Metabolic engineering of *Corynebacterium glutamicum* S9114 based on whole-genome sequencing for efficient *N*-acetylglucosamine synthesis. Synth Syst Biotechnol 4(3):120–129. https://doi.org/10.1016/j.synbio.2019.05.002

Downstream Processing in Biopharma (2022) Mettler-Toledo International Inc. All Rights Reserved. https://www.mt.com/us/en/home/applications/L1_AutoChem_Applications/fermentation/downstream-processing-in-biotechnology.html

Du Y, Wang MY, Yang L, Tong L, Guo D, Ji X (2022) Optimization and scale-up of fermentation processes driven by models. Bioengineering (Basel). https://doi.org/10.3390/bioengineering9090473

Erb TJ, Jones PR, Bar-Even A (2017) Synthetic metabolism: metabolic engineering meets enzyme design. Curr Opin Chem Biol 37:56–62

Eshawu AB, Ghalsasi VV (2024) Metabolomics of natural samples: a tutorial review on the latest technologies. J Sep Sci 47(1):2300588

Ganeshan S, Kim SH, Vujanovic V (2021) Scaling-up production of plant endophytes in bioreactors: concepts, challenges and perspectives. Bioresour Bioprocess 8(1):63. https://doi.org/10.1186/s40643-021-00417-y

Gargalo CL, Udugama I, Pontius K et al (2020) Towards smart biomanufacturing: a perspective on recent developments in industrial measurement and monitoring technologies for bio-based production processes. J Ind Microbiol Biotechnol 47:947–964. https://doi.org/10.1007/s10295-020-02308-1

Garner KL (2021) Principles of synthetic biology. Essays Biochem 65(5):791–811

Gonzalez-Covarrubias V, Martínez-Martínez E, del Bosque-Plata L (2022) The potential of metabolomics in biomedical applications. Meta 12(2):194

Hajam YA, Kumar R, Kumar A (2023) Environmental waste management strategies and vermi transformation for sustainable development. Environ Chall 100747

Himmel M (2023) Biosecurity risk assessment in the life sciences. http://books.google.ie/books?id=Mjf3zwEACAAJ&dq=Biosecurity+Risk+Assessment+in+the+Life+Sciences:+Towards+a+Toolkit+for+Individual+Practitioners+Author:%09Dr+Mirko+Himmel/SIPRI&hl=&cd=1&source=gbs_api

Johanningsmeier SD, Harris GK, Klevorn CM (2016) Metabolomic technologies for improving the quality of food: practice and promise. Annu Rev Food Sci Technol 7:413–438

Kulshreshtha A, Sharma S, Padilla CS, Mandadi KK (2022) Plant-based expression platforms to produce high-value metabolites and proteins. Front Plant Sci 13:1043478

Leong HY, Chang CK, Khoo KS, Chew KW, Chia SR, Lim JW et al (2021) Waste biorefinery towards a sustainable circular bioeconomy: a solution to global issues. Biotechnol Biofuels 14:1–15

Lydon B (2022) Optimize energy efficiency and sustainability. isa.org. https://www.isa.org/intech-home/2022/june-2022/columns/optimize-energy-efficiency-and-sustainability https://www.isa.org/intech-home/2022/june-2022/columns/optimize-energy-efficiency-and-sustainability

Magar ST (2023) Bioreactor- definition, design, principle, parts, types, applications, limitations. Microbe Notes. https://microbenotes.com/bioreactor/

Makovitzki A, Jayson A, Oren Z, Lerer E, Kafri Y, Dor E, Cherry L, Tzadok H, Levin L, Hazan O, Simon I, Tal A, Girshengorn M, Rosen O (2021) In-line monitoring of downstream purification processes for VSV based SARS-CoV-2 vaccine using a novel technique. Biotech (Basel (Switzerland)) 10(4):25. https://doi.org/10.3390/biotech10040025

Marchev AS, Vasileva LV, Amirova KM, Savova MS, Balcheva-Sivenova ZP, Georgiev MI (2021) Metabolomics and health: from nutritional crops and plant-based pharmaceuticals to profiling of human biofluids. Cell Mol Life Sci 78(19–20):6487–6503

Mattanovich D, Sauer M, Gasser B (2014) Yeast biotechnology: teaching the old dog new tricks. Microb Cell Factories 13:1–5

Muscat A, de Olde EM, Ripoll-Bosch R, Van Zanten HH, Metze TA, Termeer CJ et al (2021) Principles, drivers and opportunities of a circular bioeconomy. Nature Food 2(8):561–566

Ogenyi UE, Liu J, Yang C, Ju Z, Liu H (2021) Physical human-robot collaboration: robotic systems, learning methods, collaborative strategies, sensors, and actuators. IEEE Trans Cybernet 51(4):1888–1901. https://doi.org/10.1109/TCYB.2019.2947532

Patra P, Das M, Kundu P, Ghosh A (2021) Recent advances in systems and synthetic biology approaches for developing novel cell-factories in non-conventional yeasts. Biotechnol Adv 47:107695

Pickens LB, Tang Y, Chooi YH (2011) Metabolic engineering for the production of natural products. Annu Rev Chem Biomol Eng 2:211–236

Prasad K, Ranjan RK, Lutfi Z, Pandey H (2009) Biosensors: applications and overview in industrial automation. https://doi.org/10.18000/ijabeg.10041

Qiu S, Cai Y, Yao H, Lin C, Xie Y, Tang S, Zhang A (2023) Small molecule metabolites: discovery of biomarkers and therapeutic targets. Signal Transduct Target Ther 8(1):132

Ramawat KG, Ekiert HM, Goyal S (2020) Plant cell and tissue differentiation and secondary metabolites. Springer International Publishing

Roth MG, Westrick NM, Baldwin TT (2023) Fungal biotechnology: from yesterday to tomorrow. Front Fungal Biol 4:1135263

Sarenkova I, Sáez-Orviz S, Rendueles M, Ciprovica I, Zagorska J, Díaz M (2022) Downstream approach routes for the purification and recovery of lactobionic acid. Food Secur 11(4):583. https://doi.org/10.3390/foods11040583

Sebastián S-L, Angela M, Martin M, Anton F (2019) A review on the feedstocks for the sustainable production of bioactive compounds in biorefineries. Sustain For 11(23):6765. https://doi.org/10.3390/SU11236765

Seidel S, Mozaffari F, Maschke RW, Kraume M, Eibl R, Eibl D (2023) Automated shape and process parameter optimization for scaling up geometrically non-similar bioreactors. PRO. https://doi.org/10.3390/pr11092703

Selvaraj D (2024) 7 energy optimization techniques for your retail store. Facilio Blog. https://facilio.com/blog/energy-optimization-roadmap/

Singh R, Kumar M, Mittal A, Mehta PK (2017) Microbial metabolites in nutrition, healthcare and agriculture. 3 Biotech 7:1–14

Smith JA, Sandbrink JB (2022) Biosecurity in an age of open science. PLoS Biol 20(4):e3001600. https://doi.org/10.1371/journal.pbio.3001600

The National Academies presents: what you need to know about energy (n.d.) Copyright 2024, The National Academies of Science. http://needtoknow.nas.edu/energy/energy-efficiency/industrial-efficiency/

Trinh CT, Wlaschin A, Srienc F (2009) Elementary mode analysis: a useful metabolic pathway analysis tool for characterizing cellular metabolism. Appl Microbiol Biotechnol 81:813–826

Verma A, Verma M, Singh A (2020) Animal tissue culture principles and applications. In: Animal biotechnology. Academic Press, pp 269–293

Wang L, Birol I, Hatzimanikatis V (2004) Metabolic control analysis under uncertainty: framework development and case studies. Biophys J. https://doi.org/10.1529/biophysj.104.048090

Wang C, Zhao S, Shao X, Park JB, Jeong SH, Park HJ, Kwak WJ, Wei G, Kim SW (2019) Challenges and tackles in metabolic engineering for microbial production of carotenoids. Microb Cell Factories 18:1–8

Wenk S, Claassens NJ, Lindner SN (2022) Synthetic metabolism approaches: a valuable resource for systems biology. Curr Opin Syst Biol 30:100417

Yao T, Asayama Y (2017) Animal-cell culture media: history, characteristics, and current issues. Reprod Med Biol 16(2):99–117

16 Regulatory, Economic, and Educational Issues in Metabolomics

Hue Vu Thi, Luong Tri Duc, Hoang Duc Anh, Le Duc Anh, Ha Phuong Nguyen, Lan Anh Nguyen Hoang, Que Anh Trieu Nguyen, Vijai Singh, and Dinh-Toi Chu

16.1 Introduction

"Omics" technologies are the vehicles aiming to detect and analyze structures and functions of all components at a certain level of a particular biological system, and they are separated into four main sections based on kinds of analysis objections, including genomics, transcriptomics, proteomics, and metabolomics (Dai and Shen 2022). These technologies have great potential in biomedical, providing insight into normal physiology as well as abnormal states via screening and diagnosis, facilitating the development of responsible drugs and treatments, and are believed to contribute toward personal therapy in the future (Horgan and Kenny 2011). In this context, metabolomics terms for the systematic identification of all metabolites, which are low-molecular-weight final or intermediate products of enzyme reactions and quantity them in a certain biological sample or organism (Idle and Gonzalez 2007; Belhaj et al. 2021). Despite being the latest of the above "omics" technologies, metabolomics has been significantly developing and effectively applied in many fields (Pinu et al. 2019).

H. V. Thi · D.-T. Chu (✉)
The Interdisciplinary Research Group on Biomedicine and Health (IRGBH), International School, Vietnam National University, Hanoi, Vietnam

Faculty of Applied Sciences, International School, Vietnam National University, Hanoi, Vietnam
e-mail: toicd@vnu.edu.vn

L. T. Duc · H. D. Anh · L. D. Anh · H. P. Nguyen · L. A. N. Hoang · Q. A. T. Nguyen
The Interdisciplinary Research Group on Biomedicine and Health (IRGBH), International School, Vietnam National University, Hanoi, Vietnam

V. Singh
Department of Biosciences, School of Science, Indrashil University, Rajpur, Mehsana, India

V. Singh (ed.), *Advances in Metabolomics*,
https://doi.org/10.1007/978-981-97-7459-3_16

In particular, metabolomics can be utilized to determine biomarkers representing human diseases via consistent and accurate evaluations of metabolites (Gonzalez-Covarrubias et al. 2022a). For example, metabolomics tools are used to distinguish unhealthy, overweight, and obese people from healthy people by observing the changes in concentration of some related compounds such as branched-chain amino acids, aromatic amino acids, fatty acids, nucleotide (Rangel-Huerta et al. 2019) or detecting signals of different cancers like ratio of free fatty acids in breast cancer, colorectal cancer, and pancreatic diagnosis (Zhang et al. 2014, 2016; Gong et al. 2010). Moreover, metabolomics provides optimal strategies for the discovery, development and test of drugs, playing an important role in a plain order to treat human diseases indicated to be attributed to abnormal metabolism (Wishart 2016). Specifically, these drugs may target a particular enzyme that catalyzes for a metabolite or a group of metabolites, causing these diseases in a known process (Wishart 2016). Take Trimethylamine-N-oxidase (TMAO) as an instance, through metabolomics, they are identified as a cardiovascular disease-causing agent, generated by gut microbiota; therefore, drugs in this case can target to inhibit CntA and Cntb, an enzyme system that participates in the pathway of TMAO formation in bacteria (Brown and Hazen 2015). Apart from the current applications, metabolomics seems to have potential in new directions. Metabolomics can research exposome, a systems that include all environmental factors that affect an individual from all sources, to find the relationship between external exposures such as pollution, diet, medication, and biological elements molecules that are considered relevant during a person's life, facilitating to appreciate and improve the environment that is good for human health (Gonzalez-Covarrubias et al. 2022a). Another noticeable direction is pharmacometabolomics, a field that proposes predictions about the effects of drug treatment in a subject (Beger et al. 2020). Metabolomics technologies will provide a metabolic profile during treatment compared to a standard line to determine the molecular effects of drugs (even adverse effects), motivating personalized medicine (Beger et al. 2020).

The indispensable thing in metabolomics is the analysis of several metabolites extracted from biological samples, done by some specific tools in which nuclear magnetic resonance (NMR) and mass spectrometry (MS) are known as the main platforms (Chen et al. 2022). MS was first used to measure a molecule in 1910 by J.J. Thomson. With the continuous developments from that, MS now is widely applied as a highly sensitive measurement allowing to detect, quantity, and structure clarifying numerous metabolism via their ionization degree (Gowda and Djukovic 2014; Glish and Vachet 2003). One of the important advances is the combination of several chromatography techniques, such as capillary electrophoresis, gas chromatography or high-performance liquid chromatography, which help separate interested metabolites from a complex metabolites system in order for MS to give more accurate results (Gowda and Djukovic 2014). Moreover, some variants of MS justified that it is feasible to skip chromatography method usage, instead using ambient ionization methods, which can be faster and have no need for sample preparation (Gowda and Djukovic 2014). NMR technique was developed slower, with the first use being recognized in Rabi's experiment in 1939; however, this technique was not observed on biological samples until 1946 when Bloch tested with his finger (Becker 1993). After that, NMR started to be interesting as a potential method in analyzing metabolism systems with many involved studies (Becker 1993). In general, in the context of metabolomics, this technique is claimed to enable identity of unknown metabolites, observe the metabolic processes, and measure related substances by using atomic nuclei like 31P, 15N, and 13C (Nagana Gowda and Raftery 2021). Overall, it takes less time to measure by NMR than MS, while MS's result seems to be more sensitive and selective due to their different interaction to samples (Beisken et al. 2015). In other words, they are complementary and can be both used in some cases (Gowda and Djukovic 2014). Now, both techniques still play a key role in metabolomics and will be continuously improved and perfected.

Indeed, the potential of metabolomics cannot be undermined with many significant applications. However, not an exception, this technology has to face many challenges, not only with biological issues (Johnson and Gonzalez 2012), but also issues involving the economy, education and regulation. In fact, because the publicity of metabolomics seems pale in comparison with other "omic" technologies that earlier developed and may be more perfect, it is easier to understand why metabolomics gains less interest from funding organizations (Pinu et al. 2019). In addition, although the use of metabolomics is believed to reduce total cost (Wishart 2016), the high starting costs for facilities still is a trouble (Pinu et al. 2019), placing a financial burden for who wants to study this field. Moreover, metabolomics requires a large enough interdisciplinary knowledge to analyze the huge data that it provides about various metabolites with diverse pathways; hence, it is necessary to have a detailed and methodical education (Pinu et al. 2019). Besides, if metabolomics is used widely, especially in healthcare and biomedical fields, of course, there must be some regulation involved to limit it to a permitted ethic. This article aims to collect the available evidence and provide the most comprehensive overview of these issues.

16.2 Regulatory Issues in Metabolomics

16.2.1 The Ethical Issues in Research and Application of Metabolomics

Recently, metabolomics has a lot of benefits to molecular biomedicine as well as clinical research applications. We cannot deny the advantages that metabolics bring. However, some aspects should be given more comprehensive attention. Among them, ethical concerns need the most notice (Fig. 16.1). First of all, metabolomics has the potential to interfere with human privacy. In addition, one of the major ethical challenges is compliance with existing data-sharing policies. Furthermore, intellectual property rights and funding priorities also need to be considered (McGuire et al. 2011).

Privacy: Metabolomics collects data about your body's interactions with your environment, such as biological habits, medications, diet, and even various environmental exposures that can be tracked (Lewis Jr et al. 2020). Digitizing molecules in

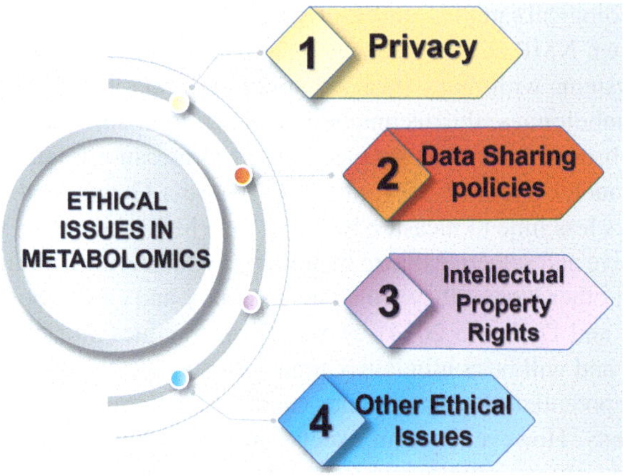

Fig. 16.1 Ethical issues in metabolomics

the environment not only reveals the contact and interaction between an individual and their environment but also reveals chemicals in that environment such as drugs, cosmetics, foods, beverages, and accompanying additives (McCall et al. 2019; Petras et al. 2016). Additionally, the collection of genetic information raises ethical questions regarding maintaining confidentiality and the risk of discrimination in insurance and employment decisions. Therefore, there is a need for privacy policies and codes of conduct regarding the appropriate use of genetic information (Reilly and Debusk 2008).

Data Sharing Policies: Policymakers' inconsistency in assessments and requirements has long been an obstacle for investigators and has since been considered challenging in clinical trials. In addition, some national policies require widespread data sharing between studies, thereby placing responsibility for reviewing and coordinating data submissions with host protection organizations and programs. This allows for community input but also invariably leads to differences in policy and practice that can confuse investigators. Because of compliance data-sharing policies, clinical trial participants may find it challenging to limit sharing or expect their data to be more tightly controlled. Data-sharing policies must be implemented to protect participants' interests while promoting scientific progress.

Intellectual Property Rights: Metabolomics-based biomarker assays can be patented. Metabolites used for therapeutic purposes may be more difficult to patent. We can only patent manufacturing processes or methods for naturally occurring and structurally characterized metabolites. However, the straightforward chemical transformation that results in the formation of a new chemical substance can be patented, which raises the question of what regulations should be in place for the owners of such substances' metabolism whose composition is not significantly changed (Kartal 2007).

Other Ethical Issues: Furthermore, the difficulties and challenges mentioned above, several ethical obstacles need to be noted. Firstly, Diagnoses and predictions based on metabolomics will change the relationship between medical service providers and patients. They will feel more worried when they have to do tests and see doctors more. Currently, the large amount of data generated leads to poor pathophysiological information, which has hindered the translation of metabolomics into clinical practice. Thus, a systematic approach is needed to determine (genetic) causation. Secondly, data on genetics and clinical or biological interactions with the surrounding environment are often collected from very large numbers of individuals from different populations and stored in their health records (response) to a certain treatment, which may belong to certain minority populations (Di Minno et al. 2022).

16.2.2 The Regulatory Issues in Research and Application of Metabolomics

16.2.2.1 Legal Problems in Metabolomics Research and Applications

Currently, particular legal actions related to metabolomics cannot be found on any platform, it is only being adjusted by legal regulations in general which include the relevant involved areas. To clarify the whole situation, we must dig into the actual metabolomics protocols to investigate the related areas of life that are being impacted (Fig. 16.2).

Environmental Issues: Metabolomics is the process of systematic research and a distinctive biological footprint in which environment samples (e.g., soil, water, etc.) are used for experiments. In that event, following the strict environmental law regulations is needed to perform precisely. For instance, researchers studying how contaminants affect metabolites in aquatic environments must abide by legal regulations of environmental preservation and water quality.

Agriculture Circumstances: Plants and animals are vastly influenced by research and application of Metabolomics. At present, these procedures are playing an important role as a foundation for other various research on animal science: Animal welfare, livestock environment, animal biotechnology, animal behavior, milk, meat,

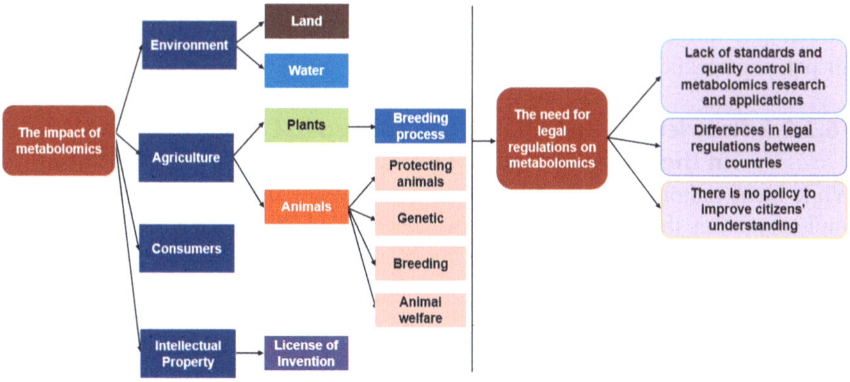

Fig. 16.2 The influence and defect of metabolomics and its legal regulatory

eggs, and their byproducts, animal genetics and breeding, physiology, nutrition, and the products of animals are all covered. Moreover, it also has a noticeable effect on the mechanism of multiplying crops, up to now 359 articles have been carried out with 160 kinds of different plants (Sakurai 2022). In contrast, it should be mentioned that the Consumer Protection Law must be followed while releasing metabolomics goods into the environment. For instance, food product safety laws are being applied in European nations and address topics including control systems, food additives, elements that come into touch with food goods, and new sorts of food products (Nezhmetdinova et al. 2020).

Animal Protection Law: Many animal-based metabolomics investigations are currently being conducted. This presents moral questions for using animals in testing, education, and research. Among these, there are several nations have legislation about animal protection, such as United States of America (Animal Welfare Act (AWA)—August 24, 1966—regulates the treatment of animals in research, teaching, testing, exhibition, and transportation activities) (Information, A.U.S.G., Animal Welfare Act 1966); The Health Research Extension Act (1985), also known as "Animals in Research," mandates that guidelines for the proper care and treatment of vertebrate specimens used in biological and behavioral studies be established and that each applicant for a funding opportunity or agreement from the National Institutes of Health or an additional national research institute be guaranteed that it follows these standards ((NIH), N.I.o.H., Health Research Extension Act 1985).

Intellectual Property Rights: The most popular IP protections utilized in biomedical research to secure ideas, inventions, methods, and materials are patents, trademarks, and copyrights. This can be applied to the study and implementation of metabolomics, where patent laws in various nations govern the following aspects of biomedicine: (1) Japan: Under the Japanese Patent Act (Government, J., Patent Act (Act No. 121 of 1959) 1959), the subject matter eligible for patenting must be an invention (Section 2(1), Japanese Patent Act. "Invention" refers to the extremely sophisticated use of natural rules in the formulation of technical concepts. After the patent reform in 2011, the amendments to the Japanese Patent Act came into effect on April 1, 2012. The "Japanese); (2) "Any invention-creation that is detrimental to the public interest or that is contrary to the laws or social morality" is not eligible for patent protection in China.

16.2.2.2 Restrictions and Challenges Related to Legal Aspects in the Study of Metabolomics

While many nations have comprehensive legal laws on metabolomics research and implementation, there are still certain obstacles and challenges in putting these policies into practice. Firstly, Inadequate quality control and calibration standards in metabolomics research and applications. Sample deterioration and standardization provide challenges. Since metabolomics is a young science, sample collection, processing, and analysis do not yet have universally recognized standards. Crucially, a sample's metabolic signature could change as soon as it is taken out of the biological system. This highlights how crucial preparing samples is and makes it challenging to repeat tests and compare findings across many studies. Secondly, Global metabolomics research and implementation may face obstacles due to variations in metabolomics-related legislative restrictions among nations. Procedures for integrating biotechnology into the healthcare industry will vary based on the national health agency of each nation. For instance, in the USA, organizations like HIPAA (Health Insurance Portability and Accountability Act) (Centers for Disease Control and Prevention 2022) provide standards and principles that are frequently followed when managing data security of medical data. Nonetheless, the General Data Protection Regulation (GDPR) governs data security in the European Union (Intersoft Consulting 2018). When we compare the two nations, we can observe that the European Union has stricter and more comprehensive laws than the USA. This may have an effect on how organizations and scientists handle and secure data in metabolomics research, particularly when it comes to participant personal information. Thirdly, there are no policies or legal regulations related to improving people's understanding of metabolomics. In particular, there is a genuine variation in the knowledge held regarding this problem among nations. There are many study publications on this topic, but only a small portion of the Vietnamese population is interested in it. However, since 2004, hundreds of people from developed nations, including Japan, Korea, Australia, and Switzerland, have participated in metabolomics research associations, networks, and forums established in Europe and other developed countries worldwide (University of Birmingham and Birmingham Metabolomics Training Centre 2017). It is evident that underdeveloped nations, like Vietnam, have not given much attention to this problem; as a result, there are still few legal restrictions on metabolomics in Vietnam.

16.3 The Economic Issues in Metabolomics

16.3.1 Metabolomics Market

The metabolomics market can be divided by application, including biomarker discovery, drug discovery, toxicology testing, nutrigenomics, functional genomics, personalized medicine, and others (Allied Market Research 2022). This market can also be analyzed from a geographic perspective, with regions such as North America, Europe, Asia-Pacific, and LAMEA being considered, or by indication, such as cancer, cardiovascular disorders, neurological disorders, inborn errors of metabolism, and others. According to a report conducted by MordorIntelligence (MordorIntelligence n.d.), the metabolomics market size is projected at USD 2.26 billion in 2024 and is predicted to reach USD 3.82 billion by 2029, increasing at a compound annual growth rate of 11.02% during the forecast period (2024–2029). Biomarkers and drug discovery are the biggest application segments in metabolomics. In terms of indications, cancer stands out as the largest market segment. Geographically, North America dominates the metabolomics market, while the Asia-Pacific region is projected to experience the highest compound annual growth rate. North America is estimated to dominate the market during the forecast period due to the existence of several key players, the rising prevalence of chronic diseases, and the increased use of metabolic applications in technological health care (MordorIntelligence n.d.).

Currently, the demand for health care services, medical examinations, and treatment is increasing, with a special emphasis on personalized medicine. According to Forbes magazine, personalized medicine is a business trend to watch in 2024 as patients get increasingly bored with traditional care, leading many to seek online alternatives and individualized approaches to suit their needs (Garcia-Ibáñez 2024). Therefore, these markets have significant growth and development prospects. Statistics show that pharmaceutical sales in emerging markets have doubled in 5 years, thanks to large populations, prosperity, and a longer life expectancy. Diseases such as cardiovascular disease, diabetes, and cancer are also on the rise in these markets, suggesting that the pharmaceutical industry can market its global products in these countries (Tannoury and Attieh 2017). Although high-income countries receive more government support for the health sector, low-income countries have higher levels of personal spending (Rahman and Schellhorn 2023). However, as for instrumentation needed for personalized medicine, especially biobanking, although most of them are accessible in high-income countries, only a few disadvantaged countries can afford biobanks due to the high expenses of equipment, sample storage, and maintenance (Rahman and Schellhorn 2023). Top-of-the-line analytical and separation machines utilized in metabolomics research can be highly expensive, making it difficult for emerging and disadvantaged countries to buy such equipment. This may slow the expansion of the metabolomics market in certain regions (Allied Market Research 2022). Figure 16.3 briefly depicts the prospects and challenges of the metabolomics market.

16.3.2 The Economics of Metabolomics

The wide array of applications present in metabolomics can facilitate translational and economic opportunities centered around clinical and biological applications

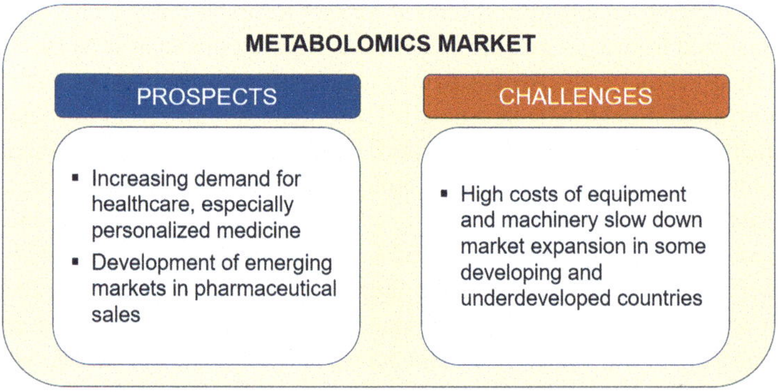

Fig. 16.3 Prospects and challenges of the metabolomics market

(Fig. 16.4) (Yang et al. 2019). The most prominent prospect of metabolomics application relates to disease detection and biomarkers. Current US figures estimate $4.5 trillion in healthcare spending (CMS 2024), with 2009 data placing diagnostics at 10% of total cost and increasing (Feldman 2009). Metabolomic approaches have been reported to improve existing diagnostic frameworks (Hirata et al. 2017; Saorin et al. 2020; Gao et al. 2021), and innovate new strategies for early diagnosis (Gonzalez-Covarrubias et al. 2022b). Better diagnosis techniques can reduce healthcare institutions' costs and the workload on healthcare personnel, which became a critical issue during COVID-19 (Biber et al. 2022). In addition, improved diagnostics provide patients with more accurate and timely results, speeding up treatment and further recovery. Another highlighted case for economic benefits is the application of metabolomics in drug development. Metabolomics can provide pertinent information on drug mechanisms (pathways, targets, interactions with other medications), drug repurposing, and personalized therapy (Pang and Hu 2023). Increasing drug development capacity will save resources, enable more options for patients, and increase profit margins, given that all other factors stay constant. Metabolomics has also been found to be used outside clinical environments. Food scientists have begun experimenting with the quantification of metabolites in products to classify, trace, and test safety criteria (Wu et al. 2022). Since food is a core consumer product, more robust methods are needed to ensure that high-quality meat and produce receives great industry and consumer attention. Should the relevant technologies be commercially available, they will be in high demand, especially from developing countries aiming to comply with international safety regulations and export to foreign, high-end consumer markets (Suanin 2022).

The nature of metabolomics, if sufficiently developed, can be highly cost-efficient and high-throughput. Metabolomics will likely be incorporated into frameworks similar to blood tests, which possess degrees of automation (Balter et al. 2018). Once the appropriate biomarkers and metabolites are identified, the process can be streamlined into sample collection, preparation, and analysis. However,

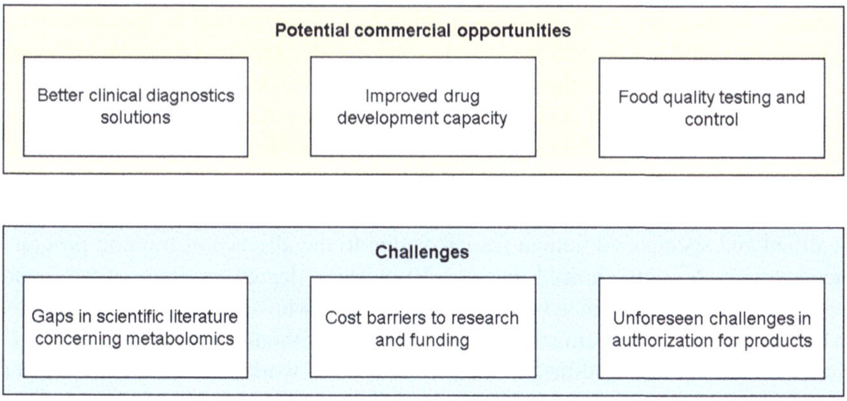

Fig. 16.4 Opportunities and challenges in the commercialization of metabolomics

significant barriers prevent metabolomics from reaching a commercially viable position. The biggest, and perhaps most challenging, hurdle to surpass is research. Metabolomics is an emerging field (the word "metabolome" was first introduced in 1998 (Kell and Oliver 2016)), and the available data does not provide adequate coverage for a wide range of diseases, thus limiting commercial potential. This hurdle will be slowly resolved in the future, opening up avenues for product development. The second hurdle is cost. Analytical equipment is expensive—nuclear magnetic resonance systems at the low end cost $35,000 (Anasazi Instruments), and high-performance liquid chromatography systems cost approximately $13,000. Cost factors may dissuade certain research groups and reduce research progress. Another key challenge is legislation. All metabolomic products must be approved by authorities, which brings difficulties unpredictable at the moment. Finally, public attention toward metabolomics is limited. Although known by academics, the average consumer likely has little knowledge of "metabolomic products." This issue will likely be fixed when there are commercial products that can be mass-marketed.

16.4 The Educational Issues in Research and Application of Metabolomics

Metabolism is the basic process of life, in which living organisms convert chemical compounds into other substances, providing energy and elements necessary to maintain biological activity. This process occurs not only at the cellular level but also at the body system level and is the basis of many biological processes, such as digestion and energy metabolism (Ellul et al. 2019; Beger et al. 2016). Understanding and researching the field of metabolism plays an important role in improving the quality of medical examination and treatment and creating new applications in the fields of medicine and pharmacy. Hence, it is vital to educate the public and raise society's awareness of the significance and relevance of metabolism in order to accomplish this. However, this would face various challenges.

Currently, the understanding of metabolism in the community is still limited and uneven. The majority of the population either lacks fundamental information about this field or could not be concerned if they are not studying in a discipline relevant to biomedicine because of the expertise in knowledge in this field. According to an epidemiological study in Australia, participants only took part in the study as respondents to the survey because they were aware that they would not be receiving a personal health evaluation for themselves or their children if they did not participate (Feher 2017). Moreover, in the current setting, there is a shortage of highly qualified and specialized human resources due to the absence of training program development. A study showed that just 50 master's degree programs were found using the search phrase "metabolomics" for students who already have a foundation in biomedicine and biochemistry and wish to study methods to advance their professional knowledge and abilities required to be able to work in metabolism (Winder et al. 2022).

There are some suggestions for resolving the challenges above. First of all, education is a method that plays the most important role in improving community awareness as well as enhancing the expertise of those who already have knowledge and experience. There are two main proposed programs: short-term and long-term programs (Table 16.1). For short-term programs, the trained subjects are people with experience and expertise. Before they participate in these short-term training courses, they must meet the basic qualifications requirements first. The purpose of this short-term training course is to help learners who want to learn more about metabolism and related industries. Regarding long-term programs, participants may be people who do not have background knowledge in the field of metabolism. Participating in this course will give them general in-depth knowledge and be oriented to different specialized research niches based on their wishes and future career orientation. Furthermore, training in the use of technology and acquiring new technical skills are essential for people who do metabolic testing due to the rapid growth of the Information Technology industry.

Presently, people can participate in online courses that offer certificates such as Biochemical Principles of Energy Metabolism provided by Korea Advanced Institute of Science and Technology Metabolism Transformation provided by the National Academy of Sports Medicine from Coursera. Besides, in the world, there is an organization called Metabolomics Society with the following missions: "To promote the growth and development of the field of metabolomics internationally; To provide the opportunity for collaboration and association among the workers in that science and in related sciences and connections between academia, government, and industry in the field of metabolomics; To provide opportunities for presentation of research achievements and creation of workshops, and To promote the publication of meritorious research in the field," and annual conferences have been held in various countries such as Japan, Canada, Spain, etc. to carry out these above missions. Additionally, several universities and training centers in various nations offer in-person and virtual training programs. These include the University of Liverpool's Liverpool Training Centre for Metabolomics, the European

Table 16.1 Main factors of short-term training and long-term education course

Factors	Short-term training	Long-term education
Participants	People with industry knowledge	People without industry knowledge
Participation duration	3 days–6 months	Longer than 6 months
Requirements	Have previous degrees, certificates, or certifications related to metabolism or biochemistry	No
Outcome	Gain more industry knowledge and knowledge of related industries	Obtain fundamental and in-depth knowledge of the industry and many related industries
Forms of training	Seminar, workshop, online course	University, college courses

Bioinformatics Institute, the West Coast Metabolomics Centre, Metabolomics South Africa, and others.

In addition to the abovementioned solutions, some of the following solutions are proposed to solve the above problems. First, making knowledge understandable and accessible is important to raise public awareness. Organizing educational exhibitions about metabolism in schools, museums, medical centers, and communities will help convey knowledge visually and dynamically. These exhibits may include demonstration models, video presentations, and educational games to help participants better understand the body's metabolic processes. Besides, programmers can convert specialized knowledge about transformation into understandable, non-technical language so everyone can access and understand. Using specific examples, simple illustrations, and images helps participants easily visualize and recall information. In addition, developing and distributing educational materials such as books, brochures, and online learning kits on transformation to provide information and guidance to the community is also necessary. This document must be written clearly and attractively to attract people's interest.

16.5 Conclusion

Nowadays, metabolomics has many benefits for molecular biomedicine and clinical research applications. The potential of metabolomics cannot be undermined with a plethora of significant applications. However, not to be an exception, this technology has to face many challenges, not only with biological issues but also issues involving ethics, legal, economy, and education. This article collected the available evidence of these issues in metabolomics. Firstly, the main ethical issues surrounding the application of metabolomics are the privacy of personal information, inconsistent data-sharing policies, and patenting standards. Secondly, particular legal actions related to metabolomics cannot be found on any platform. It is only being adjusted by legal regulations in general, which include the relevant areas involved. Thirdly, applications present in metabolomics can facilitate translational and economic opportunities, especially for disease detection and biomarkers, drug development, and food quality testing. However, significant barriers such as research, cost, legislation, and public attention prevent metabolomics from reaching a commercially viable position. Finally, education issues have also been challenged, based on the fact that understanding metabolism in the community has still been limited. Short-term and long-term programs have been suggested to raise awareness of those with knowledge and experience. At the same time, public education should also be noted through exhibitions, online information provisions, etc. In conclusion, metabolomics' ethical, legal, economic, and educational issues are understandable with a new and booming technology like metabolomics. Researchers and regulators can focus on these aspects so metabolomics can be more widely and effectively applied.

References

Allied Market Research (2022) Metabolomics market size, share, competitive landscape and trend analysis report by product and service, application and indication: global opportunity analysis and industry forecast, 2021–2030. Allied Market Research

Balter ML et al (2018) Automated end-to-end blood testing at the point-of-care: integration of robotic phlebotomy with downstream sample processing. Technology (Singap World Sci) 6(2):59–66

Becker ED (1993) A brief history of nuclear magnetic resonance. Anal Chem 65(6):295A–302A

Beger RD et al (2016) Metabolomics enables precision medicine: "a white paper, community perspective". Metabolomics 12:1–15

Beger RD, Schmidt MA, Kaddurah-Daouk R (2020) Current concepts in pharmacometabolomics, biomarker discovery, and precision medicine. Metabolites 10(4)

Beisken S, Eiden M, Salek RM (2015) Getting the right answers: understanding metabolomics challenges. Expert Rev Mol Diagn 15(1):97–109

Belhaj MR, Lawler NG, Hoffman NJ (2021) Metabolomics and lipidomics: expanding the molecular landscape of exercise biology. Meta 11(3)

Biber J et al (2022) Mental health impact on healthcare workers due to the COVID-19 pandemic: a U.S. cross-sectional survey study. J Patient Rep Outcomes 6(1):63

Brown JM, Hazen SL (2015) The gut microbial endocrine organ: bacterially derived signals driving cardiometabolic diseases. Annu Rev Med 66:343–359

Centers for Disease Control and Prevention (2022) Health Insurance Portability and Accountability Act of 1996 (HIPAA). Available from https://www.cdc.gov/phlp/publications/topic/hipaa.html#covered-entities

Chen Y, Li EM, Xu LY (2022) Guide to metabolomics analysis: a bioinformatics workflow. Metabolites 12(4)

CMS (2024) National health expenditure data. [cited 2024 20/03]; Available from: https://www.cms.gov/data-research/statistics-trends-and-reports/national-health-expenditure-data/historical

Dai X, Shen L (2022) Advances and trends in omics technology development. Front Med (Lausanne) 9:911861

Di Minno A et al (2022) Challenges in metabolomics-based tests, biomarkers revealed by metabolomic analysis, and the promise of the application of metabolomics in precision medicine. Int J Mol Sci 23(9)

Ellul S et al (2019) Metabolomics: population epidemiology and concordance in Australian children aged 11–12 years and their parents. BMJ Open 9(Suppl 3):106

Feher JJ (2017) Quantitative human physiology: an introduction. Academic Press, Amsterdam

Feldman L (2009) Managing the cost of diagnosis. Manag Care 5:43–45

Gao P et al (2021) Application of metabolomics in clinical and laboratory gastrointestinal oncology. World J Gastrointest Oncol 13(6):536–549

Garcia-Ibáñez R (2024 Why personalized medicine is a business trend to watch in 2024. [cited 2024 31/3]; Available from: https://www.forbes.com/sites/forbesbusinesscouncil/2024/03/04/why-personalized-medicine-is-a-business-trend-to-watch-in-2024/?sh=41a1adca7b52

Glish GL, Vachet RW (2003) The basics of mass spectrometry in the twenty-first century. Nat Rev Drug Discov 2(2):140–150

Gong Z et al (2010) Intake of fatty acids and antioxidants and pancreatic cancer in a large population-based case-control study in the San Francisco Bay Area. Int J Cancer 127(8):1893–1904

Gonzalez-Covarrubias V, Martinez-Martinez E, Del Bosque-Plata L (2022a) The potential of metabolomics in biomedical applications. Meta 12(2)

Gonzalez-Covarrubias V, Martínez-Martínez E, del Bosque-Plata L (2022b) The potential of metabolomics in biomedical applications. Metabolites 12. https://doi.org/10.3390/metabo12020194

Government, J., Patent Act (Act No. 121 of 1959) (1959)

Gowda GA, Djukovic D (2014) Overview of mass spectrometry-based metabolomics: opportunities and challenges. Methods Mol Biol 1198:3–12

Hirata Y et al (2017) Identification of highly sensitive biomarkers that can aid the early detection of pancreatic cancer using GC/MS/MS-based targeted metabolomics. Clin Chim Acta 468:98–104
Horgan RP, Kenny LC (2011) 'Omic' technologies: genomics, transcriptomics, proteomics and metabolomics. Obstet Gynaecol 13(3):189–195
Idle JR, Gonzalez FJ (2007) Metabolomics. Cell Metab 6(5):348–351
Information, A.U.S.G., Animal Welfare Act 1966
Intersoft Consulting (2018) General Data Protection Regulation (GDPR). Available from: https://gdpr-info.eu/
Johnson CH, Gonzalez FJ (2012) Challenges and opportunities of metabolomics. J Cell Physiol 227(8):2975–2981
Kartal M (2007) Intellectual property protection in the natural product drug discovery, traditional herbal medicine and herbal medicinal products. Phytother Res 21(2):113–119
Kell DB, Oliver SG (2016) The metabolome 18 years on: a concept comes of age. Metabolomics 12(9):148
Lewis CM Jr et al (2020) Ethical priority of the most actionable system of biomolecules: the metabolome. Am J Phys Anthropol 171(2):177–181
McCall L-I et al (2019) Analysis of university workplace building surfaces reveals usage-specific chemical signatures. Build Environ 162:106289
McGuire AL et al (2011) Ethical and practical challenges of sharing data from genome-wide association studies: the eMERGE Consortium experience. Genome Res 21(7):1001–1007
MordorIntelligence (n.d.). Global metabolomics market trends
Nagana Gowda GA, Raftery D (2021) NMR-based metabolomics. Adv Exp Med Biol 1280:19–37
Nezhmetdinova FT, Guryleva ME, Sharypova NKh, Zinurova RI, Tuzikov AR (2020) Risks of modern biotechnologies and legal aspects of their implementation in agriculture. (BIO Web Conf Volume 17, 2020 International scientific-practical conference "agriculture and food security: technology, innovation, markets, human resources" (FIES 2019)), p 6
(NIH), N.I.o.H., Health Research Extension Act 1985
Pang H, Hu Z (2023) Metabolomics in drug research and development: the recent advances in technologies and applications. Acta Pharm Sin B 13(8):3238–3251
Petras D et al (2016) Mass spectrometry-based visualization of molecules associated with human habitats. Anal Chem 88(22):10775–10784
Pinu FR, Goldansaz SA, Jaine J (2019) Translational metabolomics: current challenges and future opportunities. Meta 9(6)
Rahman M, Schellhorn HE (2023) Metabolomics of infectious diseases in the era of personalized medicine. Front Mol Biosci 10
Rangel-Huerta OD, Pastor-Villaescusa B, Gil A (2019) Are we close to defining a metabolomic signature of human obesity? A systematic review of metabolomics studies. Metabolomics 15(6):93
Reilly PR, Debusk RM (2008) Ethical and legal issues in nutritional genomics. J Am Diet Assoc 108(1):36–40
Sakurai N (2022) Recent applications of metabolomics in plant breeding. Breed Sci 72:56
Saorin A et al (2020) Emerging role of metabolomics in ovarian cancer diagnosis. Metabolites 10. https://doi.org/10.3390/metabo10100419
Suanin W (2022) Processed food exports from developing countries: the effect of food safety compliance. Eur Rev Agric Econ 50(2):743–770
Tannoury M, Attieh Z (2017) The influence of emerging markets on the pharmaceutical industry. Curr Ther Res 86:19–22
University of Birmingham and Birmingham Metabolomics Training Centre (2017) Metabolomics: understanding metabolism in the 21st century. Available from: https://www.futurelearn.com/info/courses/metabolomics/0/steps/10678
Winder CL et al (2022) Providing metabolomics education and training: pedagogy and considerations. Metabolomics 18(12):106

Wishart DS (2016) Emerging applications of metabolomics in drug discovery and precision medicine. Nat Rev Drug Discov 15(7):473–484

Wu W et al (2022) Emerging applications of metabolomics in food science and future trends. Food Chem X 16:100500

Yang Q et al (2019) Metabolomics biotechnology, applications, and future trends: a systematic review. RSC Adv 9(64):37245–37257

Zhang Y et al (2014) Decreased serum levels of free fatty acids are associated with breast cancer. Clin Chim Acta 437:31–37

Zhang Y et al (2016) Serum unsaturated free fatty acids: a potential biomarker panel for early-stage detection of colorectal cancer. J Cancer 7(4):477–483

The manufacturer's authorised representative in the EU is Springer Nature Customer Service Centre GmbH, Europaplatz 3, 69115 Heidelberg, Germany. If you have any concerns regarding our products, please contact ProductSafety@springernature.com

Printed and bound by CPI Group (UK) Ltd, Croydon, CR0 4YY
01/10/2026
02225910-0016